Chrysanthemum

Edited by

Malleshaiah SharathKumar and Ep Heuvelink

CABI

CABI is a trading name of CAB International

CABI
Nosworthy Way
Wallingford
Oxfordshire OX10 8DE
UK

Tel: +44 (0)1491 832111
E-mail: info@cabi.org
Website: www.cabi.org

CABI
200 Portland Street
Boston
MA 02114
USA

T: +1 (617)682-9015
E-mail: cabi-nao@cabi.org

A catalogue record for this book is available from the British Library, London, UK.

ISBN-13: 9781789248265 (paperback)
9781789248272 (ePDF)
9781789248289 (ePub)

DOI: 10.1079/9781789248289.0000

Commissioning Editor: Rebecca Stubbs
Editorial Assistant: Emma McCann
Production Editor: James Bishop

Typeset by Exeter Premedia Services Pvt Ltd, Chennai, India
Printed in the USA

CROP PRODUCTION SCIENCE IN HORTICULTURE SERIES

This series examines economically important horticultural crops selected from the major production systems in temperate, subtropical and tropical climatic areas. Systems represented range from open field and plantation sites to protected plastic and glass houses, growing rooms and laboratories. Emphasis is placed on the scientific principles underlying crop production practices rather than on providing empirical recipes for uncritical acceptance. Scientific understanding provides the key to both reasoned choice of practice and the solution of future problems.

Students and staff at universities and colleges throughout the world involved in courses in horticulture, as well as in agriculture, plant science, food science and applied biology at degree, diploma or certificate level will welcome this series as a succinct and readable source of information. The books will also be invaluable to progressive growers, advisers and end-product users requiring an authoritative, but brief, scientific introduction to particular crops or systems. Keen gardeners wishing to understand the scientific basis of recommended practices will also find the series very useful.

The authors are all internationally renowned experts with extensive experience of their subjects. Each volume follows a common format covering all aspects of production, from background physiology and breeding, to propagation and planting, through husbandry and crop protection, to harvesting, handling and storage. Selective references are included to direct the reader to further information on specific topics.

Titles Available:

1. **Ornamental Bulbs, Corms and Tubers** A.R. Rees
2. **Citrus** F.S. Davies and L.G. Albrigo
3. **Onions and Other Vegetable Alliums** J.L. Brewster
4. **Ornamental Bedding Plants** A.M. Armitage
5. **Bananas and Plantains** J.C. Robinson
6. **Cucurbits** R.W. Robinson and D.S. Decker-Walters
7. **Tropical Fruits** H.Y. Nakasone and R.E. Paull
8. **Coffee, Cocoa and Tea** K.C. Willson
9. **Lettuce, Endive and Chicory** E.J. Ryder
10. **Carrots and Related Vegetable *Umbelliferae*** V.E. Rubatzky, C.F. Quiros and P.W. Simon
11. **Strawberries** J.F. Hancock
12. **Peppers: Vegetable and Spice Capsicums** P.W. Bosland and E.J. Votava
13. **Tomatoes** E. Heuvelink
14. **Vegetable Brassicas and Related Crucifers** G. Dixon
15. **Onions and Other Vegetable Alliums, 2nd Edition** J.L. Brewster
16. **Grapes** G.L. Creasy and L.L. Creasy
17. **Tropical Root and Tuber Crops: Cassava, Sweet Potato, Yams and Aroids** V. Lebot

18. **Olives** I. Therios
19. **Bananas and Plantains, 2nd Edition** J.C. Robinson and V. Galán Saúco
20. **Tropical Fruits, 2nd Edition Volume 1** R.E. Paull and O. Duarte
21. **Blueberries** J. Retamales and J.F. Hancock
22. **Peppers: Vegetable and Spice Capsicums, 2nd Edition** P.W. Bosland and E.J. Votava
23. **Raspberries** R.C. Funt and H.K. Hall
24. **Tropical Fruits, 2nd Edition Volume 2** R.E. Paull and O. Duarte
25. **Peas and Beans**, A. Biddle
26. **Blackberries and Their Hybrids**, H.K. Hall & R.C. Funt
27. **Tomatoes, 2nd Edition** E. Heuvelink
28. **Grapes, 2nd Edition** G.L. Creasy and L.L. Creasy
29. **Blueberries, 2nd Edition** J. Retamales and J.F. Hancock
30. **Citrus, 2nd Edition** L.G. Albrigo, L.L. Stelinski and L.W. Timmer
31. **Tropical Root and Tuber Crops: Cassava, Sweet Potato, Yams and Aroids, 2nd Edition** V. Lebot
32. **Cucurbits, 2nd Edition** Todd C. Wehner, Rachel Naegele, James Myers, Kevin Crosby and Narinder P.S. Dhillon
33. **Carrots and Related Apiaceae Crops, 2nd Edition** Emmanuel Geoffriau and Philip W. Simon
34. **Strawberries, 2nd Edition** J.F. Hancock
35. **Sweet Cherries** Lynn E. Long, Gregory A. Lang and Clive Kaiser
36. **Cut Flowers and Foliages** James E. Faust and John M. Dole
37. **Mushrooms: *Agaricus bisporus*** Youssef Najib Sassine
38. **Date Palm** Jameel M. Al-Khayri, Shri Mohan Jain, Dennis V. Johnson and Robert R. Krueger
39. **Peach** George Manganaris, Guglielmo Costa and Carlos H. Crisosto
40. **Vegetable Brassicas and Related Crucifers, 2nd Edition** Geoffrey R. Dixon and Rachel Wells
41. **Tropical Fruits, 3rd Edition Volume 1** R.E. Paull and O. Duarte
42. **Grapes, 3rd Edition** G.L. Creasy and L.L. Creasy
43. **Chrysanthemum** Malleshaiah SharathKumar and Ep Heuvelink

Contents

Preface

This comprehensive book delves into the science, cultivation, and industry of chrysanthemums, a globally celebrated ornamental crop prized for its beauty, diversity, cultural significance, and economic impact. This book offers insights into breeding, physiology, production, protection, postharvest management, value addition, and the role of chrysanthemum in festivities and landscaping.

We begin with an introduction to the *Global Chrysanthemum Industry*, detailing its worldwide spread, production, and immense diversity in both potted and cut flower markets. This section provides a broad perspective on the economic and horticultural significance of chrysanthemums, setting the stage for the subsequent sections.

The *Breeding and Biotechnology* section explores conventional and molecular breeding approaches, including marker-assisted selection, QTL mapping, genome-wide association studies, and biotechnological advancements. Topics, such as somatic embryogenesis, embryo rescue culture, somaclonal variation, transformation protocols, viroid-free production, and AI-driven optimization of *in vitro* cultures, are discussed in depth.

The *Crop Physiology, Production, and Modeling* section examines chrysanthemum growth and development, environmental control, and intensive cultivation, from high-tech greenhouse systems to open-field farming for optimal productivity and quality. It also highlights climate-responsive crop modeling, data-driven decision support systems, and IT-based tools that are shaping the future of chrysanthemum cultivation.

The *Crop Protection and Management* section focuses on diagnosing and managing diseases and pests, detailing their symptoms, spread, life cycles, and integrated control strategies to ensure sustainable production.

The *Postharvest Management, Value Addition, and Festivities* section explores factors affecting postharvest performance, storage techniques, and commercial innovations such as tinting, dry flower technology, and edible chrysanthemums in Asian cuisine. It also highlights chrysanthemum's cultural and aesthetic role in global festivities and landscaping.

Each chapter integrates current knowledge, emerging trends, and future perspectives, enriched with high-quality photographs and illustrations to enhance understanding.

This book provides an in-depth exploration of the chrysanthemum world, making it an invaluable resource for horticulturists, researchers, students, growers, and enthusiasts. Whether you seek scientific insights or practical applications, this book serves as an essential guide to chrysanthemum cultivation, research, and industry advancements.

We, the editors, extend our deep gratitude to the researchers and experts from around the world who contributed their invaluable knowledge to this book. We are also grateful for the trust and support we received from CABI, in particular, Rebbeca Stubbs and Lucy Pritchard. We hope readers find inspiration in exploring the scientific and horticultural wonders of chrysanthemums as much as we did.

Malleshaiah SharathKumar and Ep Heuvelink
Wageningen, January 2026

Contributors

Neil O. Anderson, Department of Horticultural Science, University of Minnesota, 286 Alderman Hall, 1970 Folwell Avenue, Saint Paul, MN 55108 USA. Email: ander044@umn.edu. ORCID: https://orcid.org/0000-0001-9794-0397

N. Aswathanarayana Reddy, College of Horticulture, NH-75, Tamaka, Kolar—563 103, University of Horticultural Sciences, Bagalkot, India. Email: nanreddy5002@gmail.com. ORCID: https://orcid.org/0009-0008-7253-6038

Marcella Bovio, Plant Breeding, Department of Plant Sciences, Wageningen University and Research, 6700 AA Wageningen, The Netherlands. Email: marcella.bovio@wur.nl. ORCID: https://orcid.org/0000-0002-9390-2657

Lotte Caarls, Plant Breeding, Department of Plant Sciences, Wageningen University and Research, 6700 AA Wageningen, The Netherlands. Email: lotte.caarls@wur.nl. ORCID: https://orcid.org/0000-0003-4578-1472

Susana M.P. Carvalho, GreenUPorto-Sustainable Agrifood Production Research Centre/Inov4Agro, DGAOT, Faculty of Sciences of the University of Porto, Vairão, Portugal. Email: susana.carvalho@fc.up.pt. ORCID: https://orcid.org/0000-0001-7157-1079

Fadi Chen, State Key Laboratory of Crop Genetics and Germplasm Enhancement and Utilization, Key Laboratory of Landscaping, Ministry of Agriculture and Rural Affairs, National Forestry and Grassland Administration, College of Horticulture, Nanjing Agricultural University, 210095 Nanjing, China. Email: chenfd@njau.edu.cn. ORCID: https://orcid.org/0000-0001-9134-2353

Sumei Chen, State Key Laboratory of Crop Genetics and Germplasm Enhancement and Utilization, Key Laboratory of Landscaping, Ministry of Agriculture and Rural Affairs, National Forestry and Grassland

Administration, College of Horticulture, Nanjing Agricultural University, 210095 Nanjing, China. Email: chensm@njau.edu.cn. ORCID: https://orcid.org/0000-0003-3973-0505

Margery L. Daughtrey, Long Island Horticultural Research and Extension Center, College of Agriculture and Life Sciences, Cornell University, 3059 Sound Ave., Riverhead, NY 11901, USA. Email: mld9@cornell.edu. ORCID: NA

John M. Dole, North Carolina State University, Department of Horticultural Science, Raleigh NC 27695 USA. Email: jmdole@ncsu.edu. ORCID: https://orcid.org/0000-0001-9570-6214

James E. Faust, Clemson University, Department of Plant and Environmental Sciences E-143 Poole Agricultural Center Clemson SC 29634 USA. Email: jfaust@clemson.edu. ORCID: https://orcid.org/0000-0003-1299-5503

Vikas Kumar Galipothu, Department of Horticulture and Natural Resources, Kansas State University, Olathe, KS 66061, USA. Email: vk6150553@gmail.com. ORCID: https://orcid.org/0009-0009-2806-4536

S. Ganesh, Department of Floriculture & Landscape Architecture, Horticultural College & Research Institute,Tamil Nadu Agricultural University, Coimbatore—641 003, Tamil Nadu, India. Email: ganesh4u@gmail.com. ORCID: https://orcid.org/0000-0001-7215-9844

M. Ganga, Department of Floriculture & Landscape Architecture, Horticultural College & Research Institute,Tamil Nadu Agricultural University, Coimbatore—641 003, Tamil Nadu, India. Email: ganga.m@tnau.ac.in. ORCID: https://orcid.org/0000-0002-7525-2849

Ep Heuvelink, Horticulture and Product Physiology, Department of Plant Sciences, Wageningen University and Research, Wageningen 6700 AA, The Netherlands. Email: ep.heuvelink@wur.nl. ORCID: https://orcid.org/0000-0002-8731-7195

Oliver Körner, Leibniz-Institute of Vegetable and Ornamental Crops, Großbeeren, Germany. Email: koerner@igzev.de. ORCID: https://orcid.org/0000-0002-3241-481X

Rajiv Kumar, Division of Flower and Medicinal Crops, ICAR-Indian institute of Horticultural Research. Email: rajiv.kumar11@icar.gov.in. ORCID: NA

Naonobu Noda, Institute of Vegetable and Floriculture Science, National Agriculture and Food Research Organization, 2-1 Fujimoto, Tsukuba, Ibaraki 305-0852, Japan. Email: naonobun@affrc.go.jp. ORCID: https://orcid.org/0000-0001-7973-3984

Malleshaiah SharathKumar, Horticulture and Product Physiology, Department of Plant Sciences, Wageningen University and Research, Wageningen 6700 AA, The Netherlands; Deliflor Chrysanten B.V., Korte Kruisweg 163, 2676 BS Maasdijk, The Netherlands. Email: sharath.malleshaiah@wur.nl; sharath.malleshaiah@deliflor.nl; sharath.malleshaiah@gmail.com. ORCID: https://orcid.org/0000-0002-8736-7408

Jiangshuo Su, State Key Laboratory of Crop Genetics and Germplasm Enhancement and Utilization, Key Laboratory of Landscaping, Ministry of Agriculture and Rural Affairs, National Forestry and Grassland Administration, College of Horticulture, Nanjing Agricultural University, 210095 Nanjing, China. Email: sujiangshuo@njau.edu.cn. ORCID: https://orcid.org/0009-0001-6350-1801

N.C. Surakshitha, Plant Breeding, Wageningen University and Research, Po Box 386, 6700 AJ, Wageningen, The Netherlands. Email: surakshitha.naguvanahallychandrashekara@wur.nl. ORCID: https://orcid.org/0000-0001-7401-6513

Marie-Christine Van Labeke, Horticultural Sciences, Department of Plants and Crops, Faculty of Bioscience Engineering, Ghent University, Ghent, Belgium. Email: MarieChristine.VanLabeke@UGent.be. ORCID: https://orcid.org/0000-0002-4766-8919

M. Visalakshi, Department of Medicinal and Aromatic Crops, Horticulture college and research institute, Tamil Nadu Agricultural University, Coimbatore, India. Email: visalakshi.m@tnau.ac.in. ORCID: https://orcid.org/0000-0001-5113-9908

The Global Chrysanthemum Industry

1

Malleshaiah SharathKumar[1,2]* and Ep Heuvelink[1]

[1]Horticulture and Product Physiology, Department of Plant Sciences, Wageningen University and Research, Wageningen 6700 AA, The Netherlands
[2]Deliflor Chrysanten B.V., Korte Kruisweg 163, 2676 BS, Maasdijk, The Netherlands

Abstract

The global chrysanthemum industry has witnessed significant growth, driven by increasing demand and the widespread use of chrysanthemums as cut flowers and potted flowering plants, along with their cultural and religious significance. Chrysanthemums are a major export commodity within the floriculture industry, contributing substantially to global trade. The success of this trade is supported by an efficient supply chain, particularly cold-chain logistics, which ensures long-distance transportation. In this chapter, we will explore the global production of chrysanthemums, including recent import and export values. We will also discuss the various types of commercial chrysanthemums, the different cultivation methods employed, and the profound meaning, symbolism, and significance of chrysanthemums across cultures worldwide. Additionally, the multifaceted uses of chrysanthemums are discussed, highlighting their versatile role in ornamental horticulture, religious, and cultural practices.

Keywords

Chrysanthemum cultivation, floriculture export, global trade of chrysanthemums, ornamental horticulture

*Corresponding author: sharath.malleshaiah@wur.nl; sharath.malleshaiah@deliflor.nl; sharath.malleshaiah@gmail.com

DOI: 10.1079/9781789248289.0001

1.1 Introduction

Chrysanthemum (*Chrysanthemum morifolium*) is an herbaceous perennial species belonging to the Asteraceae family, the second largest family of flowering plants (Hao *et al.*, 2022). Chrysanthemums are a significant component of the ornamental plant industry, contributing substantially to the billion-dollar sector and ranking among the top five cut flowers in international trade (Spaargaren and van Geest, 2018). They are cultivated for their colorful flower heads (capitulum), and because of continuous breeding efforts by hobbyists and commercial breeders, a wide range of chrysanthemum cultivars is available, each displaying exquisite beauty (Fig. 1.1). They hold a prominent place in global floriculture, being highly valued as loose flowers, cut flowers, and potted plants. The ease of cultivation, coupled with high market demand and the potential for significant economic returns, has contributed to their widespread popularity. Chrysanthemums are praised for their wide variety of colors, forms, and sizes, making them suitable for numerous uses, ranging from decorative arrangements to religious ceremonies.

1.2 Types of Commercial Chrysanthemums

Commercial chrysanthemums are produced in various forms, primarily as disbud and spray types for cut flowers, and cushion types for potted plants (SharathKumar, 2023; Fig. 1.2).

- *Disbud chrysanthemums:* Disbud chrysanthemums are characterized by their large flowers, with one central flower per stem. The lateral buds are removed to enhance the growth of the central flower bud, resulting in blooms that vary in diameter from 10 to 15 cm or more, depending on the cultivar, shape, and type of flower.
- *Spray chrysanthemums:* Spray chrysanthemums have smaller flowers, with multiple blooms per stem. The central flower bud is removed to encourage the growth of lateral flowers, resulting in a spray of 10–12 equally sized flowers at a similar stage of development, attractively arranged on the stem.
- *Santini chrysanthemums:* Santini chrysanthemums are a smaller variant of cut chrysanthemums, distinguished by their small, daisy-like blooms arranged in clusters on shorter stems. These chrysanthemums have smaller flowers compared to traditional spray and disbud types, with a maximum flower diameter of around 4 cm. Despite their small size, santinis boast the same long shelf life as other cut chrysanthemums, making them a highly appealing choice for diverse floral arrangements and markets. Their compact structure and exceptional durability enhance their versatility and popularity in ornamental horticulture.
- *Potted chrysanthemums:* Potted chrysanthemums are typically small-flowered spray types cultivated as compact plants with a ball-like architecture, making them ideal for decorative purposes. In modern markets, there

Fig. 1.1. The ornamental diversity of chrysanthemums showing differences in flower color and shape across three categories: disbud chrysanthemums (A), spray chrysanthemums (B), and santini chrysanthemums (C). (Photo courtesy of Malleshaiah SharathKumar and Deliflor Chrysanten B.V.)

Fig. 1.2. Morphotypes of commercial chrysanthemums. Disbud type with single flower on a single stem (A); spray type with multiple flowers on a single stem (B); and cushion type with multiple branching (C). (Photo courtesy of Malleshaiah SharathKumar and Deliflor Chrysanten B.V.)

is increasing demand for potted disbuds, which feature vibrant color combinations and cater to consumer preferences for bold, visually striking plants. Popular for indoor decoration, potted chrysanthemums combine aesthetic appeal with versatility, making them a favorite choice during seasonal and festive occasions for indoor decoration.

1.3 Types of Chrysanthemum Cultivation

Chrysanthemums can be cultivated using several methods to suit different environmental conditions and production goals. The two primary types of chrysanthemum cultivation are greenhouse and open-field, which offer unique advantages and challenges, making them suitable for different production goals and resource availability.

1.3.1 Greenhouse cultivation

Chrysanthemum cultivation within greenhouses or protective structures offers some control over environmental conditions, depending on the low-tech to high-tech greenhouses, ensuring optimal growth and high-quality flowers throughout the year. Depending on the level of controllability required and economic feasibility, several greenhouse systems are employed, including shade-net houses, polyhouses, and glasshouses. Each type offers unique advantages and is suitable for different scales of chrysanthemum cultivation and climatic conditions. Additionally, each type provides different capabilities for manipulating factors such as temperature, humidity, light, and CO_2 levels, catering to various horticultural needs and budget considerations. Shade-net houses provide basic protection and are cost-effective, making them suitable for regions with high sunlight and temperature. Polyhouses offer a balance of cost and environmental control, suitable for diverse climates and moderate-scale operations. Glasshouses provide the highest level of control and are ideal for commercial production (Fig. 1.3; refer to Chapter 06), albeit with higher costs and maintenance requirements. By choosing the appropriate greenhouse system, growers can effectively manipulate environmental factors to optimize chrysanthemum growth, ensuring high-quality blooms and maximizing production efficiency.

1.3.2 Open-field cultivation

Open-field cultivation is the traditional method of growing chrysanthemums and remains widely practiced due to its cost-effectiveness and scalability (Fig. 1.4; refer to Chapter 07). This method involves growing plants directly in the soil, exposed to natural weather conditions. Open-field cultivation applies planting and harvesting schedules according to regional climate patterns to avoid extreme temperatures and rainfall that can negatively affect crop quality.

Fig. 1.3. Chrysanthemum cultivation in glasshouses: cut flowers (A) and pot chrysanthemum (B). (Photo courtesy of Malleshaiah SharathKumar.)

Fig. 1.4. Chrysanthemum cultivation in open-field for loose flowers. (Photo courtesy of Shambu G.B.)

1.4 Global Production of Chrysanthemums

Chrysanthemums are native to the northern hemisphere, particularly Central and East Asia, including countries such as China, Korea, and Japan. The genus comprises approximately 40 species and numerous hybrids and cultivars, most of which originate from East Asia, with China being the center of diversity (Liu *et al.*, 2012; Hao *et al.*, 2022). The introduction and subsequent cultivation in these regions have facilitated the global spread of chrysanthemums and contributed to the rich variety of genotypes available and produced worldwide. Currently, chrysanthemums are among the most important and commercially viable ornamental crops globally, holding significant market value as both cut flowers and potted plants. Chrysanthemums are the second most popular ornamental flower worldwide, following roses. However, in countries such as Japan, they hold significant market value alongside other key ornamentals such as lilies and roses (Shibata, 2008). The chrysanthemum industry is distinguished by extensive global production and trade (Fig. 1.5), positioning modern chrysanthemums as the second most traded and cultivated cut flower globally, reinforcing their prominence in the international floriculture market.

Similarly, in India, chrysanthemums hold prime importance in festivities and religious events, where they are widely used for decorations, garlands, and offerings in temples and cultural ceremonies. India has the largest production area for chrysanthemums, with 20,000 ha dedicated to cultivation. However, much of this is fragmented among smallholder farms for local uses such as

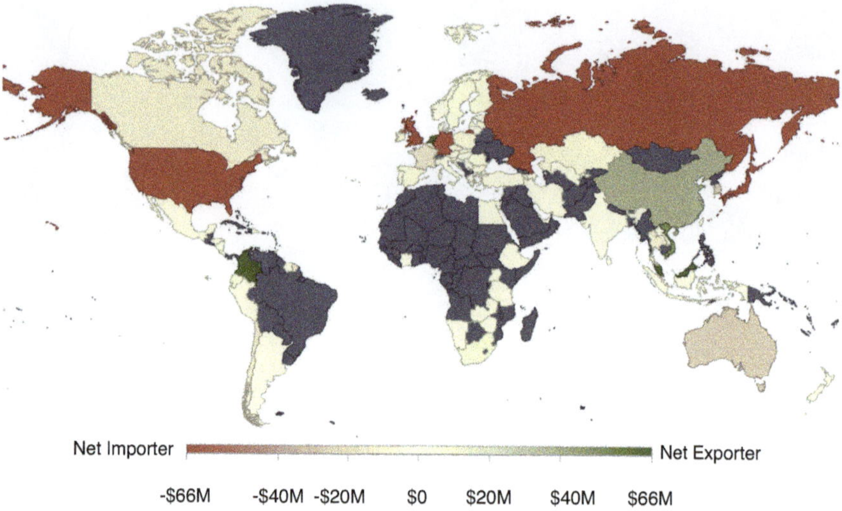

Fig. 1.5. Map showing the export and import balance of chrysanthemums by country during 2022. Each country is colored based on the difference in export and import values of chrysanthemums, measured in millions during 2022. (Adapted from OEC.world.net; Simoes and Hidalgo, 2011. CC0 public domain.)

weddings and religious ceremonies. China also has a substantial production area of 7100 ha (AIPH, 2022). Despite these extensive cultivation areas, the industry focuses on stem numbers (for cut flowers) rather than mere acreage to assess the global market. This emphasis on the number of stems produced highlights the professional and practical considerations of the market that significantly contribute to global supply chains of chrysanthemums. In terms of stem production, Colombia leads with 1.4 billion stems annually for North American and UK markets, while the Netherlands produces 1.2 billion stems per year, primarily serving the EU market (OEC World, 2022). Emerging players such as Kenya are becoming significant due to advancements in sea transportation of cut flowers. Other notable producers include Malaysia (400 million stems/year), Vietnam (300 million stems/year), and Italy (200 million stems/year), each serving different regional markets (AIPH, 2022). In 2022, the countries with the largest trade value in exports of chrysanthemums were the Netherlands (US$400M), Colombia ($180M), Malaysia ($82.8M), Vietnam ($51.3M), and China ($38.3M) (Fig. 1.6A). The countries with the largest trade value in imports were the United Kingdom ($137M), United States ($124M), Russia ($122M), Japan ($110M), and Germany ($65.9M) (Fig. 1.6B).

Additionally, mother plants for chrysanthemum cuttings are cultivated at dedicated production facilities in Africa, where they are carefully maintained to ensure the production of high-quality propagation material. These facilities

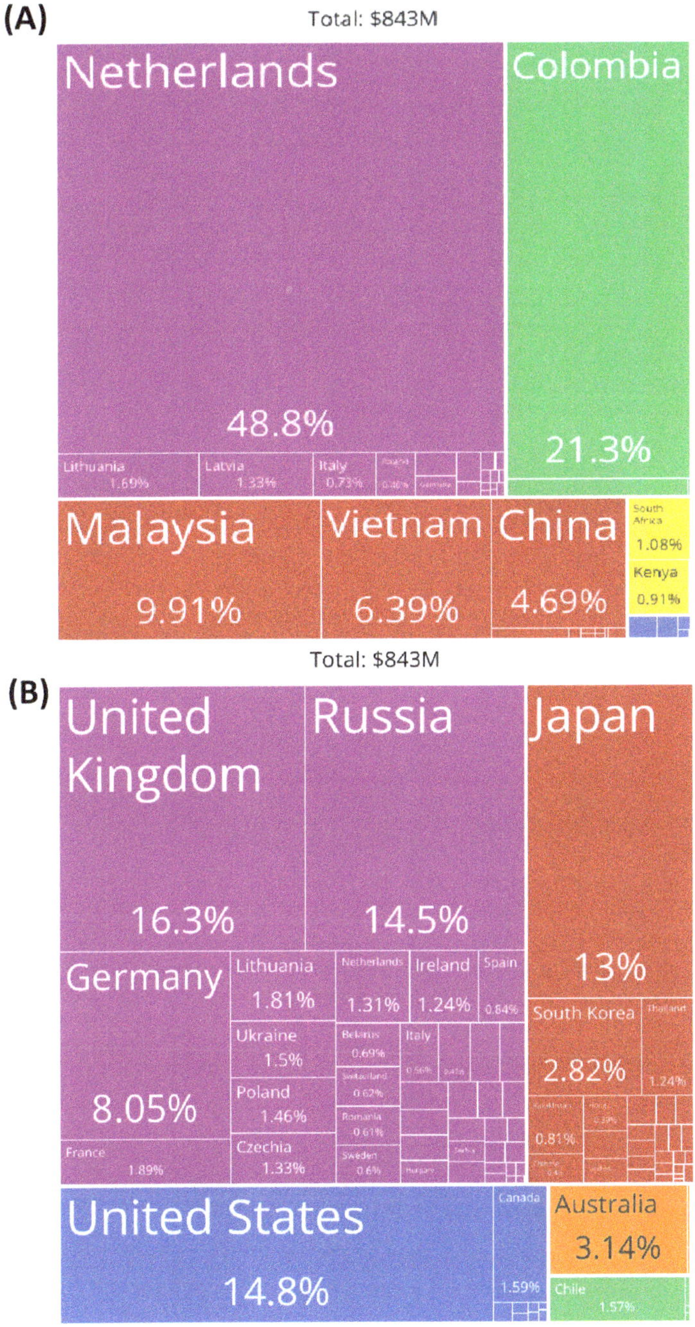

Fig. 1.6. The top countries with the percentage of export (A) and import (B) of chrysanthemum with the total value of $843M. (Adapted from OEC.world.net; Simoes and Hidalgo, 2011; CC0 public domain.)

provide optimal climatic conditions that support vigorous plant growth, enabling the continuous production of healthy, uniform cuttings. Once harvested, the unrooted cuttings are transported to the Netherlands, where they are rooted under controlled greenhouse conditions before being distributed to growers. This system combines cost-effective production in Africa with precision rooting and quality control in the Netherlands, ensuring that the final rooted cuttings meet the high standards required for commercial greenhouse cultivation in Europe and other global markets.

1.5 Meaning, Symbolism, and Significance of Chrysanthemums

1.5.1 The meaning of chrysanthemums

Chrysanthemums, commonly known as "mums," are among the most widely cultivated flowers worldwide. The name "chrysanthemum" derives from the Greek words *chrysos*, meaning "gold," and *anthemon*, meaning "flower." This etymology reflects the vibrant and rich golden hues that characterized the early varieties of the flower. The Swedish botanist Carl Linnaeus, renowned for his pioneering work in classifying and naming plants, coined the name. His choice of name highlights the enduring beauty and appeal of chrysanthemums, which have captivated cultures and gardeners globally for centuries. Popularly known as the "Queen of the East" and "Glory of the East," chrysanthemums are celebrated for their beauty and versatility.

1.5.2 Symbolism and cultural significance of chrysanthemums around the globe

Chrysanthemums are not just visually appealing; they also carry deep meanings and cultural significance in various parts of the world. Their rich history and widespread cultivation have led to diverse interpretations and uses, making them an integral part of many cultural traditions. The chrysanthemum is a flower of many faces, representing a spectrum of emotions and cultural values globally. Understanding these cultural contexts enriches our appreciation of this beautiful and resilient flower, which continues to bloom brightly in the garden of human culture.

1.5.3 China: A symbol of nobility and longevity

In China, chrysanthemums are revered for their elegance and resilience, symbolizing noble qualities and longevity. They are one of the "Four Gentlemen" in Chinese art, representing perseverance and strength by blooming in autumn as the weather cools. In traditional Chinese medicine, chrysanthemums are believed to promote longevity and good health. Chrysanthemums were believed to have medicinal properties, resulting in treating headaches with boiled roots, consuming young sprouts and petals in salads, and brewing leaves for festive drinks. Chrysanthemums are also depicted on fine Chinese porcelain, showcasing their significance. The flower symbolizes nobility in China, and even a town, Ju-Xian (now Xiaolan Town or Zhongshan City), is named after chrysanthemums, translating to "Chrysanthemum Town." A notable figure in Chinese chrysanthemum history is T'ao Ming-yang, a poet and connoisseur who cultivated a vast variety of chrysanthemums in his garden over 1500 years ago. Regarded as the greatest of all Chinese breeders, the city he lived in was renamed Chuhsein, "the city of chrysanthemums," in his honor.

1.5.4 Japan: The imperial flower

In Japan, chrysanthemums are deeply embedded in the nation's cultural and historical identity, often referred to as the "Imperial Flower" or "National Flower of Japan." Their association with Japanese royalty dates to 926 AD, when Emperor Uda ordered the finest chrysanthemum specimens for his palace in Tokyo. Introduced from China in the 5th century, chrysanthemums quickly became cherished and widely cultivated. By the Heian Period, they were a symbol of the Emperor and the Imperial Family, representing perfection, rejuvenation, and peace, and featured on the Imperial Seal of Japan. Japan honors chrysanthemums with the Chrysanthemum Festival (Kiku Matsuri) and National Chrysanthemum Day, also known as the Festival of Happiness. The Supreme Order of the Chrysanthemum, Japan's highest order of chivalry, further highlights the flower's royal association. The term "Chrysanthemum Throne" refers to the Japanese monarchy, and the 16-floret chrysanthemum is used in the Imperial Seal. Chrysanthemums also appear on Japanese passports and the 50 yen coin, symbolizing their cultural significance (Fig. 1.7). Their use spans a wide range, from everyday applications to the highest expressions of imperial honor. Chrysanthemums are prominently featured on state documents, imperial carriages, and high-ranking officials' weapons, symbolizing their enduring association with nobility, prestige, and honor.

1.5.5 India: A versatile flower in floriculture

In India, chrysanthemums are widely cultivated as loose flowers in open fields and as cut flowers in polyhouses. This dual method of cultivation highlights the flower's adaptability and its ability to thrive in diverse climatic conditions. The extensive cultivation of chrysanthemums in India extends beyond ornamental gardening to various cultural and ceremonial contexts. Chrysanthemums hold significant cultural value in India, where they are commonly used in floral arrangements and garlands integral to numerous religious and festive occasions, symbolizing purity, joy, blessing, and celebration. Known by different names across regions of India, "Guldaudi" in Hindi-speaking areas, "Chandramalika" in the eastern states, "Samanti and Sevanthige" in the southern regions, and "Shevanti" in the West, their widespread popularity and

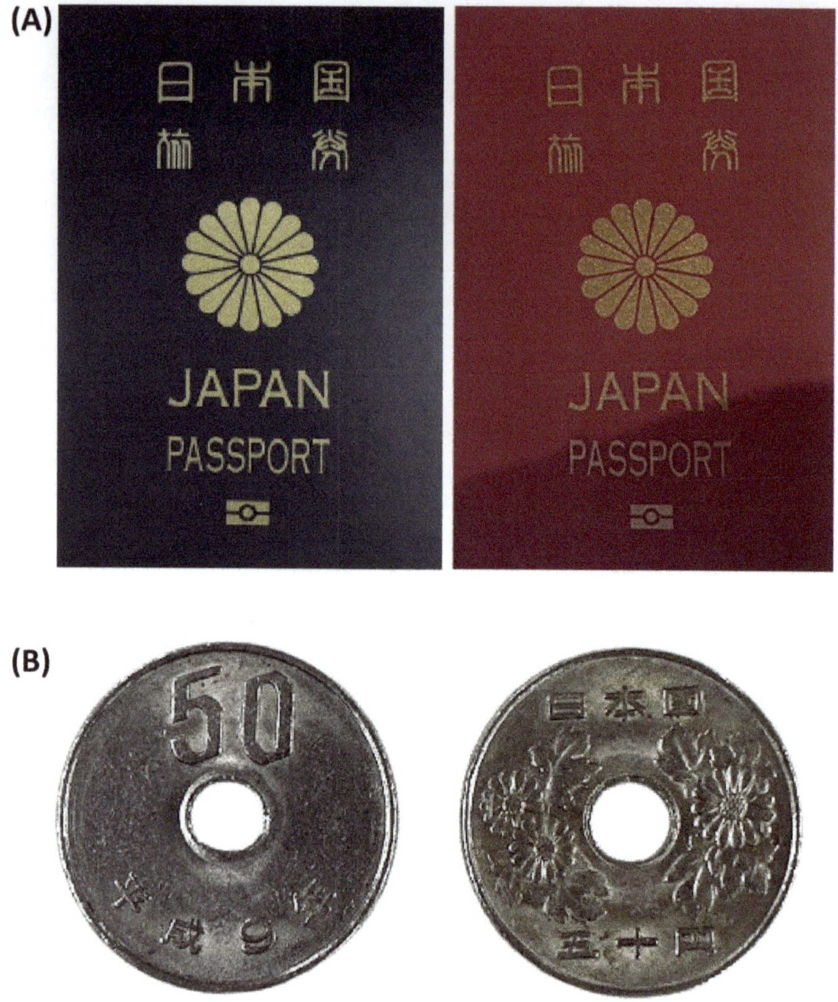

Fig. 1.7. Japanese passports featuring chrysanthemum emblem (A). The blue passport represents the five-year validity passport, often considered the "child's" passport (left). The red passport is the long-term standard passport with a ten-year validity. Both types prominently feature the chrysanthemum emblem, symbolizing Japan's cultural and diplomatic significance. Japan's 50-yen coin (B, front and back) featuring a chrysanthemum flower. First released in 1955, prominently features a chrysanthemum flower, symbolizing its cultural and historical significance in Japanese heritage. (Photo courtesy of Kavya Jayanthan.)

regional significance are evident. The growing popularity of chrysanthemums in India is driven by their aesthetic appeal and practical uses in floriculture. The flower's varied forms and colors make it a versatile choice for decorative purposes, from personal celebrations to public events. In traditional contexts, chrysanthemums are used in religious rituals and temple decorations in the form of garlands, while in contemporary settings, they feature prominently in weddings and other festive occasions. The cultivation and use of chrysanthemums continue to expand, reflecting their enduring appeal and the integral role they play in both traditional and modern Indian society.

1.5.6 Japan and western Europe: A symbol of death and mourning

In Japan, the white chrysanthemum represents purity, simplicity, and nobility of spirit. However, it is also associated with death and is commonly used in funerals and grave offerings, reinforcing its role as a flower of mourning and remembrance. Despite this association, the chrysanthemum remains deeply embedded in Japanese culture, holding significance in both solemn and celebratory contexts.

The introduction of chrysanthemums to Western Europe dates back to the 17th century, with Jacob Breynius being the first European to describe cultivated chrysanthemums in 1689. Over time, chrysanthemums have developed a rich cultural history in the West, where they are both admired for their beauty and linked to mourning rituals. In countries such as France, Belgium, Italy, Spain, Poland, Hungary, and Croatia, white chrysanthemums are strongly associated with death and remembrance. They are commonly used in funeral arrangements and placed on graves as a sign of honor, respect, and sympathy for the deceased. This tradition is especially prominent on All Saints' Day, a day dedicated to remembering the dead, when chrysanthemums are widely used for grave decorations.

Despite this somber connection, the beauty and symbolism of white chrysanthemums extend beyond mourning. These flowers, often in light tones, represent purity, sincerity, faith, and gratitude. A bouquet of white chrysanthemums is given as a gift to express deep respect for people of all ages and genders, and they are believed to heal the soul from pain and negativity, bringing peace and relaxation.

In modern floristry, chrysanthemums are increasingly considered for use in bridal bouquets and wedding decorations. Their special properties such as longevity, strong stems, and durable buds, combined with their affordable price, make them an excellent choice for wedding floral designs. The chrysanthemum's versatility and symbolism of purity, love, and innocence make it a popular choice in wedding floristry, adding elegance, and meaning to bridal bouquets and other wedding creations.

1.5.7 USA: A versatile flower of joy and cheer

Chrysanthemums were first introduced to the United States during the colonial period, and their popularity has since grown, earning them the title

"Queen of the Fall Flowers." In the United States, chrysanthemums are celebrated as symbols of happiness, positivity, and friendship. Known colloquially as "mums," these flowers are favored for their vibrant colors and durability, making them a popular choice for home gardens, public landscapes, and floral arrangements. Chrysanthemums are particularly associated with autumn, when they become a staple in seasonal decorations, symbolizing the transition of the seasons with their rich hues. They are closely associated with autumn events, such as Homecoming football games, where they are often used in corsages for girlfriends and mothers. Mums are commonly used in celebrations and as thoughtful gifts, often presented to new parents to mark the birth of a baby, symbolizing good wishes and joy. They are also a popular choice for expressing friendship and well-being, making them a versatile and cheerful addition to any floral display. Their enduring popularity in the United States highlights the cultural appreciation for their beauty and the positive emotions they evoke. This connection to tradition and celebration further cements their status as a beloved and versatile flower in the United States.

1.6 Uses of Chrysanthemums

Chrysanthemums are versatile flowers with a wide range of uses beyond their ornamental appeal, including practical applications in gardening, culinary traditions, pest control, and environmental health. Their adaptability and cultural significance make them an important plant in various aspects of human life.

1.6.1 Ornamental uses

Chrysanthemums are celebrated for their ornamental versatility. As cut flowers, they are highly regarded in floral designs, event decorations, bouquets, and wedding arrangements (Fig. 1.8). In addition to their use as cut flowers, chrysanthemums are popular as potted plants in home gardens and on balconies, adding vibrant color and charm to living spaces (Fig. 1.9). Chrysanthemum capitula in the form of garlands are part of religious rituals in temples, wedding decorations, and other festive occasions in Asian countries (Fig. 1.10). Some cultivars can be trained into various artistic forms, including bonsai, where they are shaped into miniature trees that mimic the appearance of mature deciduous or evergreen trees. Others can be molded into creative shapes, such as hemispheres, animals, furniture, or even human figures. Cascading chrysanthemums can be grown into long, pendulous drapes or shaped into large fans, pillars, or tree forms, displaying the plant's adaptability and the gardener's artistry (refer to Chapter 13).

As landscaping plants, chrysanthemums provide a stunning display of colors in the autumn, making them a popular choice for home gardens. They can accentuate entranceways, add vibrant colors to seasonal beds, or

Fig. 1.8. Chrysanthemum in floral vases (A) and bouquets (B). (Photo courtesy of Deliflor and Chrysanten B.V.)

Fig. 1.9. Indoor landscaping with chrysanthemums in potted arrangements. (Photo courtesy of Malleshaiah SharathKumar.)

dominate garden spaces with their diverse shapes, sizes, and hues. Garden-hardy chrysanthemums are perennials that can survive winter conditions in northern latitudes, producing abundant blooms without requiring extensive care. Exhibition varieties, while less hardy, are prized for their large, disbudded blooms and can be trained into elaborate forms such as "Thousand Bloom," standards (trees), fans, hanging baskets, topiary, bonsai, and cascades. These varieties often require staking and specific growing conditions such as controlled environments and additional lighting. Notable displays at locations such as Longwood Gardens in Pennsylvania and the New York Botanical Garden showcase the wide-ranging ornamental possibilities of chrysanthemums (refer to Chapter 13).

Fig. 1.10. Chrysanthemum garlands for religious and festive decorations. (Photo courtesy of Priyatharshini A.)

1.6.2 Culinary uses

In various parts of Asia and Asian cuisine, chrysanthemums are valued not only for their ornamental beauty but also for their culinary and medicinal applications. The pharmacological effects of chrysanthemums, including antibacterial, anti-inflammatory, antioxidant, antiviral, and antitumor properties, are attributed to active ingredients such as flavonoids, phenolic acids, terpenoids, volatile oils, and polysaccharides (Liu *et al.*, 2024). Yellow or white chrysanthemum flowers are often boiled to make a sweet drink known as chrysanthemum tea, popular for its refreshing taste and medicinal benefits, such as aiding recovery from influenza (The Qi, 2020). In Korea, a rice wine flavored with chrysanthemum flowers, called *gukhwaju*, is enjoyed as a traditional beverage. In China, the leaves of chrysanthemums are steamed or boiled and consumed as a vegetable. Historically, nearly all parts of the chrysanthemum were used medicinally: the roots were boiled for pain relief, petals were eaten, and leaves were brewed into a ceremonial drink believed to have life-enhancing properties. In Japan, one cherished dish is "Ohitashi of edible chrysanthemums," where chrysanthemum leaves and petals are boiled, chilled, and served with a soy sauce dressing, highlighting their delicate texture and natural flavors (refer to Chapter 12).

1.6.3 Medicinal, insecticidal, and environmental uses

Cultivated chrysanthemums and their wild relatives have a rich history of use in traditional Chinese medicine and continue to be valued for their medicinal properties (Hadizadeh *et al.*, 2022). Phytochemicals such as flavonoids, terpenoids, polysaccharides, and unsaturated fatty acids have been identified within the chrysanthemum genus (Hadizadeh *et al.*, 2022; Sharma *et al.*, 2023). Additionally, *Chrysanthemum cinerariifolium* serves as a source of pyrethrins, active compounds found in its flowers that specifically target the nervous systems of insects, causing paralysis and eventual death. This makes it a popular choice for organic pest control (Matsuo, 2019; Shimira *et al.*, 2021). Pyrethrins are extracted and used in various forms, including oleoresin suspensions and powders. Additionally, pyrethrins are biodegradable and break down easily in sunlight, making them an environmentally friendly option. Beyond their insecticidal properties, chrysanthemums have been shown to reduce indoor air pollution. According to a NASA clean air study, chrysanthemum plants can absorb toxins from the air, improving indoor air quality (Wolverton *et al.*, 1989). Incorporating chrysanthemums into both indoor and outdoor spaces not only enhances aesthetic appeal but also provides significant environmental benefits.

1.7 Conclusions and Future Prospects of Global Chrysanthemum Industry

The global chrysanthemum industry has witnessed significant growth, driven by increasing demand for ornamental plants, their cultural and religious significance, and their popularity as cut flowers and potted plants. Chrysanthemums hold substantial economic value in the floriculture sector, serving as a major export commodity. Continuous advancements in cultivation techniques, including tissue culture, genetic improvements, hybridization, and disease resistance, have enhanced both yield and flower quality, contributing to the industry's expansion. The global trade of chrysanthemums benefits from efficient supply chain logistics, particularly cold-chain systems that preserve flower quality during transportation. However, the industry faces challenges such as fluctuating demand, competition from other ornamental plants, and the impact of climate change on production. Increasing environmental concerns have also led to a greater focus on sustainable practices, addressing issues such as pesticide reduction, water efficiency, and carbon footprint reduction.

Looking ahead, the chrysanthemum industry is poised for further expansion, particularly in emerging markets such as Latin America, Asia, and Africa, where favorable climates and lower production costs present new opportunities. The global production network for mother plants and propagation materials will continue to evolve, enhancing efficiency and accessibility. Innovations in chrysanthemum breeding are expected to focus on developing new varieties with improved traits, including greater color diversity, extended vase life, and increased resistance to pests and diseases. Biotechnology and genetic engineering will play a crucial role in producing superior chrysanthemum varieties suited to evolving market demands. The future of the industry will be shaped by sustainable and eco-friendly practices, integrating energy-efficient growing systems, water and nutrient management, and pest control strategies. Additionally, the adoption of digital technologies, including artificial intelligence (AI) and the Internet of Things (IoT), will optimize operations across the entire supply chain, from precision breeding to inventory management and logistics. As the industry adapts to climate change, shifts in cultivation practices and the development of climate-resilient chrysanthemum varieties will become increasingly important to ensure long-term sustainability and growth.

References

Hadizadeh, H., Samiei, L. and Shakeri, A. (2022) *Chrysanthemum*, an ornamental genus with considerable medicinal value: A comprehensive review. *South African Journal of Botany* 144, 23–43. DOI: 10.1016/j.sajb.2021.09.007.

Hao, D.C., Song, Y., Xiao, P., Zhong, Y., Wu, P. *et al.* (2022) The genus *Chrysanthemum*: Phylogeny, biodiversity, phytometabolites, and chemodiversity. *Frontiers in Plant Science* 13, 973197. DOI: 10.3389/fpls.2022.973197.

International Association of Horticultural Producers (AIPH), (2022) The great creators of Chrysanthemums. Available at: https://aiph.org/floraculture/news/the-great-creators-of-chrysanthemums/.

Liu, P.L., Wan, Q., Guo, Y.P., Yang, J. and Rao, G.Y. (2012) Phylogeny of the genus *Chrysanthemum* L.: Evidence from single-copy nuclear gene and chloroplast DNA sequences. *PLoS One* 7, e48970. DOI: 10.1371/journal.pone.0048970.

Liu, Y., Lu, C., Zhou, J., Zhou, F., Gui, A. *et al.* (2024) *Chrysanthemum morifolium* as a traditional herb: A review of historical development, classification, phytochemistry, pharmacology and application. *Journal of Ethnopharmacology*. DOI: 10.1016/j.jep.2024.118198.

Matsuo, N. (2019) Discovery and development of pyrethroid insecticides. *Proceedings of the Japan Academy, Series B, Physical and Biological Sciences* 95(7), 378–400. DOI: 10.2183/pjab.95.027.

OEC World (2022) Chrysanthemums: Latest trends. Available at: https://oec.world/en/profile/hs/chrysanthemums?yearSelector2=2022#latest-trends

SharathKumar, M. (2023) To bloom or not to bloom: The effect of photoperiod is spectrum dependent. PhD thesis, Wageningen University, Wageningen, the Netherlands. DOI: 10.18174/589945.

Sharma, N., Radha, Kumar, M.R., Kumari, N., Puri, S. *et al.* (2023) Phytochemicals, therapeutic benefits and applications of Chrysanthemum flower: A review. *Heliyon* 9(10). DOI: 10.1016/j.heliyon.2023.e20232.

Shibata, M. (2008) Importance of genetic transformation in ornamental plant breeding. *Plant Biotechnology* 25, 3–8. DOI: 10.5511/plantbiotechnology.25.3.

Shimira, F., Senem, U., Şamil, M., Özdemir, Y. and Yalçın, M. (2021) Future and prospect use of pyrethrum (*Chrysanthemum cinerariifolium*) as part of the integrated pest and disease management (IPDM) tool in Turkey. *Turkish Journal of Agriculture - Food Science and Technology* 9(1), 150–158. DOI: 10.24925/turjaf.v9i1.150-158.3771.

Simoes, A.J.G. and Hidalgo, C.A. (2011) The economic complexity observatory: An analytical tool for understanding the dynamics of economic development. In: *Workshops at the Twenty-Fifth AAAI Conference on Artificial Intelligence*.

Spaargaren, J. and van Geest, G. (2018) Chrysanthemum. In: Huylenbroeck, J. (ed.) *Handbook of Plant Breeding. Ornamental Crops*. Springer, Cham, Switzerland, pp. 319–348. DOI: 10.1007/978-3-319-90698-0_14.

The Qi (2020) The history and healing power of chrysanthemum. Available at: https://the-qi.com/blogs/journal/the-history-and-healing-power-of-chrysanthemum?srsltid=AfmBOooWUV-pIhHyecR4SghyjLqnv-Cosw2OHiOtOLs7iq-BUoIxDLdvs (accessed January 2025).

Wolverton, B.C., Douglas, W.L. and Bounds, K. (1989) Interior landscape plants for indoor air pollution abatement [Report no.NASA-TM-101766]. In: *National Aeronautics and Space Administration (NASA) Final Report*.

2

Conventional Breeding of Chrysanthemum

Neil O. Anderson*

Department of Horticultural Science, University of Minnesota, 286 Alderman Hall, 1970 Folwell Avenue, Saint Paul, MN 55108, USA

Abstract

Chrysanthemum (*Chrysanthemum* × *grandiflorum* Tzvelv.) is a significant ornamental plant, highly valued for its aesthetic appeal and economic importance. Conventional breeding endeavors aim to develop cultivars with desirable traits, such as unique flower colors and shapes, varied plant architectures, optimized flowering times, enhanced postharvest quality, and improved resilience to biotic and abiotic stresses. This chapter provides an in-depth examination of advancements and methodologies in conventional chrysanthemum breeding, emphasizing selection strategies, hybridization techniques, and polyploidy manipulation. The concept of crop ideotypes is explored, focusing on 13 key traits that influence apical dominance, branching, flower bud initiation and development, growth rates, thermozero response, and long-day leaf numbers. Strategies to maximize seed set, including addressing self incompatibility and self compatibility mechanisms, are detailed. Additionally, the identification of reproductive barriers through seed/ovule counts, recombinant inbreeding techniques, and discussions on ploidy and inheritance modes are presented. This chapter also highlights the application of mutation breeding and cytogenetics as tools to enhance genetic variability and achieve specific breeding objectives. While limitations of conventional approaches are critically assessed, this chapter discusses the potential of integrating these traditional methods with modern molecular and biotechnological innovations to accelerate the development of superior chrysanthemum cultivars that meet the evolving demands of the market.

Keywords

Crop ideotypes, genetic variability, mutation breeding, polyploidy manipulation, reproductive barriers, seeds set, self incompatibility

*Corresponding author: ander044@umn.edu

DOI: 10.1079/9781789248289.0002

2.1 Introduction

The intense conventional breeding of floricultural crops, as documented with integrated archeological genetics (Purugganan and Fuller, 2009), is well demonstrated in chrysanthemum. The earliest reports of chrysanthemum cultivation and, presumably, the associated breeding and selection of new forms date back to the 15th century BCE (Shi *et al.*, 1994–2013). Thousands of years ago, cultivated chrysanthemums were transported from King Solomon's gardens (1000–962 BCE) to egypt by Pharaoh Thutmose III (Schweinfurth, 1919). In 385 CE, the Koreans introduced cultivated chrysanthemums to Japan (Ackerson, 1967a; b), and the flower gained prominence during the Nara, Heian, and Edo time periods (8th–12th; 17th–19th centuries CE, respectively). Depictions of the forms of cultivated chrysanthemums suggest that ancient breeding and selection occurred at an early stage of the species cultivation due to flower forms (duplex, triplex, semidoubles, doubles) and petal types (quill, spoon) that do not exist in the wild species. By the 17th century CE, chrysanthemums were introduced to Europe and America, whereupon cultivation and breeding commenced almost immediately (The New York Botanical Garden, 1797; Anderson, 2006a; van Tuyl *et al.*, 2014).

2.2 Taxa Bred for Cultivation and Commercialization

Species within the *Chrysanthemum* complex, along with its alliance of satellite genera, e.g. *Argyranthemum, Glebionis, Leucanthemopsis, Leucanthemum, Rhodanthemum,* and *Tanacetum* (Anderson, 1987), have been selected, bred, and developed for various purposes. The following taxa have been cultivated and bred for diverse uses and forms:

Pyrethrum or green pesticides, *Chrysanthemum cinerariifolium* (Trevir.) Vis. (= *C. cineriaefolium*; = *Pyrethrum cinerariifolium*; = *Tanacetum cinerariifolium*) and *Chrysanthemum coccineum* (Willd.) Grierson (= *Tanacetum coccineum*); $2n = 2x = 16$; limited $2n = 3x = 24$, $2n=4x = 36$ (Gnadinger, 1936);

Edible, vegetative shoots (annual species), *C. coronarium* L. [= *Glebionis coronaria* (L.) Tzvelev]; $2n = 2x = 16$ (Chuda *et al.*, 1998);

Salt tolerance, *Chrysanthemum arcticum* L., Arctic daisy [= *Arctanthemum arcticum*; = *Dendranthema arcticum*] and its two subspecies, *C. arcticum* L. subsp. arcticum, *C. arcticum* L. subsp. polaré Hultén; $2n = 2x = 16$, $2n = 4x = 32$ (Hultén, 1937,1960,1968);

Shasta daisy, (*Chrysanthemum* × *superbum* [Bergmans ex J. Ingram] Kent; [= *Leucanthemum* × *superbum*; = L. maximum hort. non-Ramond]); $2n = 10x = 90$ (Boase *et al.*, 1997);

The come-back mum, *Chrysanthemum* × *rubellum* Sealy (Shinn, 2012);

Groundcover mums, *Chrysanthemum weyrichii* (Maxim.) Miyabe (Dowrick, 1952);

Florists and garden chrysanthemums (cut flowers, potted plants, garden types: (*C.* × *grandiflorum* Tzvelv. [= *Dendranthema* × *grandiflora* Tzvelv.;

= *C.* × *morifolium* Ramat.], *C.* × *hybridum* Anderson; $2n = 6x = 54 \pm 9$; (Anderson *et al.*, 2008; Dowrick, 1952, 1953; Dowrick and El-Bayoumi, 1966a, b); also used for medicinal, herbal teas, and wine (Miyazawa and Hisama, 2003; Anderson, 2006a; Jeong *et al.*, 2013).

The most widely cultivated chrysanthemum crop, *C.* × *grandiflorum*, is a complex perennial geophyte, historically considered an allohexaploid of more than 10 species (*C. indicum, C. vestitum, C. zawadskii*, etc.; Ackerson, 1967a, b). Breeding this crop is complicated by self incompatibility (SI) (3 *s* loci), pseudo-self compatibility (PSC), aneuploidy, sterility, inbreeding depression, and genetic load (Anderson, 2006a). *Chrysanthemum* × *hybridum* represents a taxonomic class of unique interspecific hybrids that exhibit shrub-like growth in the second year onward, distinct from traditional garden or glasshouse chrysanthemums, which maintain smaller and consistent plant sizes irrespective of plant age (Anderson *et al.*, 2008).

2.3 Phenotypic or Biochemical Traits and Germplasm Base

2.3.1 Greenhouse/glasshouse

Chrysanthemum × *grandiflorum*, widely used as florists and garden cut flowers, potted plants, garden types, and edible forms (medicinal, herbal teas, wine), represents the most extensively bred and studied chrysanthemum species (Fig. 2.1) (Anderson, 2006a). Crop classification and ideotypes have been established for numerous phenotypic classes, all of which are vegetatively propagated (primarily as tip cuttings), with hundreds or thousands of cultivars available worldwide (Bearce *et al.*, 1964; Ackerson, 1967, b; Dole and Wilkins, 1999). Glasshouse (greenhouse) chrysanthemums are cultivated as annual crops for cut flower and/or potted plant markets, while garden or landscape types are grown as herbaceous perennials (Langevin, 1992; Anderson *et al.*, 2008). Cultivars grown for edible or medicinal purposes are produced in either glasshouse or field production environments (Bearce *et al.*, 1964). Each class is primarily distinct in its evolutionary pathway, although they are primarily cross-compatible genetic pools due to years of selection and genetic isolation.

All chrysanthemums within this species are photoperiodic, primarily quantitative short-day (SD) plants for flowering (Post, 1949; Langhans, 1964a, b; de Jong, 1981; Cockshull, 1984; Ball, 1991). However, day-neutral (DN) genotypes have also been selected (Anderson, 1991; Anderson and Ascher, 2001, 2004a, b; Anderson *et al.*, 1989a). Day neutrality results from the mis-expression of photoperiod genes affecting *FT* and its downstream pathways, enabling flowering under any day length (Caibo and Qixiang, 2010). Cultivars are tested and categorized into SD response groups, based on the number of weeks of SD required for flower bud initiation and development (FBI/FBD) (Ball, 1991; Cockshull and Kofranek, 1992). These response groups guide the

Fig. 2.1. Diversity of cultivated chrysanthemums, *Chrysanthemum* × *grandiflorum, C.* × *hybridum,* grown as flowering potted plants (multiple plants: CBR) (A); MBR (B); single stem: Fuji types (C); incurved (D); bonsai types (E); cut flowers (F); garden and landscape types (upright) (G); cushion (H); shrub, in field plantings (I); standard types grafted onto *Artemisia* rootstock, trained onto a metal frame (J); groundcover (K); and an example cut flower type "Mellow Moon" used for chrysanthemum tea (L). CBR = center bud removal, MBR = multiple bud removal (Photo courtesy of Neil Anderson.)

selection of appropriate cultivars for crop scheduling throughout the calendar year (Dole and Wilkins, 1999).

Cut flower and potted plant production systems, and to a limited extent, garden or landscape types, have distinct pinching and disbudding techniques to create various flowering phenotypes. The crop can be grown either pinched or unpinched, depending on its intended purpose (Post, 1950). When grown as a pinched crop to enhance branching from a single cutting, a soft pinch is applied to remove the terminal meristem of a rooted cutting. The crop is then grown under noninductive conditions (long days, LD) for an additional 2–3 weeks prior to the commencement of SD conditions (Ball, 1991; Dole and Wilkins, 1999). During the SD photoperiod treatment, additional pinches may be applied if the goal is to produce disbud or spray inflorescences. For a disbud crop, axillary or lateral flower buds appear during SD, they are manually removed (disbudded), leaving only the terminal or apical meristem to develop into a flower—termed "multiple bud removal" or MBR (Fig. 2.1B; Williams and Bearce, 1964; Ball, 1991). This practice often enhances the size of the apical flower as it becomes the sole "sink" for carbohydrates. In contrast, in spray crops, the apical meristem (floral bud) is removed to break apical dominance and encourage lateral shoot growth, a process known as "single bud removal" (SBR; Fig. 2.1A). This technique results in stems with all flowers opening at approximately the same time (Williams and Bearce, 1964; Ball, 1991). Examples of products within each class exhibit a wide variety of phenotypic traits, including single-stem types with MBR (Fig. 2.1C, D), bonsai (Fig. 2.1E), and cut flowers (Fig. 2.1F).

2.3.2 Landscape, garden herbaceous perennials

The breeding of garden chrysanthemums in the USA began prior to 1919 at the Arlington, VA Experimental Farm. This effort utilized glasshouse germplasm, which was subsequently adapted for garden production. Selection focused on alleles that conferred traits such as winter hardiness or cold tolerance, early flowering before the onset of frost, frost-tolerant flowers, and attractiveness to honeybee, or other pollinators (Mulford, 1937, 1939; Longley, 1939). Early studies on flowering in the garden showed that, among seedling selections, blooming occurred earlier in field conditions than in cold frames or glasshouses (Longley, 1939). Before the ploidy levels were known, flowering earliness was proposed to be a complex trait (Mulford, 1937). Soon thereafter, most university (public-sector) breeding programs across the USA began breeding garden chrysanthemums. Private-sector breeding programs started later, such as Yoder Brothers, Inc. (Barberton, OH). Currently, the only remaining public-sector chrysanthemum breeding program in the USA is at the University of Minnesota, where garden chrysanthemum breeding started in 1924, continuing their glasshouse chrysanthemum breeding, which commenced as early as 1888 (Synder, 1983; Anderson, 2006a). The germplasm for garden chrysanthemum has evolved independently from glasshouse germplasm, although the two gene pools are still cross-compatible (Anderson, 2006a).

Garden and landscape chrysanthemums (*C.* × *grandiflorum* and *C.* × *hybridum*) are popular herbaceous perennials (Morrison, 1923) and are often sold as unpinched or 1x–2x SBR pinched crops for commercial sales in the spring, such as for Mother's Day in the USA (Viehmeyer and Whitney, 1965; Hamilton, 1991; Langevin, 1992; Anderson, 2006a). When grown for autumn sales in field production, they are typically pinched at least once during production, and FBI/FBD under natural photoperiods (Hamilton, 1991; Langevin, 1992). Further pinching, whether using MBR or SBR, is typically reserved for containerized products intended for autumn displays in botanical gardens and conservatories or for events held by amateur breeding societies, such as the National Chrysanthemum Society in the USA (https://www.mums.org/) and the National Chrysanthemum Society in the UK (https://www.nationalchrysanthemumsociety.co.uk/). Such techniques are also employed for chrysanthemum festivals held in South Korea (Fig. 2.2A) (https://www.korea.net/Events/Festivals/view?articleId=7587), Taiwan (https://www.nspirement.com/2018/12/07/a-magnificent-chrysanthemum-show-in-taiwan.html), Japan, and China (Fig. 2.2B) (Ackerson, 1967a, b).

Classic garden chrysanthemum plant phenotypes include upright forms (for cut flower use, Fig. 2.1F, L; flowers are produced only at the top of the plant without the use of MBR), cushion or hemisphere types (featuring natural branching due to the lack of terminal FBD under LD; Fig. 2.1G), shrub types (Fig. 2.1H, I), and groundcover or spreading forms (exhibiting horizontal growth rather than vertical growth; Fig. 2.1K) (Viehmeyer and Whitney, 1965; Widmer, 1980; Anderson, 2006a, b; Anderson *et al.*, 2024). Some large specimens (hemispheres) or smaller bonsai forms are grafted onto *Artemisia scoparia* rootstocks (Fig. 2.1J) and trained using bamboo or metal frames to achieve specific shapes (Zhang *et al.*, 2013). In contrast, large hemispherical forms can also be created without grafting, training, or pinching with the use of the shrub cushion series, *C.* × *hybridum* "Mammoth™" series (Fig. 2.1H, I; Anderson *et al.*, 2008, 2012a, b, c, 2015a, 2017), developed at the University of Minnesota. The fall-flowering-season cultivars are divided into early-, mid-, and late-season flowering (Widmer, 1980). Example of products within each phenotype class include upright types (e.g. *C.* × *grandiflorum* "Centerpiece," "Rose Blush," "Gold Country," "Mellow Moon," or "Duluth"; Ascher *et al.*, 1993; Minnesota Agricultural Experiment Station, 2021; Widmer *et al.*, 1981, 1982), cushion types ("Minngopher," the "Minn" series; Phillips and Widmer, 1965; Widmer and Ascher, 1971, 1973, 1976), shrub cushions (*C.* × *hybridum* "Mammoth™" series; Anderson *et al.*, 2008), and wave or groundcover types (*C.* × *grandiflorum* "Cushion Ball," "Fall Color," "Snowscape," and MN Sel'n. 90-275-27, US Plant Patent No. 14,749; Anderson *et al.*, 2004; Ascher *et al.*, 1996; Hong *et al.*, 2006; Kim *et al.*, 2018). The first cushion habit garden chrysanthemums were the "Minn" series, released between the 1950s and 1970s. Other breeding programs quickly incorporated this phenotype into their product lines (Fig. 2.1G), and, as a result, the cushion plant habit now holds the majority of the market.

Fig. 2.2. Outdoor chrysanthemum festivals and exhibitions, such as in Daejeon, South Korea (A) and Xiaolan, China (B), use geometric forms and building structures in which chrysanthemums are grown for the season and flowered under natural photoperiods. (Photo courtesy of A—Neil Anderson, and B—Mark Gilquist.)

2.3.3 Salt tolerance

Salt tolerance is a trait of significant interest for use in high salinity soils or in landscapes exposed to elevated salt levels such as transportation corridors and roadway shoulders in northern climates where sodium chloride (NaCl) is applied for snow melt. In transgenic *C.* × *grandiflorum*, the heterologous expression of *AtDREB1A* has been shown to enhance salt tolerance (Hong *et al.*, 2006). Particularly important for salt tolerance within the chrysanthemum complex are species from the Arctic daisy group, including *C. arcticum* (Fig. 2.3A), *C. a.* subsp. *arcticum* (Fig. 2.3B), *C. a.* subsp. *polaré* (Fig. 2.3C) (Liu and Anderson, 2019; Liu *et al.*, 2019; Bower-Jernigan and Anderson, 2023). These species thrive exclusively in maritime environments, either in salt water directly (Fig. 2.3D) or on seaside cliffs (Fig. 2.3E) (Hultén, 1937, 1968; Porsild and Cody, 1980; Uehara *et al.*, 2017). Outstanding traits such as salt tolerance, groundcover plant habit, and winter hardiness make these species of genetic interest for conventional and molecular breeding. Breeding efforts have commenced at the University of Minnesota Flower Breeding and Genetics Laboratory to determine flowering requirements (Liu, 2021; N.O. Anderson, 2018–2019, unpublished data), the presence or absence of SI, ploidy levels, and varying levels of salt tolerance across the Arctic daisy species complex. A wide array of germplasm has been collected from the wild, making the germplasm bank of the chrysanthemum species complex more genetically diverse for germplasm enhancement and breeding than most other chrysanthemum species.

2.3.4 Miscellaneous ornamental species

The Shasta daisy (named after Mt. Shasta by Luther Burbank; Anderson and Olsen, 2015), also known as King Edward daisy, *C.* × *superbum*, is a winter-hardy perennial chrysanthemum bred for its large white, often single daisy flowers with golden central disc florets (Boase *et al.*, 1997). They are used as cut flowers as well as garden perennials. Multiple cultivars exist on the market as complex polyploids. Environmental conditions necessary for flowering have been described by Shedron and Weiler (1982) and Yahel *et al.* (1985). The come-back mum, *C.* × *rubellum*, is another herbaceous perennial (Shinn, 2012). Its flowering occurs sporadically, with blooms appearing on some but not all shoots. Common cultivars that have been on the market for decades include “Clara Curtis,” “Sheffield Pink,” and “Emperor of China.” Little is known about reproductive biology and the barriers associated with this species. Future breeding efforts should focus on achieving 100% flowering on all shoots to enhance the overall floral display and “Flower Power.” The natural groundcover chrysanthemum, *C. weyrichii*, is a hexaploid species native to Japan. Common cultivars on the market include “White Bomb” and “Pink Bomb” and have been available commercially for decades. This species also exhibits sporadic flowering among the flower stems, similar to *C.* × *rubellum*. While it is presumed to be self incompatible, there are limited data available on breeding efforts for this species. However, it has been successfully crossed with hexaploid *C.* × *grandiflorum* to produce the distinctive shrub chrysanthemum hybrids, *C.* × *hybridum* (Anderson *et al.*, 2008).

Fig. 2.3. Salt-tolerant, winter-hardy perennial American species of chrysanthemum in the Arctic daisy species complex: *Chrysanthemum arcticum* (Old Valdez, Alaska, USA) (A); *C. arcticum* ssp. *arcticum* and flower closeup—inset (Attu Island, Alaska, USA) (B), and *C. arcticum* ssp. *polaré* (Hudson Bay, Churchill, Manitoba, Canada) (C), growing in maritime conditions either directly in salt water (C, D) or adjacent to the ocean on cliff faces (E). (Photo courtesy of Neil Anderson).

2.3.5 Medicinal, tea

Edible forms of chrysanthemums, including those used for medicinal purposes, herbal teas, and wine (*C.* × *grandiflorum*), often involve fully double flower types such as incurves (football) and Fuji. Occasionally, smaller single, duplex, or triplex flower types may also be used (Miyazawa and Hisama, 2003; Jeong *et al.*, 2013). The production of liáng chá or cooling/herbal tea (Fig. 2.4) originated in the former Lingnan area of southern China for medicinal purposes, although today they are used as an herbal tea with beneficial health properties due to anthocyanin and carotenoid pigments (Liu *et al.*, 2013; Fu *et al.*, 2018a; Kan *et al.*, 2020). As many as 46 flavonoids and 17 caffeoylquinic acids have been identified, with radical scavenging, antioxidant, anti-inflammatory, and antimutagenic properties (Li *et al.*, 2019). These compounds include acacetin 7-O-galactoside and apigenin 7-O-beta-d-(4'-caffeoyl) glucuronide, which have demonstrated anti-HIV activity (Lin and Harnly, 2010). Chrysanthemum wine, traditionally consumed during Chinese celebrations such as the Double Ninth Festival (also known as the Double Yang Festival), is made by fermenting edible chrysanthemum petals with a wine-yeast ball for 24–72 hour (Five Seasons TCM, 2021). Regardless of the intended use of the chrysanthemum flowers, the optimal stage for harvesting to maximize carotenoid contents, such as caffeoylquinic acid, is when the flowers are 10% open (Chen *et al.*, 2021). Breeding programs in China and the USA (e.g. at the University of Minnesota) are working on chrysanthemum varieties for medicinal and tea uses. However, information on these breeding efforts is limited, and the germplasm diversity remains undocumented. These product lines would benefit from a clear delineation of the pedigrees and ancestry involved in generating their biochemical profiles, as well as from efforts to enhance germplasm diversity.

2.3.6 Food

Shungiku (Asian green seed) or mantilida (Crete) is an edible, annual species, *C. coronarium*, grown for its vegetative shoots and/or flower buds (Fig. 2.5), as well as for use as an ornamental bedding plant (https://www.johnnyseeds.com/vegetables/greens/asian-greens/shungiku-asian-green-seed-514.11.html). The shoots can be stir-fried for enhanced flavor and fragrance or used in salads, vegetables, pickles, and sushi (Fig. 2.5; https://en.wikipedia.org/wiki/Glebionis_coronaria). The shoots are high in potassium, while the flowers have carotenoid pigments (nutraceuticals). A short life cycle and the diploid nature of the species confer a rapid generation time for seed-propagated cultivars (open-pollinated (OP)). Many seed companies throughout the world (particularly in eastern Asia and North America) produce this as an edible chrysanthemum crop, although the genetic variation within the germplasm base is unknown and undocumented. Undoubtedly, the crop would benefit from plant collections in its native habitats throughout the Mediterranean

Fig. 2.4. Chrysanthemum tea flowers (incurved or football types) being air-dried after harvest (A), and example commercial products for consumers (B). Upon brewing in hot water (*cf.* individual teacups (B)), each dried chrysanthemum flower reopens—or reblooms—in the teacup for enjoyment. (Photo courtesy of Neil Anderson and Andrew Stobbs.)

Fig. 2.5. Edible Shungiku or mantilida shoots of *Chrysanthemum coronarium*, used for culinary purposes as stir-fried or fresh vegetable greens. (Photo courtesy of DreamsTime.com stock photo; downloaded with permission.)

region (Tutin *et al.*, 1964). Development of F_1 hybrid seed-propagated types would be beneficial as a future breeding focus.

2.3.7 Green pesticides

The species *C. cinerariifolium* and *C. coccineum* have been bred for their use in producing green pesticides or the naturally occurring compounds (Pyrethrin I and II) extracted as pyrethrum. While both species have been used as insecticides for over 1000 years, most breeding efforts are relatively recent, with severe genetic bottlenecks causing production challenges. *C. coccineum*, or painted daisy, is native to Eastern European Russia, Iran, Kazakhstan, the North Caucasus, Tadzhikistan, Transcaucasus, Turkey, and Uzbekistan (Contant, 1976). It is now primarily bred for ornamental purposes due to its colored flower petals and low levels of Pyrethrin I and II (Fig. 2.6A). In contrast, *C. cinerariifolium* (native to the Baltic states) only produces white petals (Fig. 2.6B). Painted daisy pyrethrum production began in 1828, but by 1840, it had largely been replaced by *C. cinerariifolium* (Contant, 1976). Currently, it is "reduced to a homestead crop in its native area" (Contant, 1976). Limited ornamental cultivars featuring fern-like foliage are available on the market, such as the seed-propagated "Robinson Mix" (Burpee Seed Co., Warmister, PA, USA; https://www.burpee.com) and clonal "Robinson Red" (Walters Gardens, Zeeland, MI, USA; https://www.waltersgardens.com/variety.php?ID=TANRR). Between 1939 and 1941, attempts to breed painted daisy for increased cold and drought tolerance in the USSR were largely unsuccessful (Contant, 1976), although it has been bred at Huazhong Agricultural University, Hubei, Wuhan, China, for pyrethrin production in Yunan Province (Zeng *et al.*, 2020; Zheng *et al.*, 2021). Although *C. coccineum* and *C. cinerariifolium* are both primarily diploid and can hybridize, the resulting hybrids are sterile (Contant, 1976; Tuikong, 1984). Very few breeding programs focus on this species, with a limited number of commercially available cultivars. Thus, the germplasm base of the species is narrow and would benefit from collections of germplasm in the wild, at least for further development as an ornamental or as a subject for scientific study (Zheng *et al.*, 2021).

When the insecticidal properties of *C. cinerariifolium* were discovered in 1840 (Contant, 1976), this species immediately replaced *C. coccineum* for pyrethrin production in Dalmatia (Gnadinger, 1936)—a position it has maintained in worldwide production ever since. Commercial production in Dalmatia continued until 1914, when Japan became the major producer during World War I, following the species' introduction there in 1885 (Ueyama, 2017). This dominance continued until World War II, after which production shifted to Kenya, where *C. cinerariifolium* had been introduced in 1928 (Gnadinger, 1936). Early breeding research occurred in Kashmir and India, with that germplasm serving as the basis for breeding in Tasmania (Bhat and Pandita, 1977; Gullickson, 1995). Since the 1970s, the Tasmanian germplasm base has narrowed significantly, although breeding has focused on selecting plants

Fig. 2.6. Winter-hardy, herbaceous perennials. Painted daisy, *Chrysanthemum coccineum,* has colored flowers and finely dissected leaves (A); pyrethrum, *C. cinerariifolium,* the primary source of pyrethrins, has gray-white leaves and only white flower petals (B). (Photo courtesy of Neil Anderson and Cora Rost.)

suited for mechanical harvesting of the flowers, specifically the achenes, which accumulate 94% of the pyrethrum content in the plant (Fig. 2.7) (Suraweera *et al.*, 2017). The achenes also contain the highest concentration of pyrethrin (Botanical Resources of Australia Co., Ulverstone, Australia). Production has also shifted in central Africa, with Tanzania currently being the major producer in the region, where cooperative farmers hand-harvest the crop. The Tanzanian germplasm base is broader than that of Tasmania (McLaughlin Gormley King Co., Minneapolis, MN, USA). Both production companies in Tasmania and Tanzania are now owned by Sumitomo Company (Tokyo, Japan), and coordinated pyrethrum breeding efforts are jointly conducted along with the University of Minnesota (USA) breeding program (Anderson *et al.*, 2021). Most pyrethrum crops are seed propagated (OP; no F_1 hybrid seed types exist) and direct sown (Tasmania) or transplanted (Tanzania). No cultivars are commercially available, since each company keeps its selections in-house, although at least one has been patented: "Hypy" (Bhat and Menary, 1987; Bhat, 1995; Gullickson, 1995). Elite parents and selections in the breeding programs are maintained clonally (splits) for combining ability and yield testing of progeny. Current conventional breeding objectives include disease resistance, increased pyrethrin content, annualized perennial growth, and winter hardiness (Parlevliet, 1974; Anderson *et al.*, 2021). Molecular breeding is in progress to develop marker libraries for use in marker-assisted selection; its 7.1 Gb genome has been sequenced with over 2 million scaffolds and more than 60,000 genes predicted to date (Yamashiro *et al.*, 2019). Wild-collected germplasm is being used in the University of Minnesota breeding program for integration with commercial lines to enhance genetic diversity. However, additional focused germplasm collections would benefit breeding and production efforts worldwide.

2.3.7.1 Traits of interest

"Flower power and convenience" are two critical phenotypes that are important to all flowering floricultural crops (Anderson *et al.*, 2006; Anderson and Olsen, 2015). Flower power refers to all the attributes that chrysanthemum flowers and/or plants possess that make them memorable and desirable to consumers for use in both interiorscapes and landscapes. These attributes include flower color, flower types, and whether the flowers are single or double (with extra petals), as well as their utility for medicinal purposes and chrysanthemum tea. Convenience refers to the relative ease of production for growers; long-lasting postharvest vase life for wholesalers, distributors, retailers, and consumers; adaptability of the crop to a variety of production methodologies; and responses to plant growth regulators for height control, temperature for cold and/or heat tolerance, salt tolerance, and responses to day length for year-round scheduling (programmability) (Anderson *et al.*, 2006).

Specific phenotypic and biochemical traits of interest to chrysanthemum breeding programs include a wide range of biotic and abiotic factors controlling important attributes with market value (Boase *et al.*, 1997). Other traits of

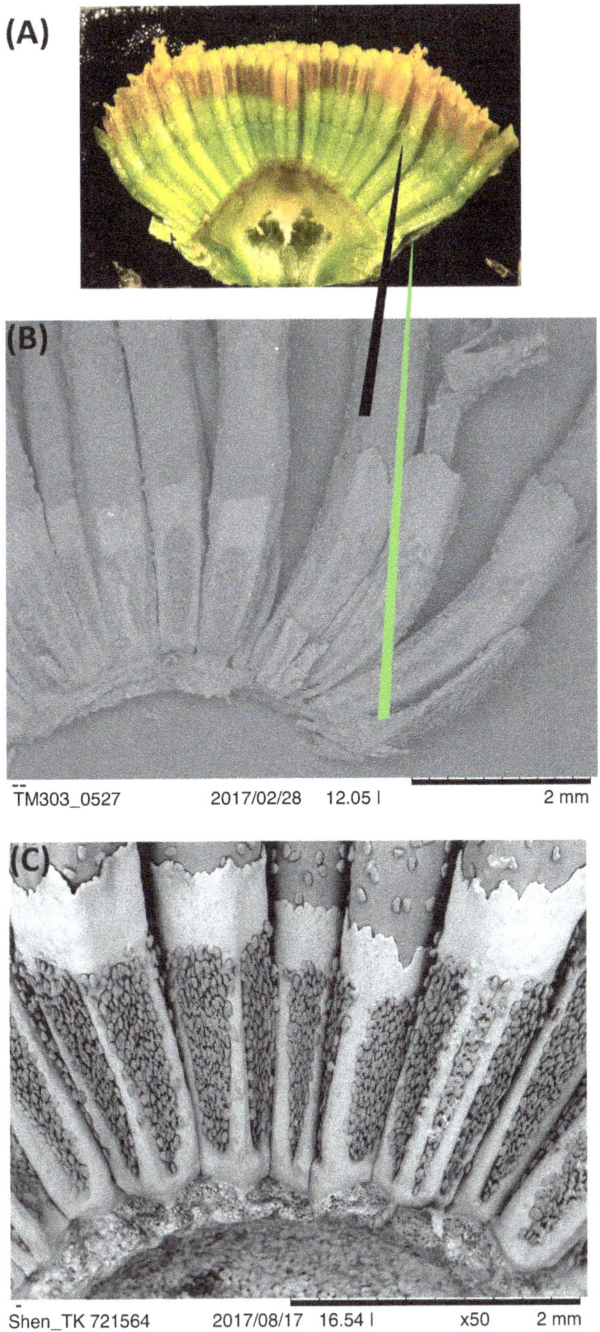

Fig. 2.7. Closeup of pyrethrum inflorescences, *Chrysanthemum cinerariifolium*, with the ray petals removed and cut in half at the midpoint diameter (A); SEM closeup of the florets showing the glandular trichomes clustered at the base of each ovule on all sides (B); red arrows indicate corollary disc florets, while orange arrows denote ray florets; greater SEM magnification of the glandular trichomes (C). SEM = scanning electron microscope. (Photo courtesy of Michele Schermann.)

interest in the crops depend on breeding objectives for market use and include phenotypic traits (flower color, flower type, flower shape, plant height, plant width, plant habit, ease of seed or clonal propagation, LD leaf number, day neutrality, perenniality, and pest and disease tolerance/resistance); biochemical compound production (pigments, medicinal components, fragrance, flavor, ethylene sensitivity, and green pesticide compounds such as pyrethrin I, II); physical attributes (postharvest vase life, rapid generation cycling (RGC), annualized perennials); abiotic (stress tolerance—heat, cold, drought); and/or molecular components for marker assisted selection (isozymes, inter-simple sequence repeats (ISSRs), simple sequence repeats (SSRs), and single-nucleotide polymorphisms (SNPs)) (Boase *et al.*, 1997; Anderson, 2006a; Anderson *et al.*, 2021).

A chrysanthemum breeder would typically focus on specific breeding objectives, selecting one or more traits that could be bred and selected for simultaneously to increase breeding efficiency. The development of parents with high combining ability (either specific or general combining ability (SCA and GCA), respectively; Allard, 1960; Bernardo, 2014), as well as the development of inbred lines, would be a priority in any breeding program. This would facilitate the creation of segregating hybrids or inbreds for parental selection. Hybrid trials involving elite lines or asexually propagated hybrids (clones) are essential to confirm breeding success by evaluating uniformity through gross phenotypic screenings, biochemical analyses, and molecular markers for gene/allele expression and trait stability. Given that chrysanthemum cultivars are selected for global production, stability is rigorously assessed through multiple trials over several years and across different locations. These trials are designed to ensure that the traits of interest are uniformly and optimally expressed in diverse environments (Allard, 1960; Anderson, 2006a, 2020, Anderson *et al.*, 2006).

2.4 Breeding Objectives and Crop Ideotype Breeding

2.4.1 Relationships between risks and economic rewards

As seen in other floricultural crops, the relationships between risks and economic rewards in breeding new chrysanthemum cultivars are dependent on numerous factors (Anderson, 2004, 2006a, b; Drew *et al.*, 2010, 2015). When possible, clonally propagated cultivars should be protected as intellectual property (IP) with plant patents (USA) and/or plant breeder's rights in Europe, Canada, etc., to provide premiums (royalties) back to the breeding program (Drew *et al.*, 2015). Due to varying levels of domestication and crop selection, chrysanthemum product classes differ in their placement along the scale of risks and rewards (Fig. 2.8) (Anderson, 2004, 2006b). Crop domestication typically begins with the release of the first cultivar into the market, establishing a new product class. For example, the introduction of the Arctic daisy (*C. arcticum*, *C. a. subsp. arcticum*, *C. a. subsp. polaré*) (Fig. 2.3) into the market as a "distinguishable product" focuses on its unique trait of salt tolerance (Liu and Anderson, 2022b).

Fig. 2.8. Relationship between risk and economic reward in relationship to research and development time a chrysanthemum breeder would spend creating new phenotypes: Within a series (relatively fast, *Chrysanthemum* × *grandiflorum* "Peach Centerpiece"; new colors can be derived from mutations or using different male parents on the female seed parent), while a new series with novel trait(s) takes more time to develop in an entire flower color palette ("Centerpiece" series). Both classes are often intraspecific hybrids in derivation. A distinguishable product with a new groundcover habit (frost-tolerant and winter-hardy perennial hybrid, MN Sel'n. 90-275-27) and a shrub chrysanthemum Innovative Product (*C.* × *hybridum* "Mammoth Red™") both take significantly more time to develop, particularly with interspecific crossing involving *C. weyrichii*. (Photo courtesy of Neil Anderson and David Hansen.)

Since chrysanthemum crop domestication status is "inversely related to increasing economic risk," after a crop is domesticated and competitive private/public-sector chrysanthemum breeding programs focus on developing new products to compete with market comparisons, it requires less breeding effort to create new cultivars within a "new series" (Fig. 2.1; Anderson, 2004). As the knowledge base increases within the crop species, it becomes easier to implement that knowledge for enhanced product development. As domestication continues, it takes less focused, but still time-consuming, breeding efforts for chrysanthemum crop species already on the market to evoke changes and create a "new series" (Fig. 2.1). The new wave or groundcover plant habit in *C.* × *grandiflorum* MN Sel'n. 90-275-27, US Plant Patent No. 14,749 (Anderson and Ascher, 2004b) is an example of a "new series" of products within the established and highly domesticated crop species (Fig. 2.1; Shinoyama *et al.*, 2006; Anderson *et al.*, 2008). While it took 10–15 years of focused breeding, accumulation of new alleles or mutations, and directed selection to create the wave or groundcover plant habit, it involved intraspecific crosses to create this "new series" (Fig. 2.1; Anderson and Gesick, 2003, 2004).

For established or "cash cow" crops with existing cultivars and series on the market (Anderson, 2004), minor changes arise from mutations or directed breeding objectives that create new products "within a series," differing by one trait or more, such as flower color changes in *C.* × *grandiflorum* (Shinoyama *et al.*, 2006; Anderson *et al.*, 2008) or leaf shape or flavor improvements in edible Shungiku or mantilida, *C. coronarium* (Fig. 2.5). These product lines realize market value and profitability based on the sheer volume of units (seed or cuttings) sold, despite a significantly lower per unit price than "distinguishable" or "innovative" products.

Dramatic alterations to crop phenotypes, often achieved through high-risk interspecific hybridizations, lead to the creation of "innovative" products (Anderson, 2004; Anderson, 2006a, b). A prime example of such alterations is the development of shrub chrysanthemums with a hemispherical, cushion plant habit, resulting from the interspecific hybridization between hexaploid *C. weyrichii* × [*C.* × *grandiflorum*] (Anderson *et al.*, 2008, 2012a, b, c, 2015a, 2017). These innovative chrysanthemums, released as the *C.* × *hybridum* "Mammoth™" series, differ significantly from standard *C.* × *grandiflorum* cushion types. From the second year of growth onward, they adopt a genetically based shrub habit, unlike standard types that require pinching, shearing, or complex grafting and wiring to achieve a similar form. The creation of this new plant form, combined with its appeal as a larger retail product and its ability to generate higher economic returns per unit, contributed to garden chrysanthemums becoming the number one herbaceous perennial in US sales since the mid-2000s, surpassing *Hemerocallis* and *Hosta* (United States Department of Agriculture, 2019).

2.5 Production Timeframes

For any new product to be successfully integrated into the market—whether it falls under an existing series, a new series, a distinguishable product, or an innovative product—cultivars must meet established specifications for crop production to seamlessly enter the horticultural distribution chain (Anderson, 2004; Anderson, 2006b; Drew *et al.*, 2010). These requirements include the total crop time to fit within scheduling requirements for year-round production, meeting aesthetic ratios, and ensuring postharvest adaptability for shipping and storage (Anderson, 2006a). For annual bedding plants, the maximum finishing time for crop production is 16 weeks, 16–18 weeks for flowering potted plants, and 18–20 weeks for cut flowers (Anderson, 2004). While long crop times are not the norm, a few exceptions may remain on the market due to narrow genetic bases and/or a lack of breeding for crop improvement. For instance, standard cyclamen potted plants, *Cyclamen persicum*, exceed the 16–18 week period (Ball, 1991; Dole and Wilkins, 1999) and spike, *Dracaena indivisa* (= *Cordyline indivisa*), a foliage bedding plant, requires at least 20 weeks or 4 weeks longer than standard bedding plants (https://www.stokeseeds.com/ca/indivisa-dracaena-895a-group). It is generally not advisable to breed new chrysanthemums that exceed the established product class range for production time.

2.6 Aesthetic Ratio

Plant height and width dimensions must conform to the aesthetic ratio of the final container size in which it is grown (Anderson, 2004). This aesthetic ratio is not applicable to the breeding, selection, and production of cut flowers, as these are grown to achieve the longest possible stem length (50–90 cm) with the support of a structure that ensures adequate stem length for multiple postharvest cuts. These cuts are made when floral hydration or preservative solutions are changed until the final stem cut is made by the floral designer for a floral arrangement (Ball, 1991; Hunter, 2000). At this point, the floral designer applies the aesthetic ratio to ensure the correct proportion of cut stems in a vase (Hunter, 2000).

The aesthetic ratio refers to the proportion, a design principle that ensures both the physical and visual balance of products. This balance allows plants to "stand on their own" without the risk of toppling over, which could damage the product or prevent it from fitting within standard shipping cart shelf heights or packing containers. Standard proportions range from 1.5 to 2 times the height or width of the container, whichever is greater (Fig. 2.9A). This design principle is routinely used in floral designing (Hunter, 2000). For example, a potted chrysanthemum that exceeds the aesthetic ratio (Fig. 2.9A) would not fit onto shelf heights of standard carts for transport prior to shipping (Fig. 2.9B). Extensive trials during the breeding process involve screening for plant height and LD leaf numbers to ensure that either the genotype is genetically capable of meeting the required aesthetic ratio or that plant growth regulators can be applied to shorten internodes (Ball, 1991; Dole and Wilkins, 1999). As Allard (1960) indirectly advised, chrysanthemum breeders must be "ruthless in selection" to avoid creating nonfunctional products that could undermine the effectiveness of a breeding program.

2.7 Crop Ideotypes

Crop ideotypes are widely used by plant breeding programs to model the desired phenotype of plant growth for a set of traits tailored to specific crop environments (Donald, 1968). Breeding objectives are incorporated into the ideotype to achieve the desired trait expression (Anderson, 2006a; Anderson and Ascher, 2001, 2004b). For chrysanthemum crops, at least four crop ideotypes have been proposed. Langton and Cockshull, 1976 developed the first *C.* × *grandiflorum* ideotype for cut spray chrysanthemums for all-year-round (AYR) production. The cut spray chrysanthemum ideotype consists of 13 traits that affect apical dominance, branching, FBI/FBD, growth rates, thermozero response, and LD leaf numbers (Langton and Cockshull, 1976). After

Fig. 2.9. The aesthetic ratio for container growing of potted chrysanthemums (indoor, garden types) is determined by multiplying the container height or width (x; whichever is greater) × 1.5–2.0 to achieve a physically/visually balanced chrysanthemum plant with a proportioned height in relationship to its container (A). Example shipping carts with flowering, garden chrysanthemum grown with the proper aesthetic ratios (B). (Photo courtesy of Neil Anderson.)

identifying the key traits for the ideotype, genotypes were tested to determine the correlation or independence of traits, as well as the broad-sense heritability, allowing for further refinement of the proposed ideotype. They tested 40 cut flower cultivars and provided specifics for each of the 13 traits, noting significant correlations between leaf and bract numbers under SD and LD conditions. They found that low SD leaf numbers were associated with correspondingly low LD leaf numbers (LDLN), resulting in undesirably shorter stems (Langton and Cockshull, 1976). As a result, the ideotype required revision, demonstrating its flexibility and adaptability to genetic variation.

Ascher (1986) proposed an ideotype for *C.* × *grandiflorum* focusing on F1 hybrid (seed-propagated) garden chrysanthemums. This ideotype was developed based on the testing of 15 F1 hybrid lines for the feasibility of glasshouse production and uniformity of flowering in both greenhouse and field conditions (Stephens *et al.*, 1983). The market comparison, "Autumn Glory," never flowered from seed under LD or SD, whereas 12 elite hybrid lines from the University of Minnesota breeding program were day neutral (Stephens *et al.*, 1983). Subsequent breeding, trialing, and selection of superior lines and inbred parents for the next 15 years resulted in a refined ideotype with built-in caveats that the phenotypic uniformity in clonally propagated cultivars could not be realized in seed-propagated hybrids, as minor variation in plant height and width and flower color persisted even after generations of parental inbreeding (Anderson *et al.*, 1995).

The *C.* × *grandiflorum* glasshouse and garden chrysanthemum DN and heat-delay-insensitive (HDI) ideotype was developed to produce flowering potted plants for glasshouse production, as well as cushion or hemispherical garden plants that would flower before the onset of a killing frost in northern latitudes (Anderson and Ascher, 2001, 2004b). This ideotype arose from studying how the cushion plant habit in garden chrysanthemums is formed under photoperiod and temperature interactions. The cushion habit, which now has the majority market share among garden chrysanthemum sales worldwide, was first developed at the University of Minnesota and first released in 1977 as the "Minn" series: "Minngopher," "Minnqueen," "Minnrose," "Minnruby," "Minnwhite," and "Minnyellow" (https://mnhardy.umn.edu/varieties/flowers/chrysanthemums). Nineteen genotypes (1 glasshouse, 18 garden) were tested in a series of five glasshouse and field environments used to identify DN plants and differentiate them for HDI. The DN and HDI ideotype consists of eight traits associated with leaf, flower, and internode development from the first to the sixth flower (Anderson and Ascher, 2001). Three critical environmental conditions are used to identify genotypes that adhere to this ideotype: SD with night temperatures of ≥ 16.7°C; LD with far-red (FR) light and night temperatures of ≥ 26.7°C; and continuous (24 hour) FR plus red (FR + R) light with day and night temperatures of ≥ 28.3°C. The DN and HDI ideotype has been validated for breeding both flowering potted chrysanthemums and garden varieties. An ideal phenotype that met the ideotype's criteria was selected for further development (Anderson, 1991; Anderson and Ascher, 2001).

In the green pesticide herbaceous perennial *C. cinerariifolium*, a crop ideotype was created for a wide array of traits (Anderson *et al.*, 2021), based on previous breeding research (Chamberlain and Clark, 1947; Parlevliet, 1969, 1970, 1974; Parlevliet and Contant, 1970; McLaughlin, 1973). Key traits included annualization (flowering within first year from seed; Wilkins and Anderson, 2006); synchronous/asynchronous flowering for mechanical vs hand harvesting, respectively, and tall plant height with strong stems for lodging resistance. Additionally, the ideotype focused on the ability to split or produce divisions for asexual clonal propagation, as well as tolerance or resistance to pathogens, insect pests, and nematodes. Other important traits included increased dried flower and pyrethrin (Py) yields per unit area, high Py content in flower dry matter (dry weights), fresh flower yield, moisture content, dry flower yield per unit area, stability across different locations and years, and winter hardiness (cold tolerance).

2.8 Conventional Breeding Strategies

Floral morphology is important when considering conventional breeding strategies to maximize seed set, strategically circumvent reproductive barriers, and implement techniques to avoid embryo abortion. Since all *Chrysanthemum* species are members of the Asteraceae, their floral morphology follows the distinct family's compound inflorescence with numerous florets (disc vs ray) in every "flower" (Smith and Laurie, 1928; Viehmeyer and Whitney, 1965; Birkeland, 1960; Cockshull, 1984). The fruit is a ribbed achene containing one anatropous ovule per disc and/or ray floret (Anderson *et al.*, 1990).

Ray florets are petaloid and gynoecious (female only), while disc florets are usually apetalous and hermaphroditic (Cockshull, 1984). However, exceptions are common in the floral anatomy of *C.* × *grandiflorum* and *C.* × *hybridum* due to years of focused breeding, mutation, and selection such that full doubles with 100% ray florets can be crossed only as female parents. Thus, multiple flower types or forms exist, including Type I—single daisy, semidouble (multiple rows of ray florets), and anemone (semi-petaloid disc florets); Type II—petals incurved or reflexed, Button or pompon (double with short petals, <3.81 cm diameter), intermediate pompon (3.81 cm to 7.62 cm diameter), large pompon (7.62 cm–11.43 cm diameter, when disbudded), standard and exhibition (>11.43 cm diameter); and Type III—tubular petals, spoon, single or semidouble (tubular petals with tips that are flattened or "spoon like"), quill (tubular petals with closed tips like a drinking straw), and spider or Fuji (tubular petals with coiled or hooked tips, respectively) (Carpenter and Shackelford, 1966). Semidouble flowers have a higher percentage of ray florets than the wild type, although varying numbers of disc florets remain as potential pollen sources. Wild-type flowers are predominantly single daisy types (although duplex and triplex may occur), with one concentric ring or row of ray florets and the remaining florets being disc type. Most other cultivated chrysanthemum species are restricted to single or duplex daisy flowers, e.g. *C. coronarium*, *C. cinerariifolium*, *C. coccineum*, and all *C. arcticum*, due to limited breeding and domestication (McLaughlin, 1973; Zhao *et al.*, 2009; Liu and Anderson, 2022b). Apetaloid (disc florets only) types are rare but do occur, similar to what is found in *Ageratum houstonianum* (Stephens, 2006). Variations in petal mutation of disc florets are common, where "promiscuous rays" may develop due to high temperatures during FBI/D, or specific flower types may have shortened or elongated "petaloid" disc florets, creating the anemone flower type (Carvalho *et al.*, 2005). These diverse flower types illustrate the wide range of genetic variation present in this complex interspecific crop.

Emasculation, pollination, and crossing procedures for conventional breeding have been thoroughly detailed in previous reviews (Smith and Laurie, 1928; Bearce *et al.*, 1964; Uhlinger and Viehmeyer, 1971; Langton, 1981; Boase *et al.*, 1997; Anderson, 2006a; Su *et al.*, 2019). The opening of each inflorescence, or "flower," commences with the outermost ring of florets (usually ray florets), followed by the successive opening of each inner, adjacent ring opens 1 day later (Smith and Laurie, 1928). The basic procedure involves trimming back the ray florets to the base of each petal. After 1–2 days, the stigmas emerge and unfold, becoming receptive to pollen. In the case of pollinating double or semidouble flowers, after emasculation, each petal meristem may continue to divide, causing the trimmed petals to elongate. This often necessitates retrimming the inner concentric rings of petals to ensure that the stigmas are exposed for pollen receipt. Pollinations are often repeated for 1–3 days following the initial pollen application to ensure successful pollen germination, pollen tube growth, and fertilization.

Pollen stainability should be assessed to ensure that male parents under consideration have adequate levels of viable pollen (*cf.* Mulford, 1937; Anderson and Ascher, 2000). Pollen can be collected and stored for brief periods at 0°C to −2°C, but it is unsuitable for long-term storage beyond a few weeks due to its trinucleate nature (Brewbaker, 1957; Martin, 1965). However, pollen can be stored for longer periods at these temperatures (0°C to −2°C) if kept on honeybee thoraxes, using "bee sticks" developed for Wisconsin Fast Plants (*Brassica;* Williams, 1980).

Following pollen germination, stigmas turn brown and remain so until embryogenesis and seed ripening are complete (Smith and Laurie, 1928). When investigating reproductive barriers, such as SI, ovule counts must also be included, as previously detailed (Anderson and Ascher, 2000; Anderson *et al.*, 1988a; Liedl and Anderson, 1993, 1994; Mulford, 1937). This allows for accurate calculation of female fertility levels, particularly in relation to reproductive barrier studies.

Pollinated inflorescences are left on the parent plant until the stem at the point of attachment starts to turn brown (Smith and Laurie, 1928; Anderson, 2006a). They are then harvested, dried at room temperature (21°C) for 1–2 weeks, followed by cleaning and counting. Chrysanthemum seeds do not have any physical dormancy (scarification) or chemical dormancy. Seed germination requirements are to sow them in 128- or 288-plug trays filled with a standard substrate or germination mix and slightly covered with a thin layer of finely milled sphagnum moss peat or fine vermiculite. Plug trays are placed under intermittent mist in a glasshouse or germination chamber (21 + 0.8/21 + 0.7°C, day/night, 16 hour, 0600–2200 hour) with high-intensity discharge lighting at a minimum set point of 150 µmol m^{-2} s^{-1}, at a mist frequency of 10 min intervals (mist nozzles, reverse osmosis water) during 0600–2200 hour with a 7 s duration. Seed germination will commence within 1–2 weeks and continue for 6–8 weeks in total. Transplanting occurs at the first true leaf stage for subsequent evaluation in either the glasshouse or field. Glasshouse conditions follow the norm of a LD photoperiod (0800–1600 hour for 3–4 weeks, supplied by 400 W high-pressure sodium lamps + 2200–0200 hour night interruption, >150 µmol m^{-2} s^{-1}), followed by SD photoperiod (0800–1600 hour) at a minimum set point of 18/20°C (day/night) with constant liquid feed of 125 mg L^{-1} N. Standard fungicide drenches and insecticides should be applied either monthly or as needed, respectively.

2.8.1 Methods to maximize seed set and incompatibility in chrysanthemum

While the mechanics of pollination, embryogenesis, and chrysanthemum seed ripening are routine, as outlined previously, genetic, cytogenetic, environmental, and physiological factors can highly influence seed set in conventional breeding. Ploidy levels of crops vary from diploid (*C. coronarium*, *C. cinerariifolium*, *C. coccineum*) to polyploid (*C.* × *grandiflorum*, *C.* × *hybridum*), which can highly influence fertility levels, the likelihood of mutations, the ability to achieve interspecific hybridization, and other important interactions to produce viable seed.

2.8.1.1 Self incompatibility

SI is a physiologically based reproductive barrier occurring in more than 70 families and 250 genera in the plant kingdom, including half of the major ornamental crop species (de Nettancourt, 1977). There are two operational SI systems, gametophytic and sporophytic, and the latter system operates in the Asteraceae and many species within the *Chrysanthemum* complex. When an SI system is in operation, self- and cross-pollinations will not produce seed when alleles of the controlling gene, *S*, are matched (Ascher, 1976; de Nettancourt, 1977; Anderson and Ascher, 2000). This prevents inbreeding of SI species and also challenges chrysanthemum breeders, such that many SI crops, particularly *C.* × *grandiflorum* and *C.* × *hybridum*, are vegetatively propagated.

In the sporophytic SI system, when *S* alleles are matched between the male and female parents in outcrosses and selfs (self-pollinations of hermaphroditic disc florets or between ray florets selfed with pollen from disc florets), pollen either fails to germinate or the pollen tube fails to penetrate the stigmatic surface to grow down the style (Ascher, 1976). Thus, SI is a prefertilization reproductive barrier with multiple *S* alleles (numbered numerically, e.g. gametes S_1, S_2, S_3, S_4, etc., with diploid denotations of the sporophyte as S_1S_2, S_1S_3, S_1S_4, S_3S_4, etc.) that encode specificities (East and Mangelsdorf, 1925; East, 1934). Matches of any *S* allele between the female and male gametes with a sporophytic SI system elicit the SI response, thereby preventing seed set. Extensive reviews of the SI systems have been previously published for further reference (Ascher, 1976; de Nettancourt, 1977; Liedl and Anderson, 1994).

Sporophytic SI systems operate in perennial chrysanthemum crops, i.e. diploid (*C. coronarium*, *C. cinerariifolium*, *C. coccineum*) (Brewer, 1968; Muturi *et al.*, 1969; McLaughlin, 1973) and hexaploid (*C.* × *grandiflorum*, *C.* × *hybridum*) (Drewlow *et al.*, 1973; Zagorski *et al.*, 1983). Annual diploid species, such as *C. coronarium*, are typically self compatible with OP seed options as inbreds, provided a clonal parent is grown in isolation.

In perennial diploid chrysanthemums, *C. coronarium*, *C. cinerariifolium*, and *C. coccineum*, SI is controlled by a single gene, *S*, whereas in polyploids multiple *S* genes control the expression of SI. For example, hexaploid *C.* × *grandiflorum* and *C.* × *hybridum* possess three epistatic genes controlling SI (Drewlow *et al.*, 1973; Zagorski *et al.*, 1983; Boase *et al.*, 1997). *S* alleles can also have dominance effect (Stephens *et al.*, 1984). As a result of dominance and epistasis, significant reductions in outcross seed set occur, producing 36–71% (Ronald, 1974) and 0.2–81.2% (Mulford, 1937) mean seed set from crossing unrelated parental genotypes. Thus, it behooves all breeding programs working with SI chrysanthemums to determine the *S* allele configuration of their parent plants, along with the usual determinations of SCA and GCA (Allard, 1960). Ascher (1976), Liedl and Anderson (1993), and Drewlow *et al.* (1973) outlined the standard procedures for determining the presence/absence of SI, the number of *S* alleles in a population and assigning *S* allele numbers.

SI is undesirable in chrysanthemum breeding programs if the number of *S* alleles is low and the goal is to create inbred lines (Anderson and Ascher, 2000). Consequently, breeding programs need to select against SI (although an operational system will be in the background) to develop expression of pseudo-self compatibility (PSC; Liedl and Anderson, 1993). In contrast, at the end of the inbred line development, the flower breeder must select for SI inbreds to prevent inbred contamination in hybrid flower seed. Strongly self incompatible inbred parents would allow for less expensive hybrid seed production, as hand emasculations and pollinations would be eliminated. Such a hybrid production method would implement the use of two SI and cross-compatible, fertile inbred and clonally propagated parents with high

GCA and SCA to produce hybrid seed cultivars. Likewise, another scenario for seed production using SI would be interplanting stock seed of multiple *S* allele genotypes that are cross-compatible and harvest OP seed.

2.8.1.2 Pseudo-self compatibility

Most reproductive barriers in plants are phenotypically plastic, including SI (Anderson and Ascher, 1996). If SI were not plastic in expression, a founding population of one genotype would become extinct. PSC is expressed when the SI system is plastic, where limited to full complements of self and/or outcross seed are produced despite matched *S* alleles. This plasticity has led to the theory that SI is part of a continuum, followed by PSC, and ending with self compatibility (SC) (Anderson *et al.*, 1989b; Liedl and Anderson, 1986a, b, 1994). Anderson and Ascher (1996) found that limited to full complements of self seed could occur in SI plants under certain conditions. PSC has been found in all SI species if enough genetic variation is screened (Ascher, 1976); it differs from SC since PSC genotypes continue to possess SI. Environmental or mechanical treatments may cause PSC to be expressed in an SI individual, e.g. application of carbon dioxide (Palloix *et al.*, 1985), high temperature stress (Ronald and Ascher, 1975), stigmatic surface mutilation (Roggen, 1982), and end-of-season stresses as the plant ages (Lawson and Williams, 1976). Percent PSC is used to calculate the level of PSC present, based on the formula (Ascher, 1976):

$$\%\text{PSC} = [(\text{mean self-seed set}) / (\text{mean outcross seed set})] \times 100$$

In the %PSC equation, decreases in fertility from selfing (inbreeding depression) are incorporated by dividing self seed set by outcross seed set (from compatible crosses; Ascher, 1976).

PSC genetics are complicated by the type of SI system operating, sensitivity to environmental and mechanical manipulations, the crop, and type of gene action (Liedl and Anderson, 1994). Qualitative PSC genes include pollen- and stylar-part mutants (Lewis, 1949; Lewis and Crowe, 1952) and the *Sf* or self-fertile gene in gametophytic SI systems, but this does not exist in chrysanthemum (Wergin, 1936). Thresholds for PSC expression have been reported in *Nemesia strumosa*, with a critical number of homozygous alleles at various loci must be present prior to PSC expression (Henny and Ascher, 1976; Robacker and Ascher, 1978). Two PSC phenotypes exist in *Nemesia*. In *C.* × *grandiflorum*, a threshold effect has also been reported (Anderson and Ascher, 1996). Two specific PSC phenotypes are expressed in *Nemesia* (Robacker and Ascher, 1982) and *Petunia* (Flaschenriem and Ascher, 1979; Dana and Ascher, 1985): pollen-mediated PSC (PMPSC) and discriminating styles (DS). These types of PSC have not been found in chrysanthemums.

PSC was first selected for and used successfully in the *C.* × *grandiflorum* breeding program at the University of Minnesota during the 1970s to develop a series of inbred families via self-, full-sib, and half-sib pollinations, and selecting for increased seed set (Anderson and Ascher, 2000). The first garden chrysanthemum cultivar released with PSC was "Golden Star" (Widmer and Ascher,

1976). Many SI flower crops, such as chrysanthemum, may be inbred to reach homozygosity, although full- or half-sib mating takes much longer to reach homozygosity than selfing. To use these inbreds for hybrid seed lines, chrysanthemum breeders must then select within inbred lines for genotypes with SI expression to be used as parents in the creation of F_1 hybrids (Ascher, 1976).

2.8.1.3 Self compatibility

SC poses the opposite of issues present in SI chrysanthemums, wherein inbreeding via selfing will occur without inhibition by relatedness or shared alleles (Ascher, 1976). In chrysanthemums, the primary SC species are annual diploids (Boase *et al.*, 1997). Either OP or F_1 hybrid seed cultivars can be readily developed in annual chrysanthemums due to the short life cycle (seed to flowering). In *C. coronarium*, numerous cultivars exist on the market for use in Chinese, Japanese, and Vietnamese cooking (Huxley *et al.*, 1992).

2.8.1.4 Identification of reproductive barriers: seed/ovule counts and FCC/MCC

Since the number of disc and/or ray florets varies considerably within and among flowers of a single genotype, across populations, growing environments, and years, reporting only seed set/inflorescence skews the data for studies of reproductive barriers (Anderson and Ascher, 1996, 2000; Anderson *et al.*, 1988a; Holland and Chamberlain, 2007). The first research to take this into account reported on the number of disc and ray florets or ovules/inflorescence to calculate % seed set (Mulford, 1937). A wide range of % seed set was found, ranging from 0% (either female/male sterile or cross-incompatible) to > 80%. Kawase and Tsukamoto (1977) also reported similar % seed set when the varying number of ovules was considered. Other researchers ignored ovule counts (Drewlow *et al.*, 1973; Zagorski *et al.*, 1983), most likely due to the laborious floret or ovule counts for every inflorescence.

Other studies on seed:ovule (S/O) and pollen:ovule (P/O) ratios found that P/O was correlated with reproductive barriers in most plants and that perennials had lower S/O ratios than annuals (Cruden, 1977; Holland and Chamberlain, 2007). The effects of accounting for S/O ratios in chrysanthemum reproductive studies were performed with complete (Type I) diallels with three diploid species: *Chrysanthemum carinatum*, *C. coronarium*, *C. coronarium* subsp. *spatiosum*, and *C. segetum* (Anderson *et al.*, 1988a). Seed set data were analyzed with and without ovule counts using the SI computer program SIGMAS (Liedl and Anderson, 1986b). S/O counts provided a realistic delineation of the SI system in operation, whereas ignoring ovule counts was misleading in all species.

A new calculation was developed and tested for statistically determining the existence of SI, SC, and PSC, from other reproductive barriers (Anderson *et al.*, 1989b). Female and male coefficients of crossability (FCC and MCC) were calculated for diallels of seed set to determine the rigor of the statistics with characterized reproductive barriers in a range of species. A four-part test to accomplish this involved: "(1) calculating male (MCC) and female (FCC) coefficients of crossability using seed set from a diallel crossing design; (2) determining significant male and female differences using a 1:1 paired Chi-square test; (3) constructing scatter diagrams of FCC vs MCC; (4) performing linear regression analysis between FCC and MCC" (Anderson *et al.*, 1989b). A complete diallel using *C. coronarium* subsp. *spatiosum* with or without ovule counts showed the significant need for determining S/O ratios to study reproductive barriers such as SI (Fig. 2.10). Of significant interest, SI could be separated from other reproductive barriers (including sterility, incongruity, and inbreeding depression) using the 1:1 Chi-square and regression analyses (Fig. 2.10; Anderson *et al.*, 1989b). The ease of using FCC/MCC along with ovule counts in chrysanthemum reproductive barriers is time-saving, removing the need for pollen tube growth studies, etc.

2.8.1.5 Inbreeding depression and genetic load

Creating SI genotypes through PSC inevitably elicits expression of inbreeding depression (IBD) and genetic load in outcrossing, SI chrysanthemums, e.g. *C. carinatum, C. coronarium, C. cinerariifolium, C. coccineum, C. coronarium* subsp. *spatiosum, C.* × *grandiflorum, C.* × *hybridum,* and *C. segetum* (Mulford, 1937; Kawase and Tsukamoto, 1977; Anderson *et al.*, 1988a, 1992a, b, 2021; YanFei *et al.*, 2009). Niwa (1931) first reported evidence for IBD in *C.* × *grandiflorum* after creating one to three inbred generations with noticeable reductions in seed set (fertility), % seed germination, plant height, and flowering. Later research compared F_1 and I_1 (first inbred generation) progenies, documenting decreased fertility and malformed flowers (Mulford, 1937; Tsukamoto *et al.*, 1964; Kawase and Tsukamoto, 1966). However, a high rate of genetic variation (standard errors) in the inbreds overlapped with outcross hybrids.

After creating advanced inbred lines via PSC, multiple noninbred, inbred, and recombinant inbred parent families from diverse hexaploid *C.* × *grandiflorum* and/or diploid *C. makinoi* backgrounds were tested for male/female changes in fertility across successive inbred generations in several studies (Anderson and Ascher, 2000; Anderson *et al.*, 1992a, b). One to three inbred generations were created using RGC techniques—laboratory seed ripening to heart stage and subsequent embryo rescue (Anderson *et al.*, 1990). Mean seed set ranges overlapped among the three groups, although the highest levels of seed set occurred in inbred and recombinant inbred parents (Anderson *et al.*, 1993). These levels steadily declined in the $I_1 - I_3$ generations due to IBD, although several inbred families had equivalent or higher fertility levels than the parents (Anderson *et al.*, 1993; Anderson and Ascher, 2000). Thus, while IBD can be severe in SI chrysanthemums, there is opportunity to select fertile inbreds to achieve homozygosity.

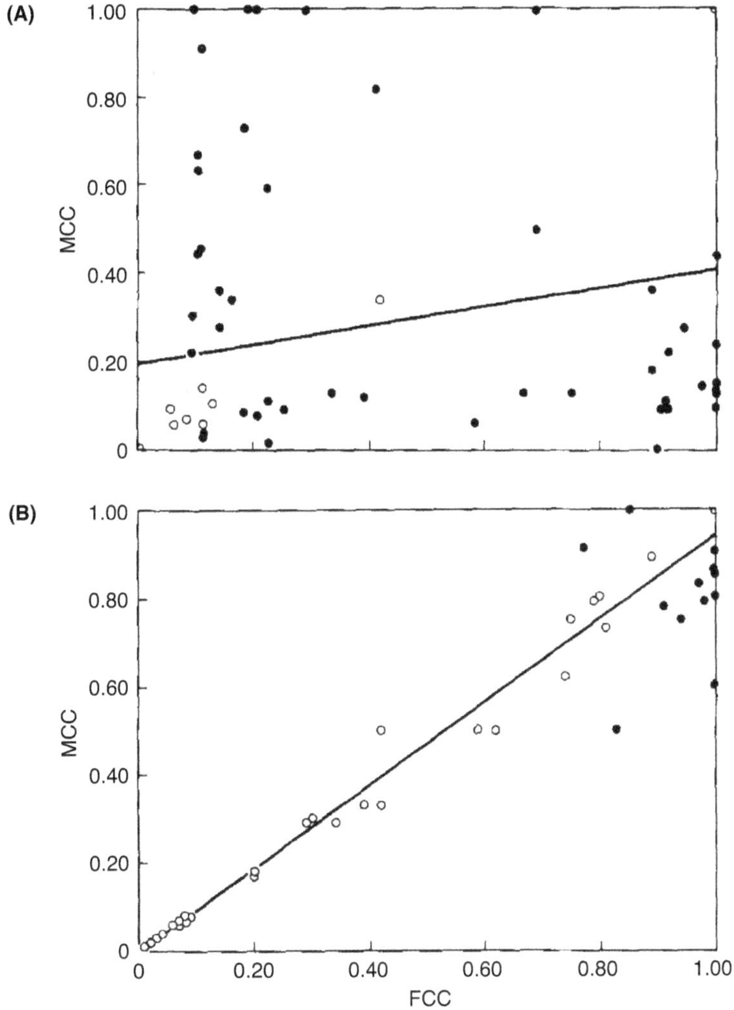

Fig. 2.10. Scatter diagrams and linear regression of FCC and MCC (A, B), illustrating the impact of ovule counts on % seed set data in *Chrysanthemum coronarium* subsp. *spatiosum* possessing a sporophytic self incompatibility system. 1:1 Chi-square deviations are shown as solid circles. FCC = female coefficient of crossability, MCC = male coefficient of crossability (Adapted from Anderson *et al.*, 1989b, reproduced with permission from Springer Nature.)

When inbreeding was performed with noninbred parents to generate I_1–I_3 generations, seed set was highly reduced in the I_1–I_2 generations (89.5–97.2%) when compared with their parents (YanFei *et al.*, 2009). Responses of other traits followed suit, and, in particular, the I_1–I_3 generation progeny flowering time was predominantly delayed from 5.9 to 11.9 days when compared with their parents (YanFei *et al.*, 2009). One exceptional I_3 inbred family, however, was 10.7 days earlier in flowering than its parent. For the remaining traits and inbred families, the effects of IBD were severe (YanFei *et al.*, 2009).

PSC expression had significant G × E (genotype × environment) interactions among and within populations and across inbred generations (Anderson *et al.*, 1992a; Anderson and Ascher, 1996). Several inbred families were eliminated by I_2 due to IBD and/or an enhanced SI system. Seed germination and survivorship vs % homozygosity (F values = Fisher's coefficient of inbreeding) had negative correlations, although correlation of % germination and % survivorship was high ($r = 0.95$) and significant ($P \leq 0.001$; Anderson *et al.*, 1992a).

The fitness of inbred vs noninbred chrysanthemums was measured using lethal equivalents to determine genetic load in *C.* × *grandiflorum* and *C. makinoi*, as well as other plant crops (Anderson *et al.*, 1992b). Lethal equivalents are defined as the number of mutant genes which, on average, cause death (lethality) and are determined using two different equations (Anderson *et al.*, 1992b). Estimates of A (−0.02 to 4.49) and B (−2.96 to 2.28), based on linear regression of % germination and % survivorship, were larger than previous reports, and negative lethal equivalents were also found. Multiplicative (epistatic) effects, in addition to additive effects, were found for the first time in chrysanthemum (Anderson *et al.*, 1992b).

2.8.1.6 Recombinant inbreeding

Due to the high levels of IBD and genetic load predominating in most inbred generations of *C.* × *grandiflorum* and *C. makinoi* (Niwa, 1931; Mulford, 1937; Kawase and Tsukamoto, 1977; Anderson *et al.*, 1988a; Anderson *et al.*, 1989b; YanFei *et al.*, 2009), the potential value of creating recombinant inbreds of hexaploid *C.* × *grandiflorum* was investigated by Anderson and Ascher (2000). Recombinant inbreeding is interspersing inbreeding with outcrossing in successive generations and reduces or avoids IBD deleterious effects (Anderson *et al.*, 1993). Unrelated inbred lines can be crossed in the process, reintroducing hybrid vigor or heterosis and, as a result, F values return to zero in the recombinant inbred progeny. These are then subsequently used as parents for hybrid seed production or further inbred and then, subsequently, intercrossed (recombined) again to repeat the process.

Average seed set ranged from 0.9 to 85.1% in recombinant inbred parents, which was within the comparative data found for inbreds (0.3–96.1%), while noninbred parents were significantly lower in fertility (24.5–38.5%) (Anderson and Ascher, 2000). As inbreeding progressed, significant decreases in fertility and seed set occurred in 56.1% of inbred families. In contrast, at least one recombinant inbred family was tolerant of an additional inbred generation, possibly due to fixed heterozygosity or allelic makeup (Anderson and Ascher, 2000). Additionally, the recombinant inbreds were more vigorous than those derived from the comparison groups. Such findings demonstrate that recombinant inbreeding may be a useful tool for achieving nearly homozygous lines without losing fertility in advanced inbred lines.

2.8.1.7 Methods to overcome incongruity and facilitate recombination for tightly linked traits

Hybrid breakdown in the F_1, F_2, or later generations, due to incongruity, frequently occurs with wide intra-/interspecific or intra-generic crosses in chrysanthemum. As defined by Hogenboom (1973, 1975), incongruity arises from one or more mismatches in genetic information among the parents used in wide crosses, due to divergent evolution and selection. Hybridizations between them result in mistakes in physiological signals and pathways for any possible plant function, such as sterility, albinism, lack of flowering, and crippling (Haghighi and Ascher, 1988).

Effective methods to overcome incongruity have been created in the past few decades, including congruity backcrossing (CBC) and hybrid bridging (HB) and are reviewed by Liedl and Anderson (1993). The basis for both CBC and HB is continual reintroduction of either each parental species or the hybrid itself into subsequent generations of individuals such that, in advanced generations incongruity and hybrid breakdown disappear, fertility is restored, linkage drag is removed, and new recombinants can be found. This has been effective in many crops, such as tomato (*Solanum lycopersicon*), beans (*Phaseolus vulgaris, Phaseolus acutifolius, Phaseolus coccineus*), wheat (*Hordeum vulgare*), etc. (Liedl and Anderson, 1993). CBC has been started in perennial chrysanthemums, and early backcross generations in the *C.* × *hybridum* "Maximum™" series are one set of products (Anderson *et al.*, 2008, 2012a, b, c, 2015a, 2017; Shinoyama *et al.*, 2006). Either technique is inexpensive to do and, after multiple generations of either CBC or HB, novel chrysanthemums may arise.

2.8.1.8 Early flowering selection techniques

The "rush to market" to capture market share and generate revenue continues to pressure private- and public-sector chrysanthemum breeding programs alike (Anderson, 2006a, 2020). The rapid pace of crop domestication, breeding/selection, and creation of new chrysanthemum products to fit within a series or for enhanced innovations in current crops throughout the horticultural distribution chain, with adequate worldwide protection of IP, is imperative to generate novelty with flower power and convenience (Aguirre, 2006; Drew *et al.*, 2010; van Tuyl *et al.*, 2014). Life history of chrysanthemum crop (annual, perennial) adjustments need to be circumvented to create annualized perennials or fast-cycling annuals in a process known as RGC (Anderson *et al.*, 1990; Wilkins and Anderson, 2006; Anderson, 2020). Breeder/producer/distributor companies may choose to enhance the efficiency of product testing through all phases, from seed/cutting to the final product, from a linear model to a condensed one as it moves through the horticultural distribution chain in the product pathway (Drew *et al.*, 2010; Anderson, 2020). Potential products commonly fast-tracked in the condensed model are innovative or distinguishable products (Fig. 2.11; Aguirre, 2006; Anderson, 2020). Differences in approaches between the linear and condensed models have been extensively reviewed (*cf.* Anderson, 2020), and breeding programs may choose to enhance the rate of new product development with careful implementation of a condensed model approach. This is particularly important since the market life of an annual cultivar is brief—1–3 + years—depending on its popularity, which benefits propagule pressure. Popular items with wide adaptability and stability over locations and years as a worldwide product may last longer on the market, sometimes for decades, e.g. garden chrysanthemum, *C.* × *grandiflorum* "Minngopher" (Anderson, 2020).

In the past 30 + years, the University of Minnesota Flower Breeding and Genetics program has focused RGC breeding and selection efforts in many herbaceous, geophyte perennial crops (monocots: *Gladiolus, Iris, Lilium;* dicots: *C. cinerariifolium, C.* × *grandiflorum, C.* × *hybridum*) to reduce the generation time from seed to flowering by developing selection techniques for conventional breeding programs (Anderson, 2005, 2013; Anderson *et al.*, 1990, 2015b, 2021). Significant reduction in the generation time of hexaploid *C.* × *grandiflorum*, from 0.75 to 1.5 years (seed to flower) to < 0.3 years, was accomplished first with a combinatory series of laboratory seed ripening and embryo rescue (at the post heart stage) techniques (Anderson *et al.*, 1990). Embryogenesis occurs at a faster rate with laboratory seed ripening than when inflorescences are left *in situ* on the plants, e.g. the heart stage occurred as soon as 2 days postpollination (vs 11 days for *in situ*) when embryo rescue could be initiated (Anderson *et al.*, 1990). Added benefits of laboratory seed ripening include removal of the source/sink interactions, adequate sucrose for seed development, optimal embryogenesis conditions, and a lack of drought, heat, insect pest, and/or disease stressors. Seeds were ready to harvest as fast as 25 days postpollination with laboratory seed ripening techniques without embryo rescue, whereas when embryo rescue was used at 2 days postpollination (heart stage), they frequently germinated *in vitro* in < 1 week (Anderson *et al.*, 1990). Using both techniques reduced the average generation time to 100 days to realize three generations/year.

A subsequent conventional RGC breeding technique, "tooth picking," was developed to enhance selection for earlier seed germination, flowering, and reduced generation times in perennials, particularly geophytes (Anderson *et al.*, 2015b; Anderson, 2019; Anderson *et al.*, 2021). The tooth-picking technique occurs in plug trays during seed germination (128, 288, etc.; https://www.agriexpo.online/agricultural-manufacturer/plug-tray-4318.html), such that each week post-sowing, germinated seedlings in each plug are tooth picked with the same color pick for the week they germinated, e.g. germination week one or G1 with red toothpicks, G2 germinating seedlings picked with green, etc. (Fig. 2.11A; Anderson, 2019; Anderson *et al.*, 2021). Upon transplanting, only the early germinating seedlings are transplanted

Fig. 2.11. A RGC technique (A), tooth picking, used with *Chrysanthemum cinerariifolium*, *C.* × *grandiflorum* to select for early germinating seedlings in a 128-plug tray, which are marked weekly after sowing in germination week 1 (G1, with red toothpicks), G2 (with pink toothpicks), etc., which, at transplant, are (B) selected for the earliest germination weeks for transplanting (G1 or red tooth-picked seedlings). Reductions in flowering and generation times are effectively realized with this RGC technique. RGC, rapid generation cycling. (Photo courtesy of Neil Anderson.)

(Fig. 2.11B), with the early germinating seedlings flowering significantly earlier than later germination weeks, resulting in progressively earlier and earlier seedling populations. The method works equally well for annuals and perennials but is the most useful for the latter to decrease generation times. Early germination weeks are correlated with the number of days to visible bud date (VBD; FBI) and flowering (FBD) (Anderson, 2019). This has been used successfully in many perennial crops, changing them from perennials that flower in year two or later to "annualized perennials" that flower in < 1 year from seed (Wilkins and Anderson, 2006), i.e. *Gladiolus* (Anderson *et al.*, 2015b; Anderson, 2019), *Lilium* (Anderson *et al.*, 2009; Anderson, 2013; Anderson *et al.*, 2015b), and *Chrysanthemum cinerariifolium* (Anderson *et al.*, 2021), etc. In *Gladiolus* × *grandiflorus*, after instituting tooth picking for ~ 20 years, the generation time from seed to flower was decreased from 5 years to < 1 year (as early as 3–4 months' time without any dormancy, cold requirements, etc.) (Anderson, 2019; Aljaser, 2021). The same was possible in *Lilium* × *formolongi* (Anderson, 2013). In pyrethrum (*C. cinerariifolium*), we have changed the normal two-year period from seed to flower (Martin and Tattersfield, 1934; Gnadinger, 1936) to < 1 year (Anderson *et al.*, 2021).

A mathematical analysis method was created for use in selecting earlier flowering chrysanthemums in conventional breeding programs (Zhang *et al.*, 2018). In the procedure, chrysanthemum genotypes are assigned to flowering groups of overlapping daylengths, followed by incorporating 15 phenotypic traits in an analytical hierarchical process and gray relational analysis coupled with a Q cluster analysis, to select early flowering types (Zhang *et al.*, 2018).

2.9 Mutation Breeding

The horticulture industry, and particularly the ornamental/floriculture commodity, thrive on novelty which drives the market (Anderson, 2004, 2006a). The need for visual appeal necessitates that breeders breed or select for novel mutations with market value to create a discernible series with at least one cultivar in each of the major flower color classes (Sampson *et al.*, 1958). Development of a color series for a particular crop phenotype in seed-propagated crops, such as a specific flower color, plant habit, flower form, or other traits, is usually accomplished by using one female parent with high GCA and SCA to cross with different males that confer a different flower color in the hybrid progeny while maintaining the characteristic traits of the series (Anderson, 2006a). For clonally propagated crops, however, selection of genotypes with the series phenotype but different flower colors, etc. takes a different form of conventional breeding methodologies, i.e. mutation breeding (Shinoyama *et al.*, 2006). There are two systems of generating new mutants to produce novel phenotypes within a series: mutations vs mitotic errors.

Normal background radiation levels of ~ 1.5–3.5 millisievert per year cause errors in cell division with every 1×10^{-5} to 10^{-6} divisions (World Nuclear Association, n.d.), most of which do not cause macroscopic changes to a phenotype as they are simple changes in the DNA sequence. Hugo de Vries advanced the theory of mutations (de Vries, 1901–1903) and their usefulness in flower breeding, despite the lack of acceptance by Luther Burbank (de Vries, 1905, 1907). For centuries, it was obvious to chrysanthemum breeders that mutations were an important source of new genotypes for commercialization (Sampson *et al.*, 1958).

Dowrick (1953) examined the generation of sports in clonally propagated garden cultivars of hexaploid *C.* × *grandiflorum*, documenting that new mutants or "sports" had different chromosome numbers. Wasscher (1956) investigated mutations in chrysanthemum, finding that almost one third (29.5%+) of the 633 cultivars examined were sports. Abnormalities in mitosis occurred at a rate of 0.89% (Sampson *et al.*, 1958) to 1.11% (Dowrick, 1953). He theorized that errors in mitosis could also be a major source of chrysanthemum sports due to the gain/loss of one or more chromosomes (aneuploidy). This was tested by Sampson *et al.* (1958), who found that ~ 30% of sports in ten sport families led to new flower color mutants. Sports varied in chromosome numbers from $2n = 45–64$, departing from the $2n = 56$ norm. Genotypes with higher chromosome counts ($2n = 58–64$) had larger inflorescences, while lower counts ($2n = 54–58$) had correspondingly smaller to medium-sized ones. Not all sport families had chromosome gain/loss to create different flower colors (Sampson *et al.*, 1958). For example, the Masterpiece sport family with five flower colors were all $2n = 57$. In contrast, the Valencia sport family also had $2n = 57$ chromosomes for all flower colors ("Valenica," "White Valencia," "Dark Valencia," "Salmon Valencia," "Orchid Valencia," "Apricot Valencia," "Dubonnet Valencia," "Crimson Valencia"), while one sport ("Yellow Valencia") deviated from this, having $2n = 56$ chromosomes (Sampson *et al.*, 1958). The other sport families studied varied more widely in chromosome numbers for their flower color mutants. Later studies by Dowrick and El-Bayoumi (1966a) showed similar trends in gain/loss of chromosomes affecting flower colors. Most new flower color mutants were determined to be periclinal chimeras (Dowrick and El-Bayoumi, 1966a). Three mitotic abnormalities can create chimeras, although anaphasic stickiness of the ends of chromosomes causes the break-fusion-bridge cycle, resulting in fragment chromosomes (Dowrick and El-Bayoumi, 1966a, b).

Creation of these mutations, while arising spontaneously, can also be achieved with treatments such as ionizing radiation and *in vitro* callus culture (Shinoyama *et al.*, 2006; Nagatomi and Degi, 2009). The first study of this with chrysanthemum "New Princess" (Dowrick and El-Bayoumi, 1966b) tested X- and γ-rays at varying dosage levels to elicit rearrangement of existing chimeras as well as creating chromosomal chimeras. They reported that levels of 1 Krad γ-rays and 1000r X-rays were optimal for whole plant treatments; higher levels killed the plant tissues (Dowrick and El-Bayoumi, 1966b). Numerous other reports have used one or both irradiation treatments to create new flower color and flower-type sports (Datta *et al.*, 2001). Floret tissue was then tissue cultured to regenerate shoot explants; the best regeneration medium was Murashige-Skoog medium with 1 mg l^{-1} 6-benzylaminopurine (BAP) plus 0.5 mg l^{-1} 1-naphthaleneacetic acid (NAA) (Datta *et al.*, 2001). Since these initial studies, a worldwide focus on creating new mutant flower colors for sport families has been a conventional breeding priority for many programs (Su *et al.*, 2019).

The United Nations Food and Agriculture Organization (UN FAO) and the International Atomic Energy Association maintain a Mutant Variety Database that contains chrysanthemums (https://www.iaea.org/resources/databases/mutant-varieties-database) and also publish the *Plant Mutation Reports* (https://www.iaea.org/publications/search/type/plant-mutation-reports) with techniques and novel creations reported in many chrysanthemum sport families (Nagatomi and Degi, 2009; Yamaguchi, 2018). Ion beam (Soliman *et al.*, 2014; Okamura *et al.*, 2015) and electromagnetic radiation (microwaves; Miler and Kulus, 2018) are newer techniques used to induce chrysanthemum mutations. Regardless of the mutagen treatment used, the high level of heterozygosity present in the species aids in quickly generating sports to fit within a series (Shinoyama *et al.*, 2006), although mutation breeding is limited due to the unpredictable nature and location of mutations affecting phenotypes, needing large numbers of mutated plantlets to test for rare mutants (0.89%+; Sampson *et al.*, 1958), having an irradiation facility, determining the correct mutagen method and optimal dosage, etc. (Yamaguchi, 2018; Su *et al.*, 2019). Choosing the correct genotype for mutation breeding is also crucial, as some flower colors (yellow) will only produce flower colors with carotenoids, due to the lack of anthocyanin pigments—eliciting the inability to obtain purples, pink, or red/orange colors (Anderson *et al.*, 1988b). Likewise, as with conventional chrysanthemum breeding, obtaining blue flowers is impossible due to the lack of the necessary pigment systems.

To complement and extend the capabilities of conventional breeding in chrysanthemums, molecular breeding techniques that employ various tools such as molecular markers, linkage maps, quantitative trait loci (QTL) mapping, and "omics" technologies are utilized. Additionally, regeneration and transformation systems are in place to support these advancements. For more detailed information, please refer to Chapters 03 and 04.

2.10 Ploidy and Inheritance Modes

The cytogenetics of chrysanthemums has been studied for over a century. Species in the *Chrysanthemum* complex constitute a polyploid series, ranging from diploid annual *C. coronarium* and perennial *C. cinerariifolium*, *C. coccineum* ($2n = 2x = 18$) species to perennial hexaploids (*C.* × *grandiflorum*, *C.* × *hybridum*; $2n = 6x = 54 \pm 9$), octoploids (*C. arcticum*; $2n = 8x = 72$), and decaploids (*C. maximum*, =*Leucanthemum maximum*; $2n = 10x = 90$) (Dowrick, 1952, 1953; Anderson, 2006a). It has long been assumed that cultivated, perennial greenhouse and garden chrysanthemums are allohexaploids due to the ≥ 10 primarily hexaploid species in their ancestry (*C. erubescens*, *C. indicum*, *C. japonense*, *C. makinoi*, *C. ornatum*, *C. sinense*, *C. vestitum*, *C. zawadskii* var. *latilobum*, etc.; Ackerson, 1967a, b; Dowrick, 1953; Ma *et al.*, 2016).

Assumptions of diploidization and a disomic inheritance mode have been made, inferring chrysanthemum to be a segmental disomic (Anderson,

2006a; Anderson *et al.*, 1992a, b; Li *et al.*, 2005, 2011; Teynor *et al.*, 1989a, b; Zhang *et al.*, 2011, 2014). Occurrence of multidose (>1) dominant alleles in segregants may also indicate a disomic nature (de Jong and Rademaker, 1986; De Backer, 2012; Klie *et al.*, 2014). In other instances, researchers postulated, tested, and proved that the hexasomic inheritance mode predominated rather than the disomic mode (Langton, 1989). The small population sizes in most of these studies, however, heighten the degree of inaccuracy. Since genomic research would provide the necessary evidence of inheritance mode, molecular analysis in this arena has commenced. In some instances, depending on the EST-SSR (expressed sequence tag–SSR) marker alleles, *C.* × *grandiflorum* has both disomic and hexasomic modes of inheritance (Fan *et al.*, 2021). Full-sib F_1 progeny had random chromosome pairing and a majority of markers (n = 204) behaving with hexasomic inheritance, while fewer (n = 150) were disomic, concluding that the species "is a complete or near-complete hexasome" (Fan *et al.*, 2021). This differential response explains the varied inheritance modes or patterns reported in previous studies. Using genomic segregation data of 73,936/183,130 SNPs in a biparental population, van Geest *et al.* (2017) did not find strong linkage for repulsion in SNP markers (disomic inheritance), thus providing evidence for "genome-wide hexasomic inheritance." Thus, both conventional and molecular future studies involving genomic and other "omic" studies in chrysanthemum must account for hexasomic inheritance complications in bioinformatic analyses.

To aid in analyzing complexities of hexasomic inheritance, Bourke *et al.* (2021) created the first bioinformatics software, polyqtlR, for QTL interval mapping. Future genomic studies focusing on the species composition of this interspecific hybrid may resolve why hexasomic inheritance persists as an autohexaploid instead of an allohexaploid. Ultimately, large progeny population sizes are of paramount importance for the expansion of both phenotyping and genotyping knowledge in chrysanthemums.

2.11 Flower Color and Types

The earliest flower color research in *C.* × *grandiflorum* identified chrysanthemin, cyanidin-3-glucoside as an anthocyanin occurring in flower petals (Willstätter and Bolton, 1916); other chemical studies soon followed, as reviewed by Kawase *et al.* (1970). The advent of paper chromatography (PC) and thin-layer chromatography (TLC) initiated the chemical discovery of flower pigments in four flower color (petal) categories, with many additional pigments identified, including flavonoids (malvin, ensatin, apigenin, acacetin, quercetin, luteolin) and carotenoids (lutein or xanthophyll; taraxanthin, flavoxanthin) (Willstätter and Bolton, 1916; Karrer and Jucker, 1943; Wada and Hattori, 1953; Kakehi *et al.*, 1965; Kawase *et al.*, 1970; Endo and Yamada, 1972; Sugiyama and Takano, 1974; Kawase and Tsukamoto, 1976; Anderson *et al.*, 1988b). Similar anthocyanins were discovered in *C. coronarium* and *C. segetum* as well (Geissman and Steelink, 1957; Anyos and Steelink, 1960).

Chrysanthemum breeders soon demonstrated the trait to be complex due to heterozygosity, with positive assortative mating (white × white parents, etc.) producing the entire range of flower colors in the seedlings (Mulford, 1938; Crook, 1942; Longley, 1942; Culbert, 1957; Howard, 1962). Flower colors were categorized into phenotypic classes using color charts, e.g. Ridgway (1912) was used first, and then the Royal Horticultural Society (1939). Four basic color classes and combinations thereof were distinguished early on: white, yellow, orange-red, and purple (Mulford, 1938; Crook, 1942; Longley, 1942). In the parents used, it appeared as though white was recessive to all flower colors, while red-purple and orange were dominant; homozygous and heterozygous parents were found (Longley, 1942). More recent studies involving heterozygous and inbred parents (F = 0.71 to F = 0.9; derived *via* PSC) showed that a dominant gene with disomic inheritance is a common a pigment inhibitor (Teynor *et al.*, 1989a). Li *et al.* (2005) found that flower color was multigenic among epistatic genes, with incomplete and mosaic dominances prevailing. Red- × yellow-flowered (female × male) hybrids showed matroclinous traits, but these did not surface if white- or yellow-flowered females were crossed.

Retroposon-like elements have been reported in decaploid *C. pacificum* (Shimizu *et al.*, 1998). A transposable element or retrotransposon in *C.* × *grandiflorum* was involved in some orange and red flower color phenotypes, with the anthocyanins located in the L1 and carotenoids in the L2 layers (Teynor *et al.*, 1989b). Li *et al.* (2005) also found red cells of transposable elements overlaid onto yellow-pigmented petals (all cells had the same number of chromosomes). In a recent sequencing of the diploid and SC species, *C. seticuspe*, for use as a reference in genome editing of cultivated species, its genome (3.05 Gb) contained an 80% frequency of inserted repeats, 72% of which were transposable elements (Nakano *et al.*, 2021). A retrotransposon family, SbdRT, was also identified. Likewise, the 7.1 Gb genome of pyrethrum, *C. cinerariifolium*, contained 33.84% of the genome sequence (Yamashiro *et al.*, 2019). Since the market demands a full flower color palette available within each series of chrysanthemums (Chandler, 2007; Anderson and Walker, 2013), conventional breeders have reached an impasse to create flower colors outside of the categories of whites, yellows, oranges/reds, and pinks/purples, since pigments evoking blue ray florets do not exist (Noda *et al.*, 2017). The lack of pigments, such as flavonoid 3′5′-hydroxylase (F3′5′H), needed for the biosynthesis of delphinidin (Su *et al.*, 2018), necessitated using transgenic approaches (*cf.* reviewed by Anderson, 2020).

In floricultural crops, GMO research often focuses on creating novel flower colors (Holton *et al.*, 1993; Clark *et al.*, 2004; Auer, 2008) in addition to fragrance, postharvest vase life, etc. (Chandler, 2012). Penna *et al.* (2017) reviewed transgenic methods used to create such transformations in floricultural crops. A small number of flower crop genetically modified organisms (GMOs) have been produced to date: nine carnations, one chrysanthemum, and one rose (Anderson and Walker, 2013; Noda *et al.*, 2017; Penna *et al.*, 2017; Su *et al.*, 2018; Anderson, 2020). Issues connected with creation, testing, restrictions, and regulatory requirements, as well as a lack of market acceptance, are often costly and limit the creation of new GMO flowers, including chrysanthemum (Clark *et al.*, 2004; Anderson, 2020). This may restrict investments by private- and public-sector flower breeding programs in the development and marketing of new GMO cultivars due to a higher price point—e.g. royalties—required to recapture some of the costs involved in their development (Anderson, 2020).

Despite previous labs' failure to produce GMO blue chrysanthemums (Su *et al.*, 2018), Noda *et al.* (2017) created the first blue chrysanthemum (refer to Chapters 03 and 04). Their success relied on overexpression of flavonoid 3′,5′-hydroxylase (F3′5′H) from *Campanula medium*, which, while resulting in purple/violet transformants, led to postulating that to get blue flowers, another step involving metal complexes with polyacylated anthocyanins would be needed. They subsequently used the butterfly pea UDP (uridine diphosphate)–glucose:anthocyanin 3′,5′-O-glucosyltransferase gene along with gene expression of F3′5′H from *Campanula medium*. This two-step co-pigmentation process with flavone glucosides created the first blue-flowered chrysanthemum (Fig. 2.12). This success demonstrates the problem-solving power of integrative molecular biology when it is coupled with the needs unsolvable by conventional chrysanthemum breeders (Kishi-Kaboshi *et al.*, 2018). Future endeavors may invoke significant changes in flower types, color patterns, stable leaf variegations, or other novel phenotypes with market value.

Fig. 2.12. Genetically modified chrysanthemum flowers with blue pigmentation created a new, "natural" flower colour class (*cf.* Anderson, 2020; https://www.nature.com/news/true-blue-chrysanthemum-flowers-produced-with-genetic-engineering-1.22365). (Reproduced with permission from Springer Nature.)

2.12 Flowering Time

Early flowering times have economic value for growers to speed the rate of production time to market and provide cost savings to consumers (Anderson, 2006a; Su *et al.*, 2018). Conventional chrysanthemum breeders have successfully reduced flowering time (SD response group) from 10 to 12 weeks in cut flower greenhouse types to 6–8 weeks in the evolution of garden or landscape types of the crop (Anderson, 2006a). While aesthetic ratios provide physiological limits to achieve the minimum required LD leaf number prior to the initiation of flower buds in SD to achieve acceptable stem lengths or plant height, molecular markers would greatly enhance the rate of change for product development. Use of RNA interference (RNAi)–suppression created genotypes that flowered almost 3 weeks earlier in LD photoperiods than wild type and *Cm-BBX24*–overexpressed *C.* × *grandiflorum* plants (Yang *et al.*, 2014). More recently, CPD, FTL, GA20, MET1, and TFL genes and their homologs have been identified which affect flowering time (Su *et al.*, 2018).

One of the genes responsible for the switch from vegetative to reproductive growth (phase change) is *CmMYB2*, a nuclear R2R3 MYB transcription factor (TF) (Zhu *et al.*, 2020). In *Arabidopsis thaliana*, when *CmMYB2* has constitutive expression, it delays flowering. However, the opposite effect was found in *C.* × *grandiflorum*, wherein plants with overexpression of *CmMYB2* flowered significantly earlier (Zhu *et al.*, 2020). Competency for phase change can be regulated by plant age to prevent precocious flowering when the plant has inadequate biomass (leaves) for photosynthetic capacity for floral sinks. When *CmNF-YB8* (a nuclear factor gene) was silenced, precocious flowering and day neutrality were found in chrysanthemum transformants (Wei *et al.*, 2017). Upregulation and downregulation of various SQUAMOSA PROMOTER BINDING-LIKE (SPL) genes were also found, leading to the discovery that *CmNF-YB8* regulated expression of the aging pathway gene, *cmo-MIR156* (Wei *et al.*, 2017). Such gene action may be the causal mechanism in the earlier creation of the precocious flowering, day-neutral garden *C.* × *grandiflorum* inbred, MN. Sel'n. 83-267-3 (Anderson *et al.*, 1989a), although confirmation awaits. The genomic sequencing of diploid *C. seticuspe* has already benefited gene discovery in the genus and is used to further gene discovery in hexaploid *C.* × *grandiflorum*. Three *FLOWERING LOCUS T* (*FT*)-like genes, *CsFTL1*, *CsFTL2*, and *CsFTL3*, were discovered to have organ-specific expression in the leaves (Oda *et al.*, 2012). Upon exposure to an SD inductive photoperiod, the genes were either upregulated or downregulated in response, which induced earlier flowering. *CsFTL3*-dependent signals could be graft transmitted, which allows for intriguing research potential of signal transmissions since the *CsFTL3* gene product(s) are proposed to be the key regulator photoperiodism in the species and, potentially, all genera and species in the chrysanthemum complex (Oda *et al.*, 2012).

While many genes and their actions still await discovery for *C.* × *grandiflorum*, corollary benefits are realized, e.g. uncovering related gibberellin synthesis genes involved in regulating flowering response times (Zhu *et al.*,

2020). Other informative side effects from such studies, such as salt/drought stress tolerance in transgenic *BBX24* mutants (Su *et al.*, 2018), illustrate many opportunities for novel gene discovery and trait associations, which may jointly aid in expanding knowledge as well as providing traits with market value.

2.13 Disease and Insect Resistance

Several papers have reviewed aspects of disease and insect resistance breeding for *C.* × *grandiflorum* and *C. cinerariifolium* that demonstrate the immense need for these breeding objectives to ensure continued, low-input (cost-effective) and sustainable production methods to avoid crop losses (Anderson, 2006a; Shinoyama *et al.*, 2006; Visser *et al.*, 2007; Teixeira da Silva *et al.*, 2013; Spaargaren and van Geest, 2018; Su *et al.*, 2018). For example, organic production of pyrethrum in Tanzania with products such as PyGanic® (https://www.mgk.com/product/pyganic-specialty/) has extended its use for organic food and flower production worldwide, as well as use in clothing and bed linens.

2.13.1 Disease resistance

Disease control for all chrysanthemum crop species includes a wide array of chemical, mechanical, physical, and/or biocontrol measures, most of which are costly and time-consuming, as well as some detrimentally affect ecosystems (Spaargaren and van Geest, 2018). Critical fungal diseases in *C.* × *grandiflorum* include soilborne fusarium (*Fusarium oxysporum* f. sp. *chrysanthemi*) and verticillium (*Verticillium dahlia*, *V. albo-atrum*), along with leaf-borne white rust (*Puccinia horiana*) (Alexander and Hall, 2011; Singh and Kumar, 2014; Shinoyama *et al.*, 2015). Transgenic chrysanthemums were found to offer cost-effective resistance to white rust, which reproduces both clonally and sexually (Singh and Kumar, 2014; Shinoyama *et al.*, 2015). While such resistance is dominant with simple inheritance, loss of resistance could be easily undermined as well (de Jong and Rademaker, 1986; Spaargaren and van Geest, 2018). Resistance of *Alternaria* leaf spot was also reported in bigeneric *Artemisia* × *Chrysanthemum* hybrids (Deng *et al.*, 2010). Integration of cost-effective and environmentally sensitive, sustainable control measures, coupled with development of disease resistances, is critical for future crop viability. Resistance breeding, using integrative conventional and molecular breeding methodologies, will be the most sustainable strategy to cope with these and other diseases in the future.

In *C. cinerariifolium*, disease resistance is needed for a wide variety of soil and leaf-borne pathogens, including wilts, *Sclerotinia minor*, *S. schlerotiiorum*; leaf/stem spots, *Pseudomonas* sp.; flower head diseases, *Botrytis cinerea*, *Ramularia belluensis*, *Coniothyrium*, *Gleosporium*; soilborne crown gall,

Agrobacterium tumefaciens or pathogens, *Fusarium*, *Phytophthora*, *Pythium*, *Rizoctonia*, and *Thielaviopsis* (Bhat, 1995; Gullickson, 1995; Pethybridge *et al.*, 2003). Conventional breeding is providing limited levels of resistance to some pathogens. However, significantly greater progress will be realized with SNPs, MAS, and GWAS in future research currently underway.

2.13.2 Insect resistance

Most commonly studied insect pest resistances of aphids (*Macrosiphoniella sanborni*), chrysanthemum leaf miners (*Liriomiza trifolii*), two-spotted spider mites (*Tetranychus urticae*), and western flower thrips (*Frankliniella occidentalis*) are multigenic, although there are commonly shared genetic constituents (Kos *et al.*, 2014; Wang *et al.*, 2014; Fu *et al.*, 2018b). Thus, conventional breeding presents finite possibilities in achieving resistances, and transgenic methods are faster and more easily obtained. Insect control is of particular interest since they can transmit diseases to chrysanthemums, e.g. tospoviruses are spread by thrips, etc. (Visser *et al.*, 2007). Thrip resistance was linked with metabolites, such as chlorogenic acid (Leiss *et al.*, 2009) and isobutylamides (Tsao *et al.*, 2005). Several genes producing *Bacillus thuringiensis* (Bt) toxins have provided effective control in transgenic chrysanthemum (Visser *et al.*, 2007). Protease inhibitors, i.e. aspartic, cysteine, and serine, effectively controlled thrips as well as other insects (Annadana *et al.*, 2002). Genes for resistance to leaf miners have been found in chrysanthemum (de Jong and Rademaker, 1991), as well as related *Ajania pacificum* (Zhao *et al.*, 2012). Transgenic chrysanthemums were found to offer cost-effective resistances to lepidopteran larvae (*Helicoverpa armigera*; Shinoyama *et al.*, 2015). Future molecular work with insect resistances involving metabolomics and other venues will undoubtedly provide novel solutions and chemical alternatives to current insect control measures exercised in chrysanthemums.

2.14 Heat and Drought Tolerance

2.14.1 Heat tolerance

High temperatures (26–32°C) during the dark period can significantly reduce or eliminate FBI/FBD in SD black cloth photoperiodic treatments in greenhouses and field conditions in the summer (Anderson and Ascher, 2001, 2004b; Anderson *et al.*, 2016; Chen *et al.*, 2001; de Jong, 1978). This is particularly crucial, given the global rise in temperatures, which increases both greenhouse and outdoor production environments. Thus, conventional breeding programs across the globe have focused on breeding and selecting HDI SD and DN genotypes (Anderson and Ascher, 2001; Anderson *et al.*, 1989a, b, 2016; de Jong, 1989; Shibata and Kawata, 1987; Wang *et al.*, 2008; Xiwang *et al.*, 2008). As discussed earlier, an HDI ideotype was created and tested to aid in the process for continued selection (Anderson and Ascher, 2001; Anderson *et al.*, 2016). Screening for heat tolerance encompasses a wide range of cell thermostability tests and identification of heat shock proteins (Du *et al.*, 2018). Transgenic lines with overexpressed *CmDREB6*, a DREB A-6 subgroup gene, had increased levels of HDI; it also enhanced expression of *CmHsfA4*, *CmHSP90*, *CmSOD*, and *CmCAT* heat shock response and ROS homeogenesis genes (Du *et al.*, 2018). Many additional genes have been identified with similar responses. Continued molecular research integrated with conventional breeding for cultivar development of cultivars with HDI responses and the genes responsible will elicit additional mechanisms of heat tolerance, as well as speed the rate of the process.

2.14.2 Drought tolerance

Water conservation, rather than drought tolerance, has been the focus for greenhouse chrysanthemum production, whereas field production and landscape or garden uses favor development of drought-tolerant forms. Screening of germplasm, interspecific hybridizations, and wild species germplasm have all been sources of drought tolerance. Interspecific hybrids of [*C.* × *grandiflorum*] × *C. indicum* had increased drought tolerance (Sun *et al.*, 2010). Genes such as *DgWRKY4* (Wang *et al.*, 2017) and *CmWRKY10* (Jaffar *et al.*, 2016) were identified to increase drought tolerance. SCoT polymorphisms have been used to study drought tolerance (Li *et al.*, 2018a). Interestingly, constitutive expression of the *C. dichrum ICE1* gene in *C.* × *grandiflorum* provided joint increases in drought tolerance as well as other abiotic stressors (low temperature and salinity) (Chen *et al.*, 2012). Dehydration-responsive element binding proteins (DREBs), acting as TFs, regulate many abiotic stresses such as drought, salt, and cold tolerance (Chen *et al.*, 2012). CmERF053, a DREB and AP2/ERF stress-induced TF, also increased drought tolerance (Chen *et al.*, 2012). Future research on this and other abiotic stressors will be increasingly successful with gene editing, genomic sequencing, QTL analysis, and MAS, any or all of which will speed creation of drought-tolerant chrysanthemum germplasm.

2.15 Cold Tolerance

Cold tolerance and/or winter hardiness is critical in higher latitudes for winter greenhouse production as a cool crop, outdoor container/direct planting field production, and garden chrysanthemum cultivation, respectively (Widmer, 1958). As reviewed by Kim and Anderson (2006), field studies involving winter hardiness are challenged with abiotic factors such as lack of low temperatures or snow cover for a "test winter," unpredictable frost heaving, soil moisture levels, or unfavorable soil types. Historic field trials to assess cold tolerance in acclimated crowns have been partially supplanted with complementary laboratory freezing tests and semilethal LT_{50} values (lethal temperatures at which 50% lethality occurs), determining that −10 and −12°C were the critical temperatures for winter survival in USDA Z4 and Z3, respectively

(Kim and Anderson, 2006). Rhizome counts were also positively correlated with cold tolerance, allowing for nondestructive screening and selection of F_1 hybrid progenies (Anderson and Gesick, 2003, 2004). Such temperatures are similar for root systems in other herbaceous perennials.

Cold acclimation, along with SD photoperiods, induces antioxidative enzymes, frost-induced gene expression, and/or proline metabolites, as has been shown in *C. dichrum*, *C. makinoi*, and *C.* × *grandiflorum* using qRT-PCR (Chen *et al.*, 2014). Electrolyte leakage tests have also been incorporated in the screening processes (Visser *et al.*, 2007). Transgenic chrysanthemums with a DREB gene (*DREB1A*) and the stress-inducible *rd29A* promoter showed increased levels of cold tolerance (Visser *et al.*, 2007). Significant research remains to further elucidate genes connected with the physiological and molecular mechanisms associated with cold tolerance and winter hardiness before easier transformation or gene editing can be used to efficiently create adaptive genotypes.

2.16 Salt Tolerance

Salt tolerance has become an increasingly critical trait for both greenhouse and field/garden production of chrysanthemums, particularly in areas with high electrical conductivity levels in irrigation water (such as California, USA) or locations where road salt applications are made during winter months (such as Minnesota, USA). Physiological research has been integrated with both conventional and molecular breeding analyses to both study and increase salt tolerances (Li *et al.*, 2018b). Trigeneric hybridizations intercrossing F_1s of ([*C.* × *grandiflorum* × *Artemisia vulgaris*] × [*C. crassum* × *Crossostephium chinense*]) enhanced salt tolerance (Deng *et al.*, 2010). Extensive screening and genotypic characterization are underway in the *Chrysanthemum arcticum* complex, all of which grow only in maritime conditions and brackish water, primarily in North America (USA and Canada) (Liu and Anderson, 2022a, b). WRKY TF genes were associated with stress responsiveness to salt levels (Li *et al.*, 2015; Liang *et al.*, 2017; Wang *et al.*, 2017). Overexpression of ClCBF1 from *Chrysanthemum lavandulifolium* increased salt tolerance (Gao *et al.*, 2018). Increasingly useful germplasm will be identified or created with conventional and molecular breeding methods to ensure that future crops are productive in and tolerant of increasingly saline conditions.

2.17 Green Pesticides, Medicinal and Edible (Herbal Teas and Wine)

Pyrethrum or Dalmatian pellitory, *C. cinerariifolium*, a primarily diploid herbaceous perennial, is the primary source for worldwide production of pyrethrin (Pyrethrum I, II), primarily in Tanzania and Tasmania, Australia, as well as Ecuador and Japan (Gnadinger, 1936; McLaughlin, 1973; Anderson *et al.*,

2021). To a lesser extent, *C. coccineum* may be produced. Uses of pyrethrin include control of human pests such as bed bugs (*Cimex lectularius*) and deer ticks (*Ixodes scapularis*), and plant pests, i.e. whiteflies (*Trialeuroides vaporariorum*) and aphids (*Myzus persicae*). For nearly two centuries, conventional pyrethrum breeding (recurrent and mass selection), coupled with physiological research, has characterized glandular trichomes in the flowers as the primary source of six monoterpene esters, i.e. three esters of chrysanthemic acid or chrysanthemates (Pyrethrin I, Cinerin I, and Jasmolin I) and three esters of pyrethric acid or pyrethrates (Pyrethrin II, Cinerin II, and Jasmolin II; Staudinger and Ružička, 1924). Selection for increased ratios of Py I/II or %Py since the 1970s (Pandita and Bhat, 1978) created seed stock that produces 1.8–2.5% Py I/II or higher (Suraweera *et al.*, 2017). Molecular breeding research is just beginning for this chrysanthemum crop. Next-generation transcriptome sequencing compared pyrethrin production in flowers and leaves of *C. cinerariifolium*, wherein 4443 and 8901 unigenes were identified, respectively (Khan *et al.*, 2017). They concluded that the major components of the pyrethrin biosynthesis pathway were in the flowers, and proteome with transcriptome comparisons would enhance understanding of secondary metabolite accumulations in the flowers (Khan *et al.*, 2017). Transcriptome and metabolite analyses have also recently been studied in *C. coccineum*, with the highest pyrethrin and terpene precursor expression also occurring in the flowers rather than the leaves (Zeng *et al.*, 2021). The complete chloroplast genome of *C. coccineum* was sequenced in 2020 (https://www.ncbi.nlm.nih.gov/nuccore/MT104464.1; Zeng *et al.*, 2020). Future molecular research on cost-benefits of metabolite production and increasing gene expression to enhance Py I/II production will be forthcoming to enhance the rate of new cultivar development.

Medicinal and edible (herbal teas, wine) uses of *C.* × *grandiflorum* involve many metabolites (Lin and Harnly, 2010; Liu *et al.*, 2013; Fu *et al.*, 2018a; Kan *et al.*, 2020; Chen *et al.*, 2021). To date, use of high-pressure liquid chromatography in *C.* × *grandiflorum* (Wang *et al.*, 2014) and ISSRs and SRAPs in *C. indicum* (Fang *et al.*, 2012) have identified the important compounds involved and the production methods. All product lines would benefit from transcriptome and metabolite analyses to complement conventional breeding programs.

2.18 Conclusion and Future Prospects

Conventional breeding of chrysanthemums, as practiced and developed over the past few thousand years, has created novel flower colors, flower types, plant habits, and a myriad of uses for the cultivated species. Propagation, production, postharvest, and marketing needs have been addressed by numerous research groups in public- and private-sector breeding programs, along with plant physiologists. Certainly, with the wide genomic base and genetic variability arising from chromosomal rearrangements and natural mutations, the pace of these conventional breeding advances and the creation of new cultivars for the worldwide markets will continue unabated. The fortunate fusion of molecular breeding has resulted in the creation of traits out of the reach of conventional breeding—e.g. blue flowers—and allowed for MAS and GWAS enhancement of selection to increase the precision of future chrysanthemum breeding and development. While much research remains, the future is definitely bright for all of these crops.

References

Ackerson, C. (1967a) Development of the chrysanthemum in China. *National Chrysanthemum Society Bulletin* 23(4), 146–155.

Ackerson, C. (1967b) Original species of the chrysanthemum. *National Chrysanthemum Society Bulletin* 23(3), 105–107.

Aguirre, P. (2006) Protection: Plant patents, utility patents, plant breeder's rights, trademarks, branding, royalties. In: Anderson, N.O. (ed.) *Flower Breeding and Genetics: Issues, Challenges and Opportunities for the 21st Century*. Springer, Dordrecht, The Netherlands, pp. 81–112.

Alexander, S. and Hall, R. (2011) Verticillium wilt of chrysanthemum: Anatomical observations on colonization of roots, stem, and leaves. *Canadian Journal of Botany* 52, 783–789. DOI: 10.1139/b74-101.

Aljaser, J. (2021) Gladiolus breeding for rapid generation cycling for potted plant production and the discovery OF gladiolus genes, UPSTREAM OF FLOWERING LOCUS C (UFC) and FLOWERING LOCUS c EXPRESSOR (FLX). PhD dissertation, University of Minnesota, St. Paul, Minnesota.

Allard, R.W. (1960) *Principles of Plant Breeding*. John Willey and Sons Inc., New York.

Anderson, N.O. (1987) Reclassifications of the genus *Chrysanthemum* L. *HortScience* 22(2), 313.

Anderson, N.O. (1991) The discovery of day neutral chrysanthemums. *Greenhouse Grower* 9, 50–53.

Anderson, N.O. (2019) Selection tools for reducing generation time of geophytic herbaceous perennials. *Acta Horticulturae* 1237, 53–66.

Anderson, N.O. (2020) Advances in marker-assisted breeding of ornamentals. In: Reid (ed.) *Achieving Sustainable Cultivation of Ornamental Crops*. BDS Publishing, Cambridge, pp. 189–230.

Anderson, N.O. and Ascher, P.D. (1996) Inheritance of pseudo-self compatibility in self-incompatible garden and greenhouse chrysanthemums, *Dendranthema grandiflora* Tzvelv. *Euphytica* 87(2), 153–164. DOI: 10.1007/BF00021888.

Anderson, N.O. and Ascher, P.D. (2000) Fertility changes in inbred families of self-incompatible chrysanthemums (*Dendranthema grandiflora*). *Journal of the American Society for Horticultural Science* 125(5), 619–625. DOI: 10.21273/JASHS.125.5.619.

Anderson, N.O. and Ascher, P.D. (2001) Selection of day-neutral, heat-delay-insensitive *Dendranthema grandiflora* genotypes. *Journal of the American Society for Horticultural Science* 126(6), 710–721. DOI: 10.21273/JASHS.126.6.710.

Anderson, N.O. and Gesick, E. (2003) Container production of prostrate garden chrysanthemums. *HortScience* 38(7), 1344–1348. DOI: 10.21273/HORTSCI.38.7.1344.

Anderson, N.O. and Gesick, E. (2004) Phenotypic markers for selection of winter hardy garden chrysanthemum (*Dendranthema* x *Grandiflora* Tzvelv.) genotypes. *Scientia Horticulturae* 101(1–2), 153–167.

Anderson, N.O. (2004) Breeding flower seed crops: Biology and technology. In: McDonald, M. and Kwong, F. (eds) *Flower Seeds*. CAB International, Wallingford, UK, pp. 53–86. Available at: https://api.elsevier.com/content/abstract/scopus_id/33744775625

Anderson, N.O. and Ascher, P.D. (2004a) Garden chrysanthemum, Mn. Sel'n. 90-275-27. U.S. Plant Patent No.14,749. U.S. Plant Patent Office, Washington, DC. Available at: https://patents.justia.com/patent/PP14749

Anderson, N.O. and Ascher, P.D. (2004b) Inheritance of seed set, germination, and day neutrality/heat delay insensitivity of garden chrysanthemums (*Dendranthema* x *grandiflora*) under glasshouse and field conditions. *Journal of the American Society for Horticultural Science* 129(4), 509–516. DOI: 10.21273/JASHS.129.4.509.

Anderson, N.O. (2005) Novel traits in the genus Lilium: Their discovery & breeding potential. *North American Lily Society Yearbook* 58, 97–104.

Anderson, N.O. (2006a) *Chrysanthemum: Dendranthema* × *Grandiflora* Tzvelv. In: Anderson, N.O. (ed.) *Flower Breeding and Genetics: Issues, Challenges and Opportunities for the 21st Century*. Springer, Dordrecht, The Netherlands, pp. 389–437. DOI: 10.1007/978-1-4020-4428-1-14.

Anderson, N.O. (2006b) Prevention of invasiveness in floricultural crops. In: Anderson, N.O. (ed.) *Flower Breeding and Genetics: Issues, Challenges and Opportunities for the 21st Century*. Springer, Dordrecht, The Netherlands, pp. 177–214. DOI: 10.1007/978-1-4020-4428-1-6.

Anderson, N.O. (2013) Discovery of novel traits in seed-propagated Lilium: Non-vernalization-requiring, day-neutral, reflowering, frost-tolerant, winter-hardy L. × *Formolongi*. I. Photoperiodism in parents and hybrids. *Floriculture and Ornamental Biotechnology* 6(Horticulture – Plants for People and Places Special Issue 2), 73–83.

Anderson, N.O. and Olsen, R.T. (2015) A vast array of beauty: The accomplishments of the father of American ornamental breeding, Luther Burbank. *HortScience* 50(2), 161–188. DOI: 10.21273/HORTSCI.50.2.161.

Anderson, N.O. and Walker, N.J. (2013) Marketing genetically modified organism carnations by future floral designers: Student-designed policy formulation. *HortTechnology* 23(5), 683–688. Available at: https://journals.ashs.org/view/journals/horttech/23/5/article-p683.xml

Anderson, N., Gomez, N. and Galatowitsch, S. (2006) A non-invasive crop ideotype to reduce invasive potential. *Euphytica* 148, 185–202. DOI: 10.1007/s10681-006-5936-6.

Anderson, N., Ascher, P., Gesick, E., Klossner, L., Eash, N. *et al.* (2008) Series garden chrysanthemums "Red Daisy," "White Daisy," and "Coral Daisy" sporting a shrub plant habit. *HortScience* 43(3), 648–654.

Anderson, N.O., Liedl, B.E., Ascher, P.D., Widmer, R.E. and Desborough, S.L. (1988a) Evaluating self incompatibility in chrysanthemum: The influence of ovule number. *Sexual Plant Reproduction* 1(3), 173–181. DOI: 10.1007/BF00193748.

Anderson, N.O., Ascher, P.D. and Widmer, R.E. (1988b) Thin-layer chromatographic analysis of flower color phenotypes in *Dendranthema grandiflorum* Ramatuelle inbreds and clonal cultivars. *Euphytica* 37(3), 229–239. DOI: 10.1007/BF00015119.

Anderson, N.O., Ascher, P.D. and Widmer, R.E. (1989a) *Chrysanthemum plant-day neutral*. United States Patent and Trademark Office, Alexandria, Virginia.

Anderson, N.O., Ascher, P.D. and Widmer, R.E. (1989b) *Greenhouse potted chrysanthemum, Mn. Sel'n. 83-267-3*. U.S. Plant Patent No. 6,884. United States Patent and Trademark Office, Washington, DC. Available at: https://patents.justia.com/patent/PP6884

Anderson, N.O., Liedl, B.E., Ascher, P.D. and Desborough, S.L. (1989b) Distinguishing between self-incompatibility and other reproductive barriers in plants using male (MCC) and female (FCC) coefficient of crossability. *Sexual Plant Reproduction* 2(2), 116–126. DOI: 10.1007/BF00192000.

Anderson, N.O., Ascher, P.D., Widmer, R.E. and Luby, J.J. (1990) Rapid generation cycling of chrysanthemums, *Dendranthema grandiflora* Tzvelv. (*Chrysanthemum morifolium* Ramat.), using laboratory seed development and embryo rescue techniques. *Journal of the American Society for Horticultural Science* 115(2), 329–336.

Anderson, N.O., Ascher, P.D. and Widmer, R.E. (1992a) Inbreeding depression in garden and glasshouse chrysanthemums: Germination and survivorship. *Euphytica* 62, 155–169. DOI: 10.1007/BF00041750.

Anderson, N.O., Ascher, P.D. and Widmer, R.E. (1992b) Lethal equivalents and genetic load. *Plant Breeding Reviews* 10, 93–127.

Anderson, N.O., Ascher, P.D. and Luby, J.J. (1993) Variations of recombinant inbreeding: Circumventing inbreeding depression in chrysanthemum populations. In: Dhir, K.K. and Sarren, T.S. (eds) *Frontiers in Plant Science Research*. Frontiers in Plant Science Research, Lausanne, Switzerland, pp. 1–16.

Anderson, N.O., Ascher, P.D. and Luby, J.J. (1995) Consumer evaluation as an adjunct to F_1 hybrid seed development of hexaploid chrysanthemums (*Dendranthema grandiflora* Tzvelv.). *HortScience* 30(4), 811.

Anderson, N.O., Widmer, R.E., Ascher, P.D. and Gesick, E. (2004) Garden chrysanthemum, Mn. Sel'n. 90-275-27. U.S. Plant Patent No.14,749. U.S. Plant Patent Office, Washington, DC.

Anderson, N.O., Younis, A. and Optiz, E. (2009) Development of colored, non-vernalization-requiring seed propagated lilies. *Acta Horticulturae* 836, 193–198. DOI: 10.17660/ActaHortic.2009.836.27.

Anderson, N.O., Gesick, E., Ascher, P.D., Poppe, S., Yao, S. *et al.* (2012a) Mammoth™ "Twilight pink daisy" garden chrysanthemum. *HortScience* 47(8), 1182–1186. DOI: 10.21273/HORTSCI.47.8.1182.

Anderson, N.O., Poppe, S., Ascher, P.D., Gesick, E., Yao, S. *et al.* (2012b) Mammoth™ "Yellow Quill" garden chrysanthemum. *HortScience* 47(2), 285–288. DOI: 10.21273/HORTSCI.47.2.285.

Anderson, N.O., Gesick, E., Ascher, P.D., Poppe, S., Yao, S. *et al.* (2012c) On the cover: Closeup of flowers for Mammoth™ Twilight Pink Daisy perennial garden chrysanthemum highlighting the eye of color in the center. *HortScience* 47(8).

Anderson, N.O., Ascher, P.D., Fritz, V., Rohwer, C., Poppe, S. *et al.* (2015a) *Chrysanthemum × hybridum* Mammoth™ "Dark Bronze Daisy" garden chrysanthemum. *HortScience* 50(8), 1260–1264. DOI: 10.21273/HORTSCI.50.8.1260.

Anderson, N.O., Carter, J., Hershman, A. and Houseright, V. (2015b) Rapid generation cycling enhances selection rate of *Gladiolus × hybridus*. *Acta Horticulturae* 1087, 429–436.

Anderson, N.O., Won, M.K. and Kim, D.C. (2016) Garden chrysanthemum cell membrane thermostability and flowering heat delay differences among U.S.

And South Korean Germplasm. *HortScience* 51(3), 216–226. DOI: 10.21273/HORTSCI.51.3.216.

Anderson, N.O., Ascher, P.D., Fritz, V., Rohwer, C., Poppe, S. *et al.* (2017) *Chrysanthemum × Hybridum* MN 98-89-7 shrub garden chrysanthemum. *HortScience* 52(2), 306–309. DOI: 10.21273/HORTSCI11203-16.

Anderson, N.O., Suranyi, R. and Gullickson, S.M. (2021) Rapid generation cycling transforms pyrethrum, *Chrysanthemum cinerariifolium*, into an annualized perennial. *Crop Science* 61(2), 1207–1227. DOI: 10.1002/csc2.20453.

Anderson, N.O., Widmer, R.E., Ascher, P.D. and Gesick, E. (2024) Garden chrysanthemum, Mn. Sel'n. 90-275-27. Plant Patent no.14,749. United States Plant Patent Office, Washington, D.C. Available at: https://ppubs.uspto.gov/api/pdf/downloadPdf/PP14749requestToken=eyJzdWIiOiJhMTg1NGFkZi0zZTVkLTQxNWYtYTZhOC1kY2JlYWQ5MjFhOWEiLCJ2ZXIiOiJkYTJmNTg4Ny1mMzdiLTRiZWItYmU3NS01NGZmZjg3NzZjZTciLCJleHAiOjB9

Annadana, S., Peters, J., Gruden, K., Schipper, A., Outchkourov, N.S. *et al.* (2002) Effects of cysteine protease inhibitors on oviposition rate of the western flower thrips, *Frankliniella occidentalis*. *Journal of Insect Physiology* 48(7), 701–706. DOI: 10.1016/S0022-1910(02)00093-8.

Anyos, T. and Steelink, C. (1960) Fluorescent petal constituents of *Chrysanthemum coronarium* L. *Archives of Biochemistry and Biophysics* 90, 63–67.

Ascher, P.D. (1976) Self incompatibility systems in floriculture crops. *Acta Horticulturae* 63, 205–215.

Ascher, P.D. (1986) Breeding garden chrysanthemums. Ornamental Plant Breeding Workshop, International Horticultural Congress, Davis, California. *HortScience* 21, 616.

Ascher, P.D., Anderson, N., Fritz, V., Poppe, S., Strasser, T. *et al.* (1993) Rose Blush™ chrysanthemum. The newest mum from the University of Minnesota. Department of Horticultural Science, Minnesota. Report No. 231-1993.

Ascher, P.D., Anderson, N.O., Fritz, V., Poppe, S., Strasser, T. *et al.* (1996) Inca™ and Snowscape™ chrysanthemums: New from the University of Minnesota's Department of Horticultural Science, St. Paul, Minnesota. Report 240-1996.

Auer, C. (2008) Ecological risk assessment and regulation for genetically-modified ornamental plants. *Critical Reviews in Plant Science* 27(4), 255–271.

Ball, V. (1991) *Ball Redbook*, 15th ed. George J. Ball Publishing, Chicago, Illinois.

Bearce, B., Bing, A., Boodley, J., Brewster, R., Dimock, A.W. *et al.* (1964) *Chrysanthemums: A Manual of the Culture, Diseases, Insects and Economics of Chrysanthemums.* New York State Extension Service Chrysanthemum School, New York State Flower Growers Association, Ithaca, New York.

Bernardo, R. (2014) *Essentials of Plant Breeding.* Stemma Press, Woodbury, Minnesota.

Bhat, B.K. and Pandita, P.N. (1977) Variation in components of yield of pyrethrins in pyrethrum (*Chrysanthemum cinerariaefolium* Vis.). In: Atal, C.K. and Kapoor, B.M. (eds) *Cultivation and Utilization of Medicinal and Aromatic Plants.* Regional Research Laboratory, Jammu-Tawi, pp. 166–169.

Bhat, B.K. and Menary, R.C. (1987) *Chrysanthemum plant named Hypy*. U.S. Plant Patent No. 5848. United States Patent and Trademark Office, Washington, DC.

Bhat, B.K. (1995) Breeding methodologies applicable to pyrethrum. In: Casida, J.E. (ed.) *Pyrethrum, the Natural Insecticide. McLaughlin Gormley King Co. and the Pyrethrum Board of Kenya.* Academic Press, New York, pp. 67–94.

Birkeland, C.J. (1960) Plant breeding as a hobby. University of Illinois, College of Agriculture and the United States Department of Agriculture cooperating, Publication No. 10M-4-60-69627. Urbana-Champaign, Illinois. Available at: https://www.scribd.com/document/274462766/Plant-Breeding-as-a-Hobby

Boase, M.R., Miller, R. and Deroles, S.C. (1997) Chrysanthemum systematics, genetics, and breeding. *Plant Breeding Reviews* 14, 321–361.

Bourke, P.M., Voorrips, R.E., Hackett, C.A., van Geest, G., Willemsen, J.H. *et al.* (2021) Detecting quantitative trait loci and exploring chromosomal pairing in autopolyploids using polyqtlR. *Bioinformatics* 37(21), 3822–3829. DOI: 10.1093/bioinformatics/btab574.

Bower-Jernigan, L. and Anderson, N. (2023) *Chrysanthemum arcticum* L. subsp. *polaré* is genetically distinct from other taxa in the *Chrysanthemum arcticum* species complex. *Acta Horticulturae* 1383, 47–63. (Proc. XXVII International EUCARPIA Symposium Section Ornamentals: From Nature to Culture – Breeding Ornamentals for Sustainability Eds.: M. Beruto *et al.*). DOI: 10.17660/ActaHortic.2023.1383.6T.

Brewbaker, J.L. (1957) Pollen cytology and self incompatibility systems in plants. *Journal of Heredity* 48, 271–277.

Brewer, J.G. (1968) Flowering and seed setting in pyrethrum (*Chrysanthemum cinerariifolium* Vis.). *Pyrethrum Post* 4, 18–21.

Caibo, P. and Qixiang, Z. (2010) Advances and prospects in molecular breeding of day-neutral chrysanthemum. *Molecular Plant Breeding* 8(2), 350–358.

Carpenter, W.J. and Shackelford, T.B. (1966) *Garden Chrysanthemums in Kansas*. Kansas Agricultural Experiment Station, Department of Horticulture. Bulletin No. 485, Contribution No. 383, Manhattan, KS.

Carvalho, S.M.P., Abi-Tarabay, H. and Heuvelink, E. (2005) Temperature affects chrysanthemum flower characteristics differently during three phases of the cultivation period. *The Journal of Horticultural Science and Biotechnology* 80(2), 209–216. DOI: 10.1080/14620316.2005.11511919.

Chamberlain, E.E. and Clark, P.J. (1947) Investigations on growing pyrethrum in New Zealand. II. Selection of clones: Their variation in habit of growth, susceptibility to root-rot, yield, and pyrethrin content of flowers. *New Zealand Journal of Science and Technology* 29(4), 215–222.

Chandler, S. (2007) Genetic modification in floriculture. *Critical Reviews in Plant Science* 26(4), 169–197.

Chandler, S. (2012) Genetic modification the development of transgenic ornamental plant varieties. *Plant Biotechnology Journal* 10(8), 891–903.

Chen, F., Chen, S., Fang, W. and Liu, J. (2001) Determining heat tolerance for *Chrysanthemum vestitum* and four *C. morifolium* cultivars with small flowers. *Acta Agriculturae Shanghai* 17(3), 80–82.

Chen, L., Chen, Y., Jiang, J., Chen, S., Chen, F. *et al.* (2012) The constitutive expression of *Chrysanthemum dichrum ICE1* in *Chrysanthemum grandiflorum* improves the level of low temperature, salinity and drought tolerance. *Plant Cell Reports* 31, 1747–1758. DOI: 10.1007/s00299-012-1288-y.

Chen, S., Liu, J., Dong, G., Zhang, X., Liu, Y. *et al.* (2021) Flavonoids and caffeoylquinic acids in *Chrysanthemum morifolium* Ramat flowers: A potentially rich source of bioactive compounds. *Food Chemistry* 344, 128733.

Chen, Y., Jiang, J., Chang, Q., Gu, C., Song, A. *et al.* (2014) Cold acclimation induces freezing tolerance via antioxidative enzymes, proline metabolism and gene expression changes in two chrysanthemum species. *Molecular Biology Reports* 41, 815–822. DOI: 10.1007/s11033-013-2921-8.

Chuda, Y., Suzuki, M., Nagata, T. and Tsushida, T. (1998) Contents and cooking loss of three quinic acid derivatives from garland (*Chrysanthemum coronarium* L.). *Journal of Agricultural and Food Chemistry* 46(4), 1437–1439. DOI: 10.1021/jf9708554.

Clark, D., Klee, H. and Dandekar, A.M. (2004) Despite benefits, commercialization of transgenic horticultural crops lags. *California Agriculture* 58(2), 89–98.

Cockshull, K.E. (1984) *Chrysanthemum morifolium* (En. Chrysanthemum; Fr. Chrysantheme; Gc. Chrsantheme; Sp. Crisantemo). In: *CRC Handbook of Flowering*, Vol. 1. CRC Press, Boca Raton, FL, pp. 238–257.

Cockshull, K.E. and Kofranek, A.M. (1992) Responses of garden chrysanthemum to day length. *HortScience* 27, 113–115.

Contant, R.B. (1976) Pyrethrum. In: Simmonds, N.W. (ed.) *Evolution of Crop Plants*. Longman, London, pp. 33–36.

Crook, C.B. (1942) Genetic studies of chrysanthemum. M.S. Thesis, Kansas State University, Manhattan, KS.

Cruden, R.W. (1977) Pollen-ovule ratios: A conservative indicator of breeding systems in flowering plants. *Evolution* 31(1), 32–46. DOI: 10.2307/2407542.

Culbert, J.R. (1957) Seeding spray type chrysanthemums. *Bulletin of the Illinois State Florists Association* 175, 8–12.

da Backer, M. (2012) Characterization and detection of *Puccinia horiana* on chrysanthemum for resistance breeding and sustainable control. PhD dissertation, Ghent University, Ghent, Belgium.

Dana, M.N. and Ascher, P.D. (1985) Discriminating styles (DS) and pollen-mediated pseudo-self compatibility (PMPSC) in *Petunia hybrida* Hort. *Euphytica* 35, 237–244.

Datta, S.K., Chakrabarty, D. and Mandal, A.K.A. (2001) Gamma ray-induced genetic manipulations in flower colour and shape in *Dendranthema grandiflorum* and their management through tissue culture. *Plant Breeding* 120, 91–92.

de Jong, J. (1978) Selection of wide temperature adaptation in *Chrysanthemum morifolium*. *Netherlands Journal of Agricultural Science* 26, 110–118.

de Jong, J. (1981) Effects of irradiance and juvenility on the selection of chrysanthemum. *Euphytica* 30, 493–500.

de Jong, J. (1989) The flowering of *Chrysanthemum morifolium* seedlings and cuttings in relation to seasonal fluctuation in light. *Scientia Horticulturae* 41, 117–124.

de Jong, J. and Rademaker, W. (1986) The reaction of *Chrysanthemum* cultivars to *Puccinia horiana* and the inheritance of resistance. *Euphytica* 35, 945–952.

de Jong, J. and Rademaker, W. (1991) Life history studies of the leafminer, *Liriomyza trifolii*, on susceptible and resistant cultivars of *Dendranthema grandiflora*. *Euphytica* 56, 47–53.

de Nettancourt, D. (1977) *Incompatibility in Angiosperms*. Springer, New York.

de Vries, H. (1907) *Plant Breeding: Comments on the Experiments of Nilsson and Burbank*. Open Court Publishing, Chicago.

de Vries, H. (1901–1903) *Die mutationstheorie. Versuche und beobachtungen über die entstehung von arten im pflanzenreich*. Veit & Co, Leipzig.

Deng, Y., Chen, S., Chen, F., Cheng, X. and Zhang, F. (2010) The embryo rescue derived intergeneric hybrid between chrysanthemum and *Ajania przewalskii* shows enhanced cold tolerance. *Plant Cell Reports* 30(12), 2177–2186.

de Vries, H. (1905) A visit to Luther Burbank. *Popular Science Monthly* 67, 329–347.

Dole, J.M. and Wilkins, H.F. (1999) *Floriculture: Principles and Species*, 2nd edn. Pearson Education and Pearson Prentice Hall, Upper Saddle River, New Jersey.
Donald, C.M. (1968) The breeding of crop ideotypes. *Euphytica* 17, 385–403.
Dowrick, G.J. (1952) The chromosomes of *Chrysanthemum*, I: The species. *Heredity* 6, 365–375.
Dowrick, G.J. (1953) The chromosomes of *Chrysanthemum*, II: Garden Varieties. *Heredity* 7, 59–72.
Dowrick, G.J. and El-Bayoumi, A. (1966a) The origin of new forms of the garden *Chrysanthemum*. *Euphytica* 15, 32–38.
Dowrick, G.J. and El-Bayoumi, A. (1966b) The induction of mutations in *Chrysanthemum* using X- and Gamma radiation. *Euphytica* 15, 204–210.
Drew, J., Anderson, N.O. and Andow, D. (2010) Conundrums of a complex vector for invasive species: A detailed examination of the horticultural industry. *Biological Invasions* 12, 2837–2851. DOI: 10.007/s10530-010-9689-8.
Drew, J., Yue, C., Anderson, N.O. and Pardey, P.G. (2015) Premiums and discounts for plant patents and trademarks used on ornamental plant cultivars: A hedonic price analysis. *HortScience* 50(6), 879–887. DOI: 10.21273/HORTSCI.50.6.879.
Drewlow, L.W., Ascher, P.D. and Widmer, R.E. (1973) Genetic studies of self incompatibility in the garden chrysanthemum, *Chrysanthemum morifolium* Ramat. *Theoretical and Applied Genetics* 43, 1–5.
Du, X., Li, W., Sheng, L., Deng, Y., Wang, W. *et al.* (2018) Over-expression of chrysanthemum *CmDREB6* enhanced tolerance of chrysanthemum to heat stress. *BMC Plant Biology* 18, 178. DOI: 10.1186/s12870-018-1400-8.
East, E.M. (1934) Norms of pollen-tube growth in incompatible matings of self-sterile plants. *Proceedings of the National Academy of Sciences* 20(4), 225–230.
East, E.M. and Mangelsdorf, A.J. (1925) New interpretation of the behavior of self-sterile plants. *Proceedings of the National Academy of Sciences* 11, 166–171.
Endo, M. and Yamada, K. (1972) Studies on the flower colours in the cultivated chrysanthemum, *Chrysanthemum morifolium* Ramat. I. On the anthocyanin pigment colouring in the red chrysanthemums. *Gifu Daigaku, Faculty of Agriculture, Research Bulletin of the Faculty of Agriculture, Gifu University* 33, 51–63.
Fan, M., Gao, Y., Wu, Z., Haider, S. and Zhang, Q. (2021) Evidence for hexasomic inheritance in *Chrysanthemum morifolium* Ramat. Based on analysis of EST-SSR markers. *Genome* (2), 75–81. DOI: 10.1139/gen-2020-0155.
Fang, H.L., Guo, Q.S., Shen, H.J. and Li, Y.C. (2012) Genetic diversity evaluation of *Chrysanthemum indicum* L. by medicinal compounds and molecular biology tools. *Biochemical Systematics and Ecology* 41, 26–34. DOI: 10.1016/j.bse.2011.12.013.
Five Seasons TCM (2021) *Chrysanthemum wine*. Available at: https://fiveseasonstcm.com/blogs/beverage/chrysanthemum-wine
Flaschenriem, D.R. and Ascher, P.D. (1979) S allele discrimination in styles of *Petunia hybrida* bearing stylar-conditioned pseudo-self compatibility. *Theoretical and Applied Genetics* 55, 23–28.
Fu, X., JiangShuo, F., KaiLi, Y., YiFan, C., Fei, Z. *et al.* (2018b) Genetic variation and association mapping of aphid (*Macrosiphoniella sanbourni*) resistance in chrysanthemum (*Chrysanthemum morifolium* Ramat.). *Euphytica* 214, 21. DOI: 10.1007/s10681-017-2085-z.
Fu, Y., Yang, J.C., Cunningham, A.B., Towns, A.M., Zhang, Y. *et al.* (2018a) A billion cups: The diversity, traditional uses, safety issues and potential of Chinese herbal teas. *Journal of Ethnopharmacology* 222, 217–228. DOI: 10.1016/j.jep.2018.04.026.

Gao, W., He, M., Liu, J., Ma, X., Zhang, Y. *et al.* (2018) Overexpression of *Chrysanthemum* lavandulifolium *ClCBF1* in *Chrysanthemum morifolium* "White Snow" improves the level of salinity and drought tolerance. *Plant Physiology and Biochemistry* 124, 50–58.

Geissman, T.A. and Steelink, C. (1957) Flavonoid petal constituents of *Chrysanthemum segetum* L. *Journal of Organic Chemistry* 22, 946–948.

Gnadinger, C.B. (1936) *Pyrethrum flowers*, 2nd edn. McLaughlin Gormley King Co., Minneapolis, Minnesota.

Gullickson, W.D. (1995) History of pyrethrum in the 1970s and 1980s. In: Casida, J.E. and Quistad, G.B. (eds) *Pyrethrum Flowers: Production, Chemistry, Toxicology, and Uses*. Oxford University Press, New York, pp. 32–46.

Haghighi, K.R. and Ascher, P.D. (1988) Fertile, intermediate hybrids between *Phaseolus vulgaris* and *P. acutifolius* from congruity backcrossing. *Sexual Plant Reproduction* 1, 51–58.

Hamilton, W. (1991) Garden mums (hardy mums). In: Ball, V. (ed.) *Ball Redbook, 15th Edn., Greenhouse Growing*, 15th edn. George J. Ball Publishing, W. Chicago, Illinois, pp. 462–468.

Henny, R.J. and Ascher, P.D. (1976) The inheritance of pseudo-self compatibility (PSC) in *Nemesia strumosa* Benth. *Theoretical and Applied Genetics* 48, 185–195.

Hogenboom, N.G. (1973) A model for incongruity in intimate partner relationships. *Euphytica* 22, 219–233.

Hogenboom, N.G. (1975) Incompatibility and incongruity: Two different mechanisms for the non-functioning of intimate partner relationships. *Proceedings of the Royal Society of London, Series B* 188, 361–375.

Holland, J.N. and Chamberlain, S.A. (2007) Ecological and evolutionary mechanisms for low seed: Ovule ratios: Need for a pluralistic approach. *Ecology* 88(3), 706–715. DOI: 10.1890/06-1283.

Holton, T.A., Brugliera, F., Lester, D.R., Tanaka, Y., Hyland, C.D. *et al.* (1993) Cloning and expression of cytochrome P450 genes controlling flower colour. *Nature* 366(6452), 276–279.

Hong, B., Tong, Z., Ma, N., Kasuga, M., Yamaguchi-Shinozaki, K. *et al.* (2006) Heterologous expression of the *AtDREB1A* gene in chrysanthemum increases drought and salt stress tolerance. *Science in China, Series C* 49, 436–445. DOI: 10.1007/s11427-006-2014-1.

Howard, G.S. (1962) The results of breeding for red flower color in garden chrysanthemums. *Proceedings of the American Society for Horticultural Science* 81, 506–509.

Hultén, E. (1937) Outline of the history of arctic and boreal biota during the Quaternary period their evolution during and after the glacial period as indicated by the equiformal progressive areas of present plant species. Bokförlags Aktiebolaget Thule. Available at: http://catalog.hathitrust.org/api/volumes/oclc/2722246.html (accessed 1 January 2020).

Hultén, E. (1960) *Flora of the Aleutian Islands and westernmost Alaska Peninsula: With Notes on the Flora of Commander Islands, 2nd edn. rev. and enl.* J. Cramer, 2nd edn. Hafner Pub. Co., New York.

Hultén, E. (1968) *Flora of Alaska and Neighboring Territories; A Manual of the Vascular Plants*. Stanford University Press, Stanford, California.

Hunter, N.T. (2000) *The art of floral design*, 2nd edn. Delmar-Publishing, Albany, New York.

Huxley, A., Griffiths, M. and Levy, M. (eds) (1992) *The New Royal dictionary of gardening*, Vol. 1. MacMillan, London, pp. 611–618.

Jaffar, A.M., Song, A., Faheem, M., Chen, S., Jiang, J. *et al.* (2016) Involvement of *CmWRKY10* in drought tolerance of chrysanthemum through the ABA-signaling pathway. *International Journal of Molecular Science* 17(5), 693.

Jeong, S.C., Kim, S.M., Jeong, Y.T. and Song, C.H. (2013) Hepatoprotective effect of water extract from *Chrysanthemum indicum* L. flower. *Chinese Medicine* 8(1), 7. DOI: 10.1186/1749-8546-8-7.

Kakehi, M., Yokota, H. and Yamaguchi, K. (1965) Studies on the color tones and the anthocyanin pigments in red chrysanthemums. *Engei Gakkai Zasshi* 34(3), 232–236.

Kan, J., Wang, M., Liu, Y., Liu, H., Chen, L. *et al.* (2020) A novel botanical formula improves eye fatigue and dry eye: A randomized, double-blind, placebo-controlled study. *The American Journal of Clinical Nutrition* 112(2), 334–342. DOI: 10.1093/ajcn/nqaa139.

Karrer, P. and Jucker, E. (1943) Carotinoide aus den blueten von winterastern. Chrysanthemaxanthin. *Helvetica Chimica Acta* 26, 626–630.

Kawase, K. and Tsukamoto, Y. (1966) Studies on the breeding of chrysanthemum 10. On the inbreeding. Abstracts of the Spring Meeting, *Japanese Journal of Horticultural Science* 289–290.

Kawase, K. and Tsukamoto, Y. (1976) Studies on flower color in *Chrysanthemum morifolium* Ramat. III. Quantitative effects of major pigments on flower color variation, and measurement of color qualities of petals with a color difference meter. *Engei Gakkai Zasshi* 45(1), 65–75.

Kawase, K. and Tsukamoto, Y. (1977) Studies on breeding in chrysanthemum, *Chrysanthemum morifolium* Ramat. I. Self- fertility. *Journal of the Japanese Society for Horticultural Science* 46, 101–112.

Kawase, K., Tsukamoto, Y., Saito, N. and Osawa, Y. (1970) Studies on flower color in *Chrysanthemum morifolium* Ramat. Anthocyanins. *Plant and Cell Physiology* 11, 349–353.

Khan, S., Upadhyay, S., Khan, F., Tandon, S., Shukla, R.K. *et al.* (2017) Comparative transcriptome analysis reveals candidate genes for the biosynthesis of natural insecticide in *Tanacetum cinerariifolium*. *BMC Genomics* 18(1), 54. DOI: 10.1186/s12864-016-3409-4.

Kim, D.C. and Anderson, N.O. (2006) Comparative analysis of laboratory freezing methods to establish cold tolerance of detached rhizomes and intact crowns in garden chrysanthemums (*Dendranthema* × *Grandiflora* Tzvelv.). *Scientia Horticulturae* 109(4), 345–352. DOI: 10.1016/j.scienta.2006.05.014.

Kim, D.C., Son, K.S., Pak, H.S., Choi, J.J., Lim, S.H. *et al.* (2018) A new landscaping garden chrysanthemum cultivar "Cushion Ball" with bicolor petals, spreading type and decorative flowers. *HortScience* 53(9), S336.

Kishi-Kaboshi, M., Aida, R. and Sasaki, K. (2018) Genome engineering in ornamental plants: Current status and future prospects. *Plant Physiology and Biochemistry* 131, 47–52. Available at: https://www.sciencedirect.com/science/article/pii/S0981942818301384

Klie, M., Schie, S., Linde, M. and Debener, T. (2014) The type of ploidy of chrysanthemum is not black or white: A comparison of a molecular approach to published cytological methods. *Frontiers in Plant Science* 5, 1–8. DOI: 10.3389/fpls.2014.00479.

Kos, S.P., Klinkhamer, P.G.L. and Leiss, K.A. (2014) Cross-resistance of chrysanthemum to western flower thrips, celery leafminer, and two-spotted spider mite. *Entomologia Experimentalis et Applicata* 151(3), 198–208.

Langevin, D. (1992) *The Growing and Marketing of Fall Mums: How You Can Turn Your Backyard into a Money-Making, Growing Machine!* Annedawn Publishing, Norton, Massachusetts.

Langhans, R. (1964a) Light and photoperiod. In: Bearce, B., Bing, A., Boodley, J., Brewster, R., Dimock, A.W. *et al.* (eds) *Chrysanthemums: A Manual of the Culture, Diseases, Insects and Economics of Chrysanthemums.* New-York State College of Agriculture, Ithaca, New York, pp. 73–85.

Langhans, R. (1964b) *Temperature.* In: Bearce, B., Bing, A., Boodley, J., Brewster, R., Dimock, A.W. *et al.* (eds) *Chrysanthemums: A Manual of the Culture, Diseases, Insects and Economics of Chrysanthemums.* New-York State College of Agriculture, Ithaca, New York, pp. 86–94.

Langton, F.A. (1981) Constraints on the selection of superior genotypes in initial trials of chrysanthemum progeny. *Acta Horticulturae* 125, 21–29.

Langton, F.A. (1989) Inheritance in *Chrysanthemum morifolium* Ramat. *Heredity* 62, 419–423. Available at: https://www.nature.com/articles/hdy198957.pdf

Langton, F.A. and Cockshull, K.E. (1976) An ideotype of chrysanthemum (*C. morifolium* Ramat). *Acta Horticulturae* 63, 165–175.

Lawson, J. and Williams, W. (1976) Environmental and genetic effects of pseudo self compatiblity in *Brassica oleracea* in relation to the production of hybrid seed. *Journal of Horticultural Science* 51, 359–365.

Leiss, K., Maltese, F., Choi, Y.H., Verpoorte, R. and Klinkhamer, P.G.L. (2009) Identification of chlorogenic acid as a resistance factor for thrips in *Chrysanthemum*. *Plant Physiology* 150, 1567–1575.

Lewis, D. (1949) Incompatibility in flowering plants. *Biological Reviews* 24, 472–496.

Lewis, D. and Crowe, L.K. (1952) Male sterility as an outbreeding mechanism in *Origanum vulgare*. *Heredity* 6, 136.

Li, C., Chen, S., Chen, F., Li, J. and Fang, W. (2011) Cytogenetic study of three edible *Chrysanthemum* cultivars. *Russian Journal of Genetics* 47, 176–181. DOI: 10.1134/S1022795411010054.

Li, F., Zhang, H., Zhao, H., Gao, T., Song, A. *et al.* (2018b) *Chrysanthemum CmHSFA4* gene positively regulates salt stress tolerance in transgenic chrysanthemum. *Journal of Plant Biotechnology* 16, 1311–1321.

Li, M.T., Yu, L.J., Wang, L.M., Liu, J.M. and Lei, C. (2005) The heredity of flower colors and the discovery of flower color chimera in chrysanthemum species [In Chinese]. *Yi Chuan* 27(6), 948–952.

Li, P., Song, A., Gao, C., Wang, L., Wang, Y. *et al.* (2015) Chrysanthemum *WRKY* gene *CmWRKY17* negatively regulates salt stress tolerance in transgenic chrysanthemum and *Arabidopsis* plants. *Plant Cell Reports* 34, 1365–1378.

Li, P., Su, J., Guan, Z., Fang, W., Chen, F. *et al.* (2018a) Association analysis of drought tolerance in cut chrysanthemum (*Chrysanthemum morifolium* Ramat.) at seedling stage. *Biotechnology* 8(5), 226. DOI: 10.1007/s13205-018-1258-3.

Li, Y., Yang, P., Luo, Y., Gao, B., Sun, J. *et al.* (2019) Chemical compositions of chrysanthemum teas and their anti-inflammatory and antioxidant properties. *Food Chemistry* 286, 8–16.

Liang, Q., Wu, Y.H., Wang, K., Bai, Z.Y. and Liu, Q.L. (2017) Chrysanthemum *WRKY* gene *DgWRKY5* enhances tolerance to salt stress in transgenic chrysanthemum. *Science Reports* 7, 4799.

Liedl, B.E. and Anderson, N.O. (1986a) Incompatibility hermeneutics: The determination of cutoff values. *Plant Cell Incompatibility Newsletter* 18, 6–12.

Liedl, B.E. and Anderson, N.O. (1986b) SIGMAS-1: A software program for self incompatibility genetic modeling. *The Journal of Heredity* 77, 480.

Liedl, B.E. and Anderson, N.O. (1993) Reproductive barriers: Identification, uses, and circumvention. *Plant Breeding Reviews* 11, 11–154.

Liedl, B.E. and Anderson, N.O. (1994) Statistical differentiation between self incompatibility and pseudo-self compatibility in *Petunia hybrida* Hort, using female and male coefficient of crossability. *Sexual Plant Reproduction* 7, 229–238. DOI: 10.1007/BF00232742.

Lin, L.Z. and Harnly, J.M. (2010) Identification of the phenolic components of chrysanthemum flower (*Chrysanthemum morifolium* Ramat). *Food Chemistry* 120(1), 319–326.

Liu, Y. and Anderson, N.O. (2019) Morphological trait variation in the Arctic daisy, *Chrysanthemum arcticum*. *HortScience* 54(9), S303.

Liu, Y. and Anderson, N.O. (2022a) Genetic structure of extant populations of *Chrysanthemum arcticum* L. and *C. a. subsp. arcticum*. *Arctic Science* 9(2), 408–429. DOI: 10.1139/as-2021-0029.

Liu, Y. and Anderson, N.O. (2022b) Phenotypic differences among and within extant populations of *Chrysanthemum arcticum* L. and *C. a. subsp. arcticum*. *BMC Plant Biology* 22. DOI: 10.1186/s12870-022-03902-4.

Liu, Y., Ahmed, S. and Long, C. (2013) Ethnobotanical survey of cooling herbal drinks from Southern China. *Journal of Ethnobiology and Ethnomedicine* 9(1), 82. DOI: 10.1186/1746-4269-9-82.

Liu, Y., Anderson, N.O. and Noyszewski, A. (2019) Phenotypic and genetic variation of *Chrysanthemum arcticum*. In: *Shanghai: Conference Proceedings, Expert Speech Manual. The 13th International Chrysanthemum Symposium*, Shanghai, China, pp. 7–8.

Longley, L.E. (1939) Variation in blooming date of chrysanthemums in garden, frame and greenhouse. *Proceedings of the American Society for Horticultural Science* 37, 1001–1002.

Longley, L.E. (1942) Color in two garden chrysanthemum crosses. *Proceedings of the American Society for Horticultural Science* 40, 583–584.

Martin, J.T. and Tattersfield, F. (1934) The effect of environmental conditions upon pyrethrum (*Chrysanthemum cinerariaefolium*). I. *Annals of Applied Biology* 21(4), 670–681.

Ma, Y.P., Chen, M.M., Wei, J.X., Zhao, L., Liu, P.L. *et al.* (2016) Origin of chrysanthemum cultivars—Evidence from nuclear low-copy *LFY* gene sequences. *Biochemical Systematics and Ecology* 65, 129–136. DOI: 10.1016/j.bse.2016.02.010.

Martin, F.W. (1965) Distyly and incompatibility in *Turnera ulmifolia*. *Bulletin of the Torrey Botanical Club* 92, 185–192.

McLaughlin, G.A. (1973) *History of pyrethrum*. In: Cassida, J.E. (ed.) *Pyrethrum, the Natural Insecticide*. McLaughlin Gormley King Co. and the Pyrethrum Board of Kenya. Academic Press, New York, pp. 3–15.

Miler, N. and Kulus, D. (2018) Microwave treatment can induce chrysanthemum phenotypic and genetic changes. *Scientia Horticulturae* 227, 223–233.

Minnesota Agricultural Experiment Station (2021) *Minnesota Hardy: New and Exciting Varieties from the U of M*. St. Paul, Minnesota. Available at: https://mnhardy.umn.edu/

Miyazawa, M. and Hisama, M. (2003) Antimutagenic activity of flavonoids from *Chrysanthemum morifolium*. *Bioscience, Biotechnology, and Biochemistry* 67(10), 2091–2099. DOI: 10.1271/bbb.67.2091.

Morrison, B.Y. (1923) *Chrysanthemums for the home*. Department of Agriculture, Farmer's Bulletin No. 1311, Washington, DC.

Mulford, F.L. (1937) Pollen studies in chrysanthemums with reference to fertility. *Proceedings of the American Society for Horticultural Science* 35, 815–817.

Mulford, F.L. (1938) Results of selfing twenty-four early blooming chrysanthemums. *Proceedings of the American Society for Horticultural Science* 35, 818–821.

Mulford, F.L. (1939) *New chrysanthemums*. U.S. Dept. of Agriculture, Circular No. 528. Beltsville, Maryland.

Muturi, S.N., Parlevliet, J.E. and Brewer, J.G. (1969) Ecological requirements of pyrethrum. I. A general review. *Pyrethrum Post* 10(1), 24–28.

Nagatomi, S. and Degi, K. (2009) Mutation Breeding of Chrysanthemum by Gamma Field Irradiation and *in vitro* Culture. Food and Agriculture Organization of the United Nations (FAO), Rome.

Nakano, M., Hirakawa, H., Fukai, E., Toyoda, A., Kajitani, R. *et al.* (2021) A chromosome-level genome sequence of *Chrysanthemum seticuspe*, a model species for hexaploid cultivated chrysanthemum. *Communications Biology* 4, 1167. DOI: 10.1038/s42003-021-02704-y.

Niwa, T. (1931) Pollination and self incompatibility in chrysanthemum (Outlines). *Japanese Association for the Advancement of Science, Report* 6, 479–487.

Noda, N., Yoshioka, S., Kishimoto, S., Nakayama, M., Douzono, M. *et al.* (2017) Generation of blue chrysanthemums by anthocyanin B-ring hydroxylation and glucosylation and its coloration mechanism. *Science Advances* 3(7), e1602785. DOI: 10.1126/sciadv.1602785.

Oda, A., Narumi, T., Li, T., Kando, T., Higuchi, Y. *et al.* (2012) *CsFTL3*, a chrysanthemum FLOWERING LOCUS t-like gene, is a key regulator of photoperiodic FLOWERING in chrysanthemums. *Journal of Experimental Botany* 63(3), 1461–1477. DOI: 10.1093/jxb/err387.

Okamura, M., Hase, Y., Furusawa, Y. and Tanaka, A. (2015) Tissue-dependent somaclonal mutation frequencies and spectra enhanced by ion beam irradiation in chrysanthemum. *Euphytica* 202, 333–343.

Palloix, A., Herve, Y., Knox, R.B. and Dumas, C. (1985) Effect of carbon dioxide and relative humidity on self-incompatibility in cauliflower, *Brassica oleracea*. *Theoretical and Applied Genetics* 70, 628–633.

Pandita, P.N. and Bhat, B.K. (1978) Effect of pollination on flower yield and pyrethrins content in pyrethrum. *Indian Journal of Genetics and Plant Breeding* 38(1), 138–141.

Parlevliet, J.E. (1969) Clonal selection for yield in pyrethrum, *Chrysanthemum cinerariaefolium* Vis. *Euphytica* 18, 21–26.

Parlevliet, J.E. (1970) The effect of picking interval and flower head development on the pyrethrins contents of different pyrethrum clones. *Pyrethrum Post* 10(4), 10–14.

Parlevliet, J.E. (1974) The genetic variability of the yield components in the Kenyan pyrethrum population. *Euphytica* 23, 377–384.

Parlevliet, J.E. and Contant, R.C. (1970) Selection for combining ability in pyrethrum, *Chrysanthemum cinerariaefolium* Vis. *Euphytica* 19, 4–11.

Penna, S., Ganapathi, T.R., Ghag, S. and Jain, T.R. (2017) Genetic modifications of horticultural plants by induced mutations and transgenic approach. *Acta Horticulturae* 1187, 22. DOI: 10.17660/ActaHortic.2017.1187.22.

Pethybridge, S.J., Hay, F.S. and Groom, T. (2003) Seasonal fluctuations in fungi associated with pyrethrum foliage in Tasmania. *Australasian Plant Pathology* 32, 223–230.

Phillips, R.A. and Widmer, R.E. (1965) *Cream Top and Minnrose—New chrysanthemums for 1966*. Report No. 64. Minnesota, Misc. St. Paul: University of Minnesota, Minnesota Agricultural Experiment Station.

Porsild, A.E. and Cody, W.J. (1980) *Vascular plants of continental Northwest Territories, Canada*. National Museum of Natural Sciences, National Museums of Canada, Ottawa, Ontario, Canada.

Post, K. (1950) Controlled photoperiod and spray formation of chrysanthemum. *Proceedings of the American Society for Horticultural Science* 55, 467–472.

Post, K. (1949) *Chrysanthemum morifolium* (chrysanthemum). In: Post, K. (ed.) *Florist Crop Production and Marketing: The Application of Scientific Facts to the Production and Marketing of Florist Cut Flowers, Potted Plants and Bulbs Grown in the Greenhouse or the Field*. Orange Judd Publishing, New York, pp. 385–417.

Purugganan, M.D. and Fuller, D.Q. (2009) The nature of selection during plant domestication. *Nature* 457, 843–848.

Ridgway, R. (1912) *Color Standards and Color Nomenclature*. The author, Washington, DC. Available at: https://library.si.edu/digital-library/book/colorstandardsc00ridg

Robacker, C.D. and Ascher, P.D. (1978) Restoration of pseudo-self compatibility (PSC) in derivatives of a high-PSC x no-PSC cross in *Nemesia strumosa* Benth. *Theoretical and Applied Genetics* 53, 135–141.

Robacker, C.D. and Ascher, P.D. (1982) Discriminating styles (DS) and pollen-mediated pseudo-self compatibility (PMPSC) in *Nemesia strumosa* Benth. Part 2. Origin of PMPSC and nature of the DS-PMPSC interaction. *Theoretical and Applied Genetics* 61, 289–296.

Roggen, H. (1982) Breaking self-incompatibility in *Brassica oleracea* with high frequency alternating electric current. *Incompatibility Newsletter* 14, 92–95.

Ronald, W.G. (1974) Genetic and high temperature control of self compatibility in *Chrysanthemum morifolium* Ramat. PhD dissertation, University of Minnesota. St. Paul, Minnesota.

Ronald, W.G. and Ascher, P.D. (1975) Effects of high temperature treatments on seed yield and self incompatibility in chrysanthemum. *Euphytica* 24(2), 317–322.

Royal Horticultural Society (1939) *Horticultural Colour Chart 1, also known as the Wilson Colour Chart (Robert F. Wilson)*. British Colour Council & Royal Horticultural Society. Henry Stone and Son, Ltd, Wisley, Woking, Surrey, UK.

Sampson, D.R., Walker, G.W.R., Hunter, A.W.S. and Bragdø, M. (1958) Investigations on the sporting process in greenhouse chrysanthemums. *Canadian Journal of Plant Science* 38, 346–356.

Schweinfurth, G. (1919) Pflanzenbilder im tempel von Karnak. In: Engler, A. (ed.) *Botanische Jahrbücher* LV. pp. 464–480.

Shedron, K.G. and Weiler, T.C. (1982) Regulation of growth and flowering in *Chrysanthemum xsuperbum* Bergmans. *Journal of the American Society for Horticultural Science* 107, 874–877.

Shibata, M. and Kawata, J. (1987) The introduction of heat tolerance for flowering from Japanese summer-flowering chrysanthemums into year round chrysanthemums. *Acta Horticulturae* 197, 77–81.

Shimizu, A., Hirai, M. and Hirochika, H. (1998) Retroposon-like elements in *Dendranthema pacificum* (Nakai) Kitam. *Acta Horticulturae* 454, 377–382. DOI: 10.17660/ActaHortic.1998.454.45.

Shinn, M. (2012) The come-Back mum: Chrysanthemum Clara Curtis. Available at: https://www.hortmag.com/plants-we-love-2/the-come-back-mum-chrysanthemum-clara-curtis

Shinoyama, H., Anderson, N., Furuta, H., Mochizuki, A., Nomura, Y. *et al.* (2006) Chrysanthemum biotechnology. In: Teixeira da Silva, J.A. (ed.) *Floriculture, Ornamental and Plant Biotechnology: Advances and Topical Issues*. Global Science Books, II, London, pp. 140–163.

Shinoyama, H., Mitsuhara, I., Ichikawa, H., Kato, K. and Mochizuki, A. (2015) Transgenic chrysanthemums (*Chrysanthemum morifolium* ramat.) carrying both insect and disease resistance. *Acta Horticulturae* 1087, 485–497. DOI: 10.17660/ActaHortic.2015.1087.66.

Shi, Z., Humphries, C.J. and Gilbert, M.G. (1994–2013) Chrysanthemum. *Flora of China (Asteraceae)* 20–21(6), 669–653. Available at: http://www.efloras.org/florataxon.aspx?flora_id=2&taxon_id=106957

Singh, P.K. and Kumar, V. (2014) Fusarium wilt of *Chrysanthemum*—Problems and prospects. *Plant Pathology & Quarantine* 4(1), 34–44. DOI: 10.5943/ppq/4/1/5.

Smith, E.D. and Laurie, A. (1928) *Chrysanthemum* breeding. Special Bulletin No. 186. Agricultural Experiment Station, Michigan State College, East Lansing, Michigan.

Soliman, T.M., Lv, S., Yang, H., Hong, B. and Ma, N. (2014) Isolation of flower color and shape mutations by gamma radiation of *Chrysanthemum morifolium* Ramat cv. Youka. *Euphytica* 199, 317–324.

Spaargaren, J. and van Geest, G. (2018) Chrysanthemum. In: Van Huylenbroeck, J. (ed.) *Ornamental Crops. Handbook of Plant Breeding*, Vol. 11. Springer, Cham. DOI: 10.1007/978-3-319-90698-0_14.

Staudinger, H. and Ružička, L. (1924) Insektentötende stoffe. *Helvetica Chemica Acta* 7, 177–201. DOI: 10.1002/hlca.19240070124.

Stephens, L. (2006) Ageratum. In: Anderson, N.O. (ed.) *Flower Breeding and Genetics: Issues, Challenges, and Opportunities for the 21st Century*. Springer, Dordrecht, The Netherlands, pp. 219–223.

Stephens, L.C., Zagorski, J., Widmer, R.E., Ascher, P.D. and Stuart, M.C. (1983) Greenhouse testing of garden chrysanthemum hybrids from seed. *Bedding Plants International (BPI) News* 14(1), 5. Available at: https://hortscans.ces.ncsu.edu/uploads/g/r/greenhou_56157b38c1c80.pdf

Stephens, L.C., Ascher, P.D. and Widmer, R.E. (1984) Interaction among sporophytic *S* loci in self-incompatible garden chrysanthemums. *Euphytica* 33, 623–631.

Su, J., Yang, X., Zhang, F., Wu, S., Xiong, S. *et al.* (2018) Dynamic and epistatic QTL mapping reveals the complex genetic architecture of waterlogging tolerance in chrysanthemum. *Planta* 247, 899–924.

Su, J., Jiang, J., Zhang, F., Liu, Y., Ding, F. *et al.* (2019) Current achievements and future prospects in the genetic breeding of chrysanthemum: A review. *Horticultural Research* 6, 109. DOI: 10.1038/s41438-019-0193-8.

Sun, C.Q., Chen, F.D., Teng, N.J., Liu, Z.L., Fang, W.M. *et al.* (2010) Interspecific hybrids between *Chrysanthemum grandiflorum* (Ramat.) Kitamura and *C. indicum* (L.) Des Moul. and their drought tolerance evaluation. *Euphytica* 174, 51–60. DOI: 10.1007/s10681-009-0117-z.

Sugiyama, A. and Takano, T. (1974) Partial characterization of anthocyanins in *Chrysanthemum corollas*. *Engei Gakkai Zasshi* 43(3), 286–294.

Suraweera, D.D., Groom, T. and Nicolas, M.E. (2017) Pattern of pyrethrin accumulation, achene and trichome development in relation to pattern of flower development in pyrethrum. In: Matsuda, K. and Takabayashi, J. (eds) *Proceedings of the II International Symposium on Pyrethrum*, International Society for Horticultural Science. Acta Horticulturae, pp. 93–100.

Synder, L.C. (1983) History of the department of horticultural science and landscape architecture, 1849-1982. Available at: https://conservancy.umn.edu/bitstream/handle/11299/93379/HistoryofDeptofHorticulturalScience.pdf;sequence=1

Teixeira da Silva, J.A., Shinoyama, H., Aida, R., Matsushita, Y., Krishna Raj, S. *et al.* (2013) Chrysanthemum biotechnology: Quo vadis? *Critical Reviews in Plant Sciences* 32(1), 21–52. DOI: 10.1080/07352689.2012.696461.

Teynor, T.T., Widmer, R.E., Ascher, P.D. and Luby, J.J. (1989a) Inheritance of flower color in *Dendranthema grandiflora* Tzvelev. (*Chrysanthemum morifolium* Ramat.) using cultivars and inbreds. I. plastid pigmentation. *Euphytica* 42(3), 199–207. DOI: 10.1007/BF00034455.

Teynor, T.T., Widmer, R.E., Ascher, P.D. and Luby, J.J. (1989b) Inheritance of flower color in *Dendranthema grandiflora* Tzvelev. (*Chrysanthemum morifolium* Ramat.) using cultivars and inbreds. II. vacuole pigmentation. *Euphytica* 42(3), 297–305. DOI: 10.1007/BF00034468.

The New York Botanical Garden (1797) *Curtis' Botanical Magazine*, Vol. 10. The New York Botanical Garden, Bronx, New York.

Tsao, R., Marvin, C.H., Broadbent, A.B., Friesen, M., Allen, W.R. *et al.* (2005) Evidence for an isobutylamide associated with host plant resistance to western flower thrips *Frankliniella occidentalis*, in chrysanthemum. *Journal of Chemical Ecology* 31(1), 103–110.

Tsukamoto, Y., Kosuge, S. and Kumo, S. (1964) Studies on the breeding of chrysanthemums. 6. Results of crossing and selfing. *Japanese Journal of Horticultural Science, Abstracts of the Autumn Meeting* 36.

Tuikong, A.R. (1984) Pyrethrum breeding in Kenya: A historical account. *Pyrethrum Post* 15(4), 113–117.

Tutin, T.G., Heywood, V.H., Burges, N.A. and Valentine, D.H. (eds) (1964) *Flora Europaea: Plantaginaceae to Compositae (and Rubiaceae)*, Vol. 4. Cambridge University Press, Cambridge.

Uehara, A., Nakata, M., Uchida, A. and Iwashina, T. (2017) Chemotaxonomic consideration of flavonoids from the leaves of *Chrysanthemum arcticum* subsp. *Arcticum* and *C. yezoense*, and related species. *Biochemical Systematics and Ecology* 73, 11–15. DOI: 10.1016/j.bse.2017.05.001.

Ueyama, N. (2017) Introduction of pyrethrum flowers (130 years in Japan). In: Matsuda, K. and Takabayashi, J. (eds) *Proceedings of the II International Symposium on Pyrethrum, International Society for Horticultural Science. Acta Horticulturae* 1169, pp. 1–6.

Uhlinger, R.D. and Viehmeyer, G. (1971) *Penstemon in your garden*. (University of Nebraska Agricultural Experiment Station Circular No. 105 rev.). Available at: https://digitalcommons.unl.edu/hcnaes/121

United States Department of Agriculture (2019) *Floriculture Crops 2018 Summary*. National Agricultural Statistics Service, Washington, DC.

van Geest, G., Voorrips, R.E., Esselink, D., Bargsten, J.W., van Tuyl, J.M. *et al.* (2017) Conclusive evidence for hexasomic inheritance in chrysanthemum based on analysis of a 183 k SNP array. *BMC Genomics* 18, 585. DOI: 10.1186/s12864-017-4003-0.

van Tuyl, J.M., Arens, P., Miller, W.B. and Anderson, N.O. (2014) The role of ornamentals in human life. In: Dixon, D.R. and Aldous, D.E. (eds) *Horticulture. Plants for*

People and Places, Vol. 1. Springer, Dordrecht, The Netherlands, chapter 11. DOI: 10.1007/978-94-017-8578-5_11.

Viehmeyer, G. and Whitney, W. (1965) *Chrysanthemum Culture in Nebraska*. EC65-1273 Rev. Extension Service, University of Nebraska, College of Agriculture and Home Economics and U.S. Department of Agriculture, Lincoln, Nebraska.

Visser, P.B., Maagd, R.A. and Jongsma, M.A. (2007) Chrysanthemum. In: Pua, E.C. and Davey, M. (eds) *Transgenic Crops VI. Biotechnology in Agriculture and Forestry*, Vol. 61. Springer, Berlin, Heidelberg. DOI: 10.1007/978-3-540-71711-9_14.

Wada, M. and Hattori, T. (1953) Studies on biogenesis of anthocyanin, 2nd. Report, Isolation of Apigenin-7-monoglucoside from white chrysanthemum. *Shigen Kagaki Kenkyusho. Iho Misc. Reports* 32, 67–70.

Wang, C., Zhang, F., Guan, Z., Chen, S., Jiang, J. *et al.* (2014) Inheritance and molecular markers for aphid (*Macrosiphoniella sanbourni*) resistance in chrysanthemum (*Chrysanthemum morifolium* Ramat.). *Scientia Horticulturae* 180, 220–226. DOI: 10.1016/j.scienta.2014.10.038.

Wang, C.H., Yeh, D.M. and Sheu, C.H. (2008) Heat tolerance and flowering-heat-delay sensitivity in relation to cell membrane thermostability in chrysanthemum. *Journal of the American Society for Horticultural Science* 133, 754–759.

Wang, K., Wu, Y.H., Tian, X.Q., Bai, Z.Y., Liang, Q.Y. *et al.* (2017) Overexpression of *DgWRKY4* enhances salt tolerance in chrysanthemum seedlings. *Frontiers in Plant Science* 8, 1592. DOI: 10.3389/fpls.2017.01592.

Wasscher, J. (1956) The importance of sports in some florist's flowers. *Euphytica* 5, 163–170.

Wei, Q., Ma, C., Xu, Y., Wang, T., Chen, Y. *et al.* (2017) Control of chrysanthemum flowering through integration with an aging pathway. *Nature Communications* 8, 829. DOI: 10.1038/s41467-017-00812-0.

Wergin, W. (1936) Cyto-genetische Untersuchungen an *Petunia hybrida* Hort. *Zeitschrift Fuer Induktion Abstammlung Vererbungslehre* 71, 120–155.

Widmer, R.E. (1958) The determination of cold resistance in the garden chrysanthemum and its relation to winter survival. *Proceedings of the American Society for Horticultural Science* 71, 537–546.

Widmer, R.E. (1980) *Garden Chrysanthemums*. University of Minnesota, Agricultural Extension Service, Horticulture Fact Sheet, No. 38 (revised), St. Paul, Minnesota.

Widmer, R.E. and Ascher, P.D. (1971) *New Cushion Mums for 1972. Minnglow and Minnyellow*. University of Minnesota, Minnesota Agricultural Experiment Station, St. Paul, Minnesota.

Widmer, R.E. and Ascher, P.D. (1973) *3 new Minnesota mums for 1974*. Miscellaneous Report No. 119, University of Minnesota, Minnesota Agricultural Experiment Station, St. Paul, Minnesota.

Widmer, R.E. and Ascher, P.D. (1976) *New mums for Minnesota '77*. Misc. Report No. 140, University of Minnesota, Minnesota Agricultural Experiment Station, St. Paul, Minnesota.

Widmer, R.E., Ascher, P.D. and Stuart, M.C. (1981) *Centerpiece, New Minnesota mum for 1982*. University of Minnesota, Minnesota Agricultural Experiment Station, Miscellaneous Report No. 172, St. Paul, Minnesota.

Widmer, R.E., Ascher, P.D. and Stuart, M.C. (1982) *Gold Country and Mellow Moon. New Minnesota mums for 1983*. Misc. Report No. 185, University of Minnesota, Minnesota Agricultural Experiment Station, St. Paul, Minnesota.

Wilkins, H. and Anderson, N.O. (2006) Creation of new floral products: Annualization of perennials-horticultural and commercial significance. In: Anderson,

N.O. (ed.) *Flower Breeding and Genetics: Issues, Challenges and Opportunities for the 21st Century*. Springer, Dordrecht, The Netherlands, pp. 49–64. DOI: 10.1007/978-1-4020-4428-1-2.

Williams, P.B. (1980) Bee-sticks, an aid in pollinating cruciferae. *HortScience* 15, 802–803.

Williams, C. and Bearce, B. (1964) *Pinching and disbudding*. In: Bearce, B., Bing, A., Boodley, J., Brewster, R. and Dimock, A.W. *et al.* (eds) *Chrysanthemums: A Manual of the Culture, Diseases, Insects and Economics of Chrysanthemums*. New-York State College of Agriculture, Ithaca, New York, pp. 38–45.

Willstätter, R. and Bolton, E.K. (1916) Über ein anthocyan er winteraster (*Chrysanthemum*). *Liebig's Ann Chern* 412, 136–148.

Xiwang, J., Qixiang, Z., Ping, Z., Chuanbin, W. and Miao, L. (2008) Heat tolerance of ground-cover chrysanthemum. In: Criley, R.A. (ed.) *Ornamentals, Now! Proc. XXVII IHC-S5 Acta Horticulturae*, pp. 89–93 (Vol. 766).

Yahel, H., Kadman-Zahavi, A. and Tsukamoto, Y. (1985) Chrysanthemum (perennial species). In: Halevy, A.H. (ed.) *CRC Handbook of Flowering*, Vol. 2. CRC Press, Boca Raton, Florida, pp. 258–264.

Yamaguchi, H. (2018) Mutation breeding of ornamental plants using ion beams. *Breeding Science* 68(1), 71–78. DOI: 10.1270/jsbbs.17086.

Yamashiro, T., Shiraishi, A., Satake, H. and Nakayama, K. (2019) Draft genome of *Tanacetum cinerariifolium*, the natural source of mosquito coil. *Scientific Reports* 9, 18249. DOI: 10.1038/s41598-019-54815-6.

YanFei, X., FaDi, C., NianJun, T., SuMei, C. and FengTong, L. (2009) Preliminary study on inbreeding depression of *Dendranthema morifolium*. *Journal of Plant Resources and Environment* 18(4), 28–32.

Yang, Y., Ma, C., Xu, Y., Wei, Q., Imtiaz, M. *et al.* (2014) A zinc finger protein regulates flowering time and abiotic stress tolerance in chrysanthemum by modulating gibberellin biosynthesis. *Plant Cell* 26, 2038–2054.

Zagorski, J.S., Ascher, P.D. and Widmer, R.E. (1983) Multigenic self incompatibility in hexaploid *Chrysanthemum*. *Euphytica* 32, 1–7.

Zeng, T., Li, J., Wang, C. and He, J. (2020) The complete chloroplast genome of *Tanacetum coccineum*. *Mitochondrial DNA Part B* 5(3), 2233–2234. DOI: 10.1080/23802359.2020.1768915.

Zeng, T., Li, J.W., Zhou, L., Xu, Z.Z., Li, J.J. *et al.* (2021) Transcriptional responses and GCMS analysis for the biosynthesis of pyrethrins and volatile terpenes in *Tanacetum coccineum*. *International Journal of Molecular Science* 222, 3005. DOI: 10.3390/ijms222313005.

Zhang, F., Jiang, J., Chen, S., Chen, F. and Fang, W. (2011) Genetic analysis and associated SRAP markers for flowering traits of chrysanthemum (*Chrysanthemum morifolium*). *Euphytica* 177, 15–24.

Zhang, J., Chen, S., Liu, R., Jiang, J., Chen, F. *et al.* (2013) Chrysanthemum cutting productivity and rooting ability are improved by grafting. *The Scientific World Journal* 286328. DOI: 10.1155/2013/286328.

Zhang, M., Huang, H., Wang, Q. and Dai, S. (2018) Cross breeding new cultivars of early-flowering multiflora *Chrysanthemum* based on mathematical analysis. *HortScience* 53(4), 421–426. DOI: 10.21273/HORTSCI12769-17.

Zhang, Y., Dai, S., Hong, Y. and Song, X. (2014) Application of genomic SSR locus polymorphisms on the identification and classification of chrysanthemum cultivars in China. *PLoS One* 9, e104856. DOI: 10.1371/journal.pone.0104856.

Zhao, H.B., Chen, F., Tang, F., Jiang, J., Li, C. *et al.* (2012) Morphological characteristics and chromosome behaviour in F_1, F_2 and BC_1 progenies between *Chrysanthemum* × *morifolium* and *Ajania pacifica*. *Russian Journal of Genetics* 48(8), 808–818.

Zhao, H.E., Liu, Z.H., Hu, X., Yin, J.L., Li, W. *et al.* (2009) Chrysanthemum genetic resources and related genera of *Chrysanthemum* collected in China. *Genetic Resources and Crop Evolution* 56, 937–946. DOI: 10.1007/s10722-009-9412-8.

Zheng, R.R., Wang, Y.Y. and Wang, C.Y. (2021) Transcriptional responses and GCMS analysis for the biosynthesis of pyrethrins and volatile terpenes in *Tanacetum coccineum*. *International Journal of Molecular Sciences* 22, 13005. DOI: 10.3390/ijms222313005.

Zhu, L., Guan, Y., Liu, Y., Zhang, Z., Abuzar Jaffar, M. *et al.* (2020) Regulation of flowering time in chrysanthemum by the *R2R3 MYB* transcription factor *CmMYB2* is associated with changes in gibberellin metabolism. *Horticultural Research* 7, 96. DOI: 10.1038/s41438-020-0317-1.

3

Molecular Breeding of Chrysanthemum

Jiangshuo Su, Sumei Chen and Fadi Chen*

State Key Laboratory of Crop Genetics & Germplasm Enhancement and Utilization, Key Laboratory of Landscaping, Ministry of Agriculture and Rural Affairs, Key Laboratory of Biology of Ornamental Plants in East China, National Forestry and Grassland Administration, College of Horticulture, Nanjing Agricultural University, 210095 Nanjing, China

Abstract

Chrysanthemum (*Chrysanthemum morifolium* Ramat.) is an important ornamental crop. Breeders aim to develop cultivars that exhibit novel characteristics such as new flower colors and shapes, plant architectures, flowering times (shorter response time), improved postharvest quality, and biotic and abiotic stress tolerance in a timely and cost-effective manner. Conventional breeding has led to the development of new chrysanthemum cultivars, but it has been slow and insufficient to meet the growing market demand for novel varieties. In recent decades, molecular breeding approaches have helped shorten the breeding process by identifying and pyramiding desirable genes into breeding lines to produce superior cultivars. This chapter provides an overview of the status of molecular breeding in chrysanthemum, including various molecular marker technologies consisting of sequence-based markers such as single-nucleotide polymorphism, next-generation sequencing, genotyping by sequencing, marker-assisted selection involving linkage maps and quantitative trait loci (QTL) mapping, genome-wide association mapping, and bulked segregant analysis. Additionally, the challenges and prospects of emerging genome editing and omics technologies are discussed.

Keywords

Chrysanthemum, genome editing, genome-wide association mapping, molecular markers, next-generation sequencing, omics technologies, QTL mapping

*Corresponding author: chenfd@njau.edu.cn

DOI: 10.1079/9781789248289.0003

3.1 Introduction

Chrysanthemum (*Chrysanthemum morifolium* Ramat.) is an important ornamental crop that represents a substantial part of the global flower industry. Developing chrysanthemum cultivars with novel flower colors and shapes, plant architectures, flowering times (shorter response time), advanced post-harvest quality, and stress tolerances in a time- and cost-efficient manner is the goal of chrysanthemum breeding. Chrysanthemum breeding is on a mission of continuously discovering and pyramiding desirable genes of interest into breeding lines to produce superior cultivars. Conventional phenotype-based breeding approaches, such as crossbreeding and mutation breeding, have made a significant contribution to the development of new chrysanthemum cultivars in the last century (see Chapter 02). However, it takes years and many generations to bring desirable traits together. A variety of environmental factors influence conventional breeding, making it insufficient to keep pace with the ever-increasing market demand for new cultivars.

Recent advances in DNA marker technology, plant genomics, and molecular biology have revolutionized our understanding of plant genetics. The emergence of a series of molecular breeding methods, including marker-assisted selection (MAS), transgenic technology, and genome editing, provides a more accurate and efficient plant breeding approach. In the last two decades, different types of molecular markers have been developed and used for identifying the linkage between markers and traits in chrysanthemum through quantitative trait loci (QTL) mapping, genome-wide association mapping (GWAS), and bulked segregant analysis (BSA) (Su *et al.*, 2019). Additionally, the progress made in high-throughput next-generation sequencing (NGS) technology and various omics platforms—especially genomics, transcriptomics, phenomics, proteomics, and metabolomics—has revolutionized plant research. Integration of these omics data with traditional genetic studies will enable identifying genes/pathways responsible for important traits and accelerating plant breeding progress (Scossa *et al.*, 2021).

Notably, the development and application of sophisticated molecular breeding techniques in chrysanthemum is a daunting task and has lagged far behind the progress made in the breeding of many field and vegetable crops, primarily due to its polyploidy nature, high heterozygosity, and lack of genomic information. The purpose of this chapter is to provide an overview of the status of molecular breeding in chrysanthemum. We summarize the merits and drawbacks of different molecular marker technologies and the recent achievements in quantitative genetic studies in chrysanthemum. In addition, we discuss the challenges and prospects of the emerging genome editing and omics technologies toward accelerating breeding programs.

3.2 Molecular Markers and Marker-Assisted Selection

Conventional plant breeding requires a significant amount of time to select and evaluate desirable traits over many generations. MAS refers to the use of molecular markers to assist phenotypic selection in crop improvement, providing a faster and more accurate selection approach in a breeding population (Tiwari *et al.*, 2022). In the past decade, several types of molecular markers have been successfully used in genomic diversity studies in chrysanthemum. In order to identify markers that are significantly associated with target traits in chrysanthemum, several molecular marker types have been used. These molecular markers are categorized as hybridization-based (non-PCR-based), polymerase chain reaction (PCR)-based, and sequencing-based methods and include restriction fragment length polymorphisms (RFLPs), random amplified polymorphic DNA (RAPDs), amplified fragment length polymorphisms (AFLPs), inter-simple sequence repeats (ISSRs), simple sequence repeats (SSRs), hybridization-based DNA fingerprinting, sequence-related amplified polymorphisms (SRAPs), sequence-characterized amplified region (SCAR), and start codon-targeted (SCoT) polymorphisms, and more recent sequence-based single-nucleotide polymorphisms (SNPs) markers have also been applied.

Molecular markers, particularly DNA markers, detecting polymorphisms at the DNA level, are the most widely used genetic markers, predominantly due to their abundance, cost-effectiveness, and stability that is not affected by environmental factors and the developmental stage of plants, compared with classical morphological, cytological, and biochemical markers. In the last three decades, various molecular marker techniques have been developed and exploited, thus leading to a significant revolution in plant breeding. Generally, molecular markers evolved continuously from low-throughput hybridization-based markers and medium-throughput PCR-based markers toward high-throughput sequence-based markers. An ideal molecular marker should be codominant, evenly distributed throughout the genome, allow for easy and automated detection, be highly repeatable and polymorphic, and distinguish allelic features. Comparisons of important characteristics of the most used molecular markers in chrysanthemum are given in Table 3.1. Continuing advances in molecular marker techniques for genetic diversity assessment, molecular phylogenetics, population genetics, and gene mapping have been witnessed in chrysanthemums.

3.2.1 Hybridization-based markers (non-PCR based)

3.2.1.1 Restriction fragment length polymorphism

RFLP is the only marker system based on hybridization known as "first-generation molecular markers", which initiated the era of molecular marker technology during the 1980s (Vos *et al.*, 1995). Mutation events, including point mutation, insertion or deletion, translocation, inversion, and duplication, may occur at restriction sites or between adjacent restriction sites in the genome. RFLP markers are powerful tools to detect the different lengths of DNA fragments after restriction digestion of genomic DNA. Using agarose or polyacrylamide gel electrophoresis (PAGE) and Southern blotting, the separated fragments of interest are identified by hybridizing with complementary

Table 3.1. Feature description and comparison of the most commonly used molecular markers.

Features	RFLP	RAPD	AFLP	SSR	ISSR	SRAP	SCAR	CAPS	SCoT	SNP
Codominant	Yes	No	No	Yes	No	No	Yes	Yes	No	Yes
Reproducibility	High	High	Medium	High	Medium-high	Medium-high	High	High	Medium-high	High
Polymorphism	Medium	Very high	High	High	High	High	High	High	High	High
Quality of DNA required	High	High	High	Low	Low	Low	Medium-high	Medium-high	Low	High
Amount of DNA required	High	Medium	Low	Low	Low	Low	Low	Low	Low	Low
Genome abundance	High	Very high	Very high	Medium	Medium	Medium	Medium	Medium	Medium	Very high
Cost	High	Less	High	High	High	Medium	Medium	High	Medium	Variable
Ease of automation	Low	Medium	High	High	Medium	Medium	High	High	Medium	High
Restriction enzyme needed	Yes	No	Yes	No	No	No	No	Yes	No	No
PCR requirement	No	Yes	Yes	Yes	Yes	Yes	Yes	Yes	Yes	No
Sequence information	Yes	No	No	Yes	No	No	Yes	Yes	No	Yes
Visualization	Radioactive	Agarose gel/PAGE	Agarose gel/PAGE	Agarose gel/PAGE	Agarose gel/PAGE	Agarose gel/PAGE	Agarose gel/PAGE	Agarose gel/PAGE	Agarose gel/PAGE	SNP-VISTA
Status	Past	Past	Past	Present	Present	Present	Present	Present	Present	Present

Note: AFLP = amplified fragment length polymorphism, CAPS = cleaved amplified polymorphic sequence, ISSR = inter-simple sequence repeats, RAPD = random amplified polymorphic DNA, SSR = simple sequence repeats, SCAR = sequence-characterized amplified region, SCoT = start codon-targeted, SNP = single-nucleotide polymorphisms, SRAP = sequence-related amplified polymorphism.

radioactively labeled probes and visualized by autoradiography (Amom and Nongdam, 2017). Usually, different probes are selected in combination with several endonucleases, each combination generating an individual unique RFLP pattern at a specific locus (Grover and Sharma, 2016). Wolff *et al.* (1994) developed probes and primers of RFLP considering the hexaploid character of chrysanthemum, suggesting methylation-insensitive restriction enzymes should be used for cultivar identification (Wolff *et al.*, 1995). Fukai *et al.* (2002) divided the 36 *Chrysanthemum yoshinaganthum* lines into two groups using cpDNA PCR-RFLP analysis. RFLP could also be used to confirm the phytoplasma in chrysanthemum (Min *et al.*, 2009). RFLP exhibits codominant inheritance, high reproducibility, and transferability among different laboratories. Limitations of the RFLP technique are high time and labor demands, high costs, and the need for high quality and large quantity of DNA (5–20 μg). Therefore, few uses of RFLP in both genetic research and plant breeding programs have been reported in the last decade.

3.2.2 PCR-based markers

3.2.2.1 Random amplified polymorphic DNA

RAPD is a PCR-based marker technique, where the whole genomic DNA of an individual is amplified using a single, short (~10 bp) and random primer (Williams *et al.*, 1990). The profile of amplification products depends on the template–primer combination, and it would be expected only when the priming site occurs twice in opposite orientation within approximately 2000 bases (Grover and Sharma, 2016). RAPD polymorphisms result from mutations or rearrangements either at or between the primer binding sites and can be detected by the presence or absence of a particular band. RAPD, as a non-radioactive assay, requires a small amount of DNA (~10 ng) and is quick and simple to conduct. Additionally, the RAPD primers are universal, i.e. non-species specific, and display a higher level of polymorphism than RFLP. Though RAPD eased most of the problems associated with RFLP, it still has some disadvantages, such as dominant inheritance, poor reproducibility, and incapability to detect allelic differences in heterozygotes (Skroch and Nienhuis, 1995). Many studies have shown that RAPD markers could be an effective tool to construct DNA fingerprinting of chrysanthemums (Wolff and Peters-van Rijn, 1993; Lema-Rumińska *et al.*, 2004; Mukherjee *et al.*, 2013). Barakat *et al.* (2010) found the similarity among chrysanthemum cultivars and mutants with a matching coefficient ranging from 0.17 to 0.90 using RAPD analyses, suggesting mutants with different flower colors could be well identified at the molecular level using RAPDs. Zhang *et al.* (2010b) detected several RAPDs associated with six ornamental traits, which might improve the selection breeding efficiency in chrysanthemum. Furthermore, the authors first attempted to apply RAPDs to genetic linkage map construction (Zhang *et al.*, 2010a).

3.2.2.2 Simple sequence repeat or microsatellite

SSRs, also known as microsatellites, short tandem repeats, or sequence-tagged microsatellite sites, are PCR-based markers targeting specific genome sites. They are randomly tandem repeats of short nucleotide motifs with 1–6 bases, correspondingly designated as mono-, di-, tri-, tetra-, penta-, and hexa-nucleotide repeats, and are abundant across genomes of every organism. Moreover, the DNA sequences flanking microsatellite regions are usually conserved. Thus, primers specific for these regions can be designed for use in the PCR reaction. The DNA polymorphism arises from the difference in the number of repeats and SSR length among individuals. SSRs are assorted based on their source into genomic SSRs (g-SSRs) and expressed sequence tag (EST)-SSRs, which are in the whole genome and coding region, respectively (Mason, 2015b). Besides nuclear microsatellites, chloroplast microsatellites (cpSSRs) and mitochondrial microsatellites (mtSSRs) are developed as well. cpSSRs typically contain mononucleotide motifs repeated 8–15 times and provide an effective tool for population genetic studies. SSRs are characterized by their codominant nature, hyper-variability, locus specificity, reproducibility, random genome-wide distribution, and the genotyping protocol that can be automated, which has made them regarded as the most breeder-friendly marker system in practical plant breeding. However, this technique requires nucleotide information for primer design, is medium throughout, and involves high costs for automated detection (Jenkins *et al.*, 2012). EST-SSRs were mainly used in chrysanthemum during the early stage with no genomic information available. Wang *et al.* (2013) developed large-scale EST-SSRs through transcriptome sequencing. They found most EST-SSR loci were readily transferable between *Chrysanthemum* and *Ajania* species. Population structure analyses showed the formation from tetraploid to hexaploid and from octoploid to decaploid might be a recent event, while the transition from diploid to tetraploid might be an ancient event in the chrysanthemum lineage (Wang *et al.*, 2014b). Similar results were obtained by Fan *et al.* (2019) using EST-SSRs in another population. Han *et al.* (2019) identified six EST-SSRs with a single band-like amplicon that could specify the genetic identities of 55 chrysanthemum varieties. These markers may be useful for cultivar protection and breeding programs in chrysanthemum.

3.2.2.3 Inter-simple sequence repeat

ISSR was developed by Zietkiewicz *et al.* (1994) and is based on the amplification of DNA segments located at an amplifiable distance between two identical microsatellite repeat regions oriented in opposite directions. The microsatellite repeats used as primers for ISSR could be di-, tri-, tetra-, or penta-nucleotide. Usually, ISSR primers are longer (15–35 bases) to enable higher annealing temperature (about 45 ~ 60°C), resulting in higher stringency. The primers used may be either unanchored or, more typically, anchored at the 3′ or 5′ end with 1 ~ 4 degenerate bases, which are extended into the flanking sequences. ISSR markers are simple and highly polymorphic, require a low quantity (~5–20 ng) of DNA, and do not need any prior knowledge of DNA sequence data—thus, they can be applied to non-model species with ease. Especially, ISSR primers can be used universally among different species, which is superior to SSRs having strong species specificity. However, ISSRs exhibit dominant inheritance and have less reproducibility due to the homology of co-migrating amplification products. To date, ISSRs have been widely used in genetic studies of chrysanthemums (Palai and Rout, 2011; Mukherjee *et al.*, 2013). More recently, Nasri *et al.* (2021) applied ISSRs to assess the genetic polymorphism among EMS-induced mutants and their parents, providing important genetic resources for the generation of new chrysanthemum cultivars.

3.2.2.4 Sequence-related amplified polymorphism

SRAP was first developed by Li and Quiros (2001), mainly for the amplification of open reading frames by targeting GC-rich exons and AT-rich promoters, introns, and spacer regions. The PCR amplification system of SRAP is based on two primers. Forward primers are 17 nucleotides in length and designed to specifically amplify the exon regions with the CCGG core sequence, whereas reverse primers are 18 nucleotides in length and designed to specifically amplify the promoters, introns, and spacer regions with the AATT core sequence (Amom and Nongdam, 2017). The key characteristic of the SRAP technique is the renaturation variable temperature. SRAPs are dominant in nature and cost-efficient, and DNA fragments are simply scored by the absence or presence of a band and thus have been widely employed in chrysanthemums. Zhang *et al.* (2011b) detected 10 and 20 SRAPs linked to initial blooming time and duration of flowering, which cumulatively explained 46 and 54% of the phenotypic variation, respectively. Based on the linkage maps constructed by SRAPs, several major QTLs related to ornamental characteristics and resistance traits were identified (Su *et al.*, 2019), the flanking markers of these QTLs might be potentially useful in marker-assisted breeding. In addition, SRAPs have been employed in population genetic analysis (Li *et al.*, 2016; Hodaei *et al.*, 2019).

3.2.2.5 Sequence-characterized amplified region

SCAR markers can be derived from PAPD, SRAP, SSR, or any other multi-locus markers by cloning and sequencing the amplified DNA fragments linked to a trait of interest and followed by designing long primers. SCAR was first reported by Paran and Michelmore (1993), where it originated from several RAPD markers linked to downy mildew resistance genes in lettuce. The functional annotation of the DNA sequence can be investigated by comparing it with the known DNA sequences available at the NCBI (National Center for Biotechnology Information) database. SCARs are more reliable and robust than RAPD and ISSR markers and can potentially be converted into codominant markers (Bhagyawant, 2016). Once it has been developed, a large number of individuals can be screened in a short time. Therefore, SCAR markers have become a good choice for direct application in breeding practice. Zhao *et al.* (2009) successfully transferred one RAPD marker associated with creep plant type in ground-cover chrysanthemum to an SCAR marker. Sang (2016) converted one AFLP marker linked to white rust of chrysanthemum into an SCAR marker, which was used in selecting resistant plants in actual chrysanthemum breeding programs.

3.2.2.6 *Cleaved amplified polymorphic sequences*

Cleaved amplified polymorphic sequences (CAPSs) were originally named PCR-RFLP marker, as it is a combination of RFLP and PCR (Maeda *et al.*, 1990), where target DNA is amplified through PCR, followed by digestion with restriction enzymes (Michaels and Amasino, 1998). The CAPS primers are designed according to the sequence information present in a genome database or cloned RAPD bands or a cDNA library. Thus, CAPS markers are versatile, codominant and do not include the laborious and technically demanding steps of Southern blotting and radioactive detection procedures needed for RFLP (Agarwal *et al.*, 2008). Recently, several SNPs loci were converted into CAPS or derived CAPS (dCAPS) to reduce the production cost in chrysanthemum MAS breeding (Su *et al.*, 2019). (Refer to Genome-wide association study (GWAS) in this chapter.)

3.2.2.7 *Start codon-targeted polymorphism*

SCoT is a gene-targeted advantageous marker that developed based on the short conserved regions flanking the translation initiation start codon ATG of plant genes (Collard and Mackill, 2009). SCoT markers involve the use of an 18-mer single primer with an annealing temperature of 50°C. SCoT markers are dominant, generally reproducible, and reveal diversity at the genes level, which gives the possibility of identifying new alleles among plant germplasms. Moreover, the information revealed by SCoT markers is more informative than RAPD and ISSR markers, and it is even simpler to develop than SSR. Feng *et al.* (2016) determined relationships and clustering between 32 medicinal chrysanthemum cultivars based on SCoTs that agreed with their origin and ecological distribution, indicating SCoT markers are informative and could be used to evaluate the genetic diversity of chrysanthemum. Li *et al.* (2016) identified several SCoTs associated with plant architecture and inflorescence-related traits in chrysanthemums. Unfortunately, all the significant SCoTs had only minor effects, explaining 4.5–7.5% of the phenotypic variation. Kobeissi *et al.* (2019) revealed the efficiency of SSR and SCoT markers was comparable in fingerprinting of chrysanthemum genotypes. However, SCoTs were deemed more useful for plant breeding because they are based on the functional region of the genome.

3.2.3 Sequence-based markers

3.2.3.1 Single-nucleotide polymorphisms

Polymorphisms specific to differences at a single-nucleotide base (substitution, insertion, or deletion) occurring in the genome sequence of individuals are referred to as SNPs (Grover and Sharma, 2016). SNPs are widely distributed within the genome and can be found in coding or non-coding regions of genes or intergenic regions with different frequencies (Xu, 2010). Although SNPs are less polymorphic than SSRs due to their biallelic nature, they easily compensate for this drawback by being abundant, ubiquitous, and amenable to high- to extra-high-throughput automation. Typically, the SNP frequency is one SNP in every 100–300 bp in plants. Advancements in sequencing technologies such as NGS and genotyping by sequencing (GBS) in the recent past have contributed to the high-throughput discovery of large numbers of SNPs, making them the most attractive potential molecular markers in genetic studies and breeding. Chong *et al.* (2016) developed >90,000 high-throughput SNP markers utilizing SLAF-seq (specific-locus amplified fragment sequencing) technology in a panel of 199 individuals to investigate the evolutionary relationship among the five cultivated types. The phylogenetic analysis divided the individuals into five groups according to the cultivated types, and they found potted and ground chrysanthemum and spray cut chrysanthemum to be closer to the wild progenitor species than disbud cut chrysanthemum and traditional chrysanthemum. Additionally, several differentiated genomic regions potentially associated with selection were further identified among different cultivated types through an analysis of selective sweeps. Van Geest *et al.* (2017b) designed an Affymetrix Axiom array with 183k SNPs generated from RNA-seq data and genotyped five populations, each including 37 ~ 405 chrysanthemum accessions. They proposed a new method for dosage scoring in hexaploids from signal intensities of the array based on mixture models. Nevertheless, high technical and equipment demands and high costs for the start-up of SNP development, especially for plants with large genomes such as chrysanthemum, limit, to some extent, the application of SNPs in practical breeding programs, although commercial services of SNP assays are available for some model species. Therefore, many SNP detection methods have been developed based on different technical principles of allelic discrimination, such as SNP-RFLP, CAPS, KASP (Kompetitive allele-specific PCR), and HRM (high-resolution melting).

3.2.3.2 Next-generation sequencing

NGS is a comprehensive expression used to describe many complementary DNA sequencing techniques, which provide a much cheaper and higher-throughput alternative than traditional Sanger sequencing (Grada and Weinbrecht, 2013). The NGS technologies and powerful computational protocols have opened new opportunities for discovering massive genome-wide markers, especially EST-SSRs and SNPs, in a cost-effective way (Stapley *et al.*, 2010). Since its commercial availability in 2005, NGS has drastically changed the nature of molecular biology and revolutionized plant breeding methods. NGS relies on massive parallel sequencing and imaging techniques to yield several hundreds of millions to several hundreds of billions of nuclear bases per run (Shendure and Ji, 2008). In the past decade, several NGS platforms, such as Roche 454, Illumina MiSeq and HiSeq series, ABI SOLID, Ion Torrent PGM, and Qiagen GeneReader, have been developed. Among them, Illumina sequencing-by-synthesis technology is the most widely utilized in plants. For example, the latest Illumina HiSeq 4000 sequencing system can produce up to 100 million reads per sample, and it has a reading length of 50, 75, and 150 bp for data yields of 210–250 Gb, 650–750 Gb, and 1300–1500 Gb per flow cell in less than 3.5 days' runtime, with an accuracy greater than 99% (Taheri *et al.*, 2018). To date, NGS is widely used in *Chrysanthemum* and its related genera to develop EST-SSRs and identify candidate genes and for various omics applications (Wang *et al.*, 2013; Jo *et al.*, 2015; Sasaki *et al.*, 2017).

All NGS strategies are based on a similar protocol, first for DNA template preparation, where universal adapters are ligated at both ends of randomly sheared DNA fragments. They also depend on the cyclic interrogation of millions of clonally amplified DNA molecules immobilized on a synthetic surface to generate several billions of sequences. Sequencing is performed in an iterative manner, where the incorporation of one or more nucleotides is followed by the emission of a signal and its detection by the sequencer. Most NGS platforms can generate reliable sequences and display near-perfect coverage on GC-rich, neutral, and moderately AT-rich genomes. However, there are key differences between the quality of that data and the applications it will support.

A major issue faced by NGS technologies is the bioinformatics analysis of the huge amounts of data generated, involving raw reads filtering, alignment to a reference sequence or *de novo* assembly, and SNP calling for polymorphism identification or genotyping (Zargar *et al.*, 2015). The analysis is much more resource-dependent, and the need for computing power, storage, and several powerful bioinformatics tools is getting urgent, especially for non-model species containing large nuclear genomes, such as polyploids including chrysanthemum. Additionally, since crop breeding is an expensive process whereby a very large population size is handled, the cost of genotyping is another important determining factor for adopting appropriate NGS technologies in plant breeding. A choice between whole and partial genome sequencing would depend on the availability and judicious use of funds.

3.2.3.3 Genotyping by sequencing

Whole-genome sequencing (WGS) offers the possibility to extract nearly all information from a genome and has been used for a wide array of crops with reference genomes to develop molecular markers. However, its application is still restricted by high costs and the impossibility of applying saturated WGS to large populations or non-model species. To overcome the latter problem, a series of NGS-based reduced-representation genome sequencing strategies were accordingly proposed, i.e. complexity reduction of polymorphic sequence (CRoPS; van Orsouw *et al.*, 2007), RAD-Seq (restriction site-associated DNA sequencing; Baird *et al.*, 2008), GBS (genotyping by sequencing; Elshire *et al.*, 2011), 2b-RAD (Wang *et al.*, 2012), dd (double digest)-RAD (Peterson *et al.*, 2012), SLAF-seq (Sun *et al.*, 2013), which are capable of generating high-throughput SNPs on a genome-wide scale, even in non-model species with little or no available genetic information (Torkamaneh *et al.*, 2018). Among these, GBS, as a robust and multiplexed genotyping method, has been applied in numerous species with large and highly repetitive genomes such as maize (Wang *et al.*, 2020), wheat (Akram *et al.*, 2021), tree peony (Zhang *et al.*, 2019), and *Dendrobium* orchids (Ryu *et al.*, 2019).

The GBS approach includes the digestion of genomic DNA using restriction enzymes, followed by the ligation of barcode adapter, PCR amplification, and sequencing of the amplified DNA pool on a single lane of flow cells (He *et al.*, 2014). Subsequently, the obtained GBS datasets are analyzed and interpreted using bioinformatic pipelines. Appropriate choice of enzymes can avoid sampling low-complexity, highly repetitive regions, simplifying downstream processing. Poland *et al.* (2012) proposed a two-enzyme (*Pst*I/*Msp*I) GBS protocol, which offers a greater degree of complexity reduction and generates a more uniform library for sequencing compared with the original protocol using *Ape*KI. GBS library preparation involving selective amplification resulted in a greater number of SNP calls with improved depth of coverage, thereby allowing for a significant reduction in per-sample cost (Sonah *et al.*, 2013). More recently, a miniaturized procedure for GBS library preparation on a nanoliter scale reduced genotyping cost per sample by 72% (Torkamaneh *et al.*, 2020). Two GBS strategies have been developed based on the Ion PGM system. First, restriction enzyme digestion, with no specific SNPs identified and ideal for discovering new markers for MAS programs. The complexity of the genome can be reduced by digesting the DNA using one or two selected restriction enzymes before adapter ligation. A second strategy is called multiplex enrichment PCR, where a set of SNPs is defined for a section of the genome using PCR primers designed to amplify only the areas of interest (Amom and Nongdam, 2017).

GBS offers substantial benefits compared with RAD in terms of lower cost, lower quantity of DNA required, no need for fragments size selection, and a shorter procedure that includes fewer DNA purification steps and PCR amplification of the pooled library (Ulaszewski *et al.*, 2021). More importantly, there is no requirement for a priori knowledge of the reference genome information, making GBS an ultimate MAS tool to accelerate molecular breeding of non-model plants. Molecular breeders are rapidly implementing GBS for high-density SNP markers's discovery, genetic diversity assessment, high-resolution linkage map construction, QTL mapping, and association analysis (He *et al.*, 2014; Nguyen and Lim, 2019; Riangwong *et al.*, 2020).

Polyploids offer unique challenges for genotyping because multi-locus alleles are generally detected when a single locus is targeted (Mason, 2015a). In addition, their complex inheritance furtherly constrains genetic studies using molecular markers. Therefore, genotyping using GBS is more difficult in polyploids than in diploids, with problems in assembling duplicated regions and challenges in distinguishing between inter- and intra-genomic polymorphisms, though these issues are encountered by all genotyping methods (Chung *et al.*, 2017). Among all the tools available, however, GBS offers a higher potential to resolve the issue by increasing the read coverage and reducing genomic representation using sophisticated computational tools for SNP discovery. Genetic studies using GBS technology have been successfully constructed in allopolyploid wheat (Alipour *et al.*, 2017) and autopolyploids such as sugarcane (Balsalobre *et al.*, 2017) and alfalfa (Adhikari *et al.*, 2018). Nguyen *et al.* (2018) first attempted to construct and optimize the *de novo* GBS pipeline in chrysanthemum. They found the *ApeKI*/*MseI* combination showed superiority in library quality, highest number of tags, and SNP generation potential compared with *ApeKI* and *ApeKI*/*MspI*. Furthermore, the authors identified 7758 SNPs in a panel of 80 wild Korean native chrysanthemum germplasms using GBS and revealed their genetic diversity. These examples suggest that GBS could be an effective tool for breeding and genetic studies in polyploids. The recent progress of genetic studies based on molecular markers in chrysanthemum, by means of genetic linkage map construction and QTL mapping, GWAS, and BSA, resulted in multiple favorable alleles for ornamental traits (Table 3.2), which greatly facilitated the use of MAS in chrysanthemums.

3.2.3.4 Linkage maps and quantitative trait loci mapping

The linkage map is a road map that describes the position and relative genetic distance between markers, which is not only a prerequisite for QTL analysis but also plays an essential role in assisting genome assembly by guiding the integration and orientation of contigs. The construction and use of linkage maps are challenging in polyploids due to the unknown linkage phases of the marker alleles. Generally, specialized statistical algorithms are required for linkage map construction in polyploids (Bourke *et al.*, 2018). Cultivated chrysanthemum is a complex hexaploid or aneuploid ($2n = 54 \pm 7{\sim}10$; Spaargaren and van Geest, 2018). However, it has long been assumed to exhibit disomic inheritance because of its presumed multispecies origin and the occurrence of bivalent chromosomal pairing in meiosis (Roxas *et al.*, 1995; Gupta *et al.*, 2013); hence, to some extent, reducing the difficulty of linkage

Table 3.2. Recent published genetic studies of chrysanthemum.

Population	Population size	Marker type	Marker number	Study objective	Methodology	References
Yuhualuoying × Aoyunhanxiao	142	RAPD, ISSR, AFLP	336	Linkage map construction	Map construction	Zhang *et al.* (2010a)
Yuhualuoying × Aoyunhanxiao	142	SRAP	675	Inflorescence traits	QTL mapping	Zhang *et al.* (2011a)
Yuhualuoying × Aoyunhanxiao	142	SRAP	675	Plant architectural traits	QTL mapping	Zhang *et al.* (2012)
Yuhualuoying × Aoyunhanxiao	142	SRAP	675	Leaf traits	QTL mapping	Zhang *et al.* (2012)
Yuhualuoying × Aoyunhanxiao	142	SRAP	346	Flowering time	One-way ANOVA	Zhang *et al.* (2011b)
Yuhualuoying × Aoyunhanxiao	142	SRAP	675	Flowering time	QTL mapping	Zhang *et al.* (2013)
Han 2 × Nannong Gongfen	133	SRAP, SSR, SCoT	262	Aphid resistance	One-way ANOVA, QTL mapping, BSA	Wang *et al.* (2014b)
QX145 × Nannongyinshan	92	SRAP, SSR	234	Branching traits	QTL mapping	Peng *et al.* (2015)
QX053 × Nanong Jingyan	160	SRAP, SSR, SCoT	497	Inflorescence traits	One-way ANOVA, QTL mapping,	Tang *et al.* (2015)
Nannong Xuefeng × Monalisa	162	SRAP, SSR	502	Waterlogging tolerance	QTL mapping	Su *et al.* (2018)
Nannong Xuefeng × Monalisa	100	SRAP, SSR	502	Cold tolerance	QTL mapping	Ao *et al.* (2019)

Nannong Xuefeng × Monalisa	162	SRAP, SSR	502	Anemone-type	QTL mapping	Yang *et al.* (2019a)
C. morifolium F_1 population	192	EST-SSR	265	Inflorescence and leaf traits	QTL mapping	Fan *et al.* (2020)
Candy × 225	305	SNP	6452	Flower-type	QTL mapping	Song *et al.* (2020)
Puma White × Dancer	94	AFLP	1779	White rust	BSA	Sang (2016)
Nannong Xuefeng × QX096	80	SRAP, SCAR	NA	Anemone-type	BSA	Yang *et al.* (2019b)
Natural population	159	SRAP, SSR, SCoT	707	Plant architecture and inflorescence	GWAS	Li *et al.* (2016)
Natural population	100	SRAP, SSR, SCoT	707	Waterlogging tolerance	GWAS	Su *et al.* (2016)
Natural population	80	SRAP, SSR, SCoT	707	Aphid resistance	GWAS	Fu *et al.* (2018)
Natural population	159	SRAP, SSR, SCoT	707	Drought tolerance	GWAS	Li *et al.* (2018)
DB36451 × DB39287	406	SNP	183,000	Flowering time and inflorescence traits	QTL mapping	van Geest *et al.* (2017a)
Natural population	199	SNP	92,830	Cultivated type and inflorescence traits	GWAS	Chong *et al.* (2016)

Continued

Table 3.2. Continued

Population	Population size	Marker type	Marker number	Study objective	Methodology	References
Natural population	88	SNP	92,811	Waterlogging tolerance	GWAS	Su *et al.* (2019)
Natural population	107	SNP	92,617	Inflorescence traits	GWAS	Chong *et al.* (2019)
Natural population	200	SNP	330,710	Leaf angle	Multi-locus GWAS	Lou *et al.* (2023)

Note: AFLP = amplified fragment length polymorphism, ANOVA = analysis of variance, BSA = bulked segregant analysis, GWAS = genome wide association study, ISSR = inter-simple sequence repeats, QTL = quantitative trait loci, RAPD = random amplified polymorphic DNA, SSR = simple sequence repeats, SCAR = sequence-characterized amplified region, SCoT = start codon-targeted, SNP = single-nucleotide polymorphisms, SRAP = sequence-related amplified polymorphism.

map construction. Although the precision and accuracy of a linkage map are affected by several factors, such as the population size, molecular marker types and numbers, as well as the methods used for ordering markers within a linkage group, considerable efforts have been devoted to constructing linkage maps in chrysanthemum for the past two decades, following a "double pseudo-testcross" mapping strategy in F_1 populations using AFLP, RAPD, ISSR, SRAP, SCoT, and SSR markers. Zhang *et al.* (2010b) constructed the first preliminary linkage map of chrysanthemum by genotyping 142 F_1 progeny using 153 RAPDs, 353 AFLPs, and 61 ISSRs. However, the authors failed to produce a referenceable integrated map, and only six pairs of homologous linkage groups based on the common intercross markers of the two parental maps were identified. Furthermore, they produced another map for the same population using SRAP markers, with an average map distance of 6.9 cM for the female map and 6.6 cM for the male map (Zhang *et al.*, 2011a). Subsequently, several mapping populations were generated to genetically localize different horticultural traits (Wang *et al.*, 2014a; Peng *et al.*, 2015; Su *et al.*, 2018; Fan *et al.*, 2020).

Biparental QTL mapping is a conventional tool for identifying markers and genes associated with desired traits in a segregating population for MAS breeding. The QTL mapping methodology is described in Fig. 3.1. Linkage map quality, phenotypic accuracy, and population size are the main factors affecting QTL detection. In chrysanthemums, QTL mapping has been applied for numerous ornamental characteristics and resistance traits based on the gel-based linkage maps, i.e. inflorescence and leaf traits (Fan *et al.*, 2020), flowering time (Zhang *et al.*, 2013), plant shape (Zhang *et al.*, 2012; Peng *et al.*, 2015), aphid resistance (Wang *et al.*, 2014a; Fu *et al.*, 2019), waterlogging tolerance (Su *et al.*, 2018), cold tolerance (Ao *et al.*, 2019), and anemone-type flowers (Yang *et al.*, 2019a). These studies successfully dissected the genetic architecture underlying the targeted traits, found several molecular markers closely linked to these traits, as well as preliminarily identified some excellent progeny as potential breeding material. However, the most reported linkage maps generated in chrysanthemums are sparse, with an average map distance of 6.6 ~ 10.2 cM, mainly due to the use of small population size (92 ~ 192) or a minimal number of traditional gel-based dominant markers (234 ~ 675). Moreover, these linkage maps are not integrated, and the number of linkage groups seriously deviated from the karyotype of chrysanthemum, which is also difficult to transfer between laboratories. Therefore, it is currently difficult to identify these major-effect QTLs' chromosomal locations and candidate genes and validate these QTLs in other populations with different genetic backgrounds, which limit their application in MAS.

Currently, advancements in NGS technologies have enabled the construction of high- or even ultrahigh-density genetic maps using SNP markers in chrysanthemum. van Geest *et al.* (2017b) provided conclusive evidence that chrysanthemums are more likely to be autohexaploid than allohexaploid based on analysis of a 183k SNP array. Subsequently, these authors reported

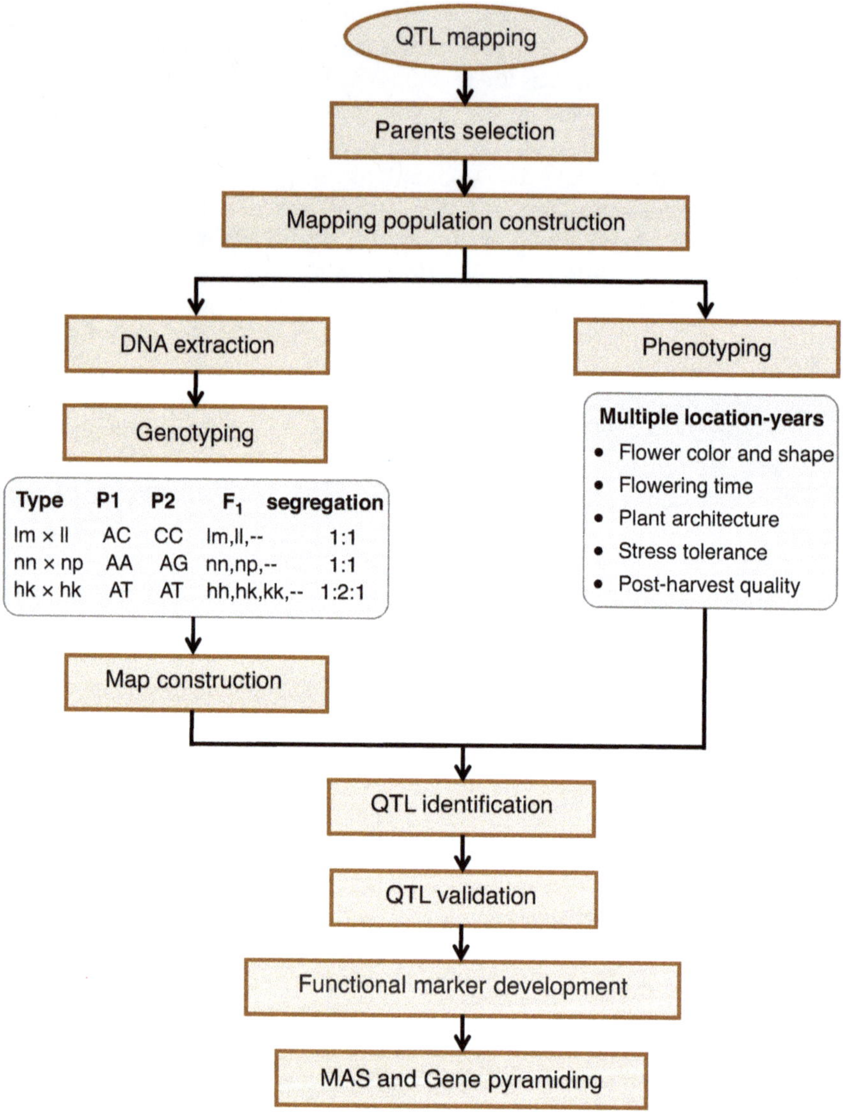

Fig. 3.1. QTL mapping methodology and its application in molecular breeding for chrysanthemum. Two parents are selected, diverse enough to exhibit an adequate level of polymorphism that is necessary to validate the initial QTL in other populations before applying the linked markers in MAS. Note: QTL = quantitative trait loci, MAS = marker-assisted selection. (Author's own image.)

the first integrated ultradense linkage map containing nine linkage groups covering 752.1 cM, with an average of 3368 markers per linkage group, in a biparental population including 406 individuals using polymapR based on marker dosage. Several QTLs associated with flowering time, flower color, ray floret number, and disc floret degreening were further detected through multiallelic QTL analyses (van Geest *et al.*, 2017a). Recently, Song *et al.* (2020) constructed an integrated high-density genetic linkage map with an average map distance of 0.76 cM using SNPs generated from SLAF-seq technology across 305 progenies in an F_1 hybrid population. Three major-effect QTLs controlling the corolla tube merging and four major-effect QTLs underlying the relative number of ray florets were repeatedly detected in different environments. Moreover, several candidate genes were identified via syntenic analysis between the genetic map and two Compositae species (sunflower and lettuce) with genomic information, and BLAST searches between the SNP markers of the QTL regions and a wild chrysanthemum transcriptome database. These results are potentially useful for MAS breeding and candidate gene exploration in chrysanthemums without a reference genome.

3.2.3.5 Genome-wide association mapping

GWAS is another effective approach based on linkage disequilibrium for decoding genotype–phenotype associations and causative locus/genes identification in plants. Successful applications of GWAS in plants have been recently reviewed by Liu *et al.* (2019) and Tibbs Cortes *et al.* (2021). The methodology of GWAS is shown in Fig. 3.2. Advantages of GWAS compared with traditional QTL mapping studies include eliminating the time and labor to construct genetic mapping populations, increasing power to detect minor-effect locus/genes, and identifying novel multiple alleles simultaneously with higher resolution (Alqudah *et al.*, 2020). In chrysanthemums, GWAS has been deployed to dissect inheritance patterns and detect causal loci underlying several important horticultural traits, such as plant architecture and inflorescence-related traits (Li *et al.*, 2016), waterlogging tolerance (Su *et al.*, 2016), aphid resistance (Fu *et al.*, 2018), and drought tolerance (Li *et al.*, 2018), using 707 SRAP, SCoT, and SSR markers. These studies found several favorable alleles closely linked to the desired traits and screened excellent germplasm resources as potential donors for future breeding. Interestingly, several common markers were found for different traits, which may be the result of linkage or pleiotropy and could be potentially used for comprehensive breeding.

The power and efficiency of GWAS are determined by marker type, marker numbers, population size, and statistical methods, which should be taken into consideration when geneticists and breeders perform GWAS for target traits. Although some progress has been made for GWAS in chrysanthemums, the currently obtained markers were impossible to target the related genomic regions and were difficult to directly use in MAS programs due to the limited number of conventional markers (<1000), low marker reliability, and high labor intensity. Nevertheless, the continuing reduction in sequencing costs makes NGS-based GWAS more affordable for non-model species with unsequenced and large genomes. Chong *et al.* (2016) identified

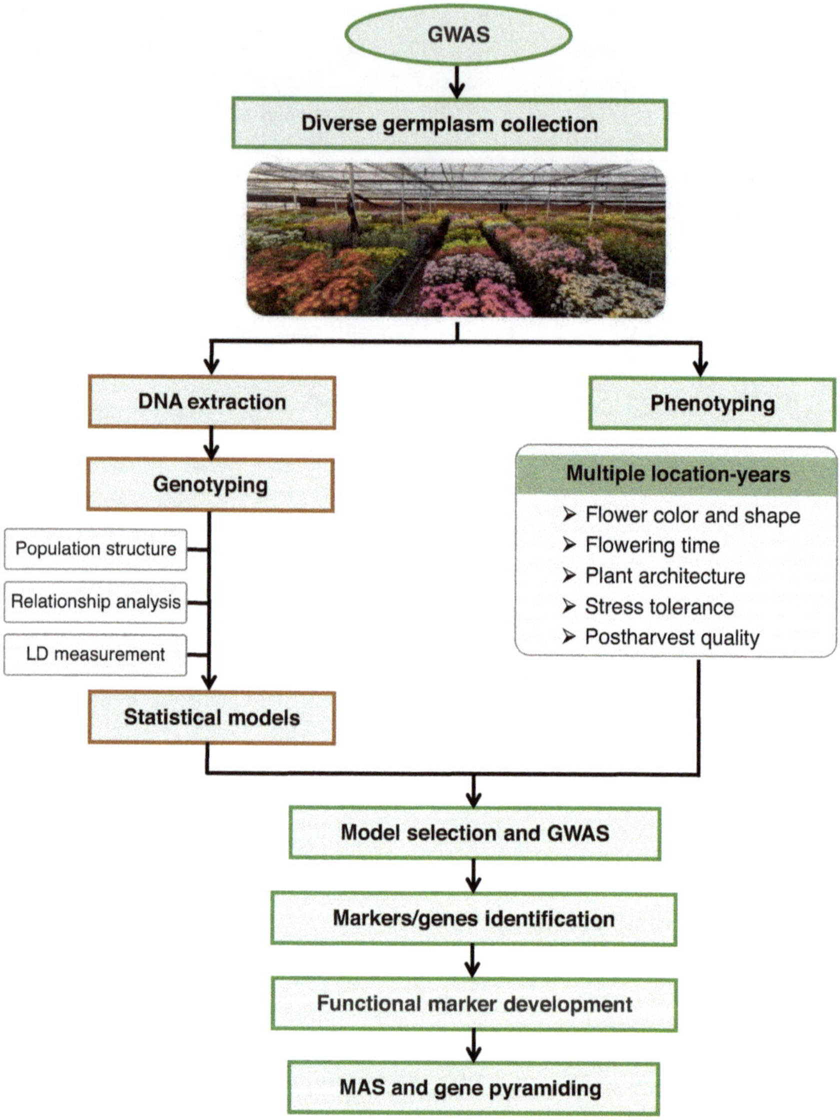

Fig. 3.2. GWAS methodology and its application in molecular breeding for chrysanthemum. Individuals from a natural population with extensive genetic diversity are used. Complete and precise phenotype data can be obtained via multiple locations and years of experiments. Population structure and kinship are generally taken into consideration to improve the accuracy of GWAS. Note: GWAS, genome-wide association study. (Author's own image.)

97 significant SNPs related to ray floret type, the cultivated type, and flower shape through GWAS. In addition, six candidate genes were predicted by aligning the SLAF-seq harboring the associations to the chrysanthemum transcriptome database, making a substantial breakthrough in integrating genetic and phenotypic data to discover candidate genes. Subsequently, several allele variants and candidate genes responsible for waterlogging tolerance (Su *et al.*, 2019) and inflorescence-related traits (Chong *et al.*, 2019) were routinely identified based on the abovementioned SNP sets in a similar way. Recently, Sumitomo *et al.* (2019) proposed a novel method to effectively develop DNA markers for petal color in hexaploid chrysanthemums using single-dose SNPs-based GWAS in the F_1 population. They identified 21 SNPs associated with *Puccinia horiana* resistance in one linkage group via GWAS, among which one SNP was proved to be an effective marker potentially used in MAS (Sumitomo *et al.*, 2021).

Markers should be technically cost-effective and straightforward enough to be applied in MAS, in contrast to those used in the research laboratory. Although SNPs have been adopted in commercial breeds for some important crop species such as rice (Yan *et al.*, 2020), maize (Chen *et al.*, 2021), and soybean (Bhat and Yu, 2021), it is still challenging to implement in chrysanthemum breeding. Recently, the major SNP locus that co-segregated with waterlogging tolerance (Su *et al.*, 2019), flowering time, and capitulum diameter (Chong *et al.*, 2019) identified via GWAS were successfully transferred into superior PCR-based dCAPS markers with an accuracy of 78.9, 88.5, and 92.3%, respectively, which present potential value for practical application in chrysanthemum MAS breeding. In addition, closer cooperation is needed between breeders and molecular geneticists to convert the markers from the laboratory to the field effectively. Indeed, the quantity and quality of phenotyping have currently become the key determinants of the accuracy of GWAS, and thus the power of the resultant MAS, particularly for complex quantitative traits that are susceptible to environmental factors. Epistasis and genotype × environment (G × E) interactions are two potential problems worth considering in post-GWAS analysis. Therefore, high-throughput phenotyping techniques for the collection of precise phenotypic data are urgently needed.

Though the GWAS has apparent advantages, it is more challenging to detect rare variants and there is a higher frequency of false positives in GWAS results than in QTL mapping. Hence, a powerful and efficient approach that combines GWAS and QTL mapping has been proposed and applied for better understanding the genetic architecture of capsaicin content in capsicum (Han *et al.*, 2018), seed vigor of rice (Guo *et al.*, 2019), and kernel water content of maize (Li *et al.*, 2020a). However, there are no reports that combine GWAS and QTL mapping in chrysanthemums, mostly due to the use of different traditional markers and the previous lack of complete genome information. Therefore, future studies are required to combine these two genetic mapping approaches and the co-identified loci for MAS breeding.

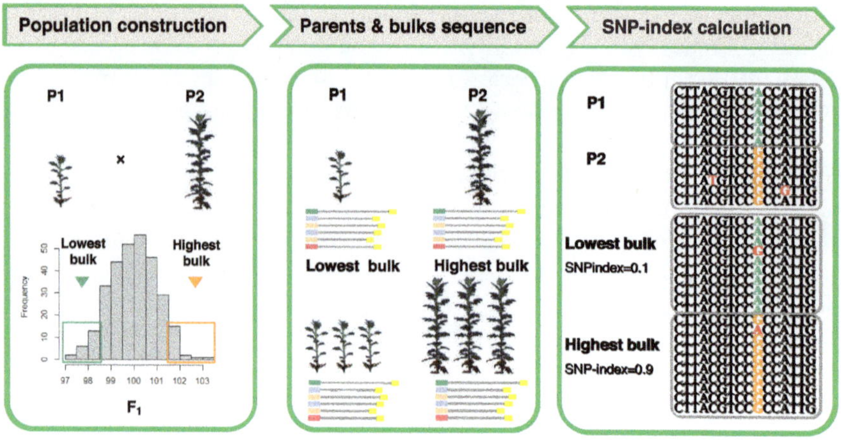

Fig. 3.3. BSA methodology for chrysanthemum using plant height as an example. Here, two DNA bulks of individuals derived from a progeny population with extreme phenotype are sequenced. Δ (SNP-index) is the most widely used algorithm for BSA. Note: BSA = bulked segregant analysis, SNP = single-nucleotide polymorphisms. (Author's own image.)

3.2.3.6 Bulked segregant analysis

The BSA approach was first introduced and used for mapping the genes of downy mildew resistance in lettuce by Michelmore *et al.* (1991). Compared with traditional QTL mapping, where progenies are individually genotyped and phenotyped, BSA enables researchers to feasibly and quickly identify markers, specific genes, or genomic regions closely linked to a target trait using two DNA bulks of individuals with extreme phenotypes (Fig. 3.3). In chrysanthemums, several BSA studies for different horticultural traits using traditional molecular markers were reported. For example, Wang *et al.* (2014b) detected two SSR markers associated with aphid resistance; Klie *et al.* (2016) identified two AFLP markers showing significant allele frequency differences in bulks with different plant architecture. The anemone type is an important flower shape of chrysanthemum with higher ornamental value. Recently, Yang *et al.* (2019b) developed an anemone-specific SRAP-SCAR marker via BSA, which was then validated in two F_1 populations with high accuracy of 87.86 and 92.5%, respectively, and had the potential to be used to discriminate anemone and non-anemone types in chrysanthemum breeding programs.

Nowadays, several studies have proved that combining the BSA strategy with NGS technologies opened a new opportunity to accelerate the process of mining candidate genes underlying the variations of important traits in numerous horticultural crops, i.e. pear (Xue *et al.*, 2017), tomato (Ruangrak *et al.*, 2018), melon (Qiao *et al.*, 2021), and sunflower (Imerovski *et al.*, 2019). However, the NGS-based BSA approach (also referred to as QTL-seq) is usually applied to species with available genome information and requires access to quantitative genetic markers that are polymorphic in the mapping population. Therefore, a modification of BSA, termed BSR-seq, that uses transcriptome sequencing data, was proposed to allow the correlation of gene expression and markers with phenotypes concurrently in non-model organisms (Liu *et al.*, 2012). This technique has been successfully applied in species with complex genomes such as wheat (Hao *et al.*, 2019) and peony (Hou *et al.*, 2018). Additionally, BSA and BSR-seq's inability to estimate allelic effect sizes or test QTL through environment interactions. Hence, a combination of BSA with traditional QTL mapping has emerged to solve the problem (Vogel *et al.*, 2021).

3.3 RNAi and Genome Editing

Genome editing (also known as targeted genome engineering) has recently emerged as an alternative to conventional breeding and transgenic methods to modified plants. This technique theoretically allows researchers to introduce mutations into any targeted genomic sequence, and mutations in genome-edited plants are heritable (Yin *et al.*, 2017). Nowadays, several available systems for genome editing have been developed, such as zinc-finger nucleases (ZFNs), TAL effector nucleases (TALENs), and the clustered regularly interspaced short palindromic repeats (CRISPR)–associated protein 9 (Cas9) (Wood *et al.*, 2011; Ran *et al.*, 2013). However, ZFNs and TALENs, which need complicated design and laborious assembly of specific DNA-binding proteins for each target gene, are gradually replaced by CRISPR/Cas9, an easier and more efficient genome-editing tool adapted from bacteria and archaea (Wiedenheft *et al.*, 2012). CRISPR/Cas9 system allows targeted cleavage of genomic DNA guided by a customizable small noncoding RNA, resulting in gene modifications by both nonhomologous end joining and homology-directed repair mechanisms (Belhaj *et al.*, 2013), and has been widely used in major crops (Moradpour and Abdulah, 2020).

Unfortunately, to date, only a few reports have demonstrated successful genome editing for petal senescence, flower color, and floral initiation in ornamental plants, mainly using the CRISPR/Cas9 method (Ahn *et al.*, 2020), because of their smaller economic importance than crop plants and complex background. Furthermore, the introduction of mutations into polyploids is difficult due to the double-strand breaks and repair events that occur independently (Zaman *et al.*, 2019). Usually, isolation of genes to target and the evaluation of mutations require special techniques for genome editing in hexaploidy chrysanthemum. Kishi-Kaboshi *et al.* (2017) first attempted to conduct gene editing using the CRISPR/Cas9 system to introduce mutations into the chrysanthemum. The resultant transgenic chrysanthemum plants expressed the yellowish-green fluorescent protein (*CpYGFP*) gene from *Chiridius poppei*. Two single-guide RNAs (sgRNAs) were used to target different positions in the *CpYGFP* and finally obtained the CRISPR-CpYGFP-chrysanthemum shoot containing a mutation in the *CpYGFP* sequence. Recently, Li *et al.* (2022)

reported a successful knockout for a TCP transcription factor gene involved in the cold stress response of chrysanthemum using a conventional CRISPR/Cas9-mediated system. Nevertheless, the low editing efficiency and high frequency of chimeric events still hindered the application of the CRISPR/Cas9 system in breeding. To resolve this issue, Liu *et al.* (2023) carried out a series of optimizations for the target sites, promoters, and Cas9 codons in the CRISPR/Cas9 gene editing system and developed a UV–visible reporter-assisted CRISPR/Cas9 gene editing system in diploid *C. indicum*, a model for cultivated chrysanthemum. They found that using mutated *eYGFPuv* as a visible marker increased the screening efficiency up to 63.6%, about four times higher than traditional antibiotic selection, and largely eliminated chimeric plants.

RNA interference (RNAi) is a well-conserved mechanism that uses small noncoding RNA to silence gene expression posttranscriptionally. To date, several RNAi transgenic chrysanthemums have been created. Higuchi *et al.* (2013) reported that loss of function of *CsAFT* by RNAi resulted in reduced sensitivity to red light night break and promoted flowering. *CmNF-YB8* silencing promotes the juvenile-to-adult transition and influences flowering time through directly regulating the expression of *cmo-MIR156* in the aging pathway (Wei *et al.*, 2017). Li *et al.* (2019) created *CmMET1* silenced transgenic plants and found they bloomed 8 days earlier than their wild type. *CmMAF2*-RNAi plants flowered 41–57 days earlier than wild type plants (Lyu *et al.*, 2022). Wang *et al.* (2022) found that *CmNF-YB8*-RNAi plants formed a thicker cuticle layer on the abaxial side of the leaf and hence improved the drought tolerance of chrysanthemum. Shinoyama *et al.* (2012) created *CmDMC1a*-RNAi transgenic chrysanthemum plants that displayed severe male sterility and a notable decrease in female fertility. However, given the hexaploid nature and self incompatibility of chrysanthemums, it is crucial to protect transgenic plants showing both male and female sterility to avoid the potential risk of transgene flow from engineered plants to other cultivars or their wild relatives. More recently, these authors simultaneously knocked out all six endogenous *CmDMC1* loci using TALEN-mediated genome-editing technology (Shinoyama *et al.*, 2020). The resulting plants exhibited both male and female sterility, suggesting knockout of all DMC1 alleles could be used to induce stable male and female sterility in chrysanthemums and thus prevent transgene flow from gene-edited plants to their wild relatives. These findings shed light on chrysanthemum genome editing and provide breeders with new tools for precise breeding and also could be a reference to other polyploid ornamental plants.

3.4 Omics Technologies

Continuing advances in omics technologies, especially genomics, transcriptomics, phenomics, proteomics, and metabolomics, enable a deeper understanding of the complex trait's mechanisms. Several attempts have been made to use emerging omics methods to accelerate chrysanthemum breeding (Fig. 3.4 for a timeline of the main developments in omics technologies).

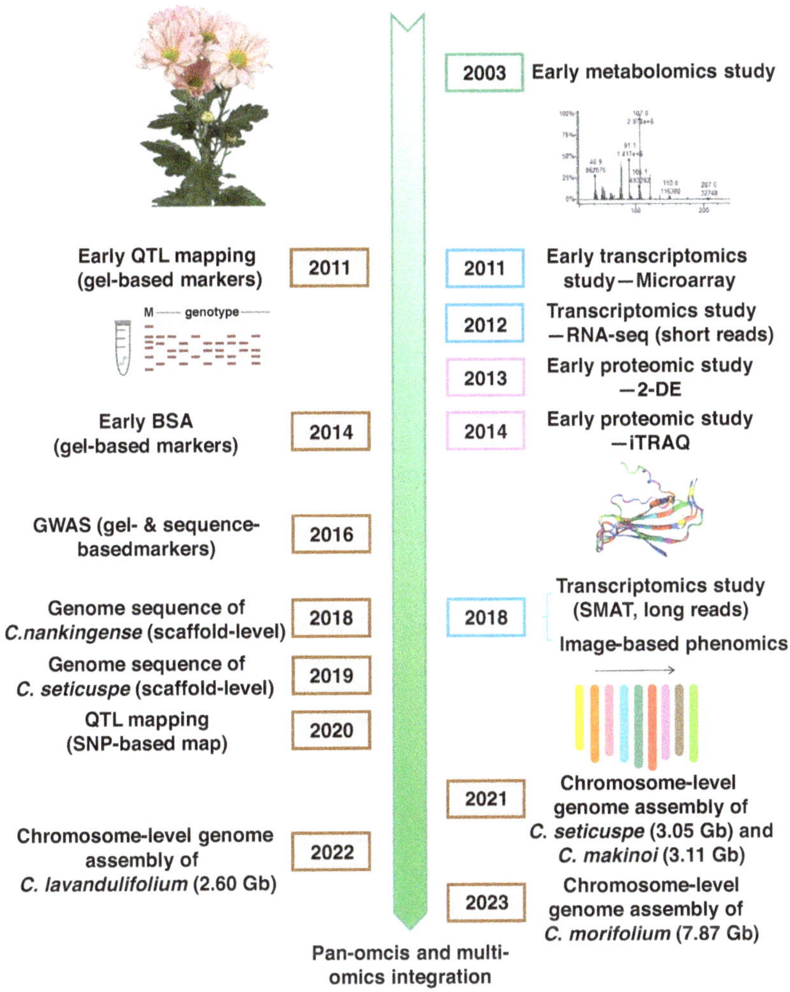

Fig. 3.4. Timeline of the main advances of genomics (brown boxes) and other omics techniques in chrysanthemum. (Author's own image.)

3.4.1 Genomics

Genomic research is the foundation of molecular biology and has a vast capability in speeding up breeding processes through MAS, genomic selection, and gene pyramiding. However, such work in chrysanthemum has lagged behind, primarily due to its large genome size, polyploidy, and high repetitiveness. Emergence of third-generation sequencing platforms, represented by PacBio single-molecule real-time sequencing and Oxford Nanopore technologies results in longer reads at low cost. In this way, the genome sequence information of two diploid *Chrysanthemum* species, *C. nankingense* (2n = 2x = 18;

2.53 Gb) and *C. seticuspe* (2n = 2x = 18; 2.72 Gb), which are potential progenitors or model species of cultivated chrysanthemums, was deciphered (Song *et al.*, 2018; Hirakawa *et al.*, 2019). Nevertheless, both reported genomes of *C. nankingense* and *C. seticuspe* have not been integrated into pseudochromosomes and are highly fragmented, with 24,051 and 354,212 scaffolds, N50 of 130.7 and 44.7 kb, respectively. The emergence and constant improvement of Hi-C (high-throughput chromosome conformation capture; Belton *et al.*, 2012) and PacBio HiFi (high-fidelity; Wenger *et al.*, 2019) technologies provide an opportunity to assemble the genomes of large complex species. More recently, two chromosome-level reference genomes of wild chrysanthemum *C. seticuspe* (Nakano *et al.*, 2021) and *C. makinoi* (2n = 2x = 18; van Lieshout *et al.*, 2022) were assembled for the first time, with N50 of 347 and 330 Mb, BUSCO (Benchmarking Universal Single-Copy Orthologs) complete score of 97.4 and 92.1%, respectively. Wen *et al.* (2022) reported a 2.60 Gb chromosome-scale reference genome of *C. lavandulifolium*, whose contig N50 and scaffold N50 values were 497.0 kb and 300.4 Mb, respectively. More recently, the genome of *C. morifolium* has been deciphered for a haploid line ($n = 3x = 27$) of pot cultivar 'Zhongshanzigui' using a combination of 126.40 × Illumina short reads, 120.70 × PacBio long reads, and 107.20 × 10X Genomics sequencing data (Song *et al.*, 2023). The generated reference genome comprises 27 pseudochromosomes (7.87 Gb) and has the contig N50 size of 1.87 Mb, scaffold N50 size of 303.69 Mb, and LAI (LTR Assembly Index) of 27.99. Multilevel evidence suggests that *C. morifolium* is likely a segmental allopolyploid with "AA'B" relationship of homoeologous chromosomes. The released high-quality genomic information contributes to understand the chrysanthemum genome evolution and provides an important step toward the application of genome editing. The cost reduction in sequencing makes it a promising prospect for future WGS and pangenome study of chrysanthemum. Furthermore, the resulting large-scale SNPs, InDels, and structural variants can be used in genome-wide analyses such as QTL mapping, GWAS, and BSA, thus accelerating the breeding progress and trait improvement of cultivated chrysanthemum.

3.4.2 Transcriptomics

Transcriptomics provides information about gene characterization and quantification in a specific plant, organ, tissue, or cell and has the potential to build a link between genotype and phenotype. A particular attraction of assaying transcription is the use of molecular markers's discovery for genetic studies. Recently, the continuing advancement of mRNA sequencing (RNA-seq) and *de novo* assembly technologies has allowed an increasing number of research communities to dissect the transcriptomes of species with limited genome information. Moreover, several public data repositories now make it possible to exchange transcriptome data among different laboratories. To date, thousands of transcriptomes of plant species have been available in NCBI (www.ncbi.nlm.nih.gov/sra). In chrysanthemum, transcriptome sequencing has been widely used for elucidating the gene regulatory mechanisms underlying complex biological processes, including plant growth and development, stress resistance, and secondary metabolism and has also led to the generation of a vast number of EST-SSR and SNP markers, as recently reviewed by Su *et al.* (2019). Nevertheless, high-throughput data integration, analysis, and interpretation have become the major bottlenecks and challenges that biologists conducting transcriptomic research face at present. With the progress in computational methodologies and bioinformatics pipelines, it becomes clear that network-focused rather than individual gene-focused strategies would be more appropriate to clarify phenotypic characters. Recently, weighted gene co-expression network analysis (WGCNA) was developed and applied to analyze highly multivariate, complex transcriptome sequencing data efficiently. In the last few years, several WGCNA research focused on carotenoid accumulation (Lu *et al.*, 2019), *Alternaria alternate* (Liu *et al.*, 2020b), axillary bud (Ahmad *et al.*, 2020), and flower development (Ding *et al.*, 2020; Sun *et al.*, 2022) has been reported in chrysanthemums. Moreover, expression GWAS (eGWAS), transcriptome-wide association studies (TWAS), and eQTL, integrating GWAS or QTL mapping with gene expression datasets as novel approaches, exhibit great potential to identify gene–trait associations (Li *et al.*, 2020b; Yao *et al.*, 2020).

3.4.3 Phenomics

Phenomics is becoming increasingly important in genetic studies and precision breeding since it allows for obtaining reliable, automatic, high-throughput phenotypic data, and it provides a better understanding of the genotype–phenotype map. During the past decade, plant phenomics has evolved from an emerging niche to a thriving research field, intending to build the extensive phenotyping database at multiple levels—from cell level, the organ level, and plant level to population level—and developing bioinformatics tools for information extraction from the overwhelming amounts of omics data (Zhao *et al.*, 2019). Currently, several high-throughput fields and greenhouse phenotypic platforms have been applied for nondestructive assessment of root morphology (Falk *et al.*, 2020), abiotic stresses (Lin *et al.*, 2020; Asif *et al.*, 2020), and flesh firmness (Alenazi *et al.*, 2020). Cultivar recognition is essential in chrysanthemum molecular breeding and commercial application. Wu *et al.* (2018) provide a powerful tool to rapidly and accurately distinguish chrysanthemum varieties using hyperspectral imaging combined with a deep convolutional neural network. Then, another image-based method to recognize large-flowered chrysanthemum cultivars using deep learning models was proposed (Liu *et al.*, 2019), both exhibiting great potential for detecting large-scale samples in industrial production. Liu *et al.* (2020a) evaluated two rapid identification approaches for automatic quality determination of chrysanthemum tea, one utilizing a regular gel imager as a hardware workbench, in association with morphological feature extraction and multivariate classification, the other applying a deep neural network to the raw image directly. The result indicated that computer vision, especially when combined with deep learning or other machine learning techniques, can be a convenient and versatile method to evaluate food quality. Qi *et al.* (2021) proposed a highly fused, lightweight detection model named the Fusion-YOLO for detecting the early flowering stage of tea chrysanthemum. The highly accurate model represented an important step toward realizing automatic harvesting of tea chrysanthemums by building a selective robot system—replacing the current manual harvesting—in the future. More recently, Lu *et al.* (2024) established an estimation model for cut chrysanthemum plant water content (PWC) prediction using leaf-based spectral reflectance (350–2500 nm), which provided the possibility for precise irrigation and nondestructive measurement of PWC in chrysanthemum. Notably, the phenomics research is still in its infancy in chrysanthemum, and several scientific and technical challenges need to be overcome to apply it in actual molecular breeding programs.

3.4.4 Proteomics

The acquisition of the genomic and transcriptomic data of plant species proceeds apace. However, much of the cell's physiology is determined by gene products (particularly, but not exclusively, proteins) rather than by nucleic acid (Witzel *et al.*, 2015). A proteomic study seeks to characterize the full protein complement present in a particular organism, organ, tissue, or cell. Proteomic methods evolved continuously from the classical two-dimensional (2-DE) gel electrophoresis to current proteomics platforms exploited, such as LC (liquid chromatography)-based methods, coupled with mass spectrometry (MS) (Sahithi *et al.*, 2020). In chrysanthemums, proteomic approaches have been successfully applied to study diverse developmental processes and environmental adaptation. Liu *et al.* (2013) identified 69 differentially accumulated protein spots between cut chrysanthemum bases of different ages using the 2-DE method and found the auxin-induced proteins PCNT115 and CmACO were tightly associated with adventitious root formation, as supported by western blot experiments. Lin *et al.* (2021) comparatively analyzed proteome-wide and crotonylation in chrysanthemum under low-temperature stress based on tandem mass tag labeling, sensitive immunoprecipitation, and high-resolution LC-MS/MS. The results provide insights into the mechanisms of crotonylation in positively regulating APX activity to reduce the oxidative damage caused by low-temperature stress. In contrast to traditional proteomics approaches, the isobaric tags for relative and absolute quantification (iTRAQ) technique enables simultaneous labeling and analysis of up to eight different samples for relative quantification with higher accuracy. To date iTRAQ was successfully applied in proteomic studies of heat stress (Zhang *et al.*, 2014b), embryo abortion (Zhang *et al.*, 2014a, 2016), pollen abortion (Wang *et al.*, 2018), and light-induced anthocyanin biosynthesis (Hong *et al.*, 2019) in chrysanthemums.

3.4.5 Metabolomics

Metabolomics is an important omics layer that can provide chemical evidence, allowing us to systematically learn the by-products of cellular processes. At least 100,000 types of chemically distinct molecules are involved in a typical plant sample, each varying in their abundance (Witzel *et al.*, 2015). Many of these compounds are of physiological importance and represent dietary sources of anti-oxidants and other health-beneficial compounds. Chrysanthemum is also an important medicinal herbaceous plant; its chemical composition is complex and is used in food, tea, and pharmaceutical products (Hadizadeh *et al.*, 2022). Miyazawa and Hisama (2003) isolated and identified the antimutagenic metabolites from chrysanthemum flower heads and pointed out that the flavonoids acacetin, apigenin, and luteolin were primary compounds and may be potent biological antimutagens that are useful as cancer chemopreventive agents. Based on a UPLC-ESI-QTOF/MS serum metabolomics approach, a total of 35 potential biomarkers were identified in chrysanthemum 'Boju' extract, which could alleviate hyperuricemia-related diseases (Peng *et al.*, 2019). Flowers produce various metabolites to attract pollinators, initiate or inhibit signaling cascades, and protect against herbivores or pathogens. Sun *et al.* (2015b) detected 193 volatile organic components from the fresh flowers of chrysanthemum using GC-MS, showing monoterpenoids and oxygenated monoterpenoids were the major scent components. Deng *et al.* (2010) inferred that higher monoterpenoid and sesquiterpenoid content enhanced aphid resistance in chrysanthemum. Sun *et al.* (2015a) found (E)-β-farnesene and artemisia ketone emitted from *Artemisia annua* might act as a potential volatile compound to resist aphids, which would be useful for future ecological control of aphids in chrysanthemum cultivation. Flower color is a major objective of ornamental plant breeding due to its strong influence on consumer choice. van Geest *et al.* (2016) showed metabolic profile differences under carbohydrate starvation, revealing that carbohydrate content alone could not explain the variation in chrysanthemum degreening sensitivity. Liu *et al.* (2019) established a metabolome-based strategy to identify metabolite markers for mutation rates associated with flower colour in chrysanthemum. Eleven metabolites were selected as features contributing to the discrimination of mutation rates; among them, taxifolin was highly accumulated in petals with high mutation rates and thus might be applicable for efficient mutation breeding. Wang *et al.* (2019) revealed the effect of flooding stress on the synthesis of flavonoids in chrysanthemum. Loh *et al.* (2022) compared the metabolite profile of chrysanthemum flower fraction with that of its detannified fraction in relation to xanthine oxidase inhibitory activity using a proton nuclear magnetic resonance (^{1}H-NMR)-based metabolomics approach coupled with multivariate data analysis. Currently, metabolome profiling is used as phenotypes to identify metabolite QTLs (mQTLs), which contribute to the breeding of traits related to the production of functionally useful metabolites in crops (Sharma *et al.*, 2021). Therefore, developing specific tools to improve the capacity to recognize metabolites and combine them with genetic studies are the main challenges researchers face in this field.

3.5 Conclusions and Future Prospects

Molecular breeding with the aid of molecular markers played a remarkable role in modern plant breeding over the last two decades. Although substantial progress has been made in MAS for important horticultural traits improvement in chrysanthemum, as mentioned in this chapter, many technical and scientific challenges that prevent the application of these technologies in actual breeding programs remain: (i) the population sizes and marker numbers are too small for accurate high-resolution mapping; (ii) the obtained QTLs/QTNs are mostly minor-effect and environment-specific with no validation, which is not robust enough for selecting phenotypes, and simple, cost-effective, high-throughput markers are urgently needed for use in commercial breeding practice; and (iii) the low genome-editing efficiency, especially for simultaneously knocking down multiple gene copies and/or alleles for hexaploidy chrysanthemum. Therefore, traditional breeding strategies, including cross and mutation breeding, are still most widely used in chrysanthemum. To accelerate the molecular breeding process and meet market demands, here we propose a comprehensive breeding scheme shown in Fig. 3.5.

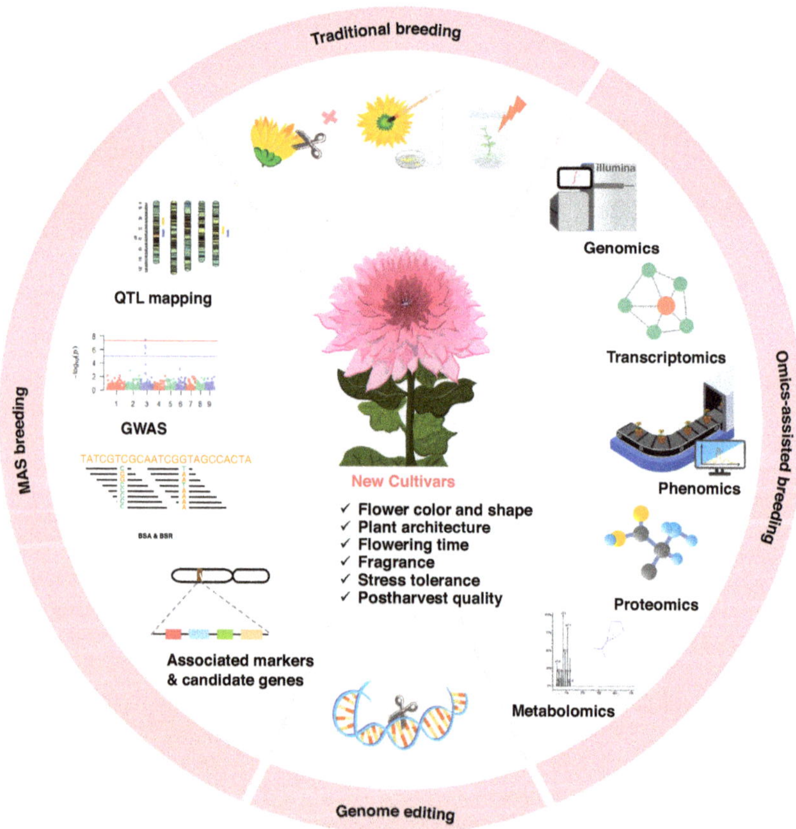

Fig. 3.5. Schematic overview of accelerating the breeding of new chrysanthemum cultivars by integrating traditional and next-generation breeding strategies. (Author's own image.)

Availability of the full-genome information of the diploid *Chrysanthemum* species (Song *et al.*, 2018; Hirakawa *et al.*, 2019; Nakano *et al.*, 2021; van Lieshout *et al.*, 2022; Wen *et al.*, 2022), as well as the cultivated chrysanthemum (Song *et al.*, 2023), provides perhaps the most important genome resource for molecular breeding programs. Moreover, high-throughput genotyping and phenotyping assays in a large-scale population should be used in future genetic studies to obtain reliable associations. Furthermore, pyramiding multiple favorable genes/QTLs through marker-assisted gene pyramiding can contribute to the development of new chrysanthemum cultivars with excellent comprehensive traits. The application of new genome-editing technologies has enormous potential to benefit the ornamental horticulture industry by supporting targeted genomic modifications. However, the development of specific experimental and analytical methods for polyploids is urgently required. Additionally, the advent and improvement of NGS and artificial intelligence (AI) technologies have accelerated the generation of multiomic data at the DNA, RNA, protein, and metabolite levels at an unprecedented rate, leading to a new era of "big biological data" (Crossa *et al.*, 2025). The marriage of these big data and genetic studies through eQTL/eGWAS, pQTL/pGWAS, and mQTL/mGWAS analyses will undoubtedly broaden and deepen our understanding of the molecular mechanisms underlying the traits of interest and generate novel opportunities for molecular breeding in chrysanthemum.

References

Adhikari, L., Lindstrom, O.M., Markham, J. and Missaoui, A.M. (2018) Dissecting key adaptation traits in the polyploid perennial *Medicago sativa* using GBS-SNP mapping. *Frontiers in Plant Science* 9, 934. DOI: 10.3389/fpls.2018.00934.

Agarwal, M., Shrivastava, N. and Padh, H. (2008) Advances in molecular marker techniques and their applications in plant sciences. *Plant Cell Reports* 27(4), 617–631. DOI: 10.1007/s00299-008-0507-z.

Ahmad, S., Yuan, C., Yang, Q., Yang, Y., Cheng, T. *et al.* (2020) Morpho-physiological integrators, transcriptome and coexpression network analyses signify the novel molecular signatures associated with axillary bud in chrysanthemum. *BMC Plant Biology* 20(1), 1–15. DOI: 10.1186/s12870-020-02336-0.

Ahn, C.H., Ramya, M., An, H.R., Park, P.M., Kim, Y.-J. *et al.* (2020) Progress and challenges in the improvement of ornamental plants by genome editing. *Plants* 9(6), 687. DOI: 10.3390/plants9060687.

Akram, S., Arif, M.A.R. and Hameed, A. (2021) A GBS-based GWAS analysis of adaptability and yield traits in bread wheat (*Triticum aestivum* L.). *Journal of Applied Genetics* 62(1), 27–41. DOI: 10.1007/s13353-020-00593-1.

Alenazi, M.M., Shafiq, M., Alsadon, A.A., Alhelal, I.M., Alhamdan, A.M. *et al.* (2020) Non-destructive assessment of flesh firmness and dietary antioxidants of greenhouse-grown tomato (*Solanum lycopersicum* L.) at different fruit maturity stages. *Saudi Journal of Biological Sciences* 27(10), 2839–2846. DOI: 10.1016/j.sjbs.2020.07.004.

Alipour, H., Bihamta, M.R., Mohammadi, V., Peyghambari, S.A., Bai, G. *et al.* (2017) Genotyping-by-sequencing (GBS) revealed molecular genetic diversity of Iranian wheat landraces and cultivars. *Frontiers in Plant Science* 8, 1293. DOI: 10.3389/fpls.2017.01293.

Alqudah, A.M., Sallam, A., Baenziger, P.S., Stephen Baenziger, P. and Börner, A. (2020) GWAS: Fast-forwarding gene identification and characterization in temperate cereals: Lessons from barley – A review. *Journal of Advanced Research* 22, 119–135. DOI: 10.1016/j.jare.2019.10.013.

Amom, T. and Nongdam, P. (2017) The use of molecular marker methods in plants: A review. *International Journal of Current Research and Review* 9(17), 1–7.

Ao, N., Ma, J., Xu, T., Su, J., Yang, X. *et al.* (2019) Genetic variation and QTL mapping for cold tolerance in a chrysanthemum F_1 population at different growth stages. *Euphytica* 215(5), 88. DOI: 10.1007/s10681-019-2412-7.

Asif, M.A., Garcia, M., Tilbrook, J., Brien, C., Dowling, K. *et al.* (2021) Identification of salt tolerance QTL in a wheat RIL mapping population using destructive and non-destructive phenotyping. *Functional Plant Biology* 48(2), 131–140. DOI: 10.1071/fp20167.

Baird, N.A., Etter, P.D., Atwood, T.S., Currey, M.C., Shiver, A.L. *et al.* (2008) Rapid SNP discovery and genetic mapping using sequenced rad markers. *PLoS ONE* 3(10), e3376. DOI: 10.1371/journal.pone.0003376.

Balsalobre, T.W.A., Pereira, G.D., Margarido, G.R.A., Gazaffi, R., Barreto, F.Z. *et al.* (2017) GBS-based single dosage markers for linkage and QTL mapping allow gene mining for yield-related traits in sugarcane. *BMC Genomics* 18(1), 1–19. DOI: 10.1186/s12864-016-3383-x.

Barakat, M.N., Fattah, R.S.A., Badr, M. and El-Torky, M.G. (2010) *In vitro* mutagenesis and identification of new variants via RAPD markers for improving *Chrysanthemum morifolium*. *African Journal of Agricultural Research* 5(8), 748–757. DOI: 10.5897/AJAR09.679.

Belhaj, K., Chaparro-Garcia, A., Kamoun, S. and Nekrasov, V. (2013) Plant genome editing made easy: Targeted mutagenesis in model and crop plants using the CRISPR/Cas system. *Plant Methods* 9(1), 39. DOI: 10.1186/1746-4811-9-39.

Belton, J.-M., McCord, R.P., Gibcus, J.H., Naumova, N., Zhan, Y. *et al.* (2012) Hi–C: A comprehensive technique to capture the conformation of genomes. *Methods* 58(3), 268–276. DOI: 10.1016/j.ymeth.2012.05.001.

Bhagyawant, S.S. (2016) RAPD-Scar markers: An interface tool for authentication of traits. *Journal of Biosciences and Medicines* 4(1), 4. DOI: 10.4236/jbm.2016.41001.

Bhat, J.A. and Yu, D. (2021) High-throughput NGS-based genotyping and phenotyping: Role in genomics-assisted breeding for soybean improvement. *Legume Science* 3(3), e81. DOI: 10.1002/leg3.81.

Bourke, P.M., Voorrips, R.E., Visser, R.G. and Maliepaard, C. (2018) Tools for genetic studies in experimental populations of polyploids. *Frontiers in Plant Science* 9, 513. DOI: 10.3389/fpls.2018.00513.

Chen, Z., Tang, D., Ni, J., Li, P., Wang, L. *et al.* (2021) Development of genic KASP SNP markers from RNA-Seq data for map-based cloning and marker-assisted selection in maize. *BMC Plant Biology* 21(1), 157. DOI: 10.1186/s12870-021-02932-8.

Chong, X., Zhang, F., Wu, Y., Yang, X., Zhao, N. *et al.* (2016) A SNP-enabled assessment of genetic diversity, evolutionary relationships and the identification of candidate genes in chrysanthemum. *Genome Biology and Evolution* 8(12), 3661–3671. DOI: 10.1093/gbe/evw270.

Chong, X., Su, J., Wang, F., Wang, H., Song, A. *et al.* (2019) Identification of favorable SNP alleles and candidate genes responsible for inflorescence-related traits via GWAS in chrysanthemum. *Plant Molecular Biology* 99(4), 407–420. DOI: 10.1007/s11103-019-00826-w.

Chung, Y.S., Choi, S.C., Jun, T.-H. and Kim, C. (2017) Genotyping-by-sequencing: A promising tool for plant genetics research and breeding. *Horticulture, Environment, and Biotechnology* 58(5), 425–431. DOI: 10.1007/s13580-017-0297-8.

Collard, B.C. and Mackill, D.J. (2009) Start codon targeted (SCoT) polymorphism: A simple, novel DNA marker technique for generating gene-targeted markers in plants. *Plant Molecular Biology Reporter* 27(1), 86–93. DOI: 10.1007/s11105-008-0060-5.

Crossa, J., Montesinos-Lopez, O.A., Costa-Neto, G., Vitale, P., Martini, J.W. *et al.* (2025) Machine learning algorithms translate big data into predictive breeding accuracy. *Trends in Plant Science* 30(2), 167–184. DOI: 10.1016/j.tplants.2024.09.011.

Deng, Y., Chen, S., Lu, A., Chen, F., Tang, F. *et al.* (2010) Production and characterisation of the intergeneric hybrids between *Dendranthema morifolium* and *Artemisia vulgaris* exhibiting enhanced resistance to chrysanthemum aphid (*Macrosiphoniella sanbourni*). *Planta* 231(3), 693–703. DOI: 10.1007/s00425-009-1081-5.

Ding, L., Song, A., Zhang, X., Li, S., Su, J. *et al.* (2020) The core regulatory networks and hub genes regulating flower development in *Chrysanthemum morifolium*. *Plant Molecular Biology* 103(6), 669–688. DOI: 10.1007/s11103-020-01017-8.

Elshire, R.J., Glaubitz, J.C., Sun, Q., Poland, J.A., Kawamoto, K. *et al.* (2011) A robust, simple genotyping-by-sequencing (GBS) approach for high diversity species. *PLoS ONE* 6(5), e19379. DOI: 10.1371/journal.pone.0019379.

Falk, K.G., Jubery, T.Z., Mirnezami, S.V., Parmley, K.A., Sarkar, S. *et al.* (2020) Computer vision and machine learning enabled soybean root phenotyping pipeline. *Plant Methods* 16(1), 5. DOI: 10.1186/s13007-019-0550-5.

Fan, M., Gao, Y., Gao, Y., Wu, Z., Liu, H. *et al.* (2019) Characterization and development of EST-SSR markers from transcriptome sequences of chrysanthemum (*Chrysanthemum* × *morifolium* Ramat.). *HortScience* 54(5), 772–778. DOI: 10.21273/hortsci13694-18.

Fan, M., Gao, Y., Wu, Z. and Zhang, Q. (2020) Linkage map development by EST-SSR markers and QTL analysis for inflorescence and leaf traits in chrysanthemum (*Chrysanthemum morifolium* Ramat.). *Plants* 9(10), 1342. DOI: 10.3390/plants9101342.

Feng, S.-G., He, R.-F., Jiang, M.-Y., Lu, J.-J., Shen, X.-X. *et al.* (2016) Genetic diversity and relationships of medicinal *Chrysanthemum morifolium* revealed by start codon targeted (SCoT) markers. *Scientia Horticulturae* 201, 118–123. DOI: 10.1016/j.scienta.2016.01.042.

Fu, X., Su, J., Yu, K., Cai, Y., Zhang, F. *et al.* (2018) Genetic variation and association mapping of aphid (*Macrosiphoniella sanbourni*) resistance in chrysanthemum (*Chrysanthemum morifolium* Ramat.). *Euphytica* 214(2), 21. DOI: 10.1007/s10681-017-2085-z.

Fu, X., Su, J.S., Li, Y.Y., Zhang, F. and Fang, W.M. (2019) Genetic variation and QTL mapping for aphid resistance in an F_1 population of chrysanthemum. *Acta Horticulturae Sinica* 46(7), 1351–1358. DOI: 10.16420/j.issn.0513-353x.2018-1031.

Fukai, S., Kamigaichi, Y., Yamasaki, N., Zhang, W. and Goi, M. (2002) Distribution, morphological variations and cpDNA PCR-RFLP analysis of *Dendranthema yoshinaganthum*. *Journal of the Japanese Society for Horticultural Science* 71(1), 114–122. DOI: 10.2503/jjshs.71.114.

Grada, A. and Weinbrecht, K. (2013) Next-generation sequencing: Methodology and application. *Journal of Investigative Dermatology* 133(8), 1–4. DOI: 10.1038/jid.2013.248.

Grover, A. and Sharma, P.C. (2016) Development and use of molecular markers: Past and present. *Critical Reviews in Biotechnology* 36(2), 290–302. DOI: 10.3109/07388551.2014.959891.

Guo, T., Yang, J., Li, D., Sun, K., Luo, L. *et al.* (2019) Integrating GWAS, QTL, mapping and RNA-Seq to identify candidate genes for seed vigor in rice (*Oryza sativa* L.). *Molecular Breeding* 39(6), 87. DOI: 10.1007/s11032-019-0993-4.

Gupta, R.C., Bala, S., Sharma, S. and Kapoor, M. (2013) Cytomorphological studies in some species of *Chrysanthemum* L. (Asteraceae). *Chromosome Botany* 8(3), 69–74. DOI: 10.3199/iscb.8.69.

Hadizadeh, H., Samiei, L. and Shakeri, A. (2022) *Chrysanthemum*, an ornamental genus with considerable medicinal value: A comprehensive review. *South African Journal of Botany* 144, 23–43. DOI: 10.1016/j.sajb.2021.09.007.

Han, J.Y., Kim, J.B., Lee, H.J., Hong, C.P. and Pak, H.S. (2019) Development of EST-SSR markers for cultivar determination and genetic diversity studies of commercial chrysanthemums in the Korean floral market. *Korean Society of Breeding Science* 51(3), 201–208. DOI: 10.9787/KJBS.2019.51.3.201.

Han, K., Lee, H.Y., Ro, N.Y., Hur, O.S., Lee, J. *et al.* (2018) QTL mapping and GWAS reveal candidate genes controlling capsaicinoid content in *Capsicum*. *Plant Biotechnology Journal* 16(9), 1546–1558. DOI: 10.1111/pbi.12894.

Hao, Z., Geng, M., Hao, Y., Zhang, Y., Zhang, L. *et al.* (2019) Screening for differential expression of genes for resistance to *Sitodiplosis mosellana* in bread wheat via BSR-Seq analysis. *Theoretical and Applied Genetics* 132(11), 3201–3221. DOI: 10.1007/s00122-019-03419-9.

He, J., Zhao, X., Laroche, A., Lu, Z.-X., Liu, H. *et al.* (2014) Genotyping-by-sequencing (GBS), an ultimate marker-assisted selection (MAS) tool to accelerate plant breeding. *Frontiers in Plant Science* 5, 484. DOI: 10.3389/fpls.2014.00484.

Higuchi, Y., Narumi, T., Oda, A., Nakano, Y., Sumitomo, K. *et al.* (2013) The gated induction system of a systemic floral inhibitor, antiflorigen, determines obligate short-day flowering in chrysanthemums. *Proceedings of the National Academy of Sciences* 110(42), 17137–17142. DOI: 10.1073/pnas.1307617110.

Hirakawa, H., Sumitomo, K., Hisamatsu, T., Nagano, S., Shirasawa, K. *et al.* (2019) *De novo* whole-genome assembly in *Chrysanthemum seticuspe*, a model species of chrysanthemums, and its application to genetic and gene discovery analysis. *DNA Research* 26(3), 195–203. DOI: 10.1093/dnares/dsy048.

Hodaei, M., Rahimmalek, M. and Arzani, A. (2019) Genetic diversity of Iranian *Chrysanthemum morifolium* cultivars using morphological traits and sequence-related amplified polymorphism (SRAP) markers. *Horticulture, Environment, and Biotechnology* 60(5), 753–765. DOI: 10.1007/s13580-019-00137-5.

Hong, Y., Li, M. and Dai, S. (2019) iTRAQ-based protein profiling provides insights into the mechanism of light-induced anthocyanin biosynthesis in chrysanthemum (*Chrysanthemum* × *morifolium*). *Genes* 10(12), 1024. DOI: 10.3390/genes10121024.

Hou, X., Guo, Q., Wei, W., Guo, L., Guo, D. *et al.* (2018) Screening of genes related to early and late flowering in tree peony based on bulked segregant RNA sequencing and verification by quantitative real-time PCR. *Molecules* 23(3), 689. DOI: 10.3390/molecules23030689.

Imerovski, I., Dedić, B., Cvejić, S., Miladinović, D., Jocić, S. *et al.* (2019) BSA-seq mapping reveals major QTL for broomrape resistance in four sunflower lines. *Molecular Breeding* 39(3), 41. DOI: 10.1007/s11032-019-0948-9.

Jenkins, J.N., McCarty, J.C., Wubben, M.J., Hayes, R., Gutierrez, O.A. *et al.* (2012) SSR markers for marker assisted selection of root-knot nematode (*Meloidogyne incognita*) resistant plants in cotton (*Gossypium hirsutum* L.). *Euphytica* 183(1), 49–54. DOI: 10.1007/s10681-011-0470-6.

Jo, K.-M., Jo, Y., Chu, H., Lian, S. and Cho, W.K. (2015) Development of EST-derived SSR markers using next-generation sequencing to reveal the genetic diversity of 50 chrysanthemum cultivars. *Biochemical Systematics and Ecology* 60, 37–45. DOI: 10.1016/j.bse.2015.03.002.

Kishi-Kaboshi, M., Aida, R. and Sasaki, K. (2017) Generation of gene-edited *Chrysanthemum morifolium* using multi-copy transgenes as targets and markers. *Plant and Cell Physiology* 58(2), 216–226. DOI: 10.1093/pcp/pcw222.

Klie, M., Menz, I., Linde, M. and Debener, T. (2016) Strigolactone pathway genes and plant architecture: Association analysis and QTL detection for horticultural traits in chrysanthemum. *Molecular Genetics and Genomics* 291(2), 957–969. DOI: 10.1007/s00438-015-1155-y.

Kobeissi, B., Saidi, A., Kobeissi, A. and Shafie, M. (2019) Applicability of SCoT and SSR molecular markers for genetic diversity analysis in *Chrysanthemum morifolium* genotypes. *Proceedings of the National Academy of Sciences, India Section B: Biological Sciences* 89(3), 1067–1077. DOI: 10.1007/s40011-018-1024-7.

Lema-Rumiñska, J., Zalewska, M. and Sadoch, Z. (2004) Radiomutants of chrysanthemum (*Dendranthema grandiflora* Tzvelev) of the lady group: RAPD analysis of the genetic diversity. *Plant Breeding* 123(3), 290–293. DOI: 10.1111/j.1439-0523.2004.00996.x.

Li, G. and Quiros, C.F. (2001) Sequence-related amplified polymorphism (SRAP), a new marker system based on a simple PCR reaction: Its application to mapping and gene tagging in Brassica. *Theoretical and Applied Genetics* 103(2–3), 2. DOI: 10.1007/s001220100570.

Li, P., Zhang, F., Chen, S., Jiang, J., Wang, H. *et al.* (2016) Genetic diversity, population structure and association analysis in cut chrysanthemum (*Chrysanthemum morifolium* Ramat.). *Molecular Genetics and Genomics* 291(3), 1117–1125. DOI: 10.1007/s00438-016-1166-3.

Li, P., Su, J., Guan, Z., Fang, W., Chen, F. *et al.* (2018) Association analysis of drought tolerance in cut chrysanthemum (*Chrysanthemum morifolium* Ramat.) at seedling stage. *3 Biotech* 8(5), 226. DOI: 10.1007/s13205-018-1258-3.

Li, S., Li, M., Li, Z., Zhu, Y., Ding, H. *et al.* (2019) Effects of the silencing of *CmMET1* by RNA interference in chrysanthemum (*Chrysanthemum morifolium*). *Plant Biotechnology Reports* 13(1), 63–72. DOI: 10.1007/s11816-019-00516-5.

Li, X., Yang, Q., Liao, X., Tian, Y., Zhang, F. *et al.* (2022) A natural antisense RNA improves chrysanthemum cold tolerance by regulating the transcription factor *DgTCP1*. *Plant Physiology* 190(1), 605–620. DOI: 10.1093/plphys/kiac267.

Li, Z., Wang, P., You, C., Yu, J., Zhang, X. *et al.* (2020b) Combined GWAS and eQTL analysis uncovers a genetic regulatory network orchestrating the initiation of secondary cell wall development in cotton. *New Phytologist* 226(6), 1738–1752. DOI: 10.1111/nph.16468.

Lin, H.-H., Lin, K.-H., Huang, M.-Y. and Su, Y.-R. (2020) Use of non-destructive measurements to identify cucurbit species (*Cucurbita maxima* and *Cucurbita moschata*) tolerant to waterlogged conditions. *Plants* 9(9), 1226. DOI: 10.3390/plants9091226.

Lin, P., Bai, H., He, L., Huang, Q., Zeng, Q.H. *et al.* (2021) Proteome-wide and lysine crotonylation profiling reveals the importance of crotonylation in chrysanthemum (*Dendranthema grandiforum*) under low-temperature. *BMC Genomics* 22(1), 51. DOI: 10.1186/s12864-020-07365-5.

Li, S., Zhang, C., Lu, M., Yang, D., Qian, Y. *et al.* (2020a) QTL mapping and GWAS for field kernel water content and kernel dehydration rate before physiological maturity in maize. *Scientific Reports* 10(1), 13114. DOI: 10.1038/s41598-020-69890-3.

Liu, C., Lu, W., Gao, B., Kimura, H., Li, Y. *et al.* (2020a) Rapid identification of chrysanthemum teas by computer vision and deep learning. *Food Science & Nutrition* 8(4), 1968–1977. DOI: 10.1002/fsn3.1484.

Liu, L., Xue, Y., Luo, J., Han, M., Liu, X. *et al.* (2023) Developing a UV–visible reporter-assisted CRISPR/Cas9 gene editing system to alter flowering time in *Chrysanthemum indicum*. *Plant Biotechnology Journal* 21(8), 1519–1521. DOI: 10.1111/pbi.14062.

Liu, R., Chen, S., Jiang, J., Zhu, L., Zheng, C. *et al.* (2013) Proteomic changes in the base of chrysanthemum cuttings during adventitious root formation. *BMC Genomics* 14(1), 919. DOI: 10.1186/1471-2164-14-919.

Liu, S., Yeh, C.-T., Tang, H.M., Nettleton, D. and Schnable, P.S. (2012) Gene mapping via bulked segregant RNA-Seq (BSR-Seq). *PLoS ONE* 7(5), e36406. DOI: 10.1371/journal.pone.0036406.

Liu, Y., Xin, J., Liu, L., Song, A., Guan, Z. *et al.* (2020b) A temporal gene expression map of chrysanthemum leaves infected with *Alternaria alternata* reveals different stages of defense mechanisms. *Horticulture Research* 7(1), 23. DOI: 10.1038/s41438-020-0245-0.

Liu, Z., Wang, J., Tian, Y. and Dai, S. (2019) Deep learning for image-based large-flowered chrysanthemum cultivar recognition. *Plant Methods* 15(1), 146. DOI: 10.1186/s13007-019-0532-7.

Loh, K.E., Chin, Y.S., Safinar Ismail, I. and Tan, H.Y. (2022) Rapid characterisation of xanthine oxidase inhibitors from the flowers of *Chrysanthemum morifolium* Ramat. using metabolomics approach. *Phytochemical Analysis* 33(1), 12–22. DOI: 10.1002/pca.3057.

Lou, L., Su, J., Ou, X., Li, C., Wang, S. *et al.* (2023) Multi-locus genome-wide association studies reveal genetic loci and candidate genes for leaf angle traits in cut chrysanthemum. *Scientia Horticulturae* 322, 112406. DOI: 10.1016/j.scienta.2023.112406.

Lu, C., Pu, Y., Liu, Y., Li, Y., Qu, J. *et al.* (2019) Comparative transcriptomics and weighted gene co-expression correlation network analysis (WGCNA) reveal potential regulation mechanism of carotenoid accumulation in *Chrysanthemum* × *morifolium*. *Plant Physiology and Biochemistry* 142, 415–428. DOI: 10.1016/j.plaphy.2019.07.023.

Lu, J., Wu, Y., Liu, H., Gou, T., Zhao, S. *et al.* (2024) Estimation of plant water content in cut chrysanthemum using leaf-based hyperspectral reflectance. *Scientia Horticulturae* 323, 112517. DOI: 10.1016/j.scienta.2023.112517.

Lyu, J., Aiwaili, P., Gu, Z., Xu, Y., Zhang, Y. *et al.* (2022) Chrysanthemum MAF2 regulates flowering by repressing gibberellin biosynthesis in response to low temperature. *The Plant Journal* 112(5), 1159–1175. DOI: 10.1111/tpj.16002.

Maeda, M., Uryu, N., Murayama, N., Ishii, H., Ota, M. *et al.* (1990) A simple and rapid method for HLA-DP genotyping by digestion of PCR-amplified DNA with allele-specific restriction endonucleases. *Human Immunology* 27(2), 111–121. DOI: 10.1016/0198-8859(90)90108-2.

Mason, A.S. (2015a) Challenges of genotyping polyploid species. In: Batley, J. (ed.) *Methods in Molecular Biology Plant Genotyping*. Humana Press, New York, pp. 161–168. DOI: 10.1007/978-1-4939-1966-6_12.

Mason, A.S. (2015b) SSR genotyping. In: Batley, J. (ed.) *Plant Genotyping. Methods in Molecular Biology*. Humana Press, New York, pp. 77–89. DOI: 10.1007/978-1-4939-1966-6_6.

Michaels, S.D. and Amasino, R.M. (1998) A robust method for detecting single-nucleotide changes as polymorphic markers by PCR. *The Plant Journal* 14(3), 381–385. DOI: 10.1046/j.1365-313x.1998.00123.x.

Michelmore, R.W., Paran, I. and Kesseli, R. (1991) Identification of markers linked to disease-resistance genes by bulked segregant analysis: A rapid method to detect markers in specific genomic regions by using segregating populations. *Proceedings of the National Academy of Sciences of the United States of America* 88(21), 9828–9832. DOI: 10.1073/pnas.88.21.9828.

Min, H., Hu, S.B., Li, Z.N., Wu, Y.F., Zhang, C.P. *et al.* (2009) A phytoplasma associated with an outbreak of an unusual disease of chrysanthemum in China in 2008. *Plant Disease* 93(8), 840–840. DOI: 10.1094/pdis-93-8-0840a.

Miyazawa, M. and Hisama, M. (2003) Antimutagenic activity of flavonoids from *Chrysanthemum morifolium*. *Bioscience, Biotechnology, and Biochemistry* 67(10), 2091–2099. DOI: 10.1271/bbb.67.2091.

Moradpour, M. and Abdulah, S.N.A. (2020) CRISPR/dCas9 platforms in plants: Strategies and applications beyond genome editing. *Plant Biotechnology Journal* 18(1), 32–44. DOI: 10.1111/pbi.13232.

Mukherjee, A.K., Dey, A., Acharya, L., Palai, S.K. and Panda, P.C. (2013) Studies on genetic diversity in elite varieties of chrysanthemum using RAPD and ISSR markers. *Indian Journal of Biotechnology* 12(2), 161–169. DOI: 10.3389/fcimb.2013.00014.

Nakano, M., Hirakawa, H., Fukai, E., Toyoda, A. and Kajitani, R. (2021) A chromosome-level genome sequence of a model chrysanthemum: Evolution and reference for hexaploid cultivated chrysanthemum. *Communications Biology* 4(1), 1–11. DOI: 10.1038/s42003-021-02704-y.

Nasri, F., Zakizadeh, H., Vafaee, Y. and Mozafari, A.A. (2021) *In vitro* mutagenesis of *Chrysanthemum morifolium* cultivars using ethylmethanesulphonate (EMS) and mutation assessment by ISSR and IRAP markers. *Plant Cell, Tissue and Organ Culture (PCTOC)* 149(3), 657–673. DOI: 10.1007/s11240-021-02163-7.

Nguyen, T.K. and Lim, J.-H. (2019) Tools for chrysanthemum genetic research and breeding: Is genotyping-by-sequencing (GBS) the best approach? *Horticulture, Environment, and Biotechnology* 60(5), 625–635. DOI: 10.1007/s13580-019-00160-6.

Nguyen, T.K., Yu, J., Choi, H.-W., In, B.-C. and Lim, J.-H. (2018) Optimization of genotyping-by-sequencing (GBS) in chrysanthemums: Selecting proper restriction enzymes for GBS library construction. *Horticultural Science and Technology* 36(1), 108–114. DOI: 10.12972/kjhst.20180012.

Palai, S.K. and Rout, G.R. (2011) Characterization of new variety of chrysanthemum by using ISSR markers. *Horticultura Brasileira* 29(4), 613–617. DOI: 10.1590/s0102-05362011000400029.

Paran, I. and Michelmore, R.W. (1993) Development of reliable PCR-based markers linked to downy mildew resistance genes in lettuce. *Theoretical and Applied Genetics* 85(8), 985–993. DOI: 10.1007/BF00215038.

Peng, A., Lin, L., Zhao, M. and Sun, B. (2019) Identifying mechanisms underlying the amelioration effect of *Chrysanthemum morifolium* Ramat. "BOJU" extract on hyperuricemia using biochemical characterization and UPLC-ESI-QTOF/MS-based metabolomics. *Food & Function* 10(12), 8042–8055. DOI: 10.1039/c9fo01821b.

Peng, H., Zhang, F., Jiang, J., Chen, S., Fang, W. *et al.* (2015) Identification of quantitative trait loci for branching traits of spray cut chrysanthemum. *Euphytica* 202(3), 385–392. DOI: 10.1007/s10681-014-1259-1.

Peterson, B.K., Weber, J.N., Kay, E.H., Fisher, H.S. and Hoekstra, H.E. (2012) Double digest RADseq: An inexpensive method for *de novo* SNP discovery and genotyping in model and non-model species. *PLoS ONE* 7(5), e37135. DOI: 10.1371/journal.pone.0037135.

Poland, J.A., Brown, P.J., Sorrells, M.E. and Jannink, J.-L. (2012) Development of high-density genetic maps for barley and wheat using a novel two-enzyme genotyping-by-sequencing approach. *PLoS ONE* 7(2), e32253. DOI: 10.1371/journal.pone.0032253.

Qi, C., Nyalala, I. and Chen, K. (2021) Detecting the early flowering stage of tea chrysanthemum using the F-YOLO model. *Agronomy* 11(5), 834. DOI: 10.3390/agronomy11050834.

Qiao, A., Fang, X., Liu, S., Liu, H., Gao, P. *et al.* (2021) QTL-seq identifies major quantitative trait loci of stigma color in melon. *Horticultural Plant Journal* 7(4), 318–326. DOI: 10.1016/j.hpj.2021.01.004.

Ran, F.A., Hsu, P.D., Wright, J., Agarwala, V., Scott, D.A. *et al.* (2013) Genome engineering using the CRISPR-Cas9 system. *Nature Protocols* 8(11), 2281–2308. DOI: 10.1038/nprot.2013.143.

Riangwong, K., Wanchana, S., Aesomnuk, W., Saensuk, C., Nubankoh, P. *et al.* (2020) Mining and validation of novel genotyping-by-sequencing (GBS)-based simple sequence repeats (SSRs) and their application for the estimation of the genetic diversity and population structure of coconuts (*Cocos nucifera* L.) in Thailand. *Horticulture Research* 7(1), 156. DOI: 10.1038/s41438-020-00374-1.

Roxas, N.J., Tashiro, Y., Miyazaki, S., Isshiki, S. and Takeshita, A. (1995) Meiosis and pollen fertility in *Higo chrysanthemum* (*Dendranthema* × *grandiflorum* (Ramat.) Kitam.). *Journal of the Japanese Society for Horticultural Science* 64(1), 161–168. DOI: 10.2503/jjshs.64.161.

Ruangrak, E., Su, X., Huang, Z., Wang, X., Guo, Y. *et al.* (2018) Fine mapping of a major QTL controlling early flowering in tomato using QTL-seq. *Canadian Journal of Plant Science* 98(3), 672–682. DOI: 10.1139/cjps-2016-0398.

Ryu, J., Kim, W.J., Im, J., Kang, K.-W., Kim, S.H. *et al.* (2019) Single nucleotide polymorphism (SNP) discovery through genotyping-by-sequencing (GBS) and genetic characterization of *Dendrobium* mutants and cultivars. *Scientia Horticulturae* 244, 225–233. DOI: 10.1016/j.scienta.2018.09.053.

Sahithi, B.M., Razi, K., Al Murad, M., Vinothkumar, A., Jagadeesan, S. *et al.* (2020) Comparative physiological and proteomic analysis deciphering tolerance and homeostatic signaling pathways in chrysanthemum under drought stress. *Physiologia Plantarum* 172(2), 289–303. DOI: 10.1111/ppl.13142.

Sang, K.P. (2016) Inheritance analysis and marker development of white rust resistance in chrysanthemum. Seoul National University Graduate School. Available at: https://hdl.handle.net/10371/121014

Sasaki, K., Mitsuda, N., Nashima, K., Kishimoto, K., Katayose, Y. *et al.* (2017) Generation of expressed sequence tags for discovery of genes responsible for floral traits of *Chrysanthemum morifolium* by next-generation sequencing technology. *BMC Genomics* 18(1), 683. DOI: 10.1186/s12864-017-4061-3.

Scossa, F., Alseekh, S. and Fernie, A.R. (2021) Integrating multi-omics data for crop improvement. *Journal of Plant Physiology* 257, 153352. DOI: 10.1016/j.jplph.2020.153352.

Sharma, V., Gupta, P., Priscilla, K., SharanKumar, S., Hangargi, B. *et al.* (2021) Metabolomics intervention towards better understanding of plant traits. *Cells* 10(2), 346. DOI: 10.3390/cells10020346.

Shendure, J. and Ji, H. (2008) Next-generation DNA sequencing. *Nature Biotechnology* 26(10), 1135–1145. DOI: 10.1038/nbt1486.

Shinoyama, H., Ichikawa, H., Saitoh-Nakashima, M., Saito, M., Aida, R. *et al.* (2012) Introduction of male sterility to gm chrysanthemum plants to prevent transgene flow. *Acta Horticulturae* (937), 337–346. DOI: 10.17660/actahortic.2012.937.41.

Shinoyama, H., Ichikawa, H., Nishizawa-Yokoi, A., Skaptsov, M. and Toki, S. (2020) Simultaneous talen-mediated knockout of chrysanthemum *DMC1* genes confers male and female sterility. *Scientific Reports* 10(1), 16165. DOI: 10.1038/s41598-020-72356-1.

Skroch, P. and Nienhuis, J. (1995) Impact of scoring error and reproducibility RAPD data on rapd based estimates of genetic distance. *Theoretical and Applied Genetics* 91(6), 1086–1091. DOI: 10.1007/BF00223923.

Sonah, H., Bastien, M., Iquira, E., Tardivel, A., Légaré, G. *et al.* (2013) An improved genotyping by sequencing (GBS) approach offering increased versatility and efficiency of SNP discovery and genotyping. *PLoS ONE* 8(1), e54603. DOI: 10.1371/journal.pone.0054603.

Song, A., Su, J., Wang, H., Zhang, Z., Zhang, X. *et al.* (2023) Analyses of a chromosome-scale genome assembly reveal the origin and evolution of cultivated chrysanthemum. *Nature Communications* 14(1), 2021. DOI: 10.1038/s41467-023-37730-3.

Song, C., Liu, Y., Song, A., Dong, G., Zhao, H. *et al.* (2018) The chrysanthemum nankingense genome provides insights into the evolution and diversification of chrysanthemum flowers and medicinal traits. *Molecular Plant* 11(12), 1482–1491. DOI: 10.1016/j.molp.2018.10.003.

Song, X., Xu, Y., Gao, K., Fan, G., Zhang, F. *et al.* (2020) High-density genetic map construction and identification of loci controlling flower-type traits in chrysanthemum (*Chrysanthemum* × *morifolium* Ramat.). *Horticulture Research* 7(1), 108. DOI: 10.1038/s41438-020-0333-1.

Spaargaren, J. and van Geest, G. (2018) Chrysanthemum. In: Van Huylenbroeck, J. (ed.) *Handbook of Plant Breeding Ornamental Crops*, Vol. 11. Springer International Publishing, Cham, pp. 319–348. DOI: 10.1007/978-3-319-90698-0_14.

Stapley, J., Reger, J., Feulner, P.G., Smadja, C., Galindo, J. *et al.* (2010) Adaptation genomics: The next generation. *Trends in Ecology & Evolution* 25(12), 705–712. DOI: 10.1016/j.tree.2010.09.002.

Su, J., Zhang, F., Li, P., Guan, Z., Fang, W. *et al.* (2016) Genetic variation and association mapping of waterlogging tolerance in chrysanthemum. *Planta* 244(6), 1241–1252. DOI: 10.1007/s00425-016-2583-6.

Su, J., Yang, X., Zhang, F., Wu, S., Xiong, S. *et al.* (2018) Dynamic and epistatic QTL mapping reveals the complex genetic architecture of waterlogging tolerance in chrysanthemum. *Planta* 247(4), 899–924. DOI: 10.1007/s00425-017-2833-2.

Su, J., Zhang, F., Chong, X., Song, A., Guan, Z. *et al.* (2019) Genome-wide association study identifies favorable SNP alleles and candidate genes for waterlogging tolerance in chrysanthemums. *Horticulture Research* 6(1), 21. DOI: 10.1038/s41438-018-0101-7.

Sumitomo, K., Shirasawa, K., Isobe, S., Hirakawa, H., Hisamatsu, T. *et al.* (2019) Genome-wide association study overcomes the genome complexity in autohexaploid chrysanthemum and tags SNP markers onto the flower color genes. *Scientific Reports* 9(1), 13947. DOI: 10.1038/s41598-019-50028-z.

Sumitomo, K., Shirasawa, K., Isobe, S.N., Hirakawa, H., Harata, A. *et al.* (2021) DNA marker for resistance to *Puccinia horiana* in chrysanthemum (*Chrysanthemum morifolium* Ramat.) "Southern Pegasus". *Breeding Science* 71(2), 261–267. DOI: 10.1270/jsbbs.20063.

Sun, D., Zhang, J., He, J., Geng, Z., Li, S. *et al.* (2022) Whole-transcriptome profiles of chrysanthemum seticuspe improve genome annotation and shed new light on mRNA–miRNA–lncRNA networks in ray florets and disc florets. *BMC Plant Biology* 22(1), 1–15. DOI: 10.1186/s12870-022-03889-y.

Sun, H., Zhang, F., Chen, S., Guan, Z., Jiang, J. *et al.* (2015a) Effects of aphid herbivory on volatile organic compounds of *Artemisia annua* and *Chrysanthemum morifolium. Biochemical Systematics and Ecology* 60, 225–233. DOI: 10.1016/j.bse.2015.04.023.

Sun, H., Zhang, T., Fan, Q., Qi, X., Zhang, F. *et al.* (2015b) Identification of floral scent in chrysanthemum cultivars and wild relatives by gas chromatography-mass spectrometry. *Molecules* 20(4), 5346–5359. DOI: 10.3390/molecules20045346.

Sun, X., Liu, D., Zhang, X., Li, W., Liu, H. *et al.* (2013) SLAF-Seq: An efficient method of large-scale *de novo* SNP discovery and genotyping using high-throughput sequencing. *PLoS ONE* 8(3), e58700. DOI: 10.1371/journal.pone.0058700.

Taheri, S., Lee Abdullah, T., Yusop, M.R., Hanafi, M., Sahebi, M. *et al.* (2018) Mining and development of novel SSR markers using next generation sequencing (NGS) data in plants. *Molecules* 23(2), 399. DOI: 10.3390/molecules23020399.

Tang, H.Q., Zhang, F., Chen, F.D., Fang, W.M. and Wang, C.C. (2015) Heterosis and mixed genetic analysis of inflorescence traits of anemone-typed chrysanthemum. *Acta Horticulturae Sinica* 42(5), 907.

Tibbs Cortes, L., Zhang, Z. and Yu, J. (2021) Status and prospects of genome-wide association studies in plants. *The Plant Genome* 14(1), e20077. DOI: 10.1002/tpg2.20077.

Tiwari, J.K., Yerasu, S.R., Rai, N., Singh, D.P., Singh, A.K. *et al.* (2022) Progress in marker-assisted selection to genomics-assisted breeding in tomato. *Critical Reviews in Plant Sciences* 41(5), 321–350. DOI: 10.1080/07352689.2022.2130361.

Torkamaneh, D., Boyle, B. and Belzile, F. (2018) Efficient genome-wide genotyping strategies and data integration in crop plants. *Theoretical and Applied Genetics* 131(3), 499–511. DOI: 10.1007/s00122-018-3056-z.

Torkamaneh, D., Boyle, B., St-Cyr, J., Légaré, G., Pomerleau, S. *et al.* (2020) NanoGBS: A miniaturized procedure for GBS library preparation. *Frontiers in Genetics* 11, 67. DOI: 10.3389/fgene.2020.00067.

Ulaszewski, B., Meger, J. and Burczyk, J. (2021) Comparative analysis of SNP discovery and genotyping in *Fagus sylvatica* L. and *Quercus robur* L. using RADseq, GBS, and ddRAD methods. *Forests* 12(2), 222. DOI: 10.3390/f12020222.

van Geest, G., Choi, Y.H., Arens, P., Post, A., Liu, Y. *et al.* (2016) Genotypic differences in metabolomic changes during storage induced-degreening of chrysanthemum disk florets. *Postharvest Biology and Technology* 115, 48–59. DOI: 10.1016/j.postharvbio.2015.12.008.

van Geest, G., Bourke, P.M., Voorrips, R.E., Marasek-Ciolakowska, A., Liao, Y. *et al.* (2017a) An ultra-dense integrated linkage map for hexaploid chrysanthemum enables multi-allelic QTL analysis. *Theoretical and Applied Genetics* 130(12), 2527–2541. DOI: 10.1007/s00122-017-2974-5.

van Geest, G., Voorrips, R.E., Esselink, D., Post, A., Visser, R.G. *et al.* (2017b) Conclusive evidence for hexasomic inheritance in chrysanthemum based on analysis of a 183 k SNP array. *BMC Genomics* 18(1), 585. DOI: 10.1186/s12864-017-4003-0.

van Lieshout, N., Kaauwen, M., Kodde, L., Arens, P., Smulders, M.J.M. *et al.* (2022) De novo whole-genome assembly of *Chrysanthemum makinoi*, a key wild ancestor to hexaploid chrysanthemum. *G3 Genes|Genomes|Genetics* 12(1), jkab358. DOI: 10.1093/g3journal/jkab358.

van Orsouw, N.J., Hogers, R.C., Janssen, A., Yalcin, F., Snoeijers, S. *et al.* (2007) Complexity reduction of polymorphic sequences (CRoPS™): A novel approach for large-scale polymorphism discovery in complex genomes. *PLoS ONE* 2(11), e1172. DOI: 10.1371/journal.pone.0001172.

Vogel, G., LaPlant, K.E., Mazourek, M., Gore, M.A. and Smart, C.D. (2021) A combined BSA-Seq and linkage mapping approach identifies genomic regions associated with phytophthora root and crown rot resistance in squash. *Theoretical and Applied Genetics* 134(4), 1015–1031. DOI: 10.1007/s00122-020-03747-1.

Vos, P., Hogers, R., Bleeker, M., Reijans, M., van de Lee, T. *et al.* (1995) AFLP: A new technique for DNA fingerprinting. *Nucleic Acids Research* 23(21), 4407–4414. DOI: 10.1093/nar/23.21.4407.

Wang, C., Zhang, F., Guan, Z., Chen, S., Jiang, J. *et al.* (2014a) Inheritance and molecular markers for aphid (*Macrosiphoniella sanbourni*) resistance in chrysanthemum (*Chrysanthemum morifolium* Ramat.). *Scientia Horticulturae* 180, 220–226. DOI: 10.1016/j.scienta.2014.10.038.

Wang, F., Zhong, X., Huang, L., Fang, W., Chen, F. *et al.* (2018) Cellular and molecular characteristics of pollen abortion in chrysanthemum cv. Kingfisher. *Plant Molecular Biology* 98(3), 233–247. DOI: 10.1007/s11103-018-0777-y.

Wang, H., Jiang, J., Chen, S., Qi, X., Peng, H. *et al.* (2013) Next-generation sequencing of the *Chrysanthemum nankingense* (Asteraceae) transcriptome permits large-scale unigene assembly and SSR marker discovery. *PLoS ONE* 8(4), e62293. DOI: 10.1371/journal.pone.0062293.

Wang, H., Qi, X., Gao, R., Wang, J., Dong, B. *et al.* (2014b) Microsatellite polymorphism among *Chrysanthemum* sp. Polyploids: The influence of whole genome duplication. *Scientific Reports* 4(1), 6730. DOI: 10.1038/srep06730.

Wang, N., Yuan, Y., Wang, H., Yu, D., Liu, Y. *et al.* (2020) Applications of genotyping-by-sequencing (GBS) in maize genetics and breeding. *Scientific Reports* 10(1), 16308. DOI: 10.1038/s41598-020-73321-8.

Wang, S., Meyer, E., McKay, J.K. and Matz, M.V. (2012) 2b-RAD: A simple and flexible method for genome-wide genotyping. *Nature Methods* 9(8), 808–810. DOI: 10.1038/nmeth.2023.

Wang, T., Zou, Q., Guo, Q., Yang, F., Wu, L. *et al.* (2019) Widely targeted metabolomics analysis reveals the effect of flooding stress on the synthesis of flavonoids in *Chrysanthemum morifolium*. *Molecules* 24(20), 3695. DOI: 10.3390/molecules24203695.

Wang, T., Wei, Q., Wang, Z., Liu, W., Zhao, X. *et al.* (2022) CmNF-YB8 affects drought resistance in chrysanthemum by altering stomatal status and leaf cuticle thickness. *Journal of Integrative Plant Biology* 64(3), 741–755. DOI: 10.1111/jipb.13201.

Wei, Q., Ma, C., Xu, Y., Wang, T., Chen, Y. *et al.* (2017) Control of chrysanthemum flowering through integration with an aging pathway. *Nature Communications* 8(1), 829. DOI: 10.1038/s41467-017-00812-0.

Wen, X., Li, J., Wang, L., Lu, C., Gao, Q. *et al.* (2022) The *Chrysanthemum lavandulifolium* genome and the molecular mechanism underlying diverse capitulum types. *Horticulture Research* 9, uhab022. DOI: 10.1093/hr/uhab022.

Wenger, A.M., Peluso, P., Rowell, W.J., Chang, P.-C., Hall, R.J. *et al.* (2019) Accurate circular consensus long-read sequencing improves variant detection and assembly of a human genome. *Nature Biotechnology* 37(10), 1155–1162. DOI: 10.1038/s41587-019-0217-9.

Wiedenheft, B., Sternberg, S.H. and Doudna, J.A. (2012) RNA-guided genetic silencing systems in bacteria and archaea. *Nature* 482(7385), 331–338. DOI: 10.1038/nature10886.

Williams, J.G., Kubelik, A.R., Livak, K.J., Rafalski, J.A. and Tingey, S.V. (1990) DNA polymorphisms amplified by arbitrary primers are useful as genetic markers. *Nucleic Acids Research* 18(22), 6531–6535. DOI: 10.1093/nar/18.22.6531.

Witzel, K., Neugart, S., Ruppel, S., Schreiner, M., Wiesner, M. *et al.* (2015) Recent progress in the use of omics technologies in brassicaceous vegetables. *Frontiers in Plant Science* 6, 244. DOI: 10.3389/fpls.2015.00244.

Wolff, K. and Peters-van Rijn, J. (1993) Rapid detection of genetic variability in chrysanthemum (*Dendranthema grandiflora* Tzvelev) using random primers. *Heredity* 71(4), 335–341. DOI: 10.1038/hdy.1993.147.

Wolff, K., Peters-Van Rijn, J. and Hofstra, H. (1994) RFLP analysis in chrysanthemum I. Probe and primer development. *Theoretical and Applied Genetics* 88(3), 472–478. DOI: 10.1007/BF00223663.

Wolff, K., Zietkiewicz, E. and Hofstra, H. (1995) Identification of chrysanthemum cultivars and stability of DNA fingerprint patterns. *Theoretical and Applied Genetics* 91(3), 439–447. DOI: 10.1007/BF00222971.

Wood, A.J., Lo, T.-W., Zeitler, B., Pickle, C.S., Ralston, E.J. *et al.* (2011) Targeted genome editing across species using ZFNs and TALENs. *Science* 333(6040), 307–307. DOI: 10.1126/science.1207773.

Wu, N., Zhang, C., Bai, X., Du, X. and He, Y. (2018) Discrimination of chrysanthemum varieties using hyperspectral imaging combined with a deep convolutional neural network. *Molecules* 23(11), 2831. DOI: 10.3390/molecules23112831.

Xu, Y. (2010) *Molecular Plant Breeding*. CAB International, Wallingford, UK.

Xue, H., Shi, T., Wang, F., Zhou, H., Yang, J. *et al.* (2017) Interval mapping for red/green skin color in Asian pears using a modified QTL-seq method. *Horticulture Research* 4, 17053. DOI: 10.1038/hortres.2017.53.

Yang, X., Wu, Y., Su, J., Ao, N., Guan, Z. *et al.* (2019b) Genetic variation and development of a scar marker of anemone-type flower in chrysanthemum. *Molecular Breeding* 39(3), 48. DOI: 10.1007/s11032-019-0958-7.

Yan, J., Zou, D., Li, C., Zhang, Z., Song, S. *et al.* (2020) SR4R: An integrative SNP resource for genomic breeding and population research in rice. *Genomics, Proteomics & Bioinformatics* 18(2), 173–185. DOI: 10.1016/j.gpb.2020.03.002.

Yang, X., Fang, X., Su, J., Ding, L., Guan, Z. *et al.* (2019a) Genetic dissection of floral traits in anemone-type chrysanthemum by QTL mapping. *Molecular Breeding* 39(9), 39. DOI: 10.1007/s11032-019-1043-y.

Yao, L., Li, Y., Ma, C., Tong, L., Du, F. *et al.* (2020) Combined genome-wide association study and transcriptome analysis reveal candidate genes for resistance to *Fusarium* ear rot in maize. *Journal of Integrative Plant Biology* 62(10), 1535–1551. DOI: 10.1111/jipb.12911.

Yin, K., Gao, C. and Qiu, J.-L. (2017) Progress and prospects in plant genome editing. *Nature Plants* 3(8), 1–6. DOI: 10.1038/nplants.2017.107.

Zaman, Q.U., Li, C., Cheng, H. and Hu, Q. (2019) Genome editing opens a new era of genetic improvement in polyploid crops. *The Crop Journal* 7(2), 141–150. DOI: 10.1016/j.cj.2018.07.004.

Zargar, S.M., Raatz, B., Sonah, H., Bhat, J.A., Dar, Z.A. *et al.* (2015) Recent advances in molecular marker techniques: Insight into QTL mapping, GWAS and genomic selection in plants. *Journal of Crop Science and Biotechnology* 18(5), 293–308. DOI: 10.1007/s12892-015-0037-5.

Zhang, F., Chen, S., Chen, F., Fang, W. and Li, F. (2010a) A preliminary genetic linkage map of chrysanthemum (*Chrysanthemum morifolium*) cultivars using RAPD, ISSR and AFLP markers. *Scientia Horticulturae* 125(3), 422–428. DOI: 10.1016/j.scienta.2010.03.028.

Zhang, F., Chen, F., Fang, W. and Chen, S. (2010b) Detection of RAPD and ISSR markers associated with ornamental traits of chrysanthemum in vegetative stage. *Acta Horticulturae Sinica* 37(8), 1345–1350.

Zhang, F., Chen, S., Chen, F., Fang, W., Chen, Y. *et al.* (2011a) SRAP-based mapping and QTL detection for inflorescence-related traits in chrysanthemum (*Dendranthema morifolium*). *Molecular Breeding* 27(1), 11–23. DOI: 10.1007/s11032-010-9409-1.

Zhang, F., Chen, S., Chen, F., Fang, W., Deng, Y. *et al.* (2011b) Genetic analysis and associated srap markers for flowering traits of chrysanthemum (*Chrysanthemum morifolium*). *Euphytica* 177(1), 15–24. DOI: 10.1007/s10681-010-0239-3.

Zhang, F., Jiang, J., Chen, S., Chen, F. and Fang, W. (2012) Mapping single-locus and epistatic quantitative trait loci for plant architectural traits in chrysanthemum. *Molecular Breeding* 30(2), 1027–1036. DOI: 10.1007/s11032-011-9686-3.

Zhang, F., Chen, S., Jiang, J., Guan, Z., Fang, W. *et al.* (2013) Genetic mapping of quantitative trait loci underlying flowering time in chrysanthemum (*Chrysanthemum morifolium*). *PLoS ONE* 8(12), e83023. DOI: 10.1371/journal.pone.0083023.

Zhang, F., Wang, Z., Dong, W., Sun, C., Wang, H. *et al.* (2014a) Transcriptomic and proteomic analysis reveals mechanisms of embryo abortion during chrysanthemum cross breeding. *Scientific Reports* 4(1), 6536. DOI: 10.1038/srep06536.

Zhang, F., Hua, L., Fei, J., Wang, F., Liao, Y. *et al.* (2016) Chromosome doubling to overcome the chrysanthemum cross barrier based on insight from transcriptomic and proteomic analyses. *BMC Genomics* 17(1), 585. DOI: 10.1186/s12864-016-2939-0.

Zhang, L., Guo, D., Guo, L., Guo, Q., Wang, H. *et al.* (2019) Construction of a high-density genetic map and QTLs mapping with GBS from the interspecific F_1 population of *P. ostii* "Fengdan Bai" and *P. suffruticosa* "Xin Riyuejin". *Scientia Horticulturae* 246, 190–200. DOI: 10.1016/j.scienta.2018.10.039.

Zhang, Y., Sun, M. and Zhang, Q. (2014b) Proteomic analysis of the heat stress response in leaves of two contrasting chrysanthemum varieties. *Plant Omics* 7(4),

229–236. Available at: https://search.informit.org/doi/abs/10.3316/INFORMIT.580054211873631

Zhao, C., Zhang, Y., Du, J., Guo, X., Wen, W. *et al.* (2019) Crop phenomics: Current status and perspectives. *Frontiers in Plant Science* 10, 714. DOI: 10.3389/fpls.2019.00714.

Zhao, J., Chen, S. and Chen, F. (2009) Conversion of RAPD marker linked to creep plant type in ground-cover chrysanthemum to SCAR marker. *Scientia Silvae Sinicae* 45(9), 147–150.

Zietkiewicz, E., Rafalski, A. and Labuda, D. (1994) Genome fingerprinting by simple sequence repeat (SSR)-anchored polymerase chain reaction amplification. *Genomics* 20(2), 176–183. DOI: 10.1006/geno.1994.1151.

Biotechnology of Chrysanthemum

4

Naonobu Noda[1]* and N.C. Surakshitha[2]*

[1]Institute of Vegetable and Floriculture Science, National Agriculture and Food Research Organization, 2-1 Fujimoto, Tsukuba, Ibaraki 305-0852, Japan
[2]Plant Breeding, Wageningen University and Research, PO Box 386, 6700 AJ, Wageningen, The Netherlands

Abstract

Chrysanthemums (*Chrysanthemum morifolium* Ramat.), highly regarded in the ornamental plant industry and second only to roses in the cut flower market worldwide, have traditionally been enhanced through crossbreeding and natural mutations, yielding a diverse array of flower morphologies and colors. Recent elucidations of the molecular mechanisms underpinning their physiology and ecology have paved the way for the development of new cultivars using advanced biotechnological tools. This chapter provides an overview of the *in vitro* methods used for germplasm conservation and describes advancements in various tissue culture techniques, including somatic embryogenesis, embryo rescue, anther culture, and protoplast culture. It also discusses genetic modifications aimed at improving flower color and floral architecture analysing, made possible by the identification of pigment biosynthetic genes and various homeotic genes. Furthermore, this chapter outlines transformation protocols for genetic engineering, producing viroid-free chrysanthemums, explores the role of artificial intelligence in optimizing *in vitro* cultural conditions, and discusses future prospects in chrysanthemum biotechnology.

Keywords

Anther and protoplast culture, cryopreservation, flower color and pigments, genetic transformation, tissue culture

*Corresponding authors: noda.naonobu737@naro.go.jp and surakshitha.naguvanahallychandrashekara@wur.nl

DOI: 10.1079/9781789248289.0004

4.1 Introduction

Chrysanthemum cultivars have long been developed through crossbreeding and spontaneous mutations, resulting in a wide range of plant morphologies and flower colors. These cultivars, including disbuds, sprays, and santinis, are now cultivated and distributed year-round. Recent research has focused on understanding the molecular mechanisms that shape the physiology and ecology of chrysanthemum. Leveraging insights from biotechnology, new methods for developing novel cultivars are emerging. Over the past decade, numerous reviews have covered key biotechnological topics, including tissue culture, with a focus on callus induction, regeneration, protoplast culture, embryo rescue, cryopreservation, and transformation (Shinoyama *et al.*, 2012a; Teixeira da Silva *et al.*, 2013, 2015a; Teixeira da Silva and Kulus 2014). Here, we briefly outline developments in tissue culture conditions, viroid-free regeneration, and the expansion of mutation techniques in chrysanthemum. Our main focus is on the analysis of gene functions and the development of chrysanthemums with desirable traits using biotechnology in recent literature. Examples include morphological modifications, such as changes in flower shape as well as the exploration of genes involved in flower pigmentation, to develop novel, bluish-colored chrysanthemums. Lastly, we describe a protocol for the transformation of chrysanthemums, which has been successfully used for various modifications. We conclude by summarizing the status of the field, highlighting the need for technologies that can introduce traits, such as sterility, to further support controlled breeding.

4.2 *In vitro* Approaches to Conserve Germplasm

In vitro approaches for conserving chrysanthemum germplasm are essential due to the highly heterozygous nature of these cultivars. To ensure genetic fidelity, chrysanthemums are typically propagated vegetatively using cuttings and maintained in greenhouses or fields, which requires significant space. Although the first generation of newly bred genotypes can be propagated by seeds, maintaining their true-to-type phenotypes is challenging due to the potential cross-pollination and wide geographical variations among populations. As a result, elite germplasm and wild species are preserved vegetatively. Although cuttings can be stored at 0°C for several months, this method only offers short-term storage. To support long-term preservation, permanent solutions, such as micropropagation, *in vitro* conservation, and cryopreservation, are increasingly being utilized to sustain these valuable germplasms.

4.2.1 Conservation of germplasm by slowing plant growth for short- or medium-term storage

Numerous studies have focused on chrysanthemum shoot-tip culture and micropropagation for germplasm conservation (Teixeira da Silva *et al.*, 2015a). Various techniques have been developed to limit plant development during *in vitro* storage, including lowering the culture temperature, modifying the culture medium, and using plant growth retardants. Among these, reducing the culture temperature is most effective for halting the growth of plantlets *in vitro*. Additionally, reducing the carbon source in the medium and adding compounds such as mannitol or benzyladenine can also significantly slow plantlet growth. Although these methods have been well documented for chrysanthemum storage under restricted growth conditions (refer to Table 4.1), they may induce somaclonal variations that could compromise the genetic integrity of the plantlets. Therefore, when maintaining true-to-type characteristics is critical, these techniques are not recommended.

4.2.2 Cryopreservation for long-term storage

Cryopreservation involves storing biological material at ultralow temperatures, typically using liquid nitrogen at −196°C. This technique is essential for preserving various plant explants—such as meristems, shoot tips, embryogenic callus, seeds, roots, and embryonic axes—and is widely regarded as the most effective method for long-term preservation of chrysanthemum and other vegetatively propagated species (Martinez-Montero and Harding, 2015; Agarwal *et al.*, 2019; Li *et al.*, 2019). Cryopreservation is both cost-effective and reliable (Kulus and Zalewska, 2014a, b), although the preparation of explants for storage can be complex (Bi *et al.*, 2016). At −196°C, cellular, metabolic, and biochemical processes are completely halted, enabling indefinite preservation without deterioration (Reed, 2017). The first successful chrysanthemum cryopreservation reports date back to the 1980s (Fukai *et al.*, 1988; Fukai and Oe, 1989; Fukai *et al.*, 1994). Since then, various techniques have been developed, including slow cooling (two-step freezing), rapid cooling (vitrification), droplet-vitrification, and encapsulation-vitrification (Martín and González-Benito, 2005; Kulus *et al.*, 2013; Teixeira da Silva *et al.*, 2015a). A pahetic qualities, although its effectivenerticularly effective protocol, known as encapsulation-vitrification-dehydration, was optimized for cryotolerance by Jeon *et al.* (2015). While the specifics of each method vary, all techniques typically use protective compounds called cryoprotectants (Halmagyi *et al.*, 2004), such as sugars (primarily sucrose), glycerol, ethylene glycol (EG), and dimethyl sulfoxide (DMSO). These cryoprotectants help dehydrate cells, stabilize cellular components, and prevent ice formation—an essential factor for successful cryostorage. Optimal concentrations and exposure time to these cryoprotectants depend on the specific technique and cultivar being preserved.

Cryopreserved material is exposed to various physicochemical and physiological stresses, posing challenges to maintaining genetic integrity. Therefore, it is crucial to periodically assess the genetic stability of the biologically conserved material. Notably, genetic disturbances in chrysanthemum chimeras were observed after utilizing a two-step freezing protocol, with 70% of recovered plants exhibiting altered inflorescence color (Fukai *et al.*, 1994). Structural damages to meristems, which affect the morphogenetic

Table 4.1. *In vitro* conservation of chrysanthemum germplasm under limiting growth conditions. Author's own table.

Modification in growth condition	Specifications	Preservation period/impact on culture	References
Reducing temperature	10°C under 1500 lux cultured on sucrose-free medium	Up to 1 year or more Restricted growth	Fukai *et al.* (1988)
	2–3°C with diffused light on normal medium		Preil and Hoffmann (1985)
	4°C		Bajaj (1986)
	2°C in the dark, 15-day-old pre-cultured shoot tips	Several months	Fukai and Oe (1989)
	In vitro cultures stored at 4°C	6 years	Bajaj (1993)
	4°C	2–6 months	Trifunović *et al.* (2007); Liu and Gao (2010)
Culture medium modification	Reduced sucrose level to 0.1 or 0%, or addition of 2–8% mannitol to the medium	Reduced growth	Fukai *et al.* (1988)
	Reduced sucrose level from 3 to 2%, addition of 2 mg l^{-1} abscisic acid to the medium	6 months	Liu and Gao (2010)
	Addition of benzyl adenine (BA) to the culture medium at room temperature	Delaying senescence	Hosoki (1989)
Use of growth retardants	Plantlets dipped in 100 mg l^{-1} uniconazol for 30 min prior to transplanting to fresh medium	Postponement of subculture for a year or more	Fukai *et al.* (1988)

response of chrysanthemum explants, have also been documented following encapsulation-dehydration storage (Kulus *et al.*, 2018). Such structural alterations in meristems can impact the stability of plantlets regenerated from cryopreserved shoot tips. Importantly, these changes are not caused directly by liquid nitrogen storage but rather by variations introduced during the preculture or recovery phases of *in vitro* procedures.

Chimeras are highly susceptible to variation following cryopreservation unless all tunica layers are adequately protected. An optimal encapsulation-dehydration technique for chrysanthemum cultivars includes a 14-day preculture with 0.09 M sucrose and 10 μM ABA, followed by 4 days of osmotic dehydration and 3 hour of desiccation (Kulus *et al.*, 2019). Although this protocol led to changes at the DNA sequence level and influenced vegetative growth, it effectively preserved the chimeric structure. Cryopreservation thus holds great promise for establishing a gene bank of chrysanthemums, including stable mutants and periclinal chimeras. However, the widespread application of cryopreservation is currently limited to a few specialized laboratories, yet it remains an efficient method to conserve elite germplasm to support future breeding efforts.

4.3 Tissue Culture

In chrysanthemum tissue culture, a variety of plant materials can serve as explants for regeneration under controlled, artificial conditions. Key *in vitro* techniques include micropropagation, somatic embryogenesis, embryo rescue, and the cultures of anthers and protoplasts, all instrumental in cultivating chrysanthemums with enhanced traits. The field of tissue culture has been extensively reviewed, providing deep insights into the methodologies and their applications in improving chrysanthemum varieties (Kulus *et al.*, 2013; Teixeira da Silva *et al.*, 2015b; Su *et al.*, 2019). Recent advancements in *in vitro* propagation techniques for chrysanthemum have been comprehensively reviewed by Eisa *et al.* (2022), who outline the current state of *in vitro* methods. This review emphasizes the selection of appropriate explant sources, optimization of medium compositions, and development of innovative methods for disinfecting culture media. It also addresses the use and optimization of phytohormones for inducing shoots, callus, somatic embryos, and roots, along with approaches to mutagenesis and acclimatization procedures. Additionally, it covers the incorporation of nanomaterials in tissue culture media and the use of arbuscular mycorrhizal fungi to enhance plant growth and resilience. Furthermore, Yan *et al.* (2022) provide an in-depth summary of virus-free chrysanthemum production through tissue culture, with a focus on meristem tip culture for producing disease-free plantlets. Refer to Yan *et al.* (2022) for the entire process, from sterilization of explants to the acclimatization of plants. The study also offers practical approaches for maintaining virus-free status in chrysanthemum cultivation. Here, in this chapter, we briefly summarize a few important techniques and their recent advancements that aid chrysanthemum research and cultivation, such as micropropagation, somatic embryogenesis, embryo rescue culture, and somaclonal variation.

4.3.1 Micropropagation

Chrysanthemums are primarily propagated vegetatively through apical cuttings, a simple and cost-effective method (refer to Chapter 06). However, it has some limitations, such as low cutting production rates, seasonal dependency, and vulnerability to viral infections (Rout and Das, 1997). These issues can be addressed using micropropagation with suitable mother plants (Datta, 2020; Miler *et al.*, 2021). The success of *in vitro* chrysanthemum propagation depends on factors such as culture medium composition, growth regulator interactions, explant type and stage, and plant genotype (Hesami *et al.*, 2014). Studies have explored tissue culture techniques for large-scale propagation of *C. morifolium* using innovative regeneration pathways (Lim *et al.*, 2012; Kulus and Zalewska, 2014b; Naing *et al.*, 2014; Chae, 2016; Tung *et al.*, 2018; Kim *et al.*, 2019; Jahan *et al.*, 2021; Miler *et al.*, 2021).

Chrysanthemums can be propagated *in vitro* using various plant parts, including seeds, cuttings, tubers, roots, anthers, pollen, and leaves. Nodal segments and shoot tips have been widely used as explants for plant regeneration through direct organogenesis and somatic embryogenesis Fig. 4.1 (Chitra *et al.*, 2006; Singh *et al.*, 2016; Ghosh *et al.*, 2018). Ray and disc florets have also been explored for culture establishment, shoot initiation, and proliferation (Datta *et al.*, 2000; Misra and Datta, 2007; Padmadevi *et al.*, 2009; Batra *et al.*,

Fig. 4.1. Micropropagation of Chrysanthemum. (A) Nodal segments used as explants and inoculated into MS medium. (B) *In vitro* shoot and root regeneration in chrysanthemum. (Photo courtesy of Floris Verbeek and NC Surakshitha.)

2011), with ray florets showing better performance in terms of organogenesis. Ray florets are particularly useful in chimera management for developing new cultivars. Moreover, alternatives, such as psyllium husk and market sugar, have been identified as cost-effective substitutes for agar and sucrose, reducing tissue culture production costs by more than six times (Pant *et al.*, 2015). A summary of recent research on the optimal cultured medium for chrysanthemum proliferation *in vitro* is provided in the study of Eisa *et al.* (2022).

4.3.2 Somatic embryogenesis

Somatic embryogenesis involves the induction of embryo-like structures from somatic cells. In chrysanthemums, this process can begin either directly from the explant's epidermal cells (Mandal and Datta, 2005) or indirectly through an intermediate callus stage (Tanaka *et al.*, 2000; Shinoyama *et al.*, 2004). Somatic embryos are essential for generating large numbers of plantlets through genetic transformation, induced mutation, or clonal propagation. However, developing a reliable protocol for somatic embryogenesis remains challenging due to the process' labor-intensive, time-consuming, and complex nature. Only a few studies have successfully generated somatic embryos in chrysanthemums (Tanaka *et al.*, 2000; Shinoyama *et al.*, 2004; Naing *et al.*, 2013; Tymoszuk *et al.*, 2014; Ali *et al.*, 2023), and these findings indicate that the efficiency of embryogenesis and plant regeneration is highly dependent on the genotype and the balance of auxins and cytokinins. (Hesami *et al.*, 2020a, b) applied computational models to optimize this process, using "multilayer perceptron (MLP)," an artificial neural network, and support vector regression. Their findings showed that support vector regression was more accurate in predicting embryogenesis outcomes. Additionally, the use of Non-dominated Sorting Genetic Algorithm-II (NSGA-II) significantly improved the rate of embryogenesis and the number of somatic embryos per explant, achieving rates as high as 99% and 56 embryos per explant, respectively. These advanced computational methods, particularly support vector regression and NSGA-II, have proven to be effective and can serve as reliable tools in future plant tissue culture studies to optimize somatic embryogenesis protocols.

In a recent study by Ali *et al.* (2023), somatic embryogenesis was used to exploit somaclonal variation for chrysanthemum improvement. Various explants from the "Dante yellow" variety were cultured on Linsmaier and Skoog (LS) medium with different 2,4-dichlorophenoxyacetic acid (2,4-D) and kinetin (KT) concentrations for callus induction. The best results were achieved with young leaf explants on 9.02 µM 2,4-D and 11.61 µM KT, resulting in larger calli. The highest callus proliferation (0.27 g ± 0.03) occurred with 0.44 µM benzylaminopurine (BAP) and 5.37 µM naphthaleneacetic acid (NAA), while 25 shootlets were regenerated with 0.45 µM 2,4-D. Clone S84 showed significant flower size improvement in field trials, making it a promising candidate for commercialization.

4.3.3 Embryo rescue culture

Most commercial chrysanthemum varieties are vulnerable to pests, diseases, and viruses, which significantly diminish their ornamental quality, as discussed under crop protection in Chapters 09 and 10. In contrast, many wild species possess traits that confer resistance to these biotic and abiotic stresses (Abraham *et al.*, 2004; Liu *et al.*, 2005). Hybridization with these wild species offers practical benefits in enhancing cultivated varieties (Eeckhaut *et al.*, 2007), though cross-incompatibility often leads to embryo abortion, even after successful fertilization.

Embryo rescue is a crucial technique in such instances, allowing viable plants to develop from weak or aborted embryos following interspecific or intergeneric hybridization. This method has enabled the transfer of multiple stress-resistance traits into chrysanthemum cultivars (refer to Table 4.2). By providing an artificial medium that substitutes for the endosperm, embryo rescue supports the development of embryos at critical stages, such as the globular and heart stages, using media developed by Murashige and Skoog (1962) and Norstog (1973). This technique also helps reduce generation time and overcome breeding barriers in interspecific and intergeneric hybrids (Sanders and Ziebur, 1963; Watanabe, 1977; Chandler and Beard, 1978; Haghighi and Ascher, 1988) and is used to culture seeds with underdeveloped endosperm (Myers and Ascher, 1982). In chrysanthemum, the timeline from pollination to seed development and from seed sowing to seedling transplantation can extend up to 21 weeks. Considering the plant's juvenility and the time from flower initiation to anthesis, the total generation time can range from 0.75 to 1.5 years (De Jong, 1981). However, rescuing and culturing heart-stage embryos can substantially shorten this period. For instance, through the rapid generation cycling technique, three generations of chrysanthemums were successfully produced within a year, demonstrating the time-saving potential of embryo rescue culture (Anderson *et al.*, 1990). Recently, Li *et al.* (2024) reported that *Crossostephium chinense*, a salt-tolerant species closely related to Chrysanthemum, was hybridized with three Chrysanthemum plant materials to develop salt-tolerant germplasm. The study found no pre-fertilization barriers in intergeneric hybridization through the examination of pollen germination and tube elongation. Using embryo rescue, the optimal inoculation time was 18 days after pollination, with the best media combination being MS +6 BA 2.0 mg L^{-1} +NAA 0.2 mg L^{-1}. True hybrids were confirmed through both morphological and cytological analyses. This approach shows promise for developing intergeneric hybrids with enhanced salt tolerance from *Crossostephium chinense*.

4.3.4 Protoplast isolation and culture

In chrysanthemum cultivation, protoplast fusion and genetic modification techniques have been utilized to overcome sexual incompatibility between distant

Table 4.2. Various studies using embryo rescue culture for chrysanthemum crop improvement. Author's own table.

Parents	Type of stress	Stress resistance	References
Chrysanthemum morifolium "Zhongshanjingui" (as female) and *Artemisia vulgaris* "Variegata" (mugwort) (as male)	Aphids (*Macrosiphoniella sanborni*)	Higher aphid tolerance compared with maternal parent	Deng *et al*. (2010)
Chrysanthemum cultivar "Zhongshanjingui" (as female) and *Ajania przewalskii* (as male—cold tolerant)	Cold (–21°C)	Tolerance from –17.7 to –21°C	Deng *et al*. (2011)
Trigeneric hybridization: (*Chrysanthemum grandiflorum* ´ *Artemisia vulgaris*) and (*Chrysanthemum crassum* ´ *Crossostephium chinense*)	Aphid (*Macrosiphoniella sanborni*) and salt stress	Higher tolerance than the parents	Deng *et al*. (2012)
Chrysanthemum morifolium "Maoyan" (as female) and *Artemisia japonica* Thunb (as male)	Salt stress	Tolerance was similar to *A. japonica*, and higher than in "Maoyan"	Zhu *et al*. (2013)
Chrysanthemum morifolium "Nannongxiaoli" (as female) and *Artemisia vulgaris* L. "Variegata" (as male)	Aphid (*Macrosiphoniella sanborni*) and Alternaria leaf spot (*Alternaria tenuissima*)	Greater resistance to aphid infestation and Alternaria leaf spot than the "Nannongxiaoli" parent, but less than the "Variegata" parent	Zhu *et al*. (2014)
Crossostephium chinense (as female) and *Chrysanthemum lavandulifolium* (as male)	Aphid (*Macrosiphoniella sanborni*)	Stronger aphid resistance than *C. lavandulifolium*	Zhong *et al*. (2021)

plant species and for precise genome editing to enhance traits such as disease resistance and stress tolerance. Protoplasts, which are single-cell plant cells with their cell wall removed, are instrumental for transient gene expression, functional characterization of plant genes, and genome editing (Subburaj *et al.*, 2016; Lin *et al.*, 2018; Xu *et al.*, 2020). Specifically, protoplast-based transformation using ribonucleoprotein (RNP) complexes helps minimize off-target effects in genetic mutants (Svitashev *et al.*, 2016; Liang *et al.*, 2017b; Zhang *et al.*, 2018).

Protoplast isolation and regeneration methods in chrysanthemum vary considerably, as reported by Sauvadet *et al.* (1990), Zhou *et al.* (2005), and Eeckhaut and Van Huylenbroeck (2011). The key steps involved and the factors influencing these processes are illustrated in Fig. 4.2. Protoplast cultures typically begin with isolation through enzymatic methods. Following isolation, purification is necessary to separate viable protoplasts from undigested material. The viable protoplasts undergo testing for viability before being cultured using

Fig. 4.2. An illustration of the protoplast isolation process, along with the factors influencing protoplast isolation and culture. (Photo courtesy of NC Surakshitha.)

techniques such as the hanging drop method, staining, agar plating, microculture, or the multidrop array technique. Furthermore, isolated protoplasts can be utilized to generate new cultivars through protoplast fusion. This technique involves the fusion of protoplasts from different species, resulting in somatic hybrids that combine desirable traits from both parent plants. This process is facilitated by polyethylene glycol (PEG) and does not require plant hormones for regeneration, as the protoplast behaves like a germinating spore (Chadipiralla *et al.*, 2020). Several factors impact chrysanthemum protoplast regeneration success, including cell density, complex additives (Fujii and Shimizu, 1990), starting material (Lindsay and Ledger, 1993), leaf age, type of osmoticum (Endo *et al.*, 1997), conditioned medium (Zhou *et al.*, 2005), culture type (Eeckhaut and Van Huylenbroeck, 2011), and the use of electrical stimulation or surfactants/antibiotics (Davey *et al.*, 2005; Eeckhaut *et al.*, 2013).

Many studies have focused on optimizing the yield and reproducibility of protoplast isolation and purification protocols (Tudses *et al.*, 2014; Cha *et al.*, 2019). For instance, Adedeji *et al.* (2020) optimized a protocol for isolating protoplasts from the chrysanthemum cv. "White ND," successfully regenerating whole plants from protoplast-derived calli. They achieved a high protoplast yield using 0.5 M mannitol, 1.5% cellulase, and a 4 hour incubation, with media supplements such as sucrose and activated charcoal facilitating rapid growth. Solid Murashige and Skoog (MS) media supplemented with 1 mg l^{-1} 6-BAP and 2 mg l^{-1} NAA promoted microcalli proliferation. However, despite the high efficiency of callus formation, shoot regeneration remained relatively low, indicating the need for further optimization. Eeckhaut *et al.* (2020) advanced chrysanthemum biotechnology by successfully isolating and regenerating protoplasts from *Chrysanthemum morifolium*. This research highlights the protoplast regeneration as a valuable tool for introducing genetic novelties into commercial varieties. Protoplasts, which are cells with their cell walls removed, provide access to somaclonal variation, a source of genetic diversity, enabling the development of improved traits, such as enhanced disease resistance or aesthetic qualities, although its effectiveness depends on the genotype.

4.4 Somaclonal Variation

Chrysanthemum breeders widely use mutation breeding techniques, including X-rays, gamma rays, and chemical mutagens to induce somatic mutations that yield new plant varieties with desirable traits, such as enhanced color and form (Kapadiya *et al.*, 2004; Teixeira da Silva *et al.*, 2015b; Yoosumran *et al.*, 2018; Asoko *et al.*, 2020). Chrysanthemums' high heterozygosity boosts mutation rates, making these techniques particularly effective (Su *et al.*, 2019). Additionally, somaclonal variation that occurs naturally or is induced during tissue culture has been a valuable tool for generating genetic diversity from explants, such as leaves, stems, or petals (Jevremovic *et al.*, 2012; Miler and Zalewska, 2014). A significant challenge in mutation breeding is chimerism, where genetic variability within the same plant leads to instability in mutant plants. This is particularly evident in traits such as floral color and shape, which can vary between the genetic layers of the flower (Vilasini and Latipah, 2000). Tissue culture techniques have been instrumental in overcoming these challenges by enabling the regeneration of plantlets from single cells, thus reducing chimeric characteristics and stabilizing traits (Ahloowalia and Maluszynski, 2001). Tissue culture not only facilitates mutagenesis and genetic transformation but also induces somaclonal variation, leading to valued variations in ploidy levels and flower color in ornamental horticulture (Bush *et al.*, 1976; Miyazaki and Tashiro, 1978; Sutter and Langhans, 1981). The type of explant used significantly impacts the extent of somaclonal variation, with shoot-tip explants typically yielding more true-to-type regenerants compared with adventitious buds (Zalewska *et al.*, 2007).

Altogether, *in vitro* propagation techniques provide an efficient means for multiplying and preserving plants with elite traits. However, regeneration *via* callus or cell suspension can induce somaclonal variation, leading to genetic and morphological changes distinct from the donor plant. Factors such as genotype, explant type, donor plant age, subculture frequency, and medium composition, particularly auxin and cytokinin balance, significantly influence this variation (Gao *et al.*, 2010; Bairu *et al.*, 2011). In chrysanthemums, somaclonal variation has been harnessed to enhance traits such as flower color and growth, with successful outcomes reported from various explants (Khalid *et al.*, 1989; Malaure *et al.*, 1991; Miler and Zalewska, 2014). Somatic mutations in regenerated plants serve as powerful breeding tools, producing high mutation rates and unique morphological changes, including novel ray floret traits derived from ovary-regenerated plants (Miler and Jedrzejczyk, 2018). Protoplast cultures from elite cultivars have also generated plants with unique flower shapes and early flowering, demonstrating the potential of somatic mutation for developing new commercial varieties (Eeckhaut *et al.*, 2020). When combined with irradiation, tissue culture further expands the spectrum of color mutants, enhancing ornamental value (Okamura *et al.*, 2015). This integrated approach highlights the utility of advanced breeding techniques to create unique cultivars with desirable traits. Therefore, the use of somaclonal variation and mutation induction through tissue culture represents a transformative advancement in chrysanthemum breeding. These methods offer a controlled and practical strategy to increase genetic diversity and produce varieties tailored to commercial and aesthetic demands, marking a significant evolution toward precision in chrysanthemum genetic enhancement.

4.5 Genetic Modification

4.5.1 Useful promoters

In an early genetic engineering experiment, Courtney-Gutterson *et al.* (1994) modified the flower color of cv. Moneymaker from pink to pale-pink or white by controlling expression of the chalcone synthase (CHS) gene. *CHS* was regulated by the expression of sense or antisense strands using a constitutive CaMV 35S promoter, combined with the 5′ untranslated region of the chlorophyll a/b binding protein gene (*Cab22L*) from *Petunia*. Despite the modification, when the resulting white-flowered transgenic lines were propagated, some of the flowers retained their original pink coloration. However, Takatsu *et al.* (2000) subsequently reported that expression of the GUS gene under the control of the CaMV 35S promoter in chrysanthemum could not be maintained for long durations, such as 12 months. To solve this problem, constitutive promoters—including those of the tobacco elongation factor 1 α-protein gene (*NtEF1α*) (Aida *et al.*, 2005), *actin* gene (Hong *et al.*, 2016; Kishi-Kaboshi *et al.*, 2019), and parsley *ubiquitin* gene (Kishi-Kaboshi *et al.*, 2019)—have been developed that can stably maintain the expression of the introduced *GUS* gene over a long period of time. For example, Ohmiya *et al.* (2006) used the *NtEF1α* promoter to repress *CmCCD4a*, which encodes carotenoid cleavage dioxygenase, in the white cultivar Sei-Marine by RNA interference (RNAi), thereby modifying the flower color to yellow.

Kishi-Kaboshi *et al.* (2017) introduced mutations using the CRISPR-Cas9 system by targeting multiple copies of a pre-transformed *CpYGFP* gene. In this case, *Cas9* was expressed with the parsley *ubiquitin* promoter and *pea3A* terminator, while the *U6* promoter from *Arabidopsis* was used to express the target sgRNA. Furthermore, Shinoyama *et al.* (2012b) used a bidirectional promoter cassette from the *mannopine synthase 1′* and *2′* genes (Pmas201) to confer sterility and prevent *GM* gene flow. This promoter was used to simultaneously express an insect-resistant gene (modified *cry1Ab* from *Bacillus thuringiensis*) and repress the meiotic homologous recombination gene *CmDMC1* via RNAi. Similarly, the Pmas201 promoter was also used for simultaneous expression of the neomycin phosphotransferase selection marker gene *NPTII* and a transcription activator-like effector nuclease (TALEN) expression vector targeting multiple *CmDMC1* genes (Shinoyama *et al.*, 2020).

Promoters that determine organ- and tissue-specific expression of transgenes are also used in chrysanthemum. For instance, the sucrose synthase gene promoter from tomato is a stamen-specific promoter, and the *InMYB1* promoter from morning glory is a petal-specific promoter used in chrysanthemum (Tanase *et al.*, 2011; Azuma *et al.*, 2016). Similarly, Noda *et al.* (2013) reported that the promoter of the petal-specific chrysanthemum flavanone 3-hydroxylase gene (*CmF3H*) is effective for expression of the flavonoid 3′,5′-hydroxylase gene (*F3′5′H*) in the ray floret petals. Lastly, the promoter of the anthocyanin 3-acyltransferase gene from *Perilla* leaves and the *CHS* promoter from rose have also been shown to be effective in chrysanthemum for the production of delphinidin by *F3′5′H* expression (Brugliera *et al.*, 2013; Noda *et al.*, 2013).

4.5.2 Application to improve flower architecture

Chrysanthemums have a variety of capitulum shapes, mainly due to differences in the length, width, shape, and number of ray florets. In Japan, breeding was actively conducted during the Edo period (1603–1868), and a wide variety of flower shapes can be seen in the classical chrysanthemum cultivars that were produced. The flower organogenesis or *ABCE* genes of chrysanthemum have been isolated by Shchennikova *et al.* (2003, 2004), including *APETALA1* (*AP1*)-like *CDM111* (Class A), *APETALA3* (*AP3*)-like *CDM19* and *CDM115* (Class B), *AGAMOUS*-like *CDM37* (Class C), and *SEPALLTA3*-like *CDM44* (Class E), as well as the *APETALA1/FRUITFUL*-like genes *CDM8* and*CDM41*. Aida *et al.* (2008) suppressed expression of the Class C *AGAMOUS*-like gene (*CAG*) by introducing a tobacco *EF1α* promoter: *CAG* antisense:*nos* terminator cassette, which resulted in a *CAG*-suppressed Sei-Marine cultivar with an increased number of small petals in addition to ray petals.

The shape of the ray florets, which greatly affects the shape of the capitulum, can be classified as flat, spoon shaped, or tubular (Song *et al.*, 2018b). Recently, candidate genes that may control the shape of ray petals have been characterized. For example, the transcription factor *BRI1-EMS-SUPPRESSOR 1* (*CmBES1*) has been shown to play an important role in the boundary formation of the ray petal. When *CmBES1* was overexpressed in chrysanthemum cv. Jimba, the morphology of the ray floret changed from flat to spoon shaped or tubular, and the expression of homeotic genes, such as *CYCLOIDEA-like 4* (*CmCYC4*) and *CUP-SHAPED COTYLEDON-like 2* (*CmCUC2*), was repressed (Cheng *et al.*, 2020). Also, *CmCYC4* and *CmTCP20*, members of the *TEOSINTE BRANCHED1/CYCLOIDEA/PROLIFERATING CELL FACTOR* (*TCP*) gene family, are thought to be involved in cell elongation. Overexpression of *CmTCP20* in cv. Jimba resulted in longer ray petals and correspondingly larger flower diameters (Wang *et al.*, 2019). Expansion of flower diameter has also been observed in *Chrysanthemum lavandulifolium* overexpressing *CmCYC2c* (Huang *et al.*, 2016). Ding *et al.* (2019) reported candidate genes involved in forming the hooked tip of ray florets. Their findings suggested that *CmYAB1* (a homolog of the organ polarity gene *YABBY*), a homolog of *ATHB15* (involved in the repression of vascular development), and the homolog of *CUC3* (a gene involved in boundary size control) are involved in hook formation at the ray petal tip.

4.5.3 Flower pigments and their biosynthetic genes

In chrysanthemum, the pigments responsible for flower color are flavonoids, including anthocyanins, carotenoids, and chlorophylls (Ohmiya, 2018). Among these, the two major anthocyanins in petals, cyanidin-3-*O*-(6″-*O*-malonyl) glucoside and cyanidin-3-*O*-(3″,6″-*O*-dimalonyl) glucoside, interact with flavone 7-malonylglucosides to produce red, pink, and magenta colors (Nakayama *et al.*, 1997; Noda *et al.*, 2017). Given their importance in flower color, transcriptome analysis has been used to identify candidate genes encoding enzymes responsible for anthocyanin biosynthesis, including CHS, chalcone isomerase, flavanone 3-hydroxylase, dihydroflavonol 4-reductase, anthocyanidin synthase, UDP-glucose:flavonoid 3-*O*-glucosyltransferase, in cv. Purple Reagan, Fenditan, Chuju, and Jinbeidahong (Hong *et al.*, 2015; Liu *et al.*, 2016; Yue *et al.*, 2018; Dong *et al.*, 2020). *Cm3MaT1* encoding malonyl-CoA:anthocyanidin 3-*O*-glucoside-6″-*O*-malonyltransferase and *Cm3MaT2* encoding malonyl-CoA:anthocyanidin 3-*O*-glucoside-3″,6″-*O*-dimalonyltrans ferase have also been identified (Suzuki *et al.*, 2004). Flower color is influenced by the interaction of anthocyanins with flavones; in this regard, a flavone synthase gene (*FNSII*) responsible for the biosynthesis of flavone aglycone has been reported in the wild species *C. indicum* (Jiang *et al.*, 2019).

Studies have also characterized transcription factors that regulate the expression of genes involved in flavonoid biosynthesis, identifying the role of MYB and basic helix-loop-helix (bHLH) transcription factors. For example, CmMYB1, a repressor of anthocyanin biosynthesis, was identified by the analysis of ARTI-Purple, a mutant line of chrysanthemum cv. Argus (Sung *et al.*, 2013). R3-type CmMYB7, which represses the natural red mutant of the white-colored cultivar Jimba, is also considered to be a negative regulator of anthocyanin biosynthesis (Xiang *et al.*, 2019). Based on the analysis of cv. Purple Reagan, CmMYB5-1, CmMYB6, and CmbHLH24 were identified as positive regulators, and CmMYB7-1 as a negative regulator of anthocyanin biosynthesis (Hong *et al.*, 2015). Lastly, CmbHLH2, cloned from cv. Zaoxiaoju, has been shown to form a complex with CmMYB6 and induce gene expression of dihydroflavonol 4-reductase (Liu *et al.*, 2015; Xiang *et al.*, 2015).

The major carotenoids in chrysanthemum petals are xanthophylls, luteins, and lutein 5,6-epoxides, which are responsible for the yellow coloration (Kishimoto *et al.*, 2004; Kishimoto and Ohmiya, 2009). An accumulation of yellow carotenoids and red anthocyanins gives rise to an orange petal color. Some genes involved in carotenoid biosynthesis and degradation have been isolated, including *LCYE* encoding lycopene ε-cyclase, which is mainly involved in the biosynthesis of luteins, and *CCD4a* encoding carotenoid cleavage dioxygenase 4a, which is involved in lutein degradation (Kishimoto and Ohmiya, 2006; Ohmiya *et al.*, 2006).

Chlorophylls are not well synthesized in petals, and when chloroplasts are lost during flowering, petals that do not synthesize anthocyanins and/or carotenoids turn white. In chrysanthemum cultivars with a green flower color, both mesophyll cells and epidermal cells retain chlorophyll-containing chloroplasts. The expression levels of several genes involved in chlorophyll biosynthesis, such as magnesium-protoporphyrin IX chelatase (*CRD*), are significantly upregulated in green petals compared with white petals. Furthermore, CONSTANS-like 16, EAR motif protein, and MYB113-like have been identified as transcription factors associated with chlorophyll content in petals; in particular, MYB113-like is specific to white petals and presumed to be a repressor of chlorophyll biosynthesis (Ohmiya *et al.*, 2017).

4.5.4 Development of novel flower colors

Hybridization and mutation breeding have resulted in increased diversity in chrysanthemum flower color. However, conventional methods cannot be

Fig. 4.3. Blue chrysanthemum produced by genetic modification. Left, wild-type cv. Sei-Elma; right, recombinant with blue flower color generated by introducing the *Campanula F3′5′H* and *Clitoria A3′5′GT* genes. (Photo courtesy of Naonobu Noda.)

used to introduce colors that are not naturally found in chrysanthemums or related wild species. Therefore, genetic engineering and genome editing technologies that can introduce, suppress, or knock out the function of specific genes may be used to obtain novel flower colors without changing any other traits. For instance, metabolic engineering has been used to introduce genes for delphinidin biosynthesis and for facilitating the 3′,5′-*O*-diglucosylation reaction of delphinidin in order to produce flowers with bluish colors, which did not previously exist in chrysanthemum (Brugliera *et al.*, 2013; Noda *et al.*, 2013, 2017). In short, if the anthocyanins in chrysanthemum petals are changed from cyanidin-type to delphinidin-type, the flower color can be modified from red or pink to purple or violet. Furthermore, delphinidin-type anthocyanins with *O*-glucosylation at the 3′ and 5′ positions of the B-ring can interact with naturally existing flavone 7-malonylglucosides to produce a blue-colored flower (Fig. 4.3). Accumulation of delphinidin-type anthocyanins in *Chrysanthemum* petals can be achieved in two ways. First, by repressing the endogenous flavonoid 3′-hydroxylase gene (*F3′H*) and expressing the pansy-derived flavonoid 3′,5′-hydroxylase gene (*F3′5′H*) under the control of the *RhCHS* promoter (Brugliera *et al.*, 2013); and second, by expressing *Campanula F3′5′H* under the control of the *CmF3H* promoter (Noda *et al.*, 2013).

4.6 Transformation Protocols

Transgenic approach can be used to transfer genes to a host plant from any source and to repress or enhance gene expression in a programmable manner. In the last two decades, the application of transgenic molecular breeding has been extensively exploited in chrysanthemum to introduce foreign genes using *Agrobacterium*-mediated transformation and biolistic transformation (rarely) (Fig. 4.4) and has led to outstanding achievements related to floral attributes, plant architecture, postharvest flower longevity, and biotic and abiotic stress tolerance (Azadi *et al.*, 2016; Noman *et al.*, 2017; Su *et al.*, 2019) (Table 4.3).

Various *Agrobacterium*-mediated transformation systems have been reported for *Chrysanthemum morifolium* (Shinoyama *et al.*, 2012a; Hesami *et al.*, 2020a). It is also possible to produce transformants of wild species such as *Chrysanthemum seticuspe* and *C. lavandulifolium* (Higuchi *et al.*, 2013; Huang *et al.*, 2016). To establish a successful transformation system, it is necessary to consider the conditions required by the *Agrobacterium tumefaciens* strain used for infection and the type and concentration of antibiotic or herbicide used both for selecting transformed cells and for subsequently eliminating *Agrobacterium*, in addition to considering the best regeneration system. The *NPTII* gene can be used as a selection marker gene, coupled with kanamycin, paromomycin, or geneticin as the selection antibiotic. Selection by hygromycin using the hygromycin phosphotransferase (HPT) gene as a marker is also possible. In addition, Brugliera *et al.* (2013) used a chlorsulfuron resistance gene encoding mutated acetolactate synthase (ALS) as a selection marker gene, selected transformants resistant to the herbicide chlorsulfuron (DuPont Glean), and eliminated *Agrobacterium* with antibiotic timentin.

Later, we describe in detail a modified version of the protocol reported by Aida *et al.* (2004) that has been successfully used to introduce blue coloration via *F3′5′H* and *A3′5′GT* expression, yellow coloration via *CmCCD4a* repression, and for genome editing of the CpYGFP transgenic chrysanthemum mentioned earlier. It has also been used to transform the wild species *C. seticuspe* to analyze the flowering regulator anti-florigen (Higuchi *et al.*, 2013).

In vitro shoots used for infection material are grown in 1/2 MS medium (pH 4.0–4.2) containing 3% sucrose solidified with 0.2% gellan gum. *Agrobacterium* transformed with a binary vector containing the transgene is cultured for 3 days at 22°C on solid Plusgrow II (Nacalai tesque, Inc.) medium containing appropriate antibiotics. Approximately 1 cm^2 of *Agrobacterium* is scraped off and suspended in 10 mM magnesium sulfate solution containing 1% Tween 20 and 0.1 mM 3′,5′-dimethoxy-4′-hydroxyacetophenone (acetosyringone). After 4–6 weeks of growth, leaves from the *in vitro* shoots are cut into 5 mm^2 pieces and soaked in the *Agrobacterium* suspension for 20 min. Excess *Agrobacterium* solution is removed, and then the leaf explants are placed on coexistence medium [MS solid medium containing 1.0 mg l^{-1} benzyladenine (BA), 2.0 mg l^{-1} of naphthalenacetic acid (NAA), 1 g l^{-1} casamino acid, and 0.1 mM acetosyringone]. Leaf explants cocultured at 22°C in the dark for 3 days are then transferred to selection medium (MS containing 3% sucrose,

Fig. 4.4. Process of chrysanthemum transformation. (A) *In vitro* shoot culture. (B) Preparation of leaf explants for infection. (C) Selection culture of infected leaf explants. (D) Shoot regeneration from leaf explants. (E) Growth of shoots confirmed to be resistant to paromomycin. (F) Acclimation of shoots. (G) Growth in greenhouse. (H) Flowering of transformant and wild-type chrysanthemum. (Photo courtesy of Naonobu Noda.)

1.0 mg l^{-1} BA, 0.5 mg l^{-1} NAA, 75 or 100 mg l^{-1} paromomycin, and 300 mg^{-1} carbenicillin) and incubated at 20°C under low light. The explants are subcultured every 2 weeks on selection medium. At 8–10 weeks after infection, small shoots regenerating from the callused explants are then transferred to shoot medium [MS containing 3% sucrose, 1.0 mg l^{-1} BA, 0.1 mg l^{-1} of NAA, 0.1 mg l^{-1} gibberellin A3 (GA_3), 300 mg l^{-1} carbenicillin, and 25 mg l^{-1} paromomycin].

Table 4.3. Summary of transgenic studies showing abiotic and biotic stress tolerance in chrysanthemum. Author's own table.

Stress tolerance	Transgene	Antibiotics	Antibiotics used for selection	Promoter	References
Abiotic stress					
Low temperature	*DREB1A*	Cefotaxime	Kanamycin	*nos*, CaMV *35S* with enhancer, *rd29A*	Hong *et al.* (2006)
Low temperature or dehydration stress	*DREB1A*	Cefotaxime	Kanamycin	*rd29A*, CaMV *35S*	Ma *et al.* (2010)
Salinity stress	*CcSOS1*	–	Hygromycin	CaMV *35S*	An *et al.* (2014)
Dehydration	*CmWRKY1*	–	Hygromycin	CaMV 35S	Fan *et al.* (2016)
Salinity and drought	*ClCBF1*	–	Kanamycin	CaMV *35S*	Gao *et al.* (2018)
Biotic stress					
Insect resistance (*Bacillus thuringiensis* -endotoxin)	*nptII, bt*	Cefotaxime	Kanamycin	CaMV *35S*	Dolgov *et al.* (1995)
Insect resistance, stress	*nptII, HPT, AFP, bt, rolC, CHS*	Cefotaxime	Kanamycin, hygromycin	*nos*, CaMV *35S*, *mas*	Dolgov *et al.* (1997)
Tomato spot wilt virus	*nptII*, TSWV *N*	Carbenicillin	Paromomycin	CaMV *35S*	Sherman *et al.* (1998)
Gray mold	*nptII, RCC2*	Cefotaxime	Kanamycin	CaMV *35S*	Takatsu *et al.* (1999)
Insect resistance	modified *cry1C*	Cefotaxime, Ticarcillin	Kanamycin	CaMV *35S*	De Jong (2001)
Thrips (*Frankliniella occidentalis*)	Potato *multicystatin*	Cefotaxime, Vancomycin	Kanamycin	*UEP1*	Annadana *et al.* (2002)

Continued

Table 4.3. Continued

Stress tolerance	Transgene	Antibiotics	Antibiotics used for selection	Promoter	References
Chrysanthemum stunt viroid	*nptII, pacI*	Cefotaxime	Kanamycin	CaMV *35S* with leader	Ishida *et al.* (2002)
	nptII, OsMADS1	Cefotaxime	Kanamycin	*nos*, CaMV *35S*	Jeong *et al.* (2002)
Tobacco budworm (*Helicoverpa armigera*)	*nptII*, non-modified *cry1Ab*	Cefotaxime	Geneticin	CaMV *35S*	Shinoyama *et al.* (2002)
Virus and viroid resistance	*nptII, pacI*	Cefotaxime	Kanamycin	CaMV *35S* with leader	Toguri *et al.* (2003)
Insect resistance	*nptII, HPT, modified cry1Ab*	Cefotaxime	Geneticin	*nos*, CaMV *35S*	Shinoyama *et al.* (2003)
Chrysanthemum stunt viroid	*pacI*	Cefotaxime	Kanamycin	CaMV *35S*	Ogawa *et al.* (2005)
tobacco budworm (*Helicoverpa armigera*) and common cutworm (*Spodoptera litura*)	*nptII, modified cry1Ab*	Cefotaxime	Geneticin	*mas*	Shinoyama and Mochizuki (2006)
Virus and viroid resistance	*ORF5 encoding the CVB coat protein*	Cefotaxime	Kanamycin	Double CaMV 35S with enhancer	Skachkova *et al.* (2006)
Insect resistance	modified *cry1Ca,* engineered *seven-domain protein inhibitor* gene, strawberry *linalool synthase* gene	–	–	*rbcS1*	Visser *et al.* (2007)

Insect resistance	*nptII*, modified *cry1Ab*	Cefotaxime	Geneticin	*mas*	Shinoyama *et al*. (2008)
Alternaria leaf spot	*hpaGX*$_{\infty}$	Carbenicillin	Kanamycin	CaMV *35S*	Xu *et al*. (2010)
Gray mold	*nptII, HPT, CaXMT1, CaMXMT1, CaDXMT1*	Cefotaxime	Kanamycin	*nos*, CaMV *35S*	Kim *et al*. (2011b)
Beet army worms (*Spodoptera exigu*) and cotton aphids (*Aphis gossypii*)	*nptII, HPT, CaXMT1, CaMXMT1, CaDXMT1*	Cefotaxime	Kanamycin	*nos*, CaMV *35S*	Kim *et al*. (2011a)
Insect resistance	*nptII, CmDMC1-RNAi, modified cry1Ab*	Cefotaxime	Geneticin	*nos, mas*	Shinoyama *et al*. (2012b)
Cucumber mosaic virus	coat protein (CP) gene of cucumber mosaic virus			CaMV *35S*	Kumar *et al*. (2012)
Septoria leaf spot	*Rice chitinase gene*	Phosphinothricin, hygromycin	Cefotaxime, hygromycin	CaMV *35S*	Sen *et al*. (2013)
Aphids	*CmWRKY48*	Carbenicillin	Carbenicillin, hygromycin	CaMV *35S*	Li *et al*. (2015)
Aphids	*Sea anemone equistatin*			*RbcS*	Valizadeh *et al*. (2013)

Shoots carrying the *NPTII* gene are selected by placing leaf explants that have grown to about 5 mm in length on leaf test medium (MS containing 3% sucrose, 1.0 mg l^{-1} BA, 2.0 mg l^{-1} NAA, 300 mg l^{-1} carbenicillin, and 100 mg l^{-1} paromomycin) for 1–2 weeks. After this period, leaf explants that are green and initiate callus formation contain the *NPTII* gene. The paromomycin-resistant shoots are transferred to 1/2 MS medium containing 3% sucrose and 400 mg l^{-1} carbenicillin and grown until they are about 4 cm long. At this stage, cuttings with three to four leaves are taken, rooting agent (oxyberon) is applied to the cut end, and the cutting is transferred to culture soil. After 3–4 weeks, cuttings with root are cold-treated at 4–5°C for a minimum period of 4 weeks and then potted for flowering in a greenhouse or incubator (Fig. 4.4).

4.6.1 Transformation for resistance to abiotic and biotic (viroids, viruses, fungi, and pests) stress

Recent advances in chrysanthemum transformation have shown promise for enhancing resistance to abiotic and biotic stresses, such as viroids, viruses, fungi, and pests. For example, overexpression of *CmHSFA4* has increased the salt tolerance of chrysanthemum by limiting Na^+ accumulation and maintaining the K^+ concentration, aligning with the function of ion transporters *CmSOS1* and *CmHKT2* (Li *et al.*, 2018). Overexpressing the *C. lavandulifolium ClCBF1* gene in chrysanthemum cv. “White Snow” has improved its salt and drought tolerance (Gao *et al.*, 2018). Furthermore, manipulation of *CmCPL1* (encodes RNAPII CTD phosphatase-like 1) expression—either overexpression or knock-down—enhanced or reduced heat tolerance of chrysanthemum, respectively (Qi *et al.*, 2018). *WRKY* transcription factor genes in chrysanthemum have also been linked to responses to salt (Li *et al.*, 2015; Liang *et al.*, 2017a; Wang *et al.*, 2017), drought (Fan *et al.*, 2016), disease (Fan *et al.*, 2015), and aphid resistance (Li *et al.*, 2015), demonstrating their broad role in stress resilience.

4.6.2 Viroid-free chrysanthemums

Chrysanthemums are almost exclusively propagated by taking cuttings from mother stock plants. If cuttings are infected by viruses and viroids, disease symptoms will appear during production and cause extensive damage. For this reason, methods for the selection and breeding of chrysanthemum resistant to chrysanthemum stunt viroid (CSVd) have been developed (Matsushita *et al.*, 2012; Nabeshima *et al.*, 2012, 2014). Tissue culture techniques have also been developed to remove CSVd from infected chrysanthemum plantlets. For example, CSVd is not detected by PCR when the shoot apical meristem of a plantlet stored at 4°C for 3 months is cultured on a medium containing 50 or 100 mg l^{-1} of the antiviral drug ribavirin (Savitri *et al.*, 2013). In addition, for chrysanthemums infected with CSVd or chrysanthemum chlorotic mottle viroid (CChMVd) and stored at 4°C for 4 weeks, preservation of the shoot tips in plant vitrification solution 2 (MS medium supplemented with 30% glycerol,

15% e.g. 15% DMSO, and 0.4 M sucrose), followed by freezing in liquid nitrogen and regeneration, leads to viroid-free chrysanthemum (Jeon *et al.*, 2016).

4.7 Artificial Intelligence to Optimize *In Vitro* Cultural Conditions

In chrysanthemum biotechnology, regeneration systems are fundamental to rapid clonal propagation and *Agrobacterium*-mediated genetic transformation. However, callus induction, somatic embryogenesis, and plantlet regeneration from various explants or isolated protoplasts depend on a combination of the type and amount of appropriate plant growth regulators (Teixeira da Silva, 2003, 2004). Furthermore, the culture conditions often vary among cultivars and breeding lines with different genetic backgrounds. Therefore, trial and error and statistical processing are generally necessary to find the optimal culture conditions for systems not investigated previously.

Modeling and optimization using mathematical models and algorithms of artificial intelligence (AI) have recently been reported as a way to optimize *in vitro* culture conditions (Hesami *et al.*, 2020a). For example, in a case study, Hesami *et al.* (2019a) used MLP-NSGA-II analysis to derive optimal conditions for the sterilization of leaf explants of chrysanthemum cv. Hornbill Dark. These authors found that, among various variables assessed, 14 min of disinfection by 1.62% sodium hypochlorite was optimal. The same authors also assessed various mathematical models combined with NSGA-II and found that medium containing 2.16 μM 6-BAP, 0.14 μM indole-3-butyric acid, 0.29 mM phloroglucinol, and 87.63 mM sucrose is optimal for shoot growth, while medium containing 1.53 mg l^{-1} 2,4-D, 1.67 mg l^{-1} BAP, 13.74g l^{-1} sucrose, 57.2g l^{-1} glucose, and 0.39g l^{-1} fructose is optimal for somatic embryogenesis (Hesami *et al.*, 2019b, c, 2020b). Firsov *et al.* (2020) attempted to produce transformants of cv. White Snowdon expressing five enzyme genes needed for the biosynthesis of artemisinin, a drug with antimalarial activity. In order to succeed in their attempt to produce valuable chemical compounds in chrysanthemum plantlets, in addition to establishing a transformation system, it was necessary to select an appropriate host and corresponding transgenes and optimize the method of controlling the transgenes by promoters and terminators. In this regard, AI technology that facilitates the optimization of these conditions would be beneficial.

4.8 Conclusions and Future Prospects

Elite chrysanthemum varieties can be propagated *in vitro* to produce sufficient quantities of planting materials that are true to type, addressing the challenges of conventional propagation methods. Newly bred chrysanthemum varieties can be introduced rapidly due to the accelerated rate of *in vitro* propagation. By optimizing media composition, using plant growth regulators effectively, and employing appropriate *in vitro* culture methodologies assisted by AI to optimize cultural conditions, the genetic engineering of chrysanthemums can be significantly improved. When using protoplasts as source explants, it is important to recognize that *in vitro* regeneration is the most common bottleneck in protoplast-related work. Improvements in protoplast regeneration protocols are crucial for advancing genome editing. Contemporary tools, such as advanced sequencing techniques, metabolite profiling, and *in situ* hybridization, offer promising prospects for gaining insights into the regeneration process. These insights could help expand the group of genotypes amenable to protoplast regeneration. Protoplasts can serve as excellent targets for genome editing techniques, such as CRISPR/Cas9, particularly when combined with DNA-free technologies, such as RNP complexes. However, the spontaneous formation of somaclones can introduce variation among the regenerants that is not due to the genome editing itself. Therefore, results from genome editing via protoplasts should be interpreted with caution to distinguish between intended edits and somaclonal variations.

For the diploid *Chrysanthemum* species *C. nankingense* (Song *et al.*, 2018a) and *C. seticuspe* (Hirakawa *et al.*, 2019), the whole genome has been sequenced. The resulting databases, Chrysanthemum Genome Database (http://www.amwayabrc.com/) and Mum Genome And Resource Database Entry (Mum GARDEN, http://mum-garden.kazusa.or.jp/), are publicly available. It is expected that further analysis of chrysanthemum genes, which are useful in horticulture, will be conducted in the future. For example, genome-wide association studies (GWAS) will facilitate the development of markers that can be used for breeding hexaploid chrysanthemums (Su *et al.*, 2019; Sumitomo *et al.*, 2019). Genetic engineering and genome editing techniques, which allow control over the function of specific genes, are also expected to advance. However, because *Chrysanthemum morifolium* is a hexaploid plant, introducing mutations in all copies of the targeted gene is more challenging than in diploid plants. In seed-propagating diploid plants, homozygous mutations can be introduced, and the genes used for "genome editing" can be removed by crossing. However, this method cannot be used in hexaploid chrysanthemums, which are vegetatively propagated. Therefore, it is necessary to develop technologies to efficiently knock out multiple copies of genes, remove the transgene used for genome editing, or introduce mutations without transformation.

Chrysanthemums with enhanced insect resistance have been developed by introducing a modified *cry1Ab* gene from *Bacillus thuringiensis*. These genetically modified chrysanthemums have been shown to cross with wild *C. japonense* var. *japonense* and other chrysanthemum cultivars (Shinoyama *et al.*, 2008). Similarly, genes responsible for blue coloration in chrysanthemum petals, such as *F3′5′H* and *A3′5′GT*, are transmitted to progeny when blue-flowered transformants are crossed with *C. japonense* var. *japonense* (Aida *et al.*, 2018, 2020). To cultivate and distribute transgenic chrysanthemums without impacting biodiversity, it is essential to introduce sterility. This would prevent the transfer of transgenes to other cultivars and wild chrysanthemum species, thereby safeguarding natural genetic diversity.

Chrysanthemum cv. "Kingfisher" has been reported to be male sterile (Wang *et al.*, 2018). However, there are no reports of cultivars exhibiting both male and female sterility. Shinoyama *et al.* (2012c) suppressed pollen maturation in cv. "Yamate-Shiro" by introducing a modified melon ethylene receptor gene (*CmETR1/H69A*). When these transformants flowered under low-temperature conditions, there was a slight restoration of pollen fertility. The authors also attempted to suppress the meiosis-specific homologous recombination gene (*CmDMC1*) using RNA interference (RNAi) to produce sterile chrysanthemums (Shinoyama *et al.*, 2012c). The resulting transformants did not form mature pollen and remained stably male- and female-sterile at temperatures between 10 and 35°C. Additionally, Shinoyama *et al.* (2020) reported a technique for introducing mutations into multiple *CmDMC1* genes using a TALEN expression vector. All six *CmDMC1* genes were disrupted in the genome-edited transformants, leading to the successful production of chrysanthemums with irreversible sterility. These advancements in sterility technologies hold promise for the future commercial production and distribution of transgenic chrysanthemums with valuable traits, without the risk of transgene transfer to other cultivars and wild species.

References

Abraham, E.M., Huang, B., Bonos, S.A. and Meyer, W.A. (2004) Evaluation of drought resistance for Texas bluegrass, Kentucky bluegrass, and their hybrids. *Crop Science* 44, 1746–1753.

Adedeji, O.S., Naing, A.H. and Kim, C.K. (2020) Protoplast isolation and shoot regeneration from protoplast-derived calli of chrysanthemum cv. White ND. *Plant Cell Tissue and Organ Culture* 141, 571–581. DOI: 10.1007/s11240-020-01816-3.

Agarwal, A., Singh, S., Malhotra, E.V., Meena, D.P.S. and Tyagi, R.K. (2019) *In vitro* conservation and cryopreservation of clonally propagated horticultural species. In: Rajasekharan, P.E. and Rao, V.R. (eds) *Conservation and Utilization of Horticultural Genetic Resources, Springer*. pp. 529–578. DOI: 10.1007/978-981-13-3669-0_18.

Ahloowalia, B.S. and Maluszynski, M. (2001) Induced mutation – a new paradigm in plant breeding. *Euphytica* 118, 167–173.

Aida, R., Ohira, K., Tanaka, Y., Yoshida, K., Kishimoto, S. *et al.* (2004) Efficient transgene expression in chrysanthemum, *Dendranthema grandiflorum* (Ramat.) kitamura, by using the promoter of a gene for chrysanthemum chlorophyll-a/b--binding protein. *Breeding Science* 54, 51–58.

Aida, R., Nagaya, S., Yoshida, K., Kishimoto, S., Shibata, M. *et al.* (2005) Efficient transgene expression in *Chrysanthemum morifolium* Ramat., with the promoter of a gene for tobacco elongation factor 1 a protein. *Japan Agricultural Research Quarterly* 39, 269–274.

Aida, R., Komano, M., Saito, M., Nakase, K. and Murai, K. (2008) Chrysanthemum flower shape modification by suppression of chrysanthemum-*AGAMOUS* gene. *Plant Biotechnology* 25, 55–59.

Aida, R., Douzono, M., Yoshioka, S., Noda, N., Tsuda, M. *et al.* (2018) Inheritance of bluish flower color of transgenic chrysanthemum by interspecific hybrids. *Japan Agricultural Research Quarterly* 52, 339–345.

Aida, R., Noda, N., Yoshioka, S., Douzono, M., Tsuda, M. *et al.* (2020) Crossability between transgenic blue chrysanthemums and the wild chrysanthemum species *Chrysanthemum japonense* var. *japonense*. *Japan Agricultural Research Quarterly* 54, 335–340.

Ali, S., Raza, S., Shahzad, S., Batool, T.S., Abdullah, A. *et al.* (2023) Regeneration of chrysanthemum (*Chrysanthemum morifolium*) via somatic embryogenesis and screening of clones for agronomic traits. *Plant Cell, Tissue and Organ Culture (PCTOC)* 153, 657–667. DOI: 10.1007/s11240-023-02506-6.

Anderson, N.O., Peter, D.A., Widmer, R.E. and Luby, J.J. (1990) Rapid generation cycling of chrysanthemum using laboratory seed development and embryo rescue techniques. *Journal of the American Society for Horticultural Science* 115(2), 329–336.

An, J., Song, A., Guan, Z., Jiang, J., Chen, F. *et al.* (2014) The over-expression of *Chrysanthemum crassum* CcSOS1 improves the salinity tolerance of chrysanthemum. *Molecular Biology Reports* 41, 4155–4162.

Annadana, S., Kuipers, G., Visser, P.B., de Kogel, W.J., Udayakumar, M. *et al.* (2002) Expression of potato multicystatin in florets of chrysanthemum and assessment of resistance to western flower thrips (*Frankliniella occidentalis*). *Acta Horticulturae* 572, 121–129.

Asoko, N., Ruamrungsri, S., Yoosumran, V. and Saetiew, K. (2020) Improvement of *Dendranthemum grandiflora* cv. Canter with colchicine *in vitro*. *International Journal of Agricultural Technology* 16, 237–246.

Azadi, P., Bagheri, H., Nalousi, A.M., Nazari, F. and Chandler, S.F. (2016) Current status and biotechnological advances in genetic engineering of ornamental plants. *Biotechnology Advances* 34, 1073–1090.

Azuma, M., Morimoto, R., Hirose, M., Morita, Y., Hoshino, A. *et al.* (2016) A petal-specific *InMYB1* promoter from Japanese morning glory: A useful tool for molecular breeding of floricultural crops. *Plant Biotechnology Journal* 14, 354–363.

Bairu, M., Aremu, A. and Van Staden, J. (2011) Somaclonal variation in plants: Causes and detection methods. *Plant Growth Regulation* 63(1), 147–173. DOI: 10.1007/s10725-010-9554-x.

Bajaj, Y.P.S. (1986) *In vitro* preservation of genetic resources - techniques and problems. In: *International Symposium on Nuclear Techniques and in Vitro Culture for Plant Improvement*, IAEA-FAO, Vienna, pp. 43–57.

Bajaj, Y.P.S. (1993) A suggested method for *in vitro* long-term storage at 4°C of chrysanthemum and petunia germplasm. *Plant Tissue Culture* 3, 57–58.

Batra, A., Verma, A.K. and Banerji, B.K. (2011) Standardization of regeneration protocol for small flower chrysanthemum (*Chrysanthemum morifolium*) cultivars. *Indian Journal of Agricultural Science* 81(11), 1078–1081.

Bi, W.L., Pan, C., Liu, J. and Wang, Q.C. (2016) Greenhouse performance, genetic stability and biochemical compounds in *Chrysanthemum morifolium* 'Hangju' plants regenerated from cryopreserved shoot tips. *Acta Physiologiae Plantarum* 38, 268. DOI: 10.1007/s11738-016-2288-2.

Brugliera, F., Tao, G.Q., Tems, U., Kalc, G., Mouradova, E. *et al.* (2013) Violet/blue chrysanthemums—metabolic engineering of the anthocyanin biosynthetic pathway results in novel petal colors. *Plant and Cell Physiology* 54, 1696–1710.

Cha, O.K., Lee, J., Lee, H.S. and Lee, H. (2019) Optimized protoplast isolation and establishment of transient gene expression system for the Antarctic flowering plant Colobanthus quitensis (Kunth) Bartl. *Plant Cell Tissue and Organ Culture* 138(3), 603–607.

Bush, S., Earle, E. and Langhans, R. (1976) Plantlets from petal segments, petal epidermis, and shoot tips of the periclinal chimera, *Chrysanthemum morifolium* 'Indianapolis'. *American Journal of Botany* 63(6), 729–737. DOI: 10.1002/j.1537-2197.1976.tb11863.x.

Chadipiralla, K., Gayathri, P., Rajani, V. and Reddy, P.V.B. (2020) Plant tissue culture and crop improvement. In: Roychowdhury, R., Choudhury, S., Hasanuzzaman, M. and Srivastava, S. (eds) *Sustainable Agriculture in the Era of Climate Change*. Springer International Publishing, Cham, Switzerland, pp. 391–412. DOI: 10.1007/978-3-030-45669-6.

Chae, S.C. (2016) Influence of auxin concentration on *in vitro* rooting of *Chrysanthemum morifolium* Ramat. *Biosciences Biotechnology Research Asia* 13, 833–837.

Chandler, J.M. and Beard, B.H. (1978) Sunflower interspecific hybridization using embryo culture. In: *The 8th International Sunflower Conference*, Minneapolis, pp. 510–515.

Cheng, P., Liu, Y., Yang, Y., Chen, H., Cheng, H. *et al.* (2020) CmBES1 is a regulator of boundary formation in chrysanthemum ray florets. *Horticulture Research* 7, 1–12.

Chitra, R., Arulmozhiyan, R., Jawaharlal, M., Indhumathi, K. and Vadivel, E. (2006) Rapid multiplication of chrysanthemum (*Dendranthema grandiflom* Tzvelev) through *in vitro* culture. *Journal of Ornamental Horticulture* 9(1), 58–60.

Datta, S.K., Mandal, A.K. and Saxena, M. (2000) Direct organogenesis from ray and disc florets of a newly evolved chlorophyll variegated chrysanthemum (*Chrysanthemum morifolium*). *Indian Journal of Agricultural Science* 71(10), 655–657.

Datta, S.K. (2020) Induced mutations: Technological advancement for development of new ornamental varieties. *Nucleus* 63, 119–129.

Courtney-Gutterson, N., Napoli, C., Lemieux, C., Morgan, A., Firoozabady, E. *et al.* (1994) Modification of flower color in florist's chrysanthemum: Production of a white-flowering variety through molecular genetics. *Bio/Technology* 12, 268–271.

Davey, M., Anthony, P., Power, J. and Lowe, K. (2005) Plant protoplasts: Status and biotechnological perspectives. *Biotechnology Advances* 23, 131–171. DOI: 10.1016/j.biotechadv.2004.09.008.

De Jong, J. (1981) Effects of irradiance and juvenility on the selection of chrysanthemums. *Euphytica* 30(2), 493–500.

De Jong, J. (2001) Transgenic Dendranthema (*Chrysanthemum*). In: Bajaj, Y.P.S. (ed.) *Transgenic Crops III (Biotechnology in Agriculture and Forestry)*, Vol. 38. Springer, Berlin, Heidelberg, New York, pp. 84–94.

Deng, Y., Chen, S., Chen, F., Cheng, X. and Zhang, F. (2011) The embryo rescue derived intergeneric hybrid between chrysanthemum and *Ajania przewalskii* shows enhanced cold tolerance. *Plant Cell Reports* 30, 2177–2186. DOI: 10.1007/s00299-011-1123-x.

Deng, Y., Jiang, J., Chen, S., Teng, N., Song, A. *et al.* (2012) Combination of multiple resistance traits from wild relative species in chrysanthemum via trigeneric hybridization. *PLoS ONE* 7(8), e44337. DOI: 10.1371/journal.pone.0044337.

Deng, Y.M., Chen, S.M., Lu, A.M., Chen, F.D., Tang, F.P. *et al.* (2010) Production and characterization of the intergeneric hybrids between *Dendranthema morifolium* and Artemisia vulgaris exhibiting enhanced resistance to chrysanthemum aphid (*Macrosiphoniella sanborni*). *Planta* 231, 693–703.

Ding, L., Zhao, K., Zhang, X., Song, A., Su, J. *et al.* (2019) Comprehensive characterization of a floral mutant reveals the mechanism of hooked petal morphogenesis in *Chrysanthemum morifolium*. *Plant Biotechnology Journal* 17, 2325–2340.

Dolgov, S.V., Mityshkina, T.U., Rukavtsova, E.B. and Buryanov, Y.I. (1995) Production of transgenic plants of *Chrysanthemum morifolium* Ramat. with the gene of Bacillus thuringiensis δ-endotoxin. *Acta Horticulturae* 441, 23–28.

Dolgov, S.V., Mitiouchkina, T.Y. and Skryabin, K.G. (1997) Agrobacterial transformation of *Chrysanthemum*. *Acta Horticulturae* 447, 329–333.

Dong, W., Li, M., Li, Z., Li, S., Zhu, Y. *et al.* (2020) Transcriptome analysis of the molecular mechanism of chrysanthemum flower color change under short-day photoperiods. *Plant Physiology and Biochemistry* 146, 315–328.

Eeckhaut, T. and Van Huylenbroeck, J. (2011) Development of an optimal culture system for callogenesis of *Chrysanthemum indicum* protoplasts. *Acta Physiologiae Plantarum* 33(4), 1547–1551. DOI: 10.1007/s11738-010-0660-1.

Eeckhaut, T., Keyser, E.D., Huylenbroeck, J.V., Riek, J.D. and Bockstaele, E.V. (2007) Application of embryo rescue after interspecific crosses in the genus Rhododendron. *Plant Cell Tissue and Organ Culture* 89, 29–35.

Eeckhaut, T., Van Houtven, W., Bruznican, S., Leus, L. and Van Huylenbroeck, J. (2020) Somaclonal variation in *chrysanthemum* x *morifolium* protoplast regenerants. *Frontiers in Plant Science* 11, 607171. DOI: 10.3389/fpls.2020.607171.

Eeckhaut, T., Lakshmanan, P., Deryckere, D., Van Huylenbroeck, J. and Bockstaele, E. (2013) Progress in plant protoplast research. *Planta* 238, 991–1003.

Eisa, E.A., Tilly-Mándy, A., Honfi, P., Shala, A.Y. and Gururani, M.A. (2022) Chrysanthemum: A comprehensive review on recent developments on *in vitro* regeneration. *Biology (Basel)* 11(12), 1774. DOI: 10.3390/biology11121774.

Endo, M., Fujii, N., Fujita, S. and Inada, I. (1997) Improvement of plating efficiency on the mesophyll protoplast culture of chrysanthemum (*Dendranthema* × *grandiflorum* (Ram.) Kitam). *Plant Biotechnology* 14(1), 81–83. DOI: 10.5511/plantbiotechnology.14.81.

Fan, Q., Song, A., Xin, J., Chen, S., Jiang, J. *et al.* (2015) *CmWRKY15* facilitates *Alternaria tenuissima* infection of chrysanthemum. *PLoS ONE* 10, e0143349. DOI: 10.1371/journal.pone.0143349.

Fan, Q., Song, A., Jiang, J., Zhang, T., Sun, H. *et al.* (2016) *CmWRKY1* enhances the dehydration tolerance of chrysanthemum through the regulation of ABA-associated genes. *PLoS ONE* 11, e0150572.

Firsov, A., Mitiouchkina, T., Shaloiko, L., Pushin, A., Vainstein, A. *et al.* (2020) *Agrobacterium*-mediated transformation of chrysanthemum with artemisinin biosynthesis pathway genes. *Plants* 9, 537.

Fujii, Y. and Shimizu, K. (1990) Callus formation from mesophyll protoplasts of pyrethrum (*Chrysanthemum coccineum*). *Plant Tissue Culture Letters* 7(2), 111–113.

Fukai, S. and Oe, M. (1989) Cold storage of shoot tips of chrysanthemum [*Dendranthema grandiflorum* (Ramat.) Kitamura]. *Plant Tissue Culture Letters* 6(1), 10–13.

Fukai, S., Goi, M. and Tanaka, M. (1994) The chimeric structure of the apical dome of chrysanthemum (*Dendranthema grandiflorum* (Ramat.) Kitam.) is affected by cryopreservation. *Scientia Horticulturae* 57, 347–351.

Fukai, S., Morii, M. and Oe, M. (1988) Storage of chrysanthemum [*Dendranthema grandiflorum* (Ramat.) Kitammura] plantlets *in vitro*. *Plant Tissue Culture Letters* 5(1), 20–25.

Gao, W., He, M., Liu, J., Ma, X., Zhang, Y. *et al.* (2018) Overexpression of *chrysanthemum lavandulifolium ClCBF1* in *Chrysanthemum morifolium* 'White Snow' improves the level of salinity and drought tolerance. *Plant Physiology and Biochemistry* 124, 50–58.

Gao, X., Yang, D., Cao, D., Ao, M., Sui, X. *et al.* (2010) *In vitro* micropropagation of Freesia hybrida and the assessment of genetic and epigenetic stability in regenerated plantlets. *Journal of Plant Growth Regulation* 29, 257–267.

Ghosh, S., Naika, M.B.N., Nishani, S., Shiragur, M. and Bhat, A. (2018) Studies on somatic embryognesis in chrysanthemum cv. Marigold using root and leaf as explants. *International Journal of Current Microbiology Applied Science* 7(8), 3965–3971.

Haghighi, K.R. and Ascher, P.D. (1988) Fertile, intermediate hybrids between *Phaseolus vulgaris* and *P. acutifolius* from congruity backcrossing. *Sexual Plant Reproduction* 1, 51–58.

Halmagyi, A., Fischer-Kluver, G., Mix-Wagner, G. and Schumacher, H.M. (2004) Cryopreservation of *Chrysanthemum morifolium* (*Dendranthema grandiflora* Ramat.) using different approaches. *Plant Cell Reports* 22, 371–375.

Hesami, M., Naderi, R., Tohidfar, M., Miler, N., Iwona, J. *et al.* (2014) Effects of growth regulators and genotypes on Pyrethrum *in vitro*. *Scientia Horticulturae* 8, 162–166.

Hesami, M., Naderi, R. and Tohidfar, M. (2019a) Modeling and optimizing *in vitro* sterilization of chrysanthemum via multilayer perceptron-non-dominated sorting genetic algorithm-II (MLP-NSGAII). *Frontiers in Plant Science* 10, 282.

Hesami, M., Naderi, R. and Tohidfar, M. (2019b) Modeling and optimizing medium composition for shoot regeneration of chrysanthemum via radial basis function-non-dominated sorting genetic algorithm-II (RBF-NSGAII). *Scientific Reports* 9, 1–11.

Hesami, M., Naderi, R., Tohidfar, M. and Yoosefzadeh-Najafabadi, M. (2019c) Application of adaptive neuro-fuzzy inference system-non-dominated sorting genetic algorithm-II (ANFIS-NSGAII) for modeling and optimizing somatic embryogenesis of chrysanthemum. *Frontiers in Plant Science* 10, 869.

Hesami, M., Alizadeh, M., Naderi, R. and Tohidfar, M. (2020a) Forecasting and optimizing Agrobacterium-mediated genetic transformation via ensemble model-fruit fly optimization algorithm: A data mining approach using chrysanthemum databases. *PLoS ONE* 15, e0239901.

Hesami, M., Naderi, R., Tohidfar, M. and Yoosefzadeh–Najafabadi, M. (2020b) Development of support vector machine-based model and comparative analysis with artificial neural network for modeling the plant tissue culture procedures: Effect of plant growth regulators on somatic embryogenesis of chrysanthemum, as a case study. *Plant Methods* 16, 112. DOI: 10.1186/s13007-020-00655-9.

Higuchi, Y., Narumi, T., Oda, A., Nakano, Y., Sumitomo, K. *et al.* (2013) The gated induction system of a systemic floral inhibitor, antiflorigen, determines obligate short-day flowering in chrysanthemums. *Proceedings of the National Academy of Sciences* 110, 17137–17142.

Hirakawa, H., Sumitomo, K., Hisamatsu, T., Nagano, S., Shirasawa, K. *et al.* (2019) De novo whole-genome assembly in *Chrysanthemum seticuspe*, a model species of chrysanthemums, and its application to genetic and gene discovery analysis. *DNA Research* 26, 195–203.

Hong, B., Tong, Z., Ma, N., Kasuga, M., Yamaguchi-Shinozuka, K. *et al.* (2006) Expression of the Arabidopsis DREB1A gene in transgenic chrysanthemum enhances tolerance to low temperature. *Journal of Horticultural Science & Biotechnology* 81, 1002–1008.

Hong, J.K., Suh, E.J., Kwon, S.J., Lee, S.B., Kim, J.A. *et al.* (2016) Promoter of chrysanthemum actin confers high-level constitutive gene expression in Arabidopsis and chrysanthemum. *Scientia Horticulturae* 211, 8–18.

Hong, Y., Tang, X., Huang, H., Zhang, Y. and Dai, S. (2015) Transcriptomic analyses reveal species-specific light-induced anthocyanin biosynthesis in chrysanthemum. *BMC Genomics* 16, 1.

Hosoki, T. (1989) *In vitro* storage of *Chrysanthemum morifolium* at room temperature. *Plant Tissue Culture Letters* 6(2), 85–87.

Huang, D., Li, X., Sun, M., Zhang, T., Pan, H. *et al.* (2016) Identification and characterization of CYC-like genes in regulation of ray floret development in *Chrysanthemum morifolium*. *Frontiers in Plant Science* 7, 1633.

Ishida, I., Tukahara, M., Yoshioka, M., Ogawa, T., Kakitani, M. *et al.* (2002) Production of anti-virus, viroid plants by genetic manipulations. *Pest Management Science* 58, 1132–1136.

Jahan, M.T., Islam, M.R., Islam, S.S., Das, P., Islam, M.M. *et al.* (2021) Clonal propagation of *Chrysanthemum morifolium* ramat using various explants obtained from field grown plants. *GSC Biological and Pharmaceutical Science* 16, 087–093.

Jeon, S.M., Arun, M., Lee, S.Y. and Kim, C.K. (2015) Application of encapsulation-vitrification in combination with air dehydration enhances cryotolerance of *Chrysanthemum morifolium* shoot tips. *Scientia Horticulturae* 194, 91–99.

Jeon, S.M., Naing, A.H., Kim, H.H., Chung, M.Y., Lim, K.B. *et al.* (2016) Elimination of chrysanthemum stunt viroid and chrysanthemum chlorotic mottle viroid from infected chrysanthemum by cryopreservation. *Protoplasma* 253, 1135–1144.

Jevremovic, S., Subotic, A., Miljkovic, D., Trifunovic, M., Petric, M. *et al.* (2012) Clonal fidelity of chrysanthemum cultivars after long term micropropagation by stem segment culture. *Acta Horticulturae* 961, 211–216.

Jeong, J.H., Chakrabarty, D., Kim, S.J. and Paek, K.Y. (2002) Transformation of Chrysanthemum (*Dendranthema grandiflorum* Kitamura) cv. Cheonsu by constitutive expression of rice OsMADS1 gene. *Journal of the Korean Society for Horticultural Science* 42, 382–386.

Jiang, Y., Ji, X., Duan, L., Ye, P., Yang, J. *et al.* (2019) Gene mining and identification of a flavone synthase II involved in flavones biosynthesis by transcriptomic analysis and targeted flavonoid profiling in *Chrysanthemum indicum* L. *Industrial Crops and Products* 134, 244–256.

Kapadiya, D.B., Chawla, S.L., Patel, A.I. and Ahlawat, T.R. (2004) Exploitation of variability through mutagenesis in chrysanthemum (*Chrysanthemum morifolium* Ramat.) var Maghi. *The Bioscan: An International Quarterly Journal of Life Sciences* 9, 1799–1804.

Khalid, N., Davey, M. and Power, J. (1989) An assessment of somaclonal variation in *Chrysanthemum morifolium*: The generation of plants of potential commercial value. *Scientia Horticulturae* 38, 287–294. DOI: 10.1016/0304-4238(89)90076-9.

Kim, S.H., Kim, Y.S., Jo, Y.D., Kang, S.Y., Ahn, J.W. *et al.* (2019) Sucrose and methyl jasmonate modulate the expression of anthocyanin biosynthesis genes and increase the frequency of flower-color mutants in chrysanthemum. *Scientia Horticulturae* 256, 108602.

Kim, Y.S., Lim, S., Kang, K.K., Jung, Y.J., Lee, Y.H. *et al.* (2011a) Resistance against beet armyworms and cotton aphids in caffeine-producing transgenic chrysanthemum. *Plant Biotechnology* 28, 393–395.

Kim, Y.S., Lim, S., Yoda, H., Choi, C.S., Choi, Y.E. *et al.* (2011b) Simultaneous activation of salicylate production and fungal resistance in transgenic chrysanthemum producing caffeine. *Plant Signaling & Behavior* 6, 409–412.

Kishi-Kaboshi, M., Aida, R. and Sasaki, K. (2017) Generation of gene-edited *Chrysanthemum morifolium* using multicopy transgenes as targets and markers. *Plant and Cell Physiology* 58, 216–226.

Kishimoto, S. and Ohmiya, A. (2006) Regulation of carotenoid biosynthesis in petals and leaves of chrysanthemum (*Chrysanthemum morifolium*). *Physiologia Plantarum* 128, 436–447.

Kishimoto, S. and Ohmiya, A. (2009) Studies on carotenoids in the petals of Compositae plants. *Journal of the Japanese Society for Horticultural Science* 78, 263–272.

Kishi-Kaboshi, M., Aida, R. and Sasaki, K. (2019) Parsley ubiquitin promoter displays higher activity than the CaMV 35S promoter and the chrysanthemum actin 2 promoter for productive, constitutive, and durable expression of a transgene in *Chrysanthemum morifolium*. *Breeding Science* 69, 536–544.

Kishimoto, S., Maoka, T., Nakayama, M. and Ohmiya, A. (2004) Carotenoid composition in petals of chrysanthemum (*Dendranthema grandiflorum* (Ramat.) Kitamura). *Phytochemistry* 65, 2781–2787.

Kulus, D. and Zalewska, M. (2014a) Cryopreservation as a tool used in long-term storage of ornamental species – a review. *Scientia Horticulturae* 168, 88–107. DOI: 10.1016/j.scienta.2014.01.014.

Kulus, D. and Zalewska, M. (2014b) *In vitro* plant recovery from alginate-encapsulated *Chrysanthemum grandiflorum* (Ramat.) Kitam. shoot tips. Propag. Ornam. *Plants* 14, 3–12.

Kulus, D., Mikuła, A. and Zalewska, M. (2013) Cryopreservation: An efficient tool to combat genetic erosion phenomenon in agriculture and horticulture. In: *Presented at the IVth International Scientific Symposium for Ph.D. Students and Students of Agricultural Colleges – Innovative researches for the future of agriculture and rural areas development*, Bydgoszcz, Poland, p. 61.

Kulus, D., Rewers, M., Serocka, M. and Mikuła, A. (2019) Cryopreservation by encapsulation-dehydration affects the vegetative growth of chrysanthemum but does not disturb its chimeric structure. *Plant Cell Tissue and Organ Culture* 138, 153–166. DOI: 10.1007/s11240-019-01614-6.

Kulus, D., Serocka, M. and Mikuła, A. (2018) Effect of various preculture and osmotic dehydration conditions on cryopreservation efficiency and morphogenetic response of chrysanthemum shoot tips. *Acta Scientiarum Polonorum Hortorum Cultus* 17(1), 139–147. DOI: 10.24326/asphc.2018.1.13.

Kumar, S., Raj, S., Sharma, A. and Varma, H. (2012) Genetic transformation and development of Cucumber mosaic virus resistant transgenic plants of *Chrysanthemum morifolium* cv. Kundan. *Scientia Horticulturae* 134, 40–45.

Li, H., Zhong, J., Guo, Z.Y., Wang, Y.X., Lu, Y.F. *et al.* (2024) Embryo age and culture medium determine the efficacy of embryo rescue in hybridizations between *Crossostephium chinense* (L.) Makino and *Chrysanthemum* spp. *Acta Horticulturae*.

In: *Proceedings of the IV Asian Horticultural Congress – AHC2023*, ISHS. DOI: 10.17660/ActaHortic.2024.1404.94.

Li, P., Song, A., Gao, C., Jiang, J., Chen, S. *et al.* (2015) The over-expression of a chrysanthemum WRKY transcription factor enhances aphid resistance. *Plant Physiology and Biochemistry* 95, 26–34.

Liang, Q., Wu, Y., Wang, K., Bai, Z., Liu, Q. *et al.* (2017a) Chrysanthemum WRKY gene *DgWRKY5* enhances tolerance to salt stress in transgenic chrysanthemum. *Scientific Reports* 7, 4799.

Liang, Z., Chen, K., Li, T., Zhang, Y., Wang, Y. *et al.* (2017b) Efficient DNA-free genome editing of bread wheat using CRISPR/Cas9 ribonucleoprotein complexes. *Nature Communications* 8, 14261.

Li, F., Zhang, H., Zhao, H., Gao, T., Song, A. *et al.* (2018) Chrysanthemum *CmHSFA4* gene positively regulates salt stress tolerance in transgenic chrysanthemum. *Plant Biotechnology Journal* 16, 1311–1321.

Li, J.W., Hosokawa, M., Nabeshima, T., Motoki, K., Yamada, H. *et al.* (2019) Cryopreservation of viroid-infected chrysanthemum shoot tips. *Scientia Horticulturae* 244, 1–9. DOI: 10.1016/j.scienta.2018.09.004.

Lim, K.B., Kwon, S.J., Lee, S.I., Hwang, Y.J. and Naing, A.H. (2012) Influence of genotype, explant source, and gelling agent on *in vitro* shoot regeneration of chrysanthemum. *Horticulture Environment and Biotechnology* 53, 329–335.

Lin, H., Chen, J.C. and Fang, S.C. (2018) A protoplast transient expression system to enable molecular, cellular, and functional studies in Phalaenopsis orchids. *Frontiers in Plant Science* 9, 843.

Lindsay, G.C. and Ledger, S.E. (1993) A protoplast to plant system for the chrysanthemum (*Dendranthema zawadskii* x *D. grandiflora*). *Plant Cell Reports* 12(5), 278–280. DOI: 10.1007/bf00237135.

Ma, C., Hong, B., Wang, T., Yang, Y.J., Tong, Z. *et al.* (2010) DREB1A regulon expression in rd29A: DREB1A transgenic chrysanthemum under low temperature or dehydration stress. *Journal of Horticultural Science & Biotechnology* 85, 503–510.

Liu, H., Sun, M., Du, D., Pan, H., Cheng, T. *et al.* (2016) Whole-transcriptome analysis of differentially expressed genes in the ray florets and disc florets of *Chrysanthemum morifolium*. *BMC Genomics* 17, 398.

Liu, J.H., Xu, X.Y. and Deng, X.X. (2005) Intergeneric somatic hybridization and its application to crop genetic improvement. *Plant Cell, Tissue and Organ Culture* 82, 19–44.

Liu, X., Xiang, L., Yin, X., Grierson, D., Li, F. *et al.* (2015) The identification of a MYB transcription factor controlling anthocyanin biosynthesis regulation in *Chrysanthemum* flowers. *Scientia Horticulturae* 194, 278–285.

Liu, Z. and Gao, S.L. (2010) The studies on germplasm conservation of chrysanthemum cinerariifolium (Trev.) Vis. *Pharmaceutical Biotechnology* 17(3), 237–239. (in Chinese with English abstract).

Malaure, R., Barclay, G., Power, J. and Davey, M. (1991) The production of novel plants from florets of *Chrysanthemum morifolium* using tissue culture 1. Shoot regeneration from ray florets and somaclonal variation exhibited by the regenerated plants. *Journal of Plant Physiology* 139, 8–13. DOI: 10.1016/s0176-1617(11)80156-2.

Mandal, A.K.A. and Datta, S.K. (2005) Direct somatic embryogenesis and plant regeneration from ray florets of *Chrysanthemum*. *Biologia Plantarum* 49(1), 29–33.

Martinez-Montero, M.E. and Harding, K. (2015) Cryobionomics: Evaluating the concept in plant cryopreservation. In: Barh, D., Khan, M. and Davies, E. (eds) *PlantOmics: The Omics of Plant Science*. Springer, New Delhi, pp. 655–682.

Matsushita, Y., Aoki, K. and Sumitomo, K. (2012) Selection and inheritance of resistance to chrysanthemum stunt viroid. *Crop Protection* 35, 1–4.

Martín, C. and González-Benito, E. (2005) Survival and genetic stability of *Dendranthema grandiflora* Tzvelev shoot apices after cryopreservation by vitrification and encapsulation-dehydration. *Cryobiology* 51, 281–289. DOI: 10.1016/j.cryobiol.2005.08.001.

Miler, N., Iwona, J., Jakubowski, S. and Winiecki, J. (2021) Ovaries of chrysanthemum irradiated with high-energy photons and high-energy electrons can regenerate plants with novel traits. *Agronomy* 11, 1111.

Misra, P. and Datta, S.K. (2007) Standardization of *in vitro* protocol in chrysanthemum cv. Madam E Roger for development of quality planting material and to induce genetic variability using gamma-radiation. *Indian Journal of Biotechnology* 6(1), 121–124.

Miyazaki, S. and Tashiro, Y. (1978) Tissue culture of *Chrysanthemum morifolium* Ramat. III. Variation in Chromosome number and flower colour of plants regenerated from different parts of shoots *in vitro*. *Bulletin of Shizuoka Agricultural Experiment Station* 44, 13–31.

Miler, N. and Zalewska, M. (2014) Somaclonal variation of chrysanthemum propagated *in vitro* from different explant types. *Acta Scientiarum Polonorum, Hortorum Cultus* 13, 69–82.

Miler, N. and Jedrzejczyk, I. (2018) Chrysanthemum plants regenerated from ovaries: A study on genetic and phenotypic variation. *Turkish Journal of Botany* 42, 289–297.

Murashige, T. and Skoog, F. (1962) A revised medium for rapid growth and bioassays with tobacco tissue cultures. *Physiologia Plantarum* 15, 473–497.

Myers, P.J. and Ascher, P.D. (1982) Culture of North American orchids from seed. *HortScience* 17(4), 550.

Nabeshima, T., Hosokawa, M., Yano, S., Ohishi, K. and Doi, M. (2012) Screening of *Chrysanthemum* cultivars with resistance to chrysanthemum stunt viroid. *Journal of the Japanese Society for Horticultural Science* 81, 285–294.

Nabeshima, T., Hosokawa, M., Yano, S., Ohishi, K. and Doi, M. (2014) Evaluation of chrysanthemum stunt viroid (CSVd) infection in newly-expanded leaves from CSVd-inoculated shoot apical meristems as a method of screening for CSVd-resistant chrysanthemum cultivars. *The Journal of Horticultural Science and Biotechnology* 89, 29–34.

Naing, A.H., Kim, C.K., Yun, B.J., Jin, J.Y. and Lim, K.B. (2013) Primary and secondary somatic embryogenesis in chrysanthemum cv. Euro. *Plant Cell, Tissue and Organ Culture* 112, 361–368.

Naing, A.H., Jeon, S.M., Han, J.S., Lim, S.H., Lim, K.B. *et al.* (2014) Factors influencing *in vitro* shoot regeneration from leaf segments of chrysanthemum. *Comptes Rendus-Biologies* 337, 383–390.

Nakayama, M., Koshioka, M., Shibata, M., Hiradate, S., Sugie, H. *et al.* (1997) Identification of cyanidin 3-O-(3″, 6″-O-dimalonyl-β-glucopyranoside) as a flower pigment of chrysanthemum (*Dendranthema grandiflorum*). *Bioscience, Biotechnology, and Biochemistry* 61, 1607–1608.

Noda, N., Aida, R., Kishimoto, S., Ishiguro, K., Fukuchi-Mizutani, M. *et al.* (2013) Genetic engineering of novel bluer-colored chrysanthemums produced by accumulation of delphinidin-based anthocyanins. *Plant and Cell Physiology* 54, 1684–1695.

Noda, N., Yoshioka, S., Kishimoto, S., Nakayama, M., Douzono, M. *et al.* (2017) Generation of blue chrysanthemums by anthocyanin B-ring hydroxylation and glucosylation and its coloration mechanism. *Science Advances* 3, e1602785.

Noman, A., Aqeel, M., Deng, J., Khalid, N., Sanaullah, T. *et al.* (2017) Biotechnological advancements for improving floral attributes in ornamental plants. *Frontiers in Plant Science* 8, 530.

Norstog, K. (1973) New synthetic medium for the culture of premature barley embryos. *In Vitro* 5(4), 307–308.

Ogawa, T., Toguri, T., Kudoh, H., Okamura, M., Momma, T. *et al.* (2005) Double-stranded RNA specific ribonuclease confers tolerance against chrysanthemum stunt viroid and tomato spotted wilt virus in transgenic chrysanthemum plants. *Breeding Science* 55, 49–55.

Ohmiya, A. (2018) Molecular mechanisms underlying the diverse array of petal colors in chrysanthemum flowers. *Breeding Science* 68, 119–127.

Ohmiya, A., Kishimoto, S., Aida, R., Yoshioka, S. and Sumitomo, K. (2006) Carotenoid cleavage dioxygenase (CmCCD4a) contributes to white color formation in chrysanthemum petals. *Plant Physiology* 142, 1193–1201.

Ohmiya, A., Sasaki, K., Nashima, K., Oda-Yamamizo, C., Hirashima, M. *et al.* (2017) Transcriptome analysis in petals and leaves of chrysanthemums with different chlorophyll levels. *BMC Plant Biology* 17, 202.

Okamura, M., Hase, Y., Furusawa, Y. and Tanaka, A. (2015) Tissue-dependent somaclonal mutation frequencies and spectra enhanced by ion beam irradiation in *Chrysanthumum*. *Euphytica* 202, 333–343. DOI: 10.1007/s10681-014-1220-3.

Padmadevi, K., Jawaharlal, M., Kannan, M. and Ganga, M. (2009) Standardization of protocol for *in vitro* plant regeneration from ray florets of chrysanthemum (*Chrysanthemum grandiflora* Tzvelev). *Journal of Ornamental. Horticulture* 12(4), 239–244.

Pant, M., Lal, A. and Jain, R. (2015) A simple cost effective method for mass propagation of *Chrysanthemum morifolium* and antibacterial activity assessment of *in vitro* raised plantlets. *Journal of Applied Pharmaceutical Science* 5(7), 103–111.

Preil, W. and Hoffmann, M. (1985) *In vitro* storage in chrysanthemum breeding and propagation. In: Schäfer-Menuhr, A. (ed.) *In Vitro Techniques*. Martinus Nijhoff, Dordrecht, pp. 161–165.

Qi, Y., Liu, Y., Zhang, Z., Gao, J., Guan, Z. *et al.* (2018) The over-expression of a chrysanthemum gene encoding an RNA polymerase II CTD phosphatase-like 1 enzyme enhances tolerance to heat stress. *Horticulture Research* 5, 37.

Reed, B.M. (2017) Plant cryopreservation: A continuing requirement for food and ecosystem security. *In Vitro Cellular & Developmental Biology - Plant* 285–288. DOI: 10.1007/s11627-017-9851-4.

Rout, G.R. and Das, P. (1997) Recent trends in the biotechnology of chrysanthemum: A critical review. *Scientia Horticulturae* 69, 239–257.

Sanders, M.E. and Ziebur, N.K. (1963) Artificial culture of embryos. In: Maheshwari, P. (ed.) *Recent Advances in the Embryology of Angiosperms*. International Society of Plant Morphology, University of Delhi, Delhi, India, pp. 297–325.

Sauvadet, M.A., Brochard, P. and Boccon-Gibod, J.A. (1990) Protoplast-to-plant system in chrysanthemum: Differential responses among several commercial clones. *Plant Cell Reports* 8, 692–695.

Savitri, W.D., Park, K.I., Jeon, S.M., Chung, M.Y., Han, J.S. *et al.* (2013) Elimination of chrysanthemum stunt viroid (CSVd) from meristem tip culture combined with prolonged cold treatment. *Horticulture, Environment, and Biotechnology* 54, 177–182.

Sen, S., Kumar, S., Ghani, M. and Thakur, M. (2013) Agrobacterium mediated genetic transformation of chrysanthemum (*Dendranthema grandiflora* Tzvelev) with rice chitinase gene for improved resistance against *Septoria obesa*. *Plant Pathology Journal* 12, 1–10.

Shchennikova, A.V., Shulga, O.A., Angenent, G.C. and Skryabin, K.G. (2003) Genetic regulation of inflorescence development in chrysanthemum. *Doklady Biological Sciences* 391, 368–370.

Shchennikova, A.V., Shulga, O.A., Immink, R., Skryabin, K.G. and Angenent, G.C. (2004) Identification and characterization of four Chrysanthemum MADS-box genes, belonging to the APETALA1/FRUITFULL and SEPALLATA3 subfamilies. *Plant Physiology* 134, 1632–1641.

Shinoyama, H. and Mochizuki, A. (2006) Insect resistant transgenic chrysanthemum [*Dendranthema* × *grandiflorum* (Ramat.) Kitamura]. *Acta Horticulturae* 714, 177–184.

Sherman, J.M., Moyer, J.W. and Daub, M.E. (1998) Tomato spot wilt virus resistance in chrysanthemum expressing the viral nucleocapsid gene. *Plant Disease* 82, 407–414.

Shinoyama, H., Mochizuki, A., Komano, M., Nomura, Y. and Nagai, T. (2003) Insect resistance in transgenic chrysanthemum [*Dendranthema* × *grandiflorum* (Ramat.) Kitamura] by the introduction of a modified δ-endotoxin gene of *Bacillus thuringiensis*. *Breeding Science* 53, 359–367.

Shinoyama, H., Komano, M., Nomura, Y. and Nagai, T. (2002) Introduction of delta-endotoxin gene of Bacillus thuringiensis to chrysanthemum [*Dendranthema* × *grandiflorum* (Ramat.) Kitamura] for insect resistance. *Breeding Science* 52, 43–50.

Shinoyama, H., Nomura, Y., Tsuchiya, T. and Kazuma, T. (2004) A simple and efficient method for somatic embryogenesis and plant regeneration from leaves of Chrysanthemum (*Dendranthema grandiflora* (Ramat.) Kitamura). *Plant Biotechnology* 21, 25–30.

Shinoyama, H., Mochizuki, A., Nomura, Y. and Kamada, H. (2008) Environmental risk assessment of genetically modified Chrysanthemums containing a modified cry1Ab gene from *Bacillus thuringiensis*. *Plant Biotechnology* 25, 17–29.

Shinoyama, H., Aida, R., Ichikawa, H., Nomura, Y. and Mochizuki, A. (2012a) Genetic engineering of chrysanthemum (*Chrysanthemum morifolium*): Current progress and perspectives. *Plant Biotechnology* 29, 323–337.

Shinoyama, H., Ichikawa, H., Saitoh-Nakashima, M., Saito, M., Aida, R. *et al.* (2012b) Introduction of male sterility to GM chrysanthemum plants to prevent transgene flow. *Acta Horticulturae* 937, 41.

Shinoyama, H., Sano, T., Saito, M., Ezura, H., Aida, R. *et al.* (2012c) Induction of male sterility in transgenic chrysanthemums (*Chrysanthemum morifolium* Ramat.) by expression of a mutated ethylene receptor gene, Cm-ETR1/H69A, and the stability of this sterility at varying growth temperatures. *Molecular Breeding* 29, 285–295.

Shinoyama, H., Ichikawa, H., Nishizawa-Yokoi, A., Skaptsov, M. and Toki, S. (2020) Simultaneous TALEN-mediated knockout of chrysanthemum *DMC1* genes confers male and female sterility. *Scientific Reports* 10, 1–14.

Singh, R., Raghunath, S., Nirmal, M., Hussain, W.S., Sarkar, H.K. *et al.* (2016) Estimating the efficiency of different explants for direct *in vitro* multiple shoots development in chrysanthemum. *International Journal of Agriculture Environment Biotechnology* 9(3), 345–352.

Skachkova, T.S., Mitiouchkina, T.Y., Taran, S.A. and Dolgov, S.V. (2006) Molecular biology approach for improving chrysanthemum resistance to virus B. *Acta Horticulturae* 714, 185–192.

Song, C., Liu, Y., Song, A., Dong, G., Zhao, H. *et al.* (2018a) The *Chrysanthemum nankingense* genome provides insights into the evolution and diversification of chrysanthemum flowers and medicinal traits. *Molecular Plant* 11, 1482–1491.

Sumitomo, K., Shirasawa, K., Isobe, S., Hirakawa, H., Hisamatsu, T. *et al.* (2019) Genome-wide association study overcomes the genome complexity in autohexaploid chrysanthemum and tags SNP markers onto the flower color genes. *Scientific Reports* 9, 1–9.

Song, X., Zhao, X., Fan, G., Gao, K., Dai, S. *et al.* (2018b) Genetic analysis of the corolla tube merged degree and the relative number of ray florets in chrysanthemum (Chrysanthemum × morifolium Ramat.). *Scientia Horticulturae* 242, 214–224.

Subburaj, S., Chung, S.J., Lee, C., Ryu, S.M., Kim, D.H. *et al.* (2016) Site-directed mutagenesis in *Petunia hybrid* protoplast system using direct delivery of purified recombinant Cas9 ribonucleoproteins. *Plant Cell Reports* 35, 1535–1544.

Su, J., Jiang, J., Zhang, F., Liu, Y., Ding, L. *et al.* (2019) Current achievements and future prospects in the genetic breeding of chrysanthemum: A review. *Horticulture Research* 6, 1–19.

Sung, S.Y., Kim, S.H., Velusamy, V., Lee, Y.M., Ha, B.K. *et al.* (2013) Comparative gene expression analysis in a highly anthocyanin pigmented mutant of colorless *Chrysanthemum*. *Molecular Biology Reports* 40, 5177–5189.

Sutter, E. and Langhans, R. (1981) Abnormalities in chrysanthemum regenerated from long-term cultures. *Annals of Botany* 48, 559–568. DOI: 10.1093/oxfordjournals.aob.a086161.

Suzuki, H., Nakayama, T., Yamaguchi, M. and Nishino, T. (2004) cDNA cloning and characterization of two *Dendranthema* × *morifolium* anthocyanin malonyltransferases with different functional activities. *Plant Science* 166, 89–96.

Svitashev, S., Schwartz, C., Lenderts, B., Young, J.K. and Cigan, A.M. (2016) Genome editing in maize directed by CRISPR-Cas9 ribonucleoprotein complexes. *Nature Communications* 7, 13274.

Takatsu, Y., Hayashi, M. and Sakuma, F. (2000) Transgene inactivation in Agrobacterium-mediated chrysanthemum (*Dendranthema grandiflorum* (Ramat.) Kitamura) transformants. *Plant Biotechnology* 17, 241–245.

Tanaka, K., Kanno, Y., Kudo, S. and Suzuki, M. (2000) Somatic embryogenesis and plant regeneration in chrysanthemum (*Dendranthema grandiflora* (Ramat.) Kitamura). *Plant Cell Reports* 19, 946–953.

Takatsu, Y., Nishizawa, Y., Hibi, T. and Akutsu, K. (1999) Transgenic chrysanthemum (*Dendranthema grandiflorum* (Ramat.) Kitamura) expressing a rice chitinase gene shows enhanced resistance to gray mold (Botrytis cinerea). *Scientia Horticulturae* 82, 113–123.

Tanase, K., Aida, R., Ohyama, A. and Onozaki, T. (2011) The promoter from tomato sucrose synthase gene TOMSF drives stamen-specific gene expression in chrysanthemum. *Plant Biotechnology* 28, 497–501.

Teixeira da Silva, J.A. (2003) Chrysanthemum: Advances in tissue culture, cryopreservation, postharvest technology, genetics and transgenic biotechnology. *Biotechnology Advances* 21, 715–766.

Teixeira da Silva, J.A. (2004) Ornamental chrysanthemums: Improvement by biotechnology. *Plant Cell, Tissue and Organ Culture* 79, 1–18.

Teixeira da Silva, J.A., Shinoyama, H., Aida, R., Matsushita, Y., Raj, S.K. *et al.* (2013) Chrysanthemum biotechnology: Quo vadis? *Critical Reviews in Plant Sciences* 32, 21–52.

Teixeira da Silva, J.A. and Kulus, D. (2014) Chrysanthemum biotechnology: Discoveries from the recent literature. *Folia Horticulturae* 26, 67–77.

Teixeira da Silva, J.A., Kim, H. and Engelmann, F. (2015a) Chrysanthemum low-temperature storage and cryopreservation: A review. *Plant Cell, Tissue and Organ Culture* 120(2), 423–440.

Teixeira da Silva, J.A., Lema-Ruminska, J., Tymoszuk, A. and Kulpa, D. (2015b) Regeneration from chrysanthemum flowers: A review. *Acta Physiologiae Plantarum* 37, 1–16.

Toguri, T., Ogawa, T., Kakitani, M., Tukahara, M. and Yoshioka, M. (2003) Agrobacterium-mediated transformation of Chrysanthemum (*Dendranthema grandiflora*) plants with a disease resistance gene (pac1). *Plant Biotechnology* 20, 121–127.

Trifunović, M., Jevremović, S., Nikolić, M., Subotić, A. and Radojević, L.J. (2007) Micropropagation of chrysanthemum cultivars—effect of cold storage on plant regeneration *in vitro*. *Acta Horticulturae* 764, 319–324.

Tudses, N., Premjet, S. and Premjet, D. (2014) Optimal conditions for high yield protoplast isolations of Jatropha curcas L. and Ricinus communis L. *American-Eurasian Journal of Agricultural & Environmental Sciences* 14, 221–230.

Tung, H.T., Nam, N.B., Huy, N.P., Luan, V.Q., Hien, V.T. *et al.* (2018) A system for large scale production of chrysanthemum using microponics with the supplement of silver nanoparticles under light-emitting diodes. *Scientia Horticulturae* 232, 153–161.

Tymoszuk, A., Zalewska, M. and Lema-Rumińska, J. (2014) Regeneration of somatic embryos from *in vitro* isolated ligulate florets of chrysanthemum. *Acta Scientiarum Polonorum Hortorum Cultus* 13(4), 13–22.

Valizadeh, M., Deraison, C., Kazemitabar, S.K., Rahbé, Y. and Jongsma, M.A. (2013) Aphid resistance in florist's chrysanthemum (*Chrysanthemum morifolium* Ramat.) induced by sea anemone equistatin overexpression. *African Journal of Biotechnology* 12, 6922–6930.

Vilasini, P. and Latipah, Z. (2000) Somaclonal variation in *Chrysanthemum morifolium* generated through petal cultures. *Journal of Tropical Agriculture and Food Science* 28, 115–120.

Visser, P.B., de Maagd, R.A. and Jongsma, M.A. (2007) *Chrysanthemum*. In: Pua, E.C. and Davey, M.R. (eds) *Transgenic Crops VI (Biotechnology in Agriculture and Forestry)*, Vol. 61. Springer-Verlag, Berlin Heidelberg, pp. 253–272.

Wang, F., Zhong, X., Huang, L., Fang, W., Chen, F. *et al.* (2018) Cellular and molecular characteristics of pollen abortion in chrysanthemum cv. Kingfisher. *Plant Molecular Biology* 98, 233–247.

Wang, J., Guan, Y., Ding, L., Li, P., Zhao, W. *et al.* (2019) The CmTCP20 gene regulates petal elongation growth in *Chrysanthemum morifolium*. *Plant Science* 280, 248–257.

Wang, K., Wu, Y.H., Tian, X.Q., Bai, Z.Y., Liang, Q.Y. *et al.* (2017) Overexpression of *DgWRKY4* enhances salt tolerance in chrysanthemum seedlings. *Frontiers in Plant Science* 8, 1592.

Watanabe, K. (1977) Successful ovary culture and production of F1 hybrids and androgenic haploids in Japanese chrysanthemum species. *Journal of Heredity* 68, 317–320.

Xiang, L., Liu, X., Li, X., Yin, X., Grierson, D. *et al.* (2015) A novel bHLH transcription factor involved in regulating anthocyanin biosynthesis in chrysanthemums (*Chrysanthemum morifolium* Ramat.). *PLoS ONE* 10, e0143892.

Xiang, L., Liu, X., Li, H., Yin, X., Grierson, D. *et al.* (2019) CmMYB# 7, an R3 MYB transcription factor, acts as a negative regulator of anthocyanin biosynthesis in chrysanthemum. *Journal of Experimental Botany* 70, 3111–3123.

Xu, G., Chen, S. and Chen, F. (2010) Transgenic chrysanthemum plants expressing a harpinXoo gene demonstrate induced resistance to *Alternaria* leaf spot and accelerated development. *Russian Journal of Plant Physiology* 57, 548–553.

Xu, J., Kang, B.C., Naing, A.H., Bae, S.J., Kim, J.S. *et al.* (2020) CRISPR/Cas9 mediated editing of 1-aminocyclopropane-1-carboxylateoxidase1 (ACO1) enhances *Petunia* flower longevity. *Plant Biotechnology Journal* 18, 287–297.

Yan, K., Du, X. and Mao, B. (2022) Production of virus-free chrysanthemum (*Chrysanthemum morifolium* Ramat) by tissue culture techniques. In: Wang, A. and Li, Y. (eds) *Methods in Molecular Biology*. Humana Press, New York, pp. 171–186. DOI: 10.1007/978-1-0716-1835-6_17.

Yoosumran, V., Ruamrungsri, S., Duangkongsan, W. and Saetiew, K. (2018) Induced mutation of *Dendranthemum grandiflora* through tissue culture by ethyl methanesulphonate (EMS). *International Journal of Agricultural Technology* 14, 73–82.

Yue, J., Zhu, C., Zhou, Y., Niu, X., Miao, M. *et al.* (2018) Transcriptome analysis of differentially expressed unigenes involved in flavonoid biosynthesis during flower development of *Chrysanthemum morifolium* 'Chuju'. *Scientific Reports* 8, 1–14.

Zalewska, M., Lema-Ruminska, J. and Miler, N. (2007) *In vitro* propagation using adventitious buds technique as a source of new variability in chrysanthemum. *Scientia Horticulturae* 113, 70–73. DOI: 10.1016/j.scienta.2007.01.019.

Zhang, Q., Xing, H.L., Wang, Z.P., Zhang, H.Y., Yang, F. *et al.* (2018) Potential high-frequency off-target mutagenesis induced by CRISPR/Cas9 in *Arabidopsis* and its prevention. *Plant Molecular Biology* 96, 445–456.

Zhong, J., Guo, Y., Chen, J., Cao, H., Liu, M. *et al.* (2021) Volatiles inheriting from *Crossostephium chinense* act as repellent weapons against aphids in *Chrysanthemum lavandulifolium* cultivars. *Industrial Crops and Products* 166, 113467. DOI: 10.1016/j.indcrop.2021.113467.

Zhou, J., Wang, B. and Zhu, L. (2005) Conditioned culture for protoplasts isolated from chrysanthemum: An efficient approach. *Colloids and Surfaces B: Biointerfaces* 45(3), 113–119. DOI: 10.1016/j.colsurfb.2005.07.012.

Zhu, W.Y., Jiang, J.F., Chen, S.M., Wang, L., Xu, L.L. *et al.* (2013) Intergeneric hybrid between *chrysanthemum* x *morifolium* and *Artemisia japonica* achieved via embryo rescue shows salt tolerance. *Euphytica* 191, 109–119. DOI: 10.1007/s10681-013-0869-3.

Zhu, W.Y., Zhang, F., Chen, S.M., Xu, L.L., Wang, L. *et al.* (2014) Intergeneric hybrids between *Chrysanthemum morifolium* 'Nannongxiaoli' and *Artemisia vulgaris* 'Variegata' show enhanced resistance against both aphids and *Alternaria* leaf spot. *Euphytica* 197(3), 399–408.

5

Developmental Physiology of Chrysanthemum

Marie-Christine Van Labeke[1], Malleshaiah SharathKumar[2,3] and Ep Heuvelink[2]*

[1]Horticultural Sciences, Department of Plants and Crops, Faculty of Bioscience Engineering, Ghent University, Ghent, Belgium
[2]Horticulture and Product Physiology, Department of Plant Sciences, Wageningen University and Research, Droevendaalsesteeg 1, 6708 PB Wageningen, The Netherlands
[3]Department of Breeding and Research, Deliflor Chrysanten B.V., Korte Kruisweg 163, 2676 BS Maasdijk, The Netherlands

Abstract

Chrysanthemum cultivation involves a dynamic interplay of factors that shape both plant development and growth. Although development and growth are intertwined, they respond to distinct environmental cues: temperature primarily influences development, while light intensity and CO_2 concentration predominantly govern growth. This chapter provides a comprehensive exploration of chrysanthemum development, elucidating phenology, plant architecture, and the intricacies of root, leaf, stem, flower, and seed formation. We discuss the environmental factors influencing the growth and development of different plant parts and delve into the nuanced relationship between environmental conditions and the developmental physiology of chrysanthemum. Furthermore, we discuss grafting techniques as a strategic approach to increase the vigor of stock plants and enhance the rooting quality of cuttings. The developmental physiology in this chapter offers insights for optimizing cultivation practices for improved productivity and quality in the chrysanthemum industry.

Keywords

Chrysanthemum development, environmental factors, flower initiation, grafting, leaf development, phenology, plant architecture, root formation, seed formation, stem growth, stock plants

*Corresponding author: ep.heuvelink@wur.nl

DOI: 10.1079/9781789248289.0005

5.1 Introduction

For understanding plant behavior, it is beneficial to distinguish between growth and development. Growth refers to the increase in size (e.g. weight, leaf area, and length), whereas development is the genetically defined progression through the life cycle stages (germination, juvenile phase, generative phase, flowering, and seed formation). Initiation of new organs (e.g. leaves, flower initiation, and formation of fruits and seeds) clearly represents plant development. Development and growth almost always occur simultaneously, and both are influenced by environmental factors. For example, temperature primarily influences development (Van der Ploeg and Heuvelink, 2006), whereas light intensity and CO_2 concentration predominantly affect growth. In this chapter, we focus on chrysanthemum development (e.g. phenology, plant architecture, root, leaf, stem, flowering, and seeds), whereas chrysanthemum growth (e.g. photosynthesis, biomass production, and partitioning) is largely presented in Chapter 08.

5.2 Plant Phenology, Rosette, and Juvenility

The cultivated chrysanthemum is an interspecific hybrid that has a long history of inbreeding, mutations, and selection (refer to Chapters 03 and 04). This leads to a highly diverse assortment, and it is not surprising that different physiological responses during the life cycle of chrysanthemum exist. The life cycle of a plant can be divided into a vegetative or juvenile phase, followed by a mature reproductive phase, where the plant is receptive to floral stimuli. The existence of a juvenile phase in chrysanthemum is illustrated in the work of De Jong (1981), who showed that the number of leaves in a seedling population from crosses between all-year-round chrysanthemums varied from 15 to 32 leaves before seedlings became receptive to short days. Therefore, the existence of a juvenile phase might be of importance for breeders but does not affect chrysanthemum production, as this starts from cuttings that have no juvenile phase.

The phenology of autumn-flowering chrysanthemums is given in Fig. 5.1 (Sumitomo *et al.*, 2008). Flower bud initiation is induced by decreasing day lengths under cooler temperatures in late summer. The main stem flowers during October–November and dies before winter after the seed set. Suckers emerge from the basal zone from mid to late autumn. These suckers become dormant and form rosettes with an inability to flower, even under short-day inductive conditions. After fulfillment of the chilling requirements, the shoots will elongate and respond again to flower-inducing triggers. Also, Cockshull (1985) reported that certain cultivars form a rosette in autumn and winter and require vernalization to flower. After passing through the rosette stage, the plants grow rapidly and flower early in a heated greenhouse at temperatures of 15–20°C. However, this is not always the case, as, for certain Japanese chrysanthemums exposed to low temperatures, shoot extension will

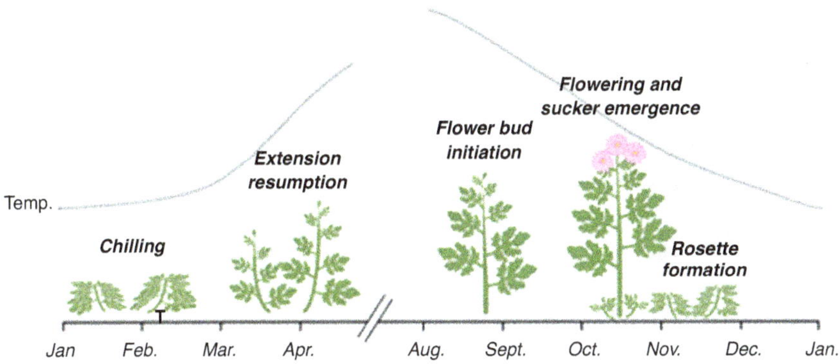

Fig. 5.1. Phenological events in herbaceous perennial chrysanthemums. (Reproduced from Sumitomo *et al.*, 2008, with permission from Oxford University Press.)

occur, but flowering is delayed compared with non-chilled plants in a heated greenhouse (Higuchi and Hara, 1974). Additionally, Kawata (1987) showed that this pattern is also present in other Japanese autumn-flowering cultivars. Likewise, exposure of mother plants of certain autumn-flowering cultivars, such as "Jinba," to temperatures lower than 10°C for an extended period did result in delayed flowering of the cuttings from these plants (Kunitake *et al.*, 2009). From these studies, it is clear that low temperatures result in different responses, depending on the genotype.

Hisamatsu *et al.* (2017) analyzed the effects of these temperature responses on rosette formation and floral induction for chrysanthemum cultivars with different genetic backgrounds. In certain genotypes, such as "Kinyo" and "City," high summer temperatures induced shorter internodes in cuttings rooted and grown under a 12 hour photoperiod leading to rosette-like types that could be considered as a dormant condition. After a sufficient chilling period, extension growth resumed. However, these chilling temperatures might result in two contrasting effects on flowering. In certain genotypes, the chilling acts as a "vernalization" and leads to flowering, while in other genotypes, the chilling has an inhibitory effect, and plants do not respond immediately to flowering stimuli (this is called a form of juvenility in older literature). The study also included a "Reagan"-type (The Netherlands), but for this cultivar, flowering was always obtained irrespective of the treatments, and it did not respond to inhibitory temperature treatments (Hisamatsu *et al.*, 2017). In Europe, all-year-round genotypes for cut flower production are required; therefore, autumn-flowering lines have been discarded during the breeding process (De Jong, 1981). Likewise, European breeding companies of autumn-flowering pot chrysanthemums discard lines with rosette formation. However, in Japan and China, flower forms are important characteristics for the floricultural market, and autumn-flowering cultivars, such as "Jinba," are still cultivated.

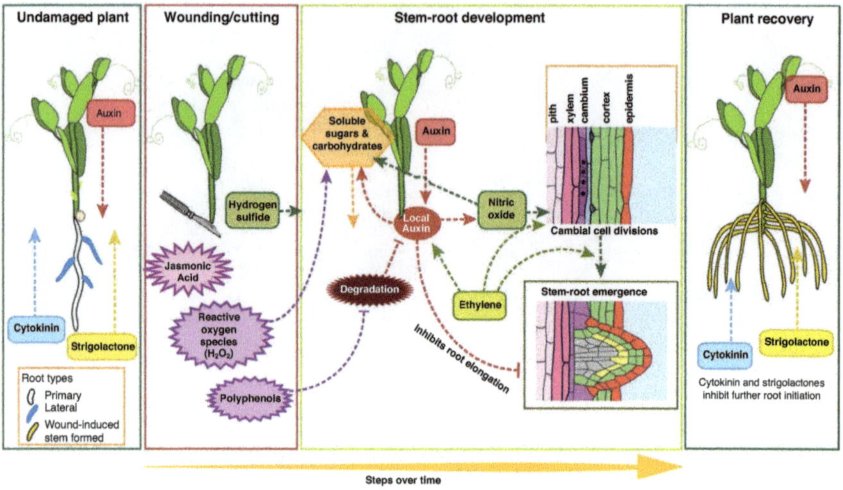

Fig. 5.2. Wound-induced adventitious root formation. (Reproduced from Steffens and Rasmussen, 2016, with permission from Oxford University Press.)

5.3 Root

Chrysanthemums are propagated asexually by leafy stem cuttings. Roots that are formed on non-root tissue, such as stem tissue in the case of chrysanthemums, are adventitious roots. Different triggers induce these roots (for an overview, see Steffens and Rasmussen, 2016; Drüge *et al.*, 2019), including wound-induced adventitious root formation, which is the basis of cutting propagation (Fig. 5.2). The process of adventitious root initiation and development is highly regulated by a suite of environmental and endogenous (hormones) factors, including physiological status of the mother plant (see Section 5.8).

In short, immediately after taking the cutting, wound response signals (jasmonic acid, hydrogen peroxidase) increase, followed by an increase of phenolics that helps to protect against this H_2O_2 burst. Adventitious root induction depends highly on auxin levels, while the role of ethylene is contradictory and depends on the investigated species. In chrysanthemum, endogenous ethylene levels increased till the early phase of adventitious root formation (Liu *et al.*, 2013), suggesting that ethylene is important in rhizogenesis. Also, a higher carbohydrate pool will enhance root initiation. Adventitious roots will be initiated in the cambial zone, and here, cell divisions will develop into new root primordia. Histological studies showed that 5 days after cutting of *Chrysanthemum* "Jinba," root primordia are already found (Fig. 5.3). Once the root system is formed, normal hormone balances are restored as both cytokinin and strigolactones are biosynthesized in the roots.

Auxins, predominantly synthesized in the shoot apical meristems, play a central role in regulating adventitious root formation; indole-3-acetic acid

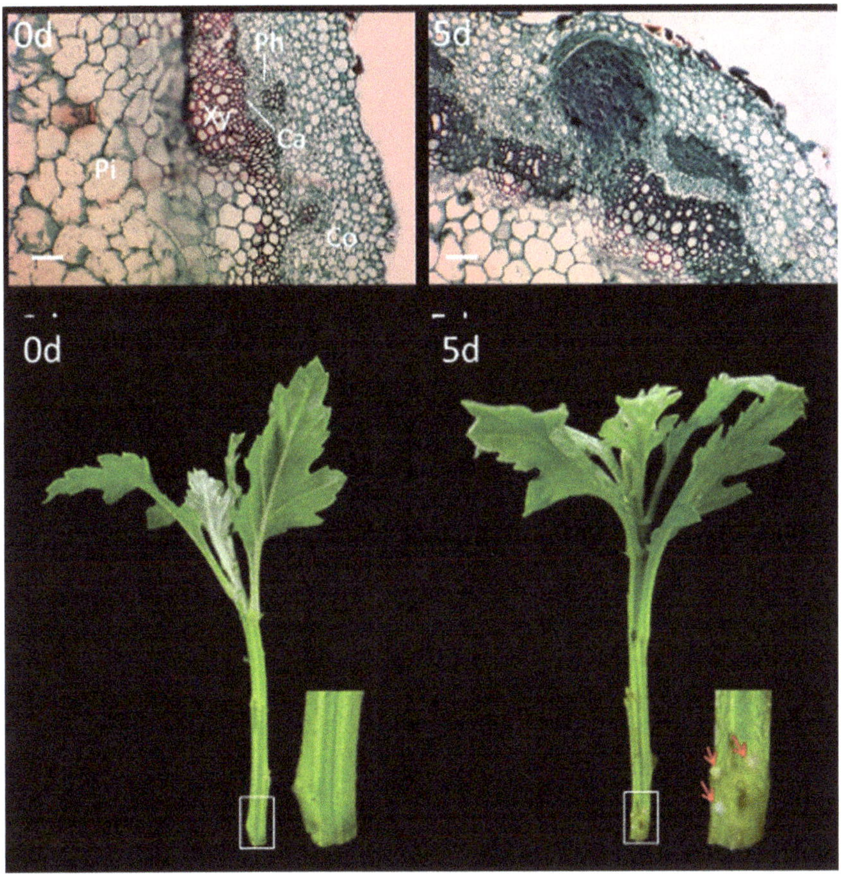

Fig. 5.3. Morphological and histological properties of adventitious root primodium formation in the stem base of *Chrysanthemum* "Jinba" cuttings. The typical stem anatomy consists of the cortex (Co), pith parenchyma (Pi), and a ring of vascular tissues comprising phloem (Ph), cambium (Ca), and xylem (Xy). Anatomy of adventitious root primordium formation at 5 days after cutting, where the first differentiated root primordia can be observed (top right). Scale bars = 500 μm. Morphology of cuttings at 0 and 5 days (bottom panel). Arrows indicate adventitious roots, and the magnified image shows the cutting position. (Adapted from Liu *et al.*, 2013, CC BY 2.0.)

(IAA), a natural auxin and synthetic analogues, and indole-3-butyric acid (IBA, which is not only an endogenous hormone but also dominantly used as rooting hormone due to its higher stability than IAA) both have a dominant and stimulatory effect on adventitious rooting (Pacurar *et al.*, 2014). In roots, the polarity of auxin transport has been described as basipetal (from apex to base) and acropetal (toward root apex) (Rashotte *et al.*, 2000). With the help of auxin influx carrier proteins (AUXIN RESISTANT1/LIKE AUX1—AUX1/LAX) and efflux carrier proteins (PIN-FORMED-PIN), auxin is transported systemically via vascular tissues (Yang *et al.*, 2006; Swarup and Péret, 2012).

The members of PIN proteins are known for their auxin-related developmental processes, including adventitious root formation. For example, in chrysanthemum, the MADS-box transcription factor *CmANR1* positively modulates the development of numerous adventitious roots and lateral roots by directly regulating the auxin transport gene *CmPIN2* (Sun *et al.*, 2018). Furthermore, the proteomic profiling in chrysanthemum enlisted the root system development proteins belonging to the TCP (TB1-CYC-PCFs) family, which influence auxin biosynthesis and transport. *CmTCP20* gene may participate in phytohormone signal transduction with improved IAA content to promote both adventitious and lateral root development in chrysanthemum (Fan *et al.*, 2019).

Environmental factors might modulate auxin levels. For example, in the shade avoidance response, a low red:far-red (R:FR) ratio increases the biosynthesis of auxin (Halliday *et al.*, 2009). The increasing use of light-emitting diodes (LEDs) in glasshouse horticulture opens the possibilities for applying different light ratios and thus influencing endogenous auxin levels. The addition of FR to red light (lower R:FR ratio) resulted in faster rooting, higher root dry mass, and an increased number of adventitious roots in chrysanthemum. However, a low R:FR ratio will also lead to stem elongation, which reduces the quality of the cuttings. This effect on internode length can be counteracted by adding blue to the spectrum (Christiaens *et al.*, 2019). Besides auxins, *LATERAL ORGAN BOUNDARIES DOMAIN* (*LBD*) genes also regulate root development. In chrysanthemum, *CmLBD* transcripts were earlier upregulated under blue light compared with red and fluorescent light, which correlated with a higher root mass under blue light (Gil *et al.*, 2020). Also, Christiaens *et al.* (2015) found a beneficial effect of blue light on the number of roots compared with red light; however, these effects were less pronounced than those observed under low R:FR ratios.

5.4 Leaf

Leaf characteristics (leaf shape, size, color, and organelles—mesophyll and palisade cells) affect growth as well as the external appearance of leaves in chrysanthemums. This is part of the ornamental value of cut chrysanthemums and contributes to the postharvest quality. Therefore, changes in leaf characteristics will affect photosynthetic capacity and ornamental value. In chrysanthemums, abundant morphological diversity in leaf phenotypes exists (leaf—base, color, shape, margin, tip, lobe, and vein distribution). Although leaf characteristics are also affected by environmental conditions, each genotype has relatively stable leaf characteristics (Ma *et al.*, 2020), which are the basis for cultivar identification and genetic variation.

The physiological status of leaves depends predominantly on the growing conditions, especially light and temperature. Leaves of plants grown under 25 and 15% irradiance were thinner with loose palisade tissue and irregularly arranged spongy mesophyll cells compared with leaves grown under full sunlight (Han *et al.*, 2017). McMahon and Kelly (1995) described the importance of light quality that influences the growth of organelles in leaf cells in chrysanthemum. Leaves grown in $CuSO_4$-filtered light (primarily absorbs FR light) had increased chlorophyll-a and -b and carotenoid contents per unit area; this appeared to be the result of more densely arranged palisade cells with reduced intercellular space. In addition, palisade cell numbers per unit area increased under $CuSO_4$ filters. Light through $CuSO_4$ filters reduced the leaf area, palisade cell length, and intercellular space but not leaf thickness. The stomatal density was not affected, but the number per leaf was reduced due to decreased leaf size by $CuSO_4$-filtered light (McMahon and Kelly, 1995).

In chrysanthemum, the commercial value depends not only on the external (visual) quality but also on postharvest longevity, primarily determined by the growing conditions. Chrysanthemum grown under monochromatic red light showed enhanced stomatal ability (the rate and the degree of stomatal closure) to regulate water loss upon desiccation (Seif *et al.*, 2021). Another critical factor that affects postharvest longevity in chrysanthemum is vapor pressure deficit (VPD). Chrysanthemums grown under lower VPD have a disturbed stomatal functionality due to a reduced sensitivity of stomata toward desiccation, leading to premature leaf wilting and leaf yellowing (Aliniaeifard and van Meeteren, 2016).

Ornamental value is affected by flower color as well as leaf color. Leaf coloration largely depends on the leaf pigments (chlorophyll, carotenoids, and anthocyanins). Mutations in the genes involved in chloroplast development can also influence leaf color. Variegated chrysanthemum can be of substantial ornamental value owing to leaf color characteristics. A recent transcriptomic study in variegated chrysanthemum variety “NAU04-1-31-1” gave molecular insights into the leaf color variation involving two genes (*CmChlH* and *CmFtsH*) responsible for yellow and green leaf tissues (Chang *et al.*, 2013). Additionally, biochemical analysis indicated that the yellow leaf tissue's chlorophyll content is lower than that in green tissue, and chloroplasts in the yellow leaf tissue were vacuolated and lacked thylakoid membrane structures (Chang *et al.*, 2013).

Temperature also influences chrysanthemum leaf characteristics. Kozáowska *et al.* (2011) observed a positive correlation between day temperature and leaf chlorophyll content. Berg and Cutter (1969) studied leaf initiation rate in chrysanthemum at 20°C. After removal of the terminal bud, the top two lateral buds produced leaves at a rate of 1.4–1.6 leaves/day for the first 9 or 10 days. Around the 10th day, there was an abrupt shift to a lower initiation rate of 0.7–0.8 leaves/day. Defoliation by periodic removal of leaves larger than 1 cm showed an increase in leaf initiation rate, but a negative effect on leaf initiation rate from maturing leaves could not explain the drastic drop in initiation rate. Berg and Cutter (1969) also concluded that the rate of leaf emergence (leaf larger than 1 cm) was equal to the rate of leaf initiation.

When the apex has been initiated to form flowers (refer to the following section), the production of true leaf primordia ceases and is initially followed by the formation of bracts. Based on data collected from chrysanthemum "Pert" plants grown in greenhouses and daylight chambers at temperatures ranging from 9 to 27°C and light integrals ranging from 0.8 to 7.8 mol m^{-2} day^{-1}, Larsen and Hidden (1995) concluded that leaf unfolding rate shows an optimum response to temperature and does not depend on light integral. This optimum temperature response was shown to be stronger when considering day temperature alone instead of the average 24 hour temperature (Carvalho *et al.*, 2002). Only when the light integral was below 2 mol m^{-2} day^{-1}, leaf unfolding rate was reduced. At temperatures between 15 and 25°C, about one leaf was unfolded (reached 3 mm length) every second day. The finding that leaf unfolding rate did not depend on light integral (when above 2 mol m^{-2} day^{-1}; Larsen and Hidden, 1995) suggests that other factors influencing assimilate availability (planting density, CO_2 concentration) will also not affect leaf unfolding rate.

5.5 Stem

Stem length depends on the number of internodes (same as the number of leaves, see Section 5.6) and their length (Fig. 5.4). Since chrysanthemum has a determinate growth pattern under short-day conditions, new internodes are formed only until the start of flower initiation. The longer the duration of the long-day period, the taller the plants, as more time exists to form new internodes (Carvalho and Heuvelink, 2003).

5.5.1 Shoot architecture and branching

Plant architecture is shaped by its branching pattern, and shoot branching depends on the formation of axillary buds at the shoot apex and their subsequent outgrowth. Therefore, the control of axillary bud outgrowth will shape the plant architecture. If axillary meristems or buds remain paradormant, new leaves will only be initiated in the apical zone, leading to a single stem. The paradormant state of the axillary bud is regulated by signals in the plant (not in the bud itself) (Lang *et al.*, 1987). Decapitation will release axillary buds from their paradormant state but with the potential to inhibit the outgrowth of axillary buds. Floral transition will also release the dormant state of other axillary buds. This phenomenon is described as apical dominance (for a review, see Cline, 1991). The regulatory system for the outgrowth of axillary buds is under both endogenous and environmental control. Endogenous control is based on the cross talk of auxins, cytokinins, and strigolactones, while endogenous sugar levels also control bud outgrowth. In particular, auxins, strigolactones, and their transport play a central role in the regulation of axillary bud outgrowth associated with apical dominance (Liang *et al.*, 2010). Branching signals from auxins and strigolactones are integrated within axillary buds by transcription factors from the TB1, CYC, and PCFs (TCP) family, *BRANCHED 1*

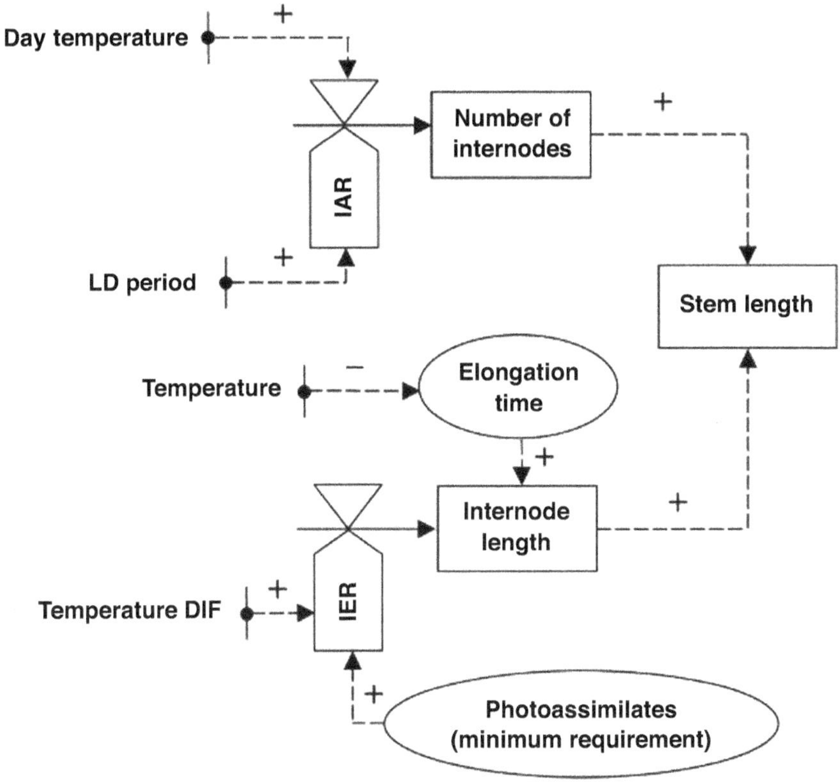

Fig. 5.4. Relational diagram for stem length in cut chrysanthemum. Boxes are state variables, ellipses are parameters, and valves are rate variables. Dashed arrows represent information flow: (+) indicates positive influence, and (-) indicates negative influence.

Note: DIF = difference between day and night temperatures, IAR = internode appearance rate, IER = internode elongation rate, LD = long day. (Reproduced from Carvalho and Heuvelink, 2004, with permission from International Society for Horticultural Science (ISHS).)

(*BRC1*) to influence bud outgrowth. *BRC1* regulates shoot branching through phyB as a response to changes in R:FR ratio (González-Grandío *et al.*, 2013). *BRC1* inhibits floral transition and stimulates outgrowth of buds by interacting with *FLOWERING LOCUS T (FT)* and *TWIN SISTER OF FT (TSF)* (Niwa *et al.*, 2013; Rameau *et al.*, 2015). In chrysanthemum, the involvement of *BRC1* in the regulation of branching was elucidated (Chen *et al.*, 2013; Dierck *et al.*, 2015). Furthermore, the sugar transporter gene, *CmSWEET17*, may be involved in the process of sucrose-induced axillary bud outgrowth by mediating the auxin transport pathway genes (Liu *et al.*, 2020). Auxin export is key to the outgrowth of buds, and Liu *et al.* (2022) showed that sucrose treatment

Fig. 5.5. Morphotypes of chrysanthemum. (A) Standard (disbud) with single flower. (B) Spray type with multiple flowers on single stem. (C) Cushion type with multiple branching. (Adapted from SharathKumar, 2023.)

of two-node stems of chrysanthemum promoted the expression of *CmPIN1* and *CmAUX1*, which induced auxin export from buds. Furthermore, plant nutrient status also influences shoot branching by influencing the phytohormones such as auxins, cytokinins, and strigolactones. For example, phosphate deficiency inhibits the outgrowth of axillary buds with the relocation of auxin and increased strigolactone level (Xi *et al.*, 2015). Also, in a chrysanthemum field experiment, shoot branching was reduced at low levels of phosphorus fertilization (Satar *et al.*, 2016).

Commercial chrysanthemums can be grouped into different morphotypes/phenotypes, varying from upright single-stem types (cut flowers, spray chrysanthemums) to highly branched growth habits (cushion types, potted chrysanthemum) (Fig. 5.5). Branching is an important quality trait in the breeding and production of chrysanthemums. Indeed, axillary bud outgrowth in cut flower types needs manual labour to disbud the stems, while in the cushion types, well-branching genotypes are needed. Therefore, knowledge of the hormonal control and genetic markers could help the breeding efforts. In chrysanthemum, markers associated with key genes in the biosynthetic pathway of strigolactones could explain the phenotypic shoot branching traits (Klie *et al.*, 2016; Dierck *et al.*, 2018). Sun *et al.* (2019) used a quantitative trait loci (QTL) approach to identify the branching pattern of chrysanthemum, grown at two planting densities, but found a strong genotype by environment (G × E) effect; thus, the search for stable QTLs linked to branching patterns is still ongoing.

5.5.2 Environmental control

In chrysanthemum production, shoot architecture is controlled by pruning, disbudding (manual decapitation or removing the axillary buds), and the use of chemical growth retardants. Recently, LED lighting technology has been explored to regulate bud outgrowth by changing the R:FR ratio. Shoot branching and outgrowth of axillary buds are interlinked with shade avoidance syndrome (SAS). A lower R:FR ratio, perceived by phyB, leads to unwanted shoot elongation and less branching. Planting density will directly influence the light intensity in the canopy and also filter wavelengths, resulting in an enrichment of FR for axillary buds. Shading and low R:FR ratio negatively regulate shoot branching in many species. Growing cut chrysanthemum in a low R:FR or high R:FR environment during their developmental stage showed that the effect of light quality was influenced by the bud position (Yuan *et al.*, 2018). In the upper stem zone, the bud length of plants grown under low R:FR was much shorter than that of the plants grown under high R:FR. Yet, this was reversed for the basal buds, indicating that a low R:FR not only inhibits bud outgrowth but this response is also influenced by bud position.

The relative progression of lateral shoot elongation from pinch to flower of chrysanthemum "Bright Golden Anne" plants grown under 2–22 mol m^{-2} day^{-1} photosynthetic photon flux, at a temperature range of 10–20°C, has been modeled using a Richards function (Karlsson and Heins, 1994). The relative time (first short day set to 0.0 and flowering to 1.0) required to attain 50% of the shoot length at flowering was lower at higher average daily temperature during plant development.

Stem length increased linearly with temperature (15–24°C) in four cultivars; however, for one cultivar, the slope of the line was steeper than that for the other three cultivars (van der Ploeg *et al.*, 2007). This increase in stem length resulted from an increase in both internode length and the number of internodes. Internode appearance rate shows an optimum around 25°C (Larsen and Hidden, 1995; Carvalho *et al.*, 2002). Chrysanthemum starts to initiate flowers soon after the start of the short-day period (Van Ruiten and De Jong, 1984), and once the plant has induced the apical flower bud, it loses the ability to initiate new internodes. Therefore, during the long-day period, only temperature influences the final leaf number. After the flower buds have been initiated, stem elongation is solely the result of internode elongation.

Besides the average temperature, the temperature regime (or temperature pattern) is very important for stem elongation. The difference between day and night temperatures is called "DIF" (short for difference). An inversed temperature regime with a higher night than day temperature is called a "negative DIF." Growers, especially of potted plants, use negative DIF to reduce elongation and obtain more compact plants. Carvalho *et al.* (2002) showed that DIF can predict the final internode length only within a temperature range, where the effects of day temperature and night temperature are equal in magnitude and opposite in sign (18–24°C). In that range, the stimulation in internode length because of higher day temperature is mitigated by the reduced internode length as a result of the higher night temperature (Fig. 5.6).

Not only the day/night temperature regime is important, but also the regime during the day. In particular, a drop in temperature during a brief period (up to a few hours; called DROP) can reduce plant elongation. The point during the day at which this DROP occurs plays a significant role. In the first hours of the light period, elongation growth is most sensitive to temperature: a low temperature during those hours reduces internode length. In the last hours of

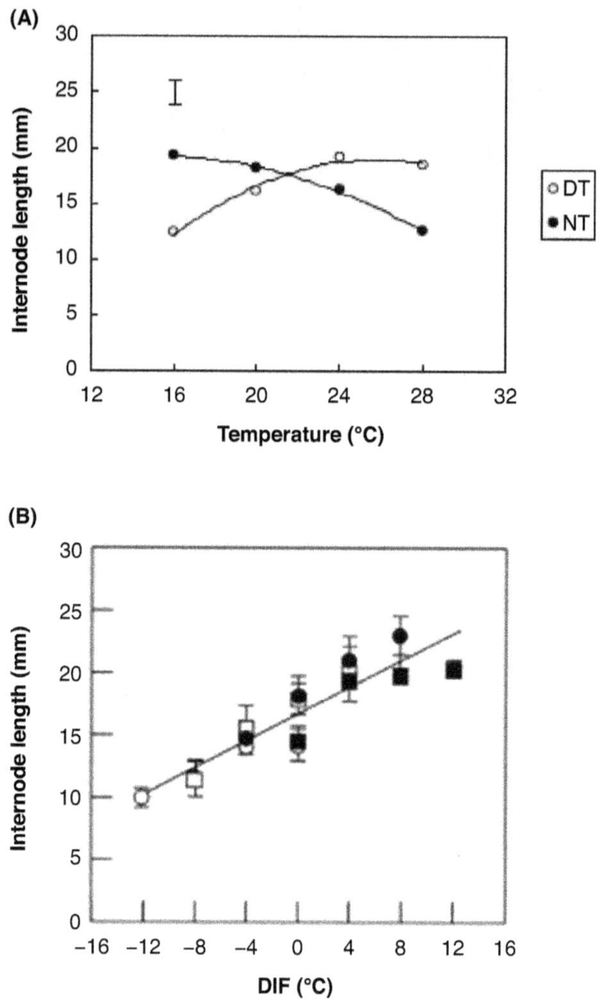

Fig. 5.6. (A) Final length of internode 10 as a function of day temperature (o: averaged over four night temperatures) or night temperature (•: averaged over 4 day temperatures) 27 days after temperature treatments started in chrysanthemum "Reagan Improved." Sixteen treatments (day length 12 h) were all combinations of 4 day temperatures (16, 20, 24, and 28°C) and four night temperatures (16, 20, 24, and 28°C). (B) Relationship between the final length of internode 10 and DIF (°C) for these 16 treatments. Symbols represent day temperatures of 16 (o), 20 (□), 24 (•), and 28°C (■). Regression line, $y = 16{\cdot}57 + 0{\cdot}547x$, $R^2 = 0{\cdot}81$. Vertical bars indicate standard error of the mean (s.e.m.) ($n = 2$) and are shown only when larger than the symbol. (Adapted from Carvalho *et al.*, 2002.)

the light period, DROP does not affect elongation. Potted plant growers use this phenomenon to keep their plants compact (Carvalho *et al.*, 2008). In a greenhouse, DROP is easier to create than a negative DIF (in late spring and summer), as day temperature only needs to be lower than night temperature during the first hours of the light period, and not throughout the whole light period. DROP in the early-morning hours often has an effect similar to that of DIF.

5.6 Flowering

Most chrysanthemums are short-day flowering plants. Nevertheless, the wide range of chrysanthemum cultivars results in a flowering range that extends from summer-flowering to autumn- and winter-flowering genotypes under natural conditions. Flower buds can thus be formed under long days, and the time to flower under long day is assessed by the number of formed leaves (long-day leaf number). Even late-flowering cultivars can start bud formation after several months under long day, although anthesis will not be reached. Therefore, flower initiation in chrysanthemum has a quantitative short-day response in early-flowering genotypes and a qualitative response in late-flowering types (Cockshull, 1985). In Asia, summer-to-autumn flowering types are commercially produced as cut and loose flowers, and in certain cultivars, unwanted flower initiation under long day occurs before the inductive photoperiod. This unwanted "premature" budding results in short-length cut flowers and the outgrowth of axillary vegetative buds. A cold pretreatment of stock plants generally reduced the incidence of premature budding in rooted cuttings (Sumitomo *et al.*, 2014). Also, in the European pot mum production, "premature budding" can occur in the early-flowering types. Here, chemical pinching (e.g. Ethrel) can be a method to avoid the negative effect of premature budding.

In Europe, for all-year-round production, autumn- and winter-flowering cultivars are used. Here, the vegetative phase continues when long days (natural light or photoperiodic lighting) are provided such that the plant can initiate a certain number of leaves before the start of the floral inductive phase under short-day conditions. The time it takes for a plant to flower (up to anthesis) from the onset of short day is called reaction time or response time. Each cultivar has its own response time, which varies between 6 and 15 weeks (see Box). Under natural conditions, cultivars with an early reaction time are induced faster than cultivars with a late response, and the duration of all stages becomes shorter with shorter reaction times. For example, Van Ruiten and De Jong (1984) showed that early response types needed 3–7 short days to transform from a vegetative to an inflorescence meristem, intermediate cultivars were induced in 10–14 days, and late response types needed 14–17 days when grown at 19/16°C (day temperature/night temperature (DT/NT)). Once the apical meristem was induced, the axillary meristems followed; the first axillary meristem needed 3 days, but the other axillary meristems followed quickly. Short day has to be maintained till commercialization to allow the floral transition in the axillary buds. In cut flowers, this leads to the spray type of flowering stems.

The *Chrysanthemum* response types (early, medium, and late response types)

Response group refers to the number of weeks it takes for the chrysanthemum to flower from the time they begin receiving short-day treatments to reach anthesis at optimum conditions. Although response groups range from 6 to 15 weeks, most potted chrysanthemums fall into the 9- or 10-week response groups, while most cultivars grown as cut chrysanthemums have a response time of 7–9 weeks although Santini types have a shorter response time, ranging from 5 to 7 weeks.

Based on their response time, chrysanthemums are classified into the following:

- Early response: flowering 7–9 weeks after the start of short days
- Medium response: flowering 10–12 weeks after the start of short days
- Late response: flowering 13–15 weeks after the start of short days

5.6.1 Floral induction

Photoperiod-induced flowering is controlled by phytochromes, with the perception of day length occurring in the leaves. Up to five phytochromes (PhyA–PhyE) exist in most dicotyledonous plants, and they largely differ in their response to spectral light quality (Bae and Choi, 2008). Three phytochromes are identified in *Chrysanthemum* spp. (PhyA, PhyB, and PhyC) (GenBank). Phytochromes exist in two forms that are reversibly interconvertible by light. Pr absorbs maximally in the red region of the spectrum (660 nm) and converts into the biologically active Pfr that has its absorption peak in the FR region (735 nm). Under darkness, Pfr gradually falls back to the inactive Pr form. PhyA is light labile and responds to FR, including the very low fluence response (VLFR) and the FR-high irradiance responses (FR-HIR). The other phytochromes (PhyB–PhyE) are responsible for the red/FR reversible low fluence response (LFR). Photoconversion of Pr to Pfr mediates an array of morphogenic responses, including the flowering response in photoperiodic plants, whereas reconversion of Pfr to Pr cancels the induction of those responses. The main photoreceptor mediating photomorphogenesis in red light is phyB.

In chrysanthemum production, day length is controlled by applying dark screens for an inductive short-day treatment or night interruption to maintain the vegetative stage and prevent flowering. It was elucidated that photoperiodic chrysanthemum flowering is primarily mediated by florigen-*FTL3* and antiflorigen-*AFT*. Under inductive short days, the flower inducer *CsFTL3* in leaves is highly expressed, which correlates with the floral evocation in the apical meristem (Oda *et al.*, 2012). On the other hand, antiflorigen-*CsAFT* is highly expressed under non-inductive long days and night break conditions with red light to prevent flowering (Higuchi *et al.*, 2013). Grafting experiments of wild-type plants on a rootstock overexpressing CsAFT clearly showed that CsAFT protein delays flowering under short days and is thus a floral repressor. Higuchi *et al.* (2012) dissected the responses of night interruption with red light at different timepoints after dusk and found that the induction peak for CsAFT occurred 8–10 hour after dusk. Dusk is thus critical for the dark-time measurement, and 8–10 hour after dusk, a light-sensitive phase or gating mechanism is present in this species. Furthermore, by evaluating different light qualities (blue, green, red, FR) for night interruption, the highest expression of CsAFT occurred under red light and was reversed by FR light, indicating the negative regulation of PhyB on floral induction (Higuchi *et al.*, 2013).

The increasing use of LED light opens possibilities to exploit the interaction between day length and light quality on floral induction in chrysanthemum. For example, when chrysanthemums were grown under 11 hour of red/blue light followed by 4 hour of blue light alone (i.e. a total of 15 hour of light, constituting a long day), they showed normal flowering—similar to what is typically seen under short-day conditions (SharathKumar *et al.*, 2021). However, an extension of the natural sunlight during the first 11 hour of the photoperiod with monochromatic red or blue LED light inhibited flowering (SharathKumar *et al.*, 2021). In addition, a 7-h day extension with green light after a 9 hour natural light period retarded flowering, although a dose–response relationship between the green photon flux density and flowering time was present (Meng and Runkle, 2019). Yet, the specific role of cryptochrome in the flowering response is still under investigation.

5.6.2 Development of the floral primordium

The flower meristem is undetermined in the first phase: reversion to the vegetative state can still occur in the absence of the inductive signals. In the second phase, when the inductive signals have acted on the apex for a sufficiently long time, flowering determination occurs, and four types of floral organs are initiated: sepals, petals, stamens, and carpels. These are formed in whorls on the flanks of the meristem. The identity of each of these organ whorls is based on the interaction of the so-called “MADS-box” genes (Theißen *et al.*, 2016).

A chrysanthemum flower head is called capitulum, and it contains many flowers. In chrysanthemum, different MADS-box genes, belonging to the APETALA1/FRUITFULL and SEPALLATA3 subfamilies, were identified (Shchennikova *et al.*, 2004). The different developmental stages of the floral transition are illustrated in Fig. 5.7.

When the plant assimilate status becomes more favorable (e.g. wider spacing and/or higher light intensity), chrysanthemum responds by producing more flowers and flower buds, rather than larger flowers (Carvalho and Heuvelink, 2003; Carvalho *et al.*, 2004). However, it has been clearly demonstrated that the reason for this response is not that flowers would not be able to grow bigger, as actual flower dry mass is only 42% of potential flower dry mass (Carvalho *et al.*, 2004). A positive linear relationship between the number of flowers per plant, including flower buds, and the total aerial plant dry mass was found (Carvalho and Heuvelink, 2003). A single linear relationship could adequately describe this number of flowers for highly diverse

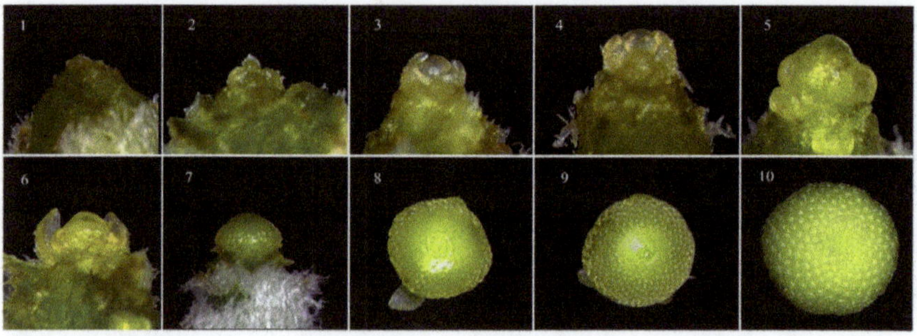

Fig. 5.7. Developmental stages of floral transition in *Chrysanthemum morifolium* cv. Radost. Stage-1 (vegetative phase)—the flat shoot apical meristem (SAM). Stage-2—between the leaf primordia, a barely perceptible bulge of SAM occurs. Stage-3—the bulge is larger, and the leaf primordia begin to deviate. Stage-4—the first bracts are visible under the leaf primordia. Stage-5–the bracts cover the dome. Stage-6 (floral initiation phase)—first floral primordia visible. Stage-7 (floral development phase)—one to four rings of floral primordia visible. Stage-8—multiple rings of floral primordia visible (less than half of area of flower head). Stage-9—multiple rings of floral primordia visible (greater than half of area of flower head). Stage-10—entire bud covered with floral primordia (visible bud emerges). Images were taken from a stereoscope with 80 × magnification. When the diameter of apex was more than 2 mm (images 8–10), a lower magnification (from 30× to 70×) was used to fit whole apex in the field of view. (Adapted from SharathKumar *et al.*, 2021, CC BY 4.0.)

growth conditions, that is, daily incident photosynthetically active radiation (4.2–18.3 mol m^{-2} day^{-1}), CO_2 concentration (345–623 µmol mol^{-1}), and duration of the LD period (0–21 days). The temperature variation was smaller but remained between 19.1 and 22.6°C (Carvalho and Heuvelink, 2003).

5.6.3 Environmental control of flower development

Besides the photoperiodic response for floral induction, flower development in chrysanthemum is dependent on both temperature and light intensity. In northern countries, the short-day phase during winter conditions also results in a relatively low daily light integral (DLI). Quite often, a macroscopic scale is used to quantify flower development, ranging from no visible bud to ray florets fully open or first row of ray florets horizontal. Using these stages, flowering development reached saturation at 6 mol m^{-2} day^{-1} (Persson and Larsen, 1998).

For all-year-round production in greenhouses, shading with blackout curtains is commonly used to achieve short-day conditions to induce flowering during long-day seasons. This action raises the air temperature in greenhouses due to reduced ventilation (Shibata, 1997) and increased absorption of irradiance. During all-year-round cultivation in greenhouses, high temperature

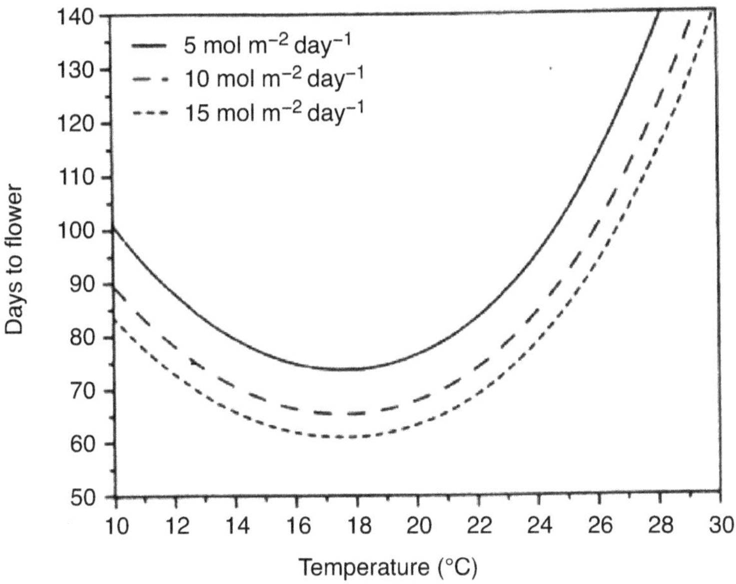

Fig. 5.8. Effects of temperature and daily light integral on flower development. (Adapted from Karlsson *et al.*, 1989, Elsevier STM agreement.)

caused by blackout shading in long-day seasons or by local climates at daytime delays the flowering of chrysanthemum, which is termed "heat-induced delay." However, higher light intensities may partially overcome the negative influence of higher temperatures on the rate of development (Fig. 5.8) (Karlsson *et al.*, 1989).

Even in short-day seasons, the air temperature in the greenhouse could raise to an unfavorable level for chrysanthemum flowering, especially in the middle of days, depending on the local climate and daily weather. The optimal temperature for chrysanthemum flowering is reported to be approximately 20°C (de Lint and Heij, 1987; Adams *et al.*, 1998). Temperature elevation (especially >26°C) could not only influence flower quality (Carvalho *et al.*, 2005) but also delay flowering (Whealy *et al.*, 1987). Although heat-tolerant cultivars have been introduced in warm regions, these cultivars are generally of reduced quality and still cannot guarantee to flower on time when the temperature reaches above 30°C (e.g. Willits and Bailey, 2000; Nakano *et al.*, 2013). Recent molecular studies revealed that the heat-induced delay is also associated with the florigenic and antiflorigenic genes that regulate flowering under photoperiod. Furthermore, high temperature during capitulum development (after visible bud stage) has a larger contribution to flowering delay compared with floral induction and initiation phases (Nozaki and Fukai, 2008; Nakano *et al.*, 2013). The possible explanation could be that during the early phases, the inhibited florigen-*FTL3* expression by high temperature is still sufficient to convert the shoot apical meristem into inflorescence meristems but becomes insufficient to promote the further stages of capitulum development and flower anthesis (Nakano *et al.*, 2015). However, Vaid and Runkle (2013) stated that high temperature during the first 2 weeks of short days has greater potential to delay flowering than later during production. Therefore, differentiation of the sensitivity of floral phases to heat-induced delay in chrysanthemum is still under investigation.

5.7 Seeds

Chrysanthemum is considered a self incompatible outcrossing species. Both pre-and postfertilization barriers are described (refer to Chapter 02). Wang *et al.* (2014) screened 24 chrysanthemum cultivars and found that most of the cultivars were self incompatible, but a small portion of cultivars was self compatible, demonstrating that chrysanthemum self incompatibility is cultivar dependent. Chrysanthemum flowers comprise ray and disk florets, and peripheral ray florets are largely responsible for ornamental value. However, decorative chrysanthemums contain a higher number of ray florets with very low fertility due to reduced receptivity of stigmatic tissues, and lack of fertilization results in poor seed set. Therefore, the ratio of disk to ray florets and the quality of pistils determines fertility and seed set in chrysanthemum. Seeds are only used to produce new cultivars, as seedlings do not come true or bear all the same characteristics as the parents. After successful pollination, it takes 10–15 days to reach the stage of the globular embryo. In most crosses, seeds become ripe within 5 weeks after self- or artificial pollination.

5.8 Developmental Physiology of Stock (Mother) Plants

5.8.1 Physiological background

Chrysanthemum production, whether cut flowers or cushion types, is a global industry with firms specialized in cutting production. The climate of the (sub) tropical highlands is excellent for producing the most popular ornamentals. Also, for chrysanthemum, the majority of the commercial cutting production takes place in Africa or Latin America, where light conditions during winter are more favorable, resulting in more vigorous cuttings compared with those in temperate climates. Also, the lower wages and other production costs are an advantage. This globalization results in extra phases in the production chain, namely the long-distance transport by air and trucks and a storage period of unrooted cuttings up to several weeks. Immediately after severing the cuttings from the mother plant, leaves maintain a high photosynthetic capacity (Hegewald *et al.*, 1999), so if this is followed immediately by rooting under sufficient light intensity, carbohydrates should not be the limiting factor. However, transport and dark storage will affect the carbohydrate pool of the cuttings.

The quality of apical cuttings starts with the vigor of the stock plants, which is highly dependent on irradiance level and nitrogen availability (Zerche *et al.*, 1999). Drüge *et al.* (2000) studied the effect of nitrogen supply to stock plants (0.6, 1.5, or 4.0 g N m^{-2} $week^{-1}$) on cutting quality, and this for increasing periods (2, 3, or 4 weeks) of dark cold storage (0.5 or 5°C) on the adventitious rooting of chrysanthemum. Internal nitrogen concentrations ranged from 2.7 to 6.5% of dry matter. Here, a positive relationship between the number of roots and root length and increasing nitrogen content was found, and this positive effect was found for both stored and unstored cuttings. In addition, the carbohydrate content was measured in this experiment; however, the conclusions were unclear as high nitrogen content correlated with high sucrose levels and decreasing starch levels. It is, however, clear that dark storage of cuttings results in depletion of carbohydrate reserves after ± 2 weeks (Drüge *et al.*, 2000).

A survey of chrysanthemum stock plants grown in Central America showed that the total nitrogen content of harvested cuttings ranged from 3.6 to 5.8% DM (average of 20 cultivars) (Zerche *et al.*, 2019). However, after shipping the cuttings to Germany, no correlation between nitrogen (total nitrogen ranged between 4.7 and 5.7% DM) and carbohydrates in leaves or stems was found, with an exception for glucose that positively correlated with N proteins. Shifts among the soluble and insoluble N pools after transport result from unpredictable events that affect carbohydrate content of the cuttings at harvest (DLI and temperature before harvest), followed by partial or total depletion of carbohydrates depending on the transport temperature and duration of dark storage. This can lead to a postharvest shift from N bound in proteins to soluble N pools, which can maintain respiration if carbohydrates are depleted. So, a sufficient and relatively high N content in the growing axillary shoots of the mother plant is needed before harvesting them as cuttings. A minimal 4.2% nitrogen content in chrysanthemum is advised to secure a qualitative rooted cutting, while lower values will increase the risk of reduced rooting (Drüge *et al.*, 2000; Zerche *et al.*, 2019).

5.8.2 Grafting to enhance stock plant's vigor and rooting quality of cuttings

Chrysanthemum apical cuttings are taken from healthy and vigorously growing stock plants. However, cutting vigor tends to decline over time after repeated harvests (multiple collections) from the same stock plants. Furthermore, the cutting homogeneity and quality (e.g. ease of root formation) also decline with the stock plant's age. This decline in vigor and quality may be associated with an internal imbalance in the hormone and metabolite levels and may also be associated with epigenetic changes (Drüge *et al.*, 2019); epigenetic changes are modifications arising from the growing environment that can potentially alter the genetic (inherent) properties of elite genotypes over time. The success of chrysanthemum propagation relies on adventitious rooting ability and homogeneity of cuttings. Homogeneity of cuttings helps crop scheduling with the predictability of flowering time and harvest date (De Ruiter, 1996). The reduced number of cuttings due to lower vigor and insufficient rooting causes a significant economic loss for breeding companies and rooted cutting suppliers, despite intensive control over the growth factors (such as photoperiod, temperature, and relative humidity) in the chrysanthemum propagation industry. It is expensive and laborious to establish a new batch of stock plants to meet the need for vigorous and well-rootable cuttings; the development of techniques to extend the period for cutting collection is of significant economic interest (Zhang *et al.*, 2013).

To mitigate reduced productivity and rooting ability of cuttings, grafting of mother stock plants has been proposed. Grafting increased water and nutrient uptake, translocation and utilization efficiency with increased synthesis of endogenous hormones to improve physiological processes associated with growth, signal transduction, and development (Nawaz *et al.*, 2016). Hence, grafting is a way to increase the stock plant's cutting productivity and rooting ability. For example, the wild species used belongs to the herbaceous genus *Artemisia*, which, like chrysanthemum, belongs to the *Asteraceae* family and serves as a useful rootstock with a strong root system. Plants grafted by "hole insertion" grafting (described by Lee, 1994) on *Artemisia* showed higher vigor with enhanced rooting ability and tolerance to various biotic (*Alternaria* leaf spot—Deng *et al.*, 2012) and abiotic stresses (drought—Chen *et al.*, 2018). Cuttings from mother stock plants grafted onto *Artemisia scoparia* showed improved rooting ability, caused by higher auxin (IAA/IBA) levels, and improved growth compared with cuttings from non-grafted mother stock plants (Zhang *et al.*, 2013). Therefore, grafting to generate stock plants of elite genotypes can be expected in commercial practice eventually. However, grafting techniques and rootstock–scion relationships need to be optimized by selecting compatible rootstocks for thousands of chrysanthemum genotypes, in order to improve the vigor, cutting productivity, and rooting ability in chrysanthemum.

5.9 Conclusion and Future Prospects

The developmental physiology of chrysanthemums is shaped by a broad array of factors that influence their growth and development. The interplay between environmental cues—temperature, light intensity, and CO_2 concentration—and the plant's physiological responses critically determines the overall productivity and quality of chrysanthemums. Notably, different chrysanthemum genotypes exhibit distinct responses to temperature variations. The photoperiod plays a critical role in floral induction, while hormonal processes regulate adventitious rooting and shoot branching. These insights not only summarize current understandings but also pave the way for future research focused on optimizing these responses to environmental variables, thereby enhancing cultivation practices and genetic selection for resilience and productivity. Utilizing grafting techniques has proven to enhance the vigor of stock plants and improve the rooting quality of cuttings. Optimizing these techniques and identifying compatible rootstock–scion combinations could further enhance propagation efficiency, leading to plants with improved health and stress tolerance. Long-term studies on grafting's effects are essential for refining commercial cultivation practices. Future research should focus on unraveling the genetic and molecular bases of chrysanthemum development. By understanding the regulation of key genes involved in phenology, root formation, and flowering, breeders can develop new varieties with desirable traits. Advances in genome editing, such as CRISPR/Cas9, offer opportunities for precise genetic enhancements to improve stress resistance, floral characteristics, and adaptability to environmental changes. As climate change continues to reshape agricultural practices, developing resilient chrysanthemum varieties is imperative. Breeding programs should prioritize traits that enhance tolerance to extreme temperatures and fluctuating photoperiods. Integrating advanced horticultural technologies, such as precision agriculture and controlled environment agriculture (CEA), can significantly improve the efficiency and sustainability of chrysanthemum production. Moreover, employing LED lighting to manipulate light quality and photoperiods could further optimize growth and flowering while minimizing energy consumption.

References

Adams, S.R., Pearson, S. and Hadley, P. (1998) An appraisal of the use of reciprocal transfer experiments: Assessing the stages of photoperiod sensitivity in chrysanthemum cv. Snowdon (*Chrysanthemum morifolium* Ramat.). *Journal of Experimental Botany* 49(325), 1405–1411. DOI: 10.1093/jxb/49.325.1405.

Aliniaeifard, S. and van Meeteren, U. (2016) Stomatal characteristics and desiccation response of leaves of cut chrysanthemum (*Chrysanthemum morifolium*) flowers grown at high air humidity. *Scientia Horticulturae* 205, 84–89. DOI: 10.1016/j.scienta.2016.04.025.

Bae, G. and Choi, G. (2008) Decoding of light signals by plant phytochromes and their interacting proteins. *Annual Review of Plant Biology* 59, 281–311. DOI: 10.1146/annurev.arplant.59.032607.092859.

Berg, A.R. and Cutter, E.G. (1969) Leaf initiation rates and volume growth rates in the shoot apex of chrysanthemum. *American Journal of Botany* 56, 153–159. DOI: 10.1002/j.1537-2197.1969.tb07518.x.

Carvalho, S.M.P. and Heuvelink, E. (2003) Effect of assimilate availability on flower characteristics and plant height of cut chrysanthemum: An integrated study. *Journal of Horticultural Science and Biotechnology* 78, 711–720. DOI: 10.1080/14620316.2003.11511688.

Carvalho, S.M.P. and Heuvelink, E. (2004) Modelling external quality of cut chrysanthemum: Achievements and limitations. *Acta Hortic* 654, 287–294. DOI: 10.17660/ActaHortic.2004.654.33.

Carvalho, S.M.P., Heuvelink, E., Cascais, R. and Kooten, O. (2002) Effect of day and night temperature on internode and stem length in chrysanthemum:Is everything explained by DIF. *Annals of Botany* 90, 111–118. DOI: 10.1093/aob/mcf154.

Carvalho, S.M.P., Heuvelink, E., Harbinson, J. and Kooten, O. (2004) Role of sink–source relationships in chrysanthemum flower size and total biomass production. *Physiologia Plantarum* 128, 263–273. DOI: 10.1111/j.1399-3054.2006.00733.x.

Carvalho, S.M.P., Abi-Tarabay, H. and Heuvelink, E. (2005) Temperature affects chrysanthemum flower characteristics differently during three phases of the cultivation period. *Journal of Horticultural Science & Biotechnology* 80, 209–216.

Carvalho, S.M.P., Noort, F.R., Postma, R. and Heuvelink, E. (2008) Possibilities for producing compact floricultural crops. Wageningen UR Greenhouse Horticulture Report 173. Wageningen. Available at: https://edepot.wur.nl/11685

Chang, Q., Chen, S., Chen, Y., Deng, Y., Chen, F. *et al.* (2013) Anatomical and physiological differences and differentially expressed genes between the green and yellow leaf tissue in a variegated chrysanthemum variety. *Molecular Biotechnology* 54, 393–411. DOI: 10.1007/s12033-012-9578-8.

Chen, X., Zhou, X., Xi, L., Li, J., Zhao, R. *et al.* (2013) Roles of DgBRC1 in Regulation of lateral branching in chrysanthemum (*Dendranthema × grandiflora* cv. Jinba). *PLoS One* 8, e61717. DOI: 10.1371/journal.pone.0061717.

Chen, Y., Sun, X., Zheng, C., Zhang, S. and Yang, J. (2018) Grafting onto artemisia annua improves drought tolerance in chrysanthemum by enhancing photosynthetic capacity. *Horticultural Plant Journal* 4(3), 117–125. DOI: 10.1016/j.hpj.2018.03.008.

Christiaens, A., Labeke, M.C., Gobin, B. and Huylenbroeck, J. (2015) Rooting of ornamental cuttings affected by spectral light quality. *Acta Horticulturae* 1104, 219–224. DOI: 10.17660/ActaHortic.2015.1104.33.

Christiaens, A., Gobin, B., Huylenbroeck, J. and Labeke, M.C. (2019) Adventitious rooting of *Chrysanthemum* is stimulated by a low red:far-red ratio. *Journal of Plant Physiology* 236, 117–123. DOI: 10.1016/j.jplph.2019.03.008.

Cline, M.G. (1991) Apical dominance. *Botanical Review* 57, 318–358. DOI: 10.1007/BF02858771.

Cockshull, K.E. (1985) *Chrysanthemum morifolium*. In: Halevy, A.H. (ed.) *Handbook of Flowering*, Vol. 4. CRC Press, Boca Raton, Fla, pp. 238–257.

De Jong, J. (1981) Flower initiation of chrysanthemum seedlings grown continuously in short days at four levels of irradiance. *Scientia Horticulturae* 14, 277–284. DOI: 10.1016/0304-4238(81)90023-6.

De Lint, P.J.A.L. and Heij, G. (1987) Effects of day and night temperature on growth and flowering of chrysanthemum. *Acta Horticulturae* 197, 53–62. DOI: 10.17660/ActaHortic.1987.197.5.

Deng, Y., Chen, S., Chang, Q., Wang, H. and Chen, F. (2012) The *Chrysanthemum × Artemisia vulgaris* intergeneric hybrid has better rooting ability and higher resistance to alternaria leaf spot than its chrysanthemum parent. *Scientia Horticulturae* 134, 185–190. DOI: 10.1016/j.scienta.2011.11.012.

De Ruiter, H.A. (1996) Development of chrysanthemum cuttings: The influence of age and position of the axillary buds. *Annals of Botany* 77, 99–104. DOI: 10.1006/anbo.1996.0012.

Dierck, R., Keyser, E., Riek, J., Dhooghe, E., Huylenbroeck, J. *et al.* (2015) Hormonal and genetic regulation of axillary bud outgrowth in *Chrysanthemum morifolium* during floral initiation. *Acta Horticulturae* 1087, 179–185. DOI: 10.17660/ActaHortic.2015.1087.21.

Dierck, R., Leus, L., Dhooghe, E., Huylebroeck, J., Riek, J. *et al.* (2018) Branching gene expression during chrysanthemum axillary bud outgrowth regulated by strigolactone and auxin transport. *Plant Growth Regulation* 86, 23–36. DOI: 10.1007/s10725-018-0408-2.

Drüge, U., Zerche, S., Kadner, R. and Ernst, M. (2000) Relation between nitrogen status, carbohydrate distribution and subsequent rooting of chrysanthemum cuttings as affected by pre-harvest nitrogen supply and cold-storage. *Annals of Botany* 85(5), 687–701. DOI: 10.1006/anbo.2000.1132.

Drüge, U., Hilo, A., Pérez-Pérez, J.M., Klopotek, Y., Acosta, M. *et al.* (2019) Molecular and physiological control of adventitious rooting in cuttings: Phytohormone action meets resource allocation. *Annals of Botany* 123, 929–949. DOI: 10.1093/aob/mcy234.

Fan, H.M., Liu, B.W., Ma, F.F., Sun, X. and Zheng, C.S. (2019) Proteomic profiling of root system development proteins in chrysanthemum overexpressing the CmTCP20 gene. *Plant Science* 287, 110175. DOI: 10.1016/j.plantsci.2019.110175.

Gil, C.S., Jung, H.J., Lee, C. and Eom, S.H. (2020) Blue light and NAA treatment significantly improve rooting on single leaf-bud cutting of *Chrysanthemum* via upregulated rooting-related genes. *Scientia Horticulturae* 274, 109650. DOI: 10.1016/j.scienta.2020.109650.

González-Grandío, E., Poza-Carrión, C., Sorzano, C.O. and Cubas, P. (2013) BRANCHED1 promotes axillary bud dormancy in response to shade in *Arabidopsis*. *The Plant Cell* 25(3), 834–850. DOI: 10.1105/tpc.112.108480.

Halliday, K.J., Martínez-García, J.F. and Josse, E.M. (2009) Integration of light and auxin signaling. *Cold Spring Harbor Perspectives in Biology* 1(6), 11. DOI: 10.1101/cshperspect.a001586.

Han, S., Chen, S.M., Song, A.P., Liu, R.X., Li, H.Y. *et al.* (2017) Photosynthetic responses of *Chrysanthemum morifolium* to growth irradiance: Morphology, anatomy and chloroplast ultrastructure. *Photosynthetica* 55(1), 184–192. DOI: 10.1007/s11099-016-0219-5.

Hegewald, J., Kadner, R., Drüge, U. and Zerche, S. (1999) Photosyntheserate von Chrysanthemenstecklingen während der Bewurzelung unter dem Einfluss unterschiedlicher Stickstofversorgung der Mutterpfanzen und Lichtbedingungen während der Vermehrung. *BDGL Schriftenreihe* 17, 159.

Higuchi, H. and Hara, M. (1974) Effect of chilling treatment of cuttings on the growth and flowering of chrysanthemum. *Bulletin of Aichi Agricultural Research Center* B-6, 7–22.

Higuchi, Y., Sumitomo, K., Oda, A., Shimizu, H. and Hisamatsu, T. (2012) Day light quality affects the night-break response in the short-day plant chrysanthemum, suggesting differential phytochrome-mediated regulation of flowering. *Journal of Plant Physiology* 169(18), 1789–1796. DOI: 10.1016/j.jplph.2012.07.003.

Higuchi, Y., Narumi, T., Oda, A., Nakano, Y., Sumitomo, K. *et al.* (2013) The gated induction system of a systemic floral inhibitor, antiflorigen, determines obligate short-day flowering in chrysanthemums. *Proceedings of the National Academy of Sciences of the United States of America* 110(42), 17137–17142. DOI: 10.1073/pnas.1307617110.

Hisamatsu, T., Sumitomo, K., Shibata, M. and Koshioka, M. (2017) Seasonal variability in dormancy and flowering competence in *Chrysanthemum*: Chilling impacts on shoot extension growth and flowering capacity. *Japan Agricultural Research Quarterly* 51, 343–350. DOI: 10.6090/jarq.51.343.

Karlsson, M.G. and Heins, R.D. (1994) A model of chrysanthemum stem elongation. *Journal of the American Society for Horticultural Science* 119, 403–407. DOI: 10.21273/JASHS.119.3.403.

Karlsson, M.G., Heins, R.D., Erwin, J.E., Berghage, R.D., Carlson, W.H. *et al.* (1989) Irradiance and temperature effects on time of development and flower size in chrysanthemum. *Scientia Horticulturae* 39(3), 257–267. DOI: 10.1016/0304-4238(89)90138-6.

Kawata, J. (1987) The phasic development of chrysanthemum as a basis for the regulation of vegetative growth and flowering in Japan. *Acta Horticulturae* 197, 115–123. DOI: 10.17660/ActaHortic.1987.197.14.

Klie, M., Menz, I., Linde, M. and Debener, T. (2016) Strigolactone pathway genes and plant architecture: Association analysis and QTL detection for horticultural traits in chrysanthemum. *Molecular Genetics and Genomics* 291(2), 957–969. DOI: 10.1007/s00438-015-1155-y.

Kozáowska, A., Breĝ, W., Krzesiêski, W. and Trelka, W. (2011) The effect of amount of light and the temperature on biomorphological characteristic of chrysanthemum during all-year culture. *Acta Scientiarum Polonorum Hortorum Cultus* 10(3), 235–246.

Kunitake, T., Tanigawa, T., Matsuno, T., Yamada, A. and Suyama, T. (2009) Exposure of rooted cuttings to long days, or mother plants to high temperature, can avoid delayed flowering of *Chrysanthemum morifolium* Ramat. "Jinba" for winter and spring cut-flower production. *Journal of the Japanese Society for Horticultural Science* 78(3), 363–368. DOI: 10.2503/jjshs1.78.363.

Lang, G., Early, J., Martin, G. and Darnell, R. (1987) Endo-, para-, and ecodormancy: Physiological terminology and classification for dormancy research. *Hortscience* 22, 377. DOI: 10.21273/HORTSCI.22.5.701b.

Larsen, R.U. and Hidden, C. (1995) Predicting leaf unfolding in flower induced shoots of greenhouse grown chrysanthemum. *Scientia Horticulturae* 63, 225–239. DOI: 10.1016/0304-4238(95)00801-Y.

Lee, J.M. (1994) Cultivation of grafted vegetables I. Current status, grafting methods, and benefits. *Hortscience* 29, 235–239. DOI: 10.21273/HORTSCI.29.4.235.

Liang, J., Zhao, L., Challis, R. and Leyser, O. (2010) Strigolactone regulation of shoot branching in chrysanthemum (*Dendranthema grandiflorum*). *Journal of Experimental Botany* 61(11), 3069–3078. DOI: 10.1093/jxb/erq133.

Liu, R., Chen, S., Jiang, J., Zhu, L., Zheng, C. *et al.* (2013) Proteomic changes in the base of chrysanthemum cuttings during adventitious root formation. *BMC Genomics* 14, 919. DOI: 10.1186/1471-2164-14-919.

Liu, W., Peng, B., Song, A., Jiang, J. and Chen, F. (2020) Sugar transporter, CmSWEET17, promotes bud outgrowth in *Chrysanthemum morifolium*. *Genes* 11(1), 26. DOI: 10.3390/genes11010026.

Liu, W., Peng, B., Song, A., Zhang, Y., Jiang, J. *et al.* (2022) Sucrose-induced bud outgrowth in *Chrysanthemum morifolium* involves changes of auxin transport and gene expression. *Scientia Horticulturae* 296, 110904. DOI: 10.1016/j.scienta.2022.110904.

Ma, Y.P., Zhao, L., Zhang, W.J., Zhang, Y.H., Xing, X. *et al.* (2020) Origins of cultivars of *Chrysanthemum*—evidence from the chloroplast genome and nuclear LFY gene. *Journal of Systematics and Evolution* 58, 925–944. DOI: 10.1111/jse.12682.

McMahon, M.J. and Kelly, J.W. (1995) Anatomy and developed pigments of chrysanthemum leaves under spectrally selective filters. *Scientia Horticulturae* 64, 203–209. DOI: 10.1016/0304-4238(95)00838-1.

Meng, Q. and Runkle, E.S. (2019) Regulation of flowering by green light depends on its photon flux density and involves cryptochromes. *Physiologia Plant* 166(3), 762–771. DOI: 10.1111/ppl.12832.

Nakano, Y., Higuchi, Y., Sumitomo, K. and Hisamatsu, T. (2013) Flowering retardation by high temperature in chrysanthemums: Involvement of FLOWERING LOCUS T-like 3 gene repression. *Journal of Experimental Botany* 64(4), 909–920. DOI: 10.1093/jxb/ers370.

Nakano, Y., Higuchi, Y., Sumitomo, K., Oda, A. and Hisamatsu, T. (2015) Delay of flowering by high temperature in chrysanthemum: Heat-sensitive time-of-day and heat effects on CsFTL3 and CsAFT gene expression. *The Journal of Horticultural Science and Biotechnology* 90(2), 143–149. DOI: 10.1080/14620316.2015.11513165.

Nawaz, M.A., Imtiaz, M., Kong, Q., Cheng, F., Ahmed, W. *et al.* (2016) Grafting: A technique to modify ion accumulation in horticultural crops. *Frontiers in Plant Science* 7, 1457. DOI: 10.3389/fpls.2016.01457.

Niwa, M., Daimon, Y., Kurotani, K., Higo, A., Pruneda-Paz, J.L. *et al.* (2013) BRANCHED1 interacts with FLOWERING LOCUS to repress the floral transition of the axillary meristems in *Arabidopsis*. *The Plant Cell* 25(4), 1228–1242. DOI: 10.1105/tpc.112.109090.

Nozaki, K. and Fukai, S. (2008) Effects of high temperature on floral development and flowering in spray chrysanthemum. *Journal of Applied Horticulture* 10(1), 8–14. DOI: 10.37855/jah.2008.v10i01.02.

Oda, A., Narumi, T., Li, T., Kando, T., Higuchi, Y. *et al.* (2012) CsFTL3, a chrysanthemum FLOWERING LOCUS T-like gene, is a key regulator of photoperiodic flowering in chrysanthemums. *Journal of Experimental Botany* 63(3), 1461–1477. DOI: 10.1093/jxb/err387.

Pacurar, D.I., Perrone, I. and Bellini, C. (2014) Auxin is a central player in the hormone cross-talks that control adventitious rooting. *Physiologia Plant* 151, 83–96. DOI: 10.1111/ppl.12171.

Persson, L. and Larsen, R.U. (1998) Adapting a prediction model for flower development in chrysanthemum to new cultivars. *Acta Horticulturae* 456, 143–150. DOI: 10.17660/ActaHortic.1998.456.16.

Rameau, C., Bertheloot, J., Leduc, N., Andrieu, B., Foucher, F. *et al.* (2015) Multiple pathways regulate shoot branching. *Frontiers in Plant Science* 5, 741. DOI: 10.3389/fpls.2014.00741.

Rashotte, A.M., Brady, S.R., Reed, R.C., Ante, S.J. and Muday, G.K. (2000) Basipetal auxin transport is required for gravitropism in roots of *Arabidopsis*. *Plant Physiology* 122, 481–490. DOI: 10.1104/pp.122.2.481.

Satar, V.P., Panchbhai, D.M., Badge, S. and Shivankar, S. (2016) Growth and flower yield of annual chrysanthemum as influenced by nitrogen and phosphorus levels. *Plant Archives* 16, 523–526.

Seif, M., Aliniaeifard, S., Arab, M., Mehrjerdi, M.Z., Shomali, A. *et al.* (2021) Monochromatic red light during plant growth decreases the size and improves the functionality of stomata in chrysanthemum. *Functional Plant Biology* 48, 515–528. DOI: 10.1071/FP20280.

SharathKumar, M. (2023) *To Bloom or not to Bloom: The Effect of Photoperiod Is Spectrum Dependent*. Doctor of Philosophy, Wageningen University, Wageningen. DOI: 10.18174/589945.

SharathKumar, M., Heuvelink, E., Marcelis, L.F.M. and Ieperen, W. (2021) Floral induction in the short-day plant chrysanthemum under blue and red extended long-days. *Frontiers in Plant Science* 11, 610041. DOI: 10.3389/fpls.2020.610041.

Shchennikova, A.V., Shulga, O., Immink, R., Skryabin, K.G. and Angenent, G.C. (2004) Identification and characterization of four chrysanthemum MADS-Box genes, belonging to the APETALA1/FRUITFULL and SEPALLATA3 subfamilies. *Plant Physiology* 134(4), 1632–1641. DOI: 10.1104/pp.103.036665.

Shibata, M. (1997) Studies on responses to temperature and photoperiod and breeding of spray type chrysanthemums with summer-to-autumn flowering. *Bulletin of the National Research Institute of Vegetables, Ornamental Plants and Tea* 12, 1–71.

Steffens, B. and Rasmussen, A. (2016) The physiology of adventitious roots. *Plant Physiology* 170(2), 603–617. DOI: 10.1104/pp.15.01360.

Sumitomo, K., Narumi, T., Satoh, S. and Hisamatsu, T. (2008) Involvement of the ethylene response pathway in dormancy induction in chrysanthemum. *Journal of Experimental Botany* 59(15), 4075–4082. DOI: 10.1093/jxb/ern247.

Sumitomo, K., Yamagata, A., Oda, A. and Hisamatsu, T. (2014) Identification of high long-day leaf number cultivars and prevention of premature budding by cold pre-treatment for fine control of flowering in summer-to-autumn-flowering chrysanthemum (*Chrysanthemum morifolium* Ramat.). *The Journal of Horticultural Science and Biotechnology* 89(6), 647–654. DOI: 10.1080/14620316.2014.11513133.

Sun, C.H., Yu, J.Q., Duan, X., Wang, J.H., Zhang, Q.Y. *et al.* (2018) The MADS transcription factor CmANR1 positively modulates root system development by directly regulating CmPIN2 in chrysanthemum. *Horticuture Research* 5, 52. DOI: 10.1038/s41438-018-0061-y.

Sun, W., Yang, X., Su, J., Guan, Z., Jiang, J. *et al.* (2019) The genetics of planting density-dependent branching in chrysanthemum. *Scientia Horticulturae* 256, 108598. DOI: 10.1016/j.scienta.2019.108598.

Swarup, R. and Péret, B. (2012) AUX/LAX family of auxin influx carriers-an overview. *Frontiers Plant Science* 3, 225. DOI: 10.3389/fpls.2012.00225.

Theißen, G., Melzer, R. and Rümpler, F. (2016) MADS-domain transcription factors and the floral quartet model of flower development: Linking plant development and evolution. *Development* 143, 3259–3271. DOI: 10.1242/dev.134080.

Vaid, T.M. and Runkle, E.S. (2013) Developing flowering rate models in response to mean temperature for common annual ornamental crops. *Scientia Horticulturae* 161, 15–23. DOI: 10.1016/j.scienta.2013.06.032.

Van der Ploeg, A. and Heuvelink, E. (2006) The influence of temperature on growth and development of chrysanthemum cultivars: A review. *Journal of Horticultural Science and Biotechnology* 81, 174–182. DOI: 10.1080/14620316.2006.11512047.

Van der Ploeg, A., Kularathne, R.J., Carvalho, S.M. and Heuvelink, E. (2007) Variation between cut chrysanthemum cultivars in response to suboptimal temperature. *Journal of the American Society for Horticultural Science* 132(1), 52–59. DOI: 10.21273/JASHS.132.1.52.

Van Ruiten, J.E.M. and De Jong, J. (1984) Speed of flower induction in *Chrysanthemum morifolium* depends on cultivar and temperature. *Scientia Horticulturae* 23, 287–294. DOI: 10.1016/0304-4238(84)90073-6.

Wang, F., Zhang, F.J., Chen, F.D., Fang, W.J. and Teng, N.J. (2014) Identification of chrysanthemum (*Chrysanthemum morifolium*) self-incompatibility. *The Scientific World Journal* 625658. DOI: 10.1155/2014/625658.

Whealy, C.A., Nell, T.A., Barret, J.E. and Larson, R.A. (1987) High temperature effects on growth and floral development of chrysanthemum. *Journal of the American Society for Horticultural Science* 112, 464–468.

Willits, D.H. and Bailey, D.A. (2000) The effect of night temperature on chrysanthemum flowering: Heat-tolerant versus heat-sensitive cultivars. *Scientia Horticulturae* 83(3–4), 325–330. DOI: 10.1016/S0304-4238(99)00091-6.

Xi, L., Wen, C., Fang, S., Chen, X., Nie, J. *et al.* (2015) Impacts of strigolactone on shoot branching under phosphate starvation in chrysanthemum (*Dendranthema grandifolium* cv. Jinba). *Frontiers Plant Science* 6, 694. DOI: 10.3389/fpls.2015.00694.

Yang, Y., Hammes, U.Z., Taylor, C.G., Schachtman, D.P. and Nielsen, E. (2006) High-affinity auxin transport by the AUX1 influx carrier protein. *Current Biology* 16, 1123–1127. DOI: 10.1016/j.cub.2006.04.029.

Yuan, C., Ahmad, S., Cheng, T., Wang, J., Pan, H. *et al.* (2018) Red to far-red light ratio modulates hormonal and genetic control of axillary bud outgrowth in Chrysanthemum (*Dendranthema grandiflorum* "Jinba"). *International Journal of Molecular Science* 19, 1590. DOI: 10.3390/ijms19061590.

Zerche, S., Kadner, R. and Drüge, U. (1999) Effect of cultivar, nitrogen nutrition and cultivating system of chrysanthemum mother plants on cutting yield, nitrogen concentration and subsequent rooting of cuttings. *Gartenbauwissenschaft* 64, 272–278.

Zerche, S., Lohr, D., Meinken, E. and Drüge, U. (2019) Metabolic nitrogen and carbohydrate pools as potential quality indicators of supply chains for ornamental young plants. *Scientia Horticulturae* 247, 449–462. DOI: 10.1016/j.scienta.2018.12.029.

Zhang, J., Chen, S., Liu, R., Jiang, J., Chen, F. *et al.* (2013) Chrysanthemum cutting productivity and rooting ability are improved by grafting. *Scientific World Journal* 2013, 286328. DOI: 10.1155/2013/286328.

6

High-tech Greenhouse Cultivation of Chrysanthemum

Malleshaiah SharathKumar[1,2]*, Susana M.P. Carvalho[3] and Ep Heuvelink[1]

[1]Horticulture and Product Physiology, Department of Plant Sciences, Wageningen University and Research, Wageningen 6700 AA, The Netherlands
[2]Department of Breeding and Research, Deliflor Chrysanten B.V., Korte Kruisweg 163, 2676 BS, Maasdijk, The Netherlands
[3]GreenUPorto-Sustainable Agrifood Production Research Centre/Inov4Agro, DGAOT, Faculty of Sciences of the University of Porto, Vairão, Portugal

Abstract

High-tech greenhouse cultivation enables year-round production of cut and potted chrysanthemums under controlled conditions, ensuring consistent yields of high-quality flowers. This chapter highlights the specific practices and successful strategies employed in Dutch intensive greenhouse cultivation, renowned for its technological sophistication. It provides an overview of protected chrysanthemum cultivation, emphasizing key environmental requirements, including light management (photoperiod, intensity, and spectrum), temperature control, relative humidity, ventilation, and CO_2 regulation. The chapter also explores cultural practices for both cut and potted chrysanthemum production, covering planting, irrigation, fertilization, and the regulation of photoperiod and temperature to optimize growth and flowering. Additionally, essential horticultural techniques, such as pinching and disbudding, are detailed to enhance plant architecture and bloom quality across different chrysanthemum types, including sprays and disbuds. Furthermore, recent advancements such as multilayer rooting systems for enhanced rooting efficiency, the integration of robotics-driven approaches, and the adoption of modern technologies such as plug-based rooting systems and hydroponics are discussed. These innovations contribute to sustainability and efficiency in chrysanthemum cultivation. However, while the techniques presented in this chapter reflect the state-of-the-art Dutch greenhouse practices, they may not be

*Corresponding author: sharath.malleshaiah@wur.nl; sharath.malleshaiah@deliflor.nl; sharath.malleshaiah@gmail.com

DOI: 10.1079/9781789248289.0006

universally applicable. Region-specific adaptations are essential, accounting for local environmental conditions, infrastructure, and resource availability, as these factors significantly influence the feasibility and suitability of these advanced cultivation methods worldwide.

Keywords

Climate control, cultivation, environmental factors, high-tech cultivation, photoperiod, smart farming, year-round production

6.1 Introduction

Chrysanthemums are a beloved group of flowering plants, a staple in both ornamental gardening and the floral industry, prized for their wide range of hues and variety of bloom shapes. Their versatility and ability to thrive in various climatic conditions make chrysanthemums a popular choice for growers worldwide. Chrysanthemums can be cultivated in open fields as well as under protected cultivation structures, such as shade-net houses, polyhouses, or glasshouses/greenhouses. The cultivation and production scheduling of chrysanthemums in greenhouses or other protected environments offers significant advantages over open-field cultivation. One key advantage is the ability to control climatic conditions, creating an optimal environment that positively influences the quality and the yield of the blooms. Key factors that must be carefully controlled and monitored include light, temperature, relative humidity (RH), and CO_2 dosing. In greenhouses, growers can manage the light environment, adjusting the intensity, duration, and spectral composition using light-emitting diodes (LEDs) throughout the cultivation period. This controlled environment not only improves the quality of the blooms but also extends the growing season, enabling year-round production. Moreover, to synchronize crop flowering with peaks in market demand, ensuring that blooms are available at the most advantageous festivities (such as Valentine's Day, Mother's Day, and Christmas season), and adequate climate control is essential. Additionally, high-tech greenhouse cultivation ensures a consistent availability of homogeneous products in terms of both quantity and quality while providing protection against adverse weather conditions, pests, and diseases, which are more challenging to manage in open-field cultivation (refer to Chapter 07).

Climate conditions inside the greenhouse do depend on outdoor climate conditions, which vary with the season. The level of control over the greenhouse climate depends largely on the type of protected cultivation structure used and equipment installed, and the local climate and seasonal conditions where the production takes place. In regions with extreme weather conditions, more advanced greenhouse technologies may be required to maintain optimal growing conditions year-round. Conversely, in areas with more subtropical climates, simpler structures may suffice. In all cases, understanding and integrating the local weather patterns and seasonal variations into the greenhouse management strategy is essential for maximizing the success of chrysanthemum cultivation.

In the subsequent sections, the critical environmental factors influencing greenhouse chrysanthemum cultivation are discussed in detail. These include light management (photoperiod, light intensity, and light quality), temperature regulation, RH control, heating, cooling, ventilation, and CO_2 dosing. This chapter also provides an in-depth discussion of specific horticultural and cultivation practices tailored for both cut and potted chrysanthemums. This includes soil preparation (ploughing, sterilization, and liming), soil composition, planting block-rooted cuttings, spacing, and planting methods. Additionally, cultivation management practices such as irrigation strategies, fertilizer management, and advancements in sustainable drip irrigation systems are explored. The role of growth regulators in controlling plant development is also examined. Furthermore, essential horticultural techniques, such as pinching and disbudding, are detailed, highlighting their role in shaping plant architecture and enhancing bloom quality. Finally, this chapter also discusses innovations in sustainable growing practices, ensuring efficient resource utilization while maintaining high production standards.

6.2 Critical Environmental Factors: Relevance and Management

6.2.1 Light duration (photoperiod)

Light can be pointed out as the most critical factor in chrysanthemum cultivation, influencing both vegetative growth and the flowering phase. Chrysanthemum is a short-day plant: it needs long nights (13 hour or longer) for flower initiation (refer to Chapter 05). Long days are applied to promote vegetative growth and short days to induce flowering. Additionally, night break strategies are often employed to precisely control flowering and schedule the date of harvest. To optimize growth, besides light duration (photoperiod), the intensity and spectral quality of light must be carefully adjusted to meet the plant's needs throughout its growth cycle. This is particularly important in regions with limited natural light during a part of the year.

6.2.1.1 Night breaks/night interruptions

Growers can manipulate flowering time by interrupting the long-night phase with a short period of light exposure, a technique known as a night break or night interruption. This method effectively inhibits flowering, when the required long-night phase is disrupted by short pulses of light. Chrysanthemum growers commonly apply multiple night breaks of 10–20 min using available light systems at very low light intensity to prevent the plants from entering the flowering stage (Higuchi *et al.*, 2012). This practice enables growers to regulate flowering time or inhibit flowering during short winter days.

6.2.1.2 Long days for vegetative growth

During the vegetative growth phase, chrysanthemum requires long days, typically 14–16 hour of light. In intensive greenhouse cultivation, such as in the Netherlands, the photoperiod is sometimes extended to 20 hour of light, with only 4 hour of darkness. This extended photoperiod promotes robust leaf and stem growth, enabling the plants to accumulate sufficient leaf area, biomass, and plant height (the latter being particularly relevant for cut chrysanthemum) before transitioning to the flowering stage. During winter or in regions with limited daylight, supplemental lighting is often used to extend the photoperiod and support vegetative growth. This can be provided either as high-intensity photosynthetic light (using high-pressure sodium lamps or LEDs) or as low-intensity light (using incandescent lamps or LEDs) to regulate day length.

6.2.1.3 Short days for flowering

Growers often use blackout screens to artificially shorten the day length (Fig. 6.1), since these screens block daylight from entering the greenhouse when natural daylight conditions do not meet the cultivation requirements. This method is needed in regions with long daylight hours during summer, where natural photoperiods would otherwise inhibit flowering. The screens are crafted from high-tech fabrics designed to either reflect or absorb light, achieving up to 99.9% blackout. This ability to control light exposure within the greenhouse creates a consistent and controlled environment, facilitating year-round chrysanthemum production regardless of external light conditions. In addition to their role in light management, blackout screens offer significant energy-saving benefits, potentially reducing energy use by 20–30% annually (A1 Group, 2024; Dutch Greenhouses, 2024). Blackout screens prevent artificial light from escaping greenhouses, enhancing energy efficiency and ensuring compliance with environmental regulations. They are particularly vital in regions with strict light pollution standards, such as the Netherlands and Canada.

6.2.2 Light intensity and spectrum

Daily light integral (DLI; mol m^{-2} day^{-1}), which is the product of light intensity and photoperiod, directly influences plant growth and flower quality. Adequate light is essential for photosynthesis, and at high latitudes with low DLI in winter, supplementary lighting is used (Fig. 6.2A and B). In the Netherlands, growers have traditionally used HPS lamps for supplementary lighting at

Fig. 6.1. Blackout screens to regulate photoperiod by totally unfolding the screens while sunlight is present, creating short-day conditions to induce flowering. (Photo courtesy of Malleshaiah SharathKumar.)

about 80 µmol m^{-2} s^{-1} light intensity at canopy level during the winter period, to ensure year-round chrysanthemum production. This level of light intensity was sufficient in cut chrysanthemum to produce flowering stems weighing 60–65 g and measuring around 65 cm. However, as market demands shifted toward heavier stems (75–80 g), growers increased the supplementary lighting intensity to 160 µmol m^{-2} s^{-1} to achieve these desired outcomes. Growers are increasingly adopting advanced LED lighting systems to enhance energy efficiency and sustainability in greenhouse cultivation (Fig. 6.2B). These systems

Fig. 6.2. Supplementary lighting using a hybrid system with HPS and LED lamps (A); and LED lighting in high-tech greenhouses, utilizing spectra rich in red and low in blue (B), to increase daily light integral. Note: HPS = high-pressure sodium, LED, light-emitting diode. (Photo courtesy of Malleshaiah SharathKumar.)

not only reduce energy consumption but also provide precise control over light intensity and spectrum, enabling growers to tailor lighting conditions to meet the specific needs of their crops. Replacing traditional HPS fixtures with advanced dimmable LED systems, equipped with an optional far-red light channel for end-of-day far-red treatment, can be especially beneficial in winter for promoting sufficient stem elongation. Lighting levels can be dynamically adjusted within the range of 150–200 μmol m^{-2} s^{-1}, ensuring cost-effective and efficient cultivation with LED technology. These adaptive LED systems allow modifications within the installed 200 μmol m^{-2} s^{-1} capacity, depending on planting density and energy costs, enabling growers to optimize both plant growth and economic feasibility. Despite the advantages of LED lighting, a complete switch to LEDs involves high investment costs and is generally more feasible in new greenhouse constructions. The expense and structural modifications required for existing facilities can be prohibitively expensive, making a hybrid (HPS + LEDs) approach a more practical solution for many growers (Fig. 6.2A).

6.2.3 Temperature

Temperature regulation is a critical factor in the successful cultivation of chrysanthemums within a greenhouse, significantly influencing every stage of the plant's growth, flowering process, and the timing of flowering (Carvalho, 2003). Chrysanthemums thrive within a specific temperature range of 20–22°C during the day, with slightly cooler temperatures at night. Maintaining this optimal temperature range is essential for promoting chrysanthemum development, ensuring plant health and high-quality bloom production. Maintaining cooler finishing temperatures (typically around 15–18°C) can improve the overall commercial quality of chrysanthemum plants, leading to larger and more vibrant blooms without negatively affecting the flowering schedule.

6.2.3.1 Daytime temperatures

The ideal daytime temperature for chrysanthemums is between 20 and 22°C, depending on the cultivar. This temperature range supports healthy vegetative growth and prepares the plants for the transition to flowering.

6.2.3.2 Nighttime temperatures

At night, temperatures should be maintained between 15 and 18°C. These cooler nighttime temperatures are crucial for slowing down plant respiration and conserving energy, which is redirected toward flower development. Consistent cool nights are also vital for the proper differentiation of flower buds, ensuring that flowering is uniform and occurs at the desired time.

6.2.3.3 Impact of high temperature on chrysanthemum flowering

The ideal temperature for inducing flowering in chrysanthemums is approximately 20°C. Flowering is often delayed when temperature exceeds 26°C, a phenomenon known as heat delay (Nozaki and Fukai, 2008; Nakano *et al.*, 2015). When temperature exceeds 26°C during the critical first 2 weeks of short day, the delay is more pronounced (Fig. 6.3). This heat delay is especially problematic in regions such as Japan, where summer and early autumn temperatures often remain above 20°C during both day and night, leading to significant disruptions in the timing of chrysanthemum flowering. Heat-induced flowering delay has posed a significant challenge for chrysanthemum production, particularly with the advent of year-round production techniques in countries such as Japan and the Netherlands. While Western countries, especially those at higher latitudes, have successfully extended the chrysanthemum flowering season through manipulation of light and darkness, Japan faces unique challenges due to its high summer temperatures, which exceed the optimal conditions for growing chrysanthemums. To mitigate this issue, Japanese breeders have developed early-flowering cultivars, known as summer-to-autumn-flowering chrysanthemums, capable of blooming under longer daylight periods and higher temperatures than traditional autumn-flowering varieties (Kawata *et al.*, 1987; Shibata, 1997). Despite these efforts, flowering delays remain problematic due to the considerable variation in heat tolerance among cultivars. This issue is further exacerbated by the increasing interannual variation in summer temperatures, amplified by global warming.

Fig. 6.3. Uneven flowering under high temperature (heat delay) in chrysanthemum "Aristotle" at 30°C. (Photo courtesy of Malleshaiah SharathKumar.)

In greenhouses, blackout screens are a crucial tool to reduce the photoperiod in natural long days, for promoting flower induction, enabling growers to produce year-round. However, during the summer months, when days are longer and sunlight is more intense, closing blackout screens can trap heat within the greenhouse. This would lead to a significant temperature increase, which could create conditions that are detrimental to chrysanthemum flowering and overall plant health. To minimize this heat buildup, growers use blackout curtains strategically, i.e. after the intense sun radiation, when ambient temperatures are naturally lower. Some growers also keep blackout screens closed in the early morning to shorten day length while radiation levels remain low.

During sunny periods with long days, blackout screens may be partially closed (up to 80%, Fig. 6.1) while sunlight is still present and later fully closed as radiation decreases, helping to reduce heat buildup within greenhouse. While this approach minimizes excessive heat accumulation, it can lead to uneven light and humidity distribution within the greenhouse, resulting in inconsistent plant growth, flower development, and coloration. Despite these challenges, blackout screens remain an effective tool for photoperiod manipulation while mitigating heat stress risks. By carefully timing their use, growers can maintain a stable greenhouse environment, ensuring plants receive the required short-day conditions without compromising on the growth and flowering quality.

6.2.4 Relative humidity

The ideal RH levels for chrysanthemum cultivation range between 60 and 70%. Humidity levels in the greenhouse are primarily influenced by crop transpiration and ventilation.

- **High humidity:** Elevated humidity levels, especially above 80 and 85%, can create conditions conducive to the development of fungal diseases, such as powdery mildew and botrytis. These diseases thrive in moist environments and can quickly spread, leading to plant damage and reduced flower quality. Excessive moisture on the leaves due to high humidity also increases the risk of bacterial infections, further compromising plant health. RH tends to rise at night due to lower temperatures and crop transpiration. To avoid very high humidity, growers employ strategies such as proper ventilation to remove excess moisture and adjusting irrigation practices to avoid overwatering, which can create overly humid conditions.
- **Low humidity:** Conversely, if humidity levels are too low (40–60%), chrysanthemums may suffer from dehydration, leading to stunted growth and poor flower development. Low humidity can also cause physiological disorders, such as leaf curl and tip burn, particularly during dry winter months, when heating systems are in use, further reducing air moisture

levels. Light misting can help stabilize low humidity levels in greenhouses. For instance, applying clean water at a rate of $0.5\,l{\cdot}m^{-2}{\cdot}h^{-1}$ on bright sunny days via misting helps maintain adequate RH levels, reducing plant stress and supporting healthy growth and development.

6.2.5 Heating, cooling, and ventilation

To successfully manage temperature in protected chrysanthemum cultivation, growers must invest in reliable climate control systems, which include heating and an adequate ventilation. These systems are essential for creating a consistent and suitable environment, thereby avoiding abrupt temperature fluctuations that could stress the plants. Moreover, such stress can delay flowering or result in bud abortion, significantly impacting yield and quality of the crop. Greenhouse climate control is crucial for optimizing growth, particularly when dealing with large variations in radiation levels between sunny and cloudy days. Sudden transitions from cloudy to sunny weather can cause rapid changes in greenhouse conditions, demanding careful adjustments to heating, cooling, and ventilation systems.

- **Heating:** In cooler regions, maintaining a warm environment inside the greenhouse, especially during the night, is essential to prevent cold stress and ensure optimal plant growth. One effective method is the installation of pipes that circulate hot water or steam throughout the greenhouse. This system provides consistent and even warmth, particularly near the plants, helping to maintain the desired temperatures.
- **Cooling:** In warmer regions, cooling systems are vital to prevent heat stress and maintaining a stable growing environment. Fans and cooling pads, installed during the construction of the greenhouse, are effective in circulating air and lowering the internal temperature. This evaporative cooling method pulls warm air through wet pads, cooling it as it passes through, and then circulates the cooler air using fans. This system effectively reduces the temperature inside the greenhouse, even during hot days, creating a more conducive environment for chrysanthemum cultivation. Nonetheless, an undesirable gradient can be created in the greenhouse.
- **Ventilation:** Proper ventilation is essential in a greenhouse to regulate both temperature and RH. Good air circulation prevents the buildup of excess heat and moisture, reducing the risk of diseases and promoting robust plant development. Early-morning ventilation can help reduce the stress caused by transitions from low nighttime transpiration to high daytime transpiration.

  - **Temperature regulation:** Ventilation helps maintain optimal temperatures by allowing hot air to escape and cooler air to circulate within the greenhouse. This is particularly crucial during warmer months, when temperatures can rise rapidly, potentially stressing the plants and hindering their growth and flowering. Proper ventilation, combined with cooling via misting or fogging, prevents overheating, ensuring that the temperature inside greenhouse remains within the ideal range. During sunny spring days, greenhouse temperatures can rise 8–10°C above outside conditions, making ventilation essential to dissipate excess heat. However, increased ventilation can also result in water vapor loss, impacting humidity levels within the greenhouse. Effectively balancing temperature control and humidity management is crucial for maintaining a stable and optimal greenhouse climate for plant growth.

- **Humidity control:** Ventilation is also key in managing humidity levels. By promoting air exchange, ventilation dissipates excess moisture, helping to maintain humidity within a safe range. Using fans or strategically placed vents helps maintain a more uniform temperature and humidity within the greenhouse. Proper ventilation minimizes the risk of stagnant air pockets, which can harbor pathogens and pests, while also promoting continuous air movement that evenly distributes carbon dioxide, enhancing photosynthesis and plant health. Additionally, effective ventilation reduces excess moisture, lowering the likelihood of disease outbreaks and creating a healthier growing environment.
- **Misting:** To regulate humidity, misting systems with nozzles installed throughout the greenhouse are recommended. For effective misting, the water pressure should be at least 3.45 megapascals (MPa). This high-pressure system produces a fine mist that evaporates quickly, cooling the air without wetting the leaves, thereby reducing the risk of disease. Proper misting helps maintain the optimal microclimate inside the greenhouse, particularly during hot periods.

6.2.6 CO_2 dosing

CO_2 level is critical for photosynthesis, making its management essential in greenhouse settings for optimizing chrysanthemum growth and achieving higher yields with consistent quality. CO_2 from flue gases is often used to enrich the greenhouse environment, particularly in the morning, when heating is required to maintain optimal temperatures. Flue gases can only be used directly for CO_2 enrichment when natural gas is the energy source; in other cases, cleaning of flue gases is needed or liquid CO_2 is used. Minimal ventilation can cause CO_2 levels inside the greenhouse to drop below ambient levels, necessitating additional CO_2 dosing to sustain plant growth. To address this, a minimum pipe temperature can be maintained, allowing CO_2 dosing even when heating demand is low. However, this approach essentially results in heat destruction to obtain CO_2. A more efficient strategy is to use the boiler for CO_2 supply during the day while storing the generated heat in a heat buffer for nighttime heating, optimizing energy use and reducing waste. Recent work at Wageningen University & Research indicates that effective chrysanthemum cultivation is achievable with lower CO_2 supply than commonly used by growers (de Visser, 2024), with benefits in terms of reduced energy consumption and associated costs. Implementing optimized CO_2 management strategies, such as adjusting dosages based on plant developmental stages and the environmental conditions can enhance sustainability in greenhouse chrysanthemum production. Following are the recommendations (de Visser, 2024), for optimizing CO_2 management in greenhouses, promoting healthy growth, maximizing photosynthetic efficiency, and minimizing resource use.

6.2.6.1 Recommended practices

- **Maintain optimal CO_2 levels:** Ensure a CO_2 concentration of 800 ppm for effective plant growth.
- **Follow light-dependent dosing:** Maintain the 800 ppm set point but adjust dosing according to available light intensity.
- **Adjust CO_2 based on ventilation:** Under high ventilation rates (window positions > 20%) significantly reduce CO_2 dosing to prevent excessive loss. Follow the guideline of 60 kg $CO_2 \cdot ha^{-1} \cdot h^{-1}$ (approximately 770–800 ppm), which aligns with the maximum uptake capacity of the crop, ensuring efficient CO_2 utilization while minimizing waste.
- **Lower doses during certain growing phases:** Apply lower doses (500–700 ppm) during the long-day phase (vegetative) and the first week of the short-day phase (generative/flowering) to match plant requirements.
- **Distribute CO_2 evenly during the day:** Inject CO_2 evenly throughout the day, except when windows are wide open.
- **Timing dosing:** Start dosing 30 min after the light period begins and stop 30 min earlier than the end of the light period.
- **Utilize fogging:** Incorporate fogging systems to enhance CO_2 utilization by the plants while reducing the need for ventilation, thereby improving overall efficiency.

6.3 Crop Installation in Cut Chrysanthemum

6.3.1 Soil preparation (ploughing, sterilization, liming)

Proper soil preparation is essential for creating optimal growing conditions for cut chrysanthemum planting by enhancing soil structure, fertility, and aeration. The process involves key steps, such as soil loosening, sterilization, and liming. Loosening the soil, typically using a small hand tractor, is the first step. It breaks up compacted soil layers, improving aeration, root penetration, and water infiltration. This step also helps remove weeds, past crop roots, and debris, ensuring a clean planting environment. After loosening, soil conditioners, such as compost or organic matter, can be incorporated to improve soil texture, increase organic content, and enhance nutrient availability.

Soil sterilization is the next step in the soil preparation process for chrysanthemum cultivation. This targets the control of soilborne pests, diseases, and weeds, ensuring a healthy growing environment. The two common methods used are steaming and fumigation, each with its own advantages and considerations.

Soil steaming is a nonchemical, environmentally friendly method that uses heat to sterilize the soil. Steam is injected into the soil, raising its temperature to around 80–100°C for 30–45 min (Fig. 6.4). The soil is typically covered with a tarp to retain the steam and ensure uniform heating. Soil steaming helps to effectively eliminate pathogens, weed seeds, nematodes, and soilborne fungi, such as *Fusarium* and *Pythium*. It also helps improve soil structure and nutrient availability and promotes healthier root development. However, steaming may not always be feasible due to the cost, logistical challenges, or unavailability of equipment.

When steaming is not a viable option, soil fumigation is an effective alternative for managing soilborne pests, diseases, and weeds. This method involves the application of chemical agents such as dichloropropane or chloropicrin to sterilize the soil. Soil fumigation begins with thorough tilling and moistening of the soil to ensure even distribution of the fumigant and to create optimal conditions for its penetration. Once prepared, the fumigant is either injected

Fig. 6.4. Soil steaming in greenhouses at 80–100°C to eliminate pathogens and pests. (Photo courtesy of Malleshaiah SharathKumar.)

Fig. 6.5. Prepared beds for planting rooted chrysanthemum cuttings. (Photo courtesy of Malleshaiah SharathKumar.)

into the soil or applied as a liquid or gas. The soil is then immediately covered with a tarp to contain the fumigant, minimize evaporation, and allow deep penetration into the soil profile. This containment ensures that the fumigant effectively eliminates soilborne pests, pathogens, and weeds. After the required exposure period, which varies based on the fumigant used, the tarp is removed, and the soil is aerated to eliminate any residual chemicals, ensuring the soil is safe for planting. While fumigation is highly effective, it necessitates strict adherence to safety protocols due to the toxic nature of the chemicals involved. Proper handling, the use of protective equipment, and compliance with local regulations are essential to minimize risks to workers, crops, and the environment.

6.3.2 Growing bed preparation

Proper bed preparation is also a critical step in chrysanthemum cultivation to create optimal growth conditions. Growing beds should be well structured to promote drainage, aeration, and nutrient availability. Beds can be prepared manually or using rotary hoes attached to tractors or small machines, which are particularly effective for incorporating soil correctives and organic matter. The ideal bed dimensions are 15–25 cm in height, 1–1.2 m in width, and approximately 25 m in length, with spacing between beds of 20–40 cm (Fig. 6.5). This spacing facilitates proper air circulation and allows sufficient room for plant maintenance, ensuring a productive and healthy crop.

6.3.3 Soil composition

A well-drained sandy loam soil with good texture and aeration is considered ideal for chrysanthemum cultivation. The soil should be rich in organic matter, nutrient-rich, and capable of supporting healthy root development. However, optimal soil conditions are seldom naturally available, prompting growers to use substrates and soil conditioners to enhance growing conditions. Organic materials, such as peat, coconut coir, rice husks, coffee husks, pine bark, and wood chips, are often incorporated to improve soil structure. Inorganic components, such as sand, vermiculite, and expanded clay, are added to enhance aeration and water retention. Since organic-rich substrates can decompose over time, potentially affecting porosity and water retention, using materials with slower decomposition rates, such as pine bark, helps maintain substrate stability and ensures consistent growth conditions.

Regular soil testing, conducted at least once every two production cycles, is essential for monitoring soil fertility and making necessary adjustments. Maintaining a soil pH between 5.5 and 6.5 is critical for nutrient availability and overall plant health, with the optimal range being 5.5–6.0. Adjustments to soil pH can be made using correctives or by modifying the soil composition. The electrical conductivity (EC) should range between 1 and 1.5 $dS \cdot m^{-1}$ (decisiemens per meter) to ensure adequate nutrient levels. High cation exchange capacity (CEC) substrates are particularly valuable, as they store and release nutrients as needed, providing a consistent nutrient supply for the plants.

Achieving the right balance between water retention and aeration is vital to prevent root diseases and promote efficient nutrient uptake.

6.3.4 Liming

The need for this step depends on the pH of the soil. In acidic soils, liming is essential to correct acidity and to improve soil health for optimal chrysanthemum cultivation. By raising the soil pH to the ideal range of 5.5–6.0, liming reduces toxic aluminum activity, enhances nutrient availability, and improves water and nutrient absorption efficiency. It should be applied about 30 days before planting to allow sufficient time for the lime to react with the soil. The amount of lime application depends on soil analysis, and dolomitic limestone is commonly used as it provides both calcium and magnesium, which are vital for plant growth.

6.3.5 Planting block-rooted cuttings

Block-rooted cuttings are commonly used as planting material (Fig. 6.6). These 5–7 cm-long cuttings are taken from healthy mother plants, typically grown in African countries, and then shipped to the Netherlands for rooting. To stimulate root development, the cuttings are treated with indole-3-butyric

Fig. 6.6. Well-rooted cuttings ready for planting, either manually or using planting machines. (Photo courtesy of Malleshaiah SharathKumar.)

acid (IBA) rooting powder. Rooting usually takes 2 weeks, after which the cuttings are delivered to growers for planting in greenhouses.

Block-rooted cuttings are planted in prepared beds with soil or substrate that is moist but not waterlogged to avoid compaction and lack of oxygen at root level. The rooted cuttings should be planted at the center of each square in the metal support frame (Fig. 6.7), at a depth of 2–3 cm. The block should be gently pressed around the base to ensure the plants remain upright and to eliminate air pockets near the root system. In high-tech greenhouses, planting is increasingly mechanized. Since 2010, machines have been used in Dutch greenhouses to plant chrysanthemum cuttings, covering vast areas with precision. Today, over 80% of all cut flower chrysanthemums in the Netherlands are planted using automated systems (Fig. 6.8). These technologies improve efficiency, reduce labor costs, and ensure uniformity, setting a benchmark for large-scale chrysanthemum production globally.

6.3.6 Spacing

A standard spacing of 12 × 15 cm is recommended, resulting in a planting density of approximately 60 plants·m^{-2}. This spacing allows adequate room

Fig. 6.7. Chrysanthemum bed view after planting block-rooted cuttings in the greenhouse. (Photo courtesy of Malleshaiah SharathKumar.)

for plant growth and airflow while optimizing the use of greenhouse space. Nonetheless, adjusting spacing based on seasonal conditions is common practice in greenhouses:

- *Higher density (summer)*: During the summer months, when ample sunlight is available, plant spacing can be reduced to 12 × 12 cm, accommodating approximately 70 plants·m^{-2}.
- *Lower density (winter)*: In the winter months, spacing is widened to 15 × 15 cm, accommodating approximately 50 plants per square meter. This adjustment ensures each plant receives sufficient light under lower DLI conditions, preventing light deprivation and maintaining plant quality. However, when supplementary lighting is used, higher planting densities can still be implemented, even during winter.

Alternative spacing configurations, such as 15 × 15 cm (yielding approximately 50 plants· per square meter), are ideal for maximizing air circulation in disease-prone environments, reducing the risk of outbreaks in high-humidity conditions. Conversely, 10 × 15 cm spacing accommodates about 67 plants

Fig. 6.8. Fully automatic transplanter operating in a high-tech Dutch greenhouse. (Photo courtesy of Malleshaiah SharathKumar)

per square meter, optimizing space utilization in controlled environments with lower disease risks. These flexible configurations enable growers to adapt effectively to specific environmental conditions, cultivar requirements, and crop management objectives, promoting optimal growth and yield across different seasons and greenhouse settings.

6.4 Cultivation Practices in Cut Chrysanthemum

6.4.1 Irrigation management

An effective irrigation management is essential for chrysanthemum growth, disease prevention, and optimizing water use efficiency. During the early growth phase, particularly in the first 2 weeks under long-day photoperiod conditions, overhead irrigation should be applied twice a week to ensure successful rooting and establishment. Micro-sprinkler systems are ideal for providing uniform moisture distribution while preventing overwatering. Typically, 15–20 $mm{\cdot}m^{-2}$ of water is recommended after planting, with adjustments based on soil type, radiation levels, season, and plant transpiration rate. Overwatering should be avoided, as excessive soil moisture creates favorable conditions for root rot infections, such as *Pythium*, which can severely impact plant health and productivity.

After the initial 2 weeks, drip irrigation is applied weekly to maintain adequate soil moisture levels. Watering from below, using two watering hoses or drip pipes per bed, ensures even water distribution across the growing beds. Irrigation should be conducted early in the morning to allow the crop to dry during the day, minimizing disease risks, particularly *Botrytis*, which thrives in humid conditions and wet foliage. Insufficient water can cause wilting, delayed flowering, or even plant death.

During hot periods, water demand may rise significantly, requiring up to 6 $l{\cdot}m^{-2}{\cdot}day^{-1}$, with weekly irrigation exceeding 40 $l{\cdot}m^{-2}$. To improve precision irrigation, growers utilize soil moisture sensors to monitor water levels in the root zone. These sensors also measure temperature and EC, though variations in soil composition after each cultivation cycle present challenges for consistent moisture readings. However, by setting irrigation thresholds per sensor section, growers can achieve reliable and comparable water content across different sections and cultivation cycles.

6.4.1.1 Factors influencing water consumption and management in chrysanthemum cultivation

Water consumption is closely related to the transpiration rate, which in turn depends on radiation level and the leaf area. Therefore, all the factors that influence leaf area are directly impacting the irrigation practices. These include:

- *Crop phase, weight, and plant height*: The plant's stage of development, weight, and height significantly influence water needs. Water demand peaks when plants reach higher than 25 cm in height, above this, crop stage is characterized by maximum transpiration. Additionally, the vegetative and flowering phases dictate varying water requirements, with flowering plants typically requiring more water.

- *Heating*: Greenhouse heating *per se* affects plant transpiration rates, especially during colder months. Depending on the intensity of heating, transpiration can increase by 1–2 $mm \cdot m^{-2} \cdot day^{-1}$, necessitating more frequent watering to compensate for moisture loss.
- *Seasonal variations, solar radiation, and supplemental lighting*: Radiation plays a crucial role in determining water usage. In greenhouses with high light transmission or supplementary lighting, particularly during the early growth stages, transpiration rates increase significantly. This heightened transpiration necessitates higher water inputs to prevent plant stress and ensure optimal growth. Seasonal variations influence these dynamics, as higher light levels in spring and summer demand more irrigation.
- *Mitigating excess moisture risks*: Excessive watering can lead to oxygen depletion in the root zone, creating favorable conditions for soil pathogens such as *Pythium* and *Fusarium*, which cause root rot and potentially plant death. Gradual and controlled watering helps maintain homogeneous soil moisture without waterlogging. Drip irrigation is particularly effective in the later stages of cultivation or during flowering, as it minimizes the risk of diseases such as *Botrytis*. Overhead or micro-sprinkler systems, while useful in early stages, should be avoided in flowering stages to prevent water contact with flowers and subsequent disease outbreaks.

6.4.2 Fertilizer management

Fertilizer management is largely influenced by soil or substrate characteristics; however, certain general practices have proven to be widely effective. A basal application of 50 $g \cdot m^{-2}$ of di-ammonium phosphate (DAP) before planting provides essential phosphorus and nitrogen for early root establishment and initial growth. From the third week after planting, weekly fertilizer applications are recommended using a balanced N–P–K fertilizer, such as 20–10–20 at 21.4–28.6 $mmol \cdot l^{-1}$ N, supplemented with micronutrients to support healthy plant development. Advanced subirrigation, drip systems, or fertigation methods that recirculate nutrient-rich water can significantly reduce fertilizer usage by 25–50%, minimize salt accumulation, and lower environmental contamination risks compared to conventional direct soil application. Along with macronutrients, balanced fertilizers enriched with micronutrients, such as calcium (Ca), magnesium (Mg), zinc (Zn), and copper (Cu), further enhance nutrient use efficiency, preventing deficiencies that could compromise plant health and flower quality. While these general fertilizer recommendations provide an effective framework, nutrient management must be tailored to specific cultivation conditions, considering factors such as crop development stage, soil moisture levels, and irradiation intensity, to optimize growth and maximize chrysanthemum yield and quality.

Drip irrigation (subirrigation) systems are increasingly favored for their ability to optimize nutrient use efficiency in greenhouse-grown chrysanthemums. By reducing nutrient supply during the vegetative phase and eliminating fertilization during the reproductive phase, growers can achieve similar yields and quality with significantly lower input costs. This strategy relies on three key physiological processes: nutrient storage during early growth, reduced nutrient uptake during flowering, and high-affinity nutrient absorption at moderate supply levels. Compared to direct soil bed fertilization, drip irrigation allows a 75–88% reduction in nitrogen, phosphorus, and sulfur inputs.

The Dutch greenhouse industry is further advancing sustainability in chrysanthemum cultivation by exploring hydroponic or water culture systems as part of the Future-proof Cut Flower Cultivation on Water project (www.vertify.nl). This innovative approach replaces traditional soil-based methods with closed-water systems, eliminating fertilizer and pesticide emissions into soil and groundwater, aligning with environmental sustainability goals. Additionally, hydroponics removes the need for energy-intensive soil disinfection practices, such as steaming, further reducing resource consumption. Research indicates that hydroponically cultivated chrysanthemums perform on par with soil-grown plants in branch structure, plant health, flower quality, and plant weight. By maintaining high production and quality standards while significantly reducing environmental impact, hydroponics is emerging as a future-oriented approach for chrysanthemum cultivation, offering a sustainable solution for high-quality flower production. However, economic feasibility remains a challenge due to the high initial investment costs, limiting widespread commercial adoption.

6.4.3 Growth regulators

Under long days with high light intensity, the continuous differentiation of nodes and leaves, coupled with increased internodal elongation, significantly affects plant height in a genotype-specific manner. In tall chrysanthemum cultivars, reducing the duration of long days is an effective strategy for controlling height while maintaining plant uniformity and flower quality. Alternatively, the application of growth regulators is a widely used and effective method to achieve consistent crop uniformity in both cut flowers and potted chrysanthemums. Growth regulators—such as B-Nine® or Dazide (daminozide), A-Rest® (ancymidol), and paclobutrazol—are commonly applied, with B-Nine® being the most common. These growth regulators inhibit cell elongation, enhance stem strength, and promote a darker green leaf color, collectively improving overall plant quality. To address specific growth requirements in cut chrysanthemums, growth regulators are applied at targeted stages of development: A 2500 ppm spray of B-Nine® is applied to the upper foliage 5 weeks after the start of short days. This application helps reduce neck stretching, ensuring stronger stems and better flower presentation.

A 25–60 ppm spray of Florgib®/ProGibb® T&O is used to elongate peduncles in pompon-type flowers. The spray is applied as a single treatment, 4–5 weeks after short-day initiation, with the solution directed toward the flower buds to achieve optimal elongation (Whipker, 2018). For optimal results, daminozide is applied one to four times during the growth cycle, with the first application occurring 2 weeks after planting to prevent internode stretching. Tall varieties and summer-grown crops, exposed to increased light intensity, may require a second application at concentrations up to 0.5%. Late applications should be avoided, as they can lead to flower discoloration, cream-colored petals in white varieties, or reduced vibrancy in pink varieties, and delays in blooming. To maintain flower quality, B-Nine® is generally not applied once flower buds are visible.

Growth regulators are most effective when applied during periods of low light, such as early morning or late evening, ensuring plants are fully turgid. This careful timing minimizes potential issues while ensuring the desired growth control and plant uniformity. Since the use of chemical growth regulators is not a sustainable practice, growers are increasingly adopting nonchemical strategies to control stem elongation and produce compact plants. One effective approach is the application of negative difference between day and night temperatures (DIF), where a higher night temperature than day temperature is maintained (refer to Chapter 05). Another promising strategy involves the use of high red-to-far-red light ratios, applied either continuously or specifically at the end of the light period (Carvalho *et al.*, 2008).

6.5 Disbudding in Spray and Disbud Chrysanthemums

Two types of cut chrysanthemums, spray types and disbud (standard) types, require disbudding as a crucial process to maintain high product quality. Disbudding is optimally performed when flower buds reach a size of 2–3 cm, typically occurring 3–4 weeks after starting the short-day phase. This labor-intensive process, currently without mechanized alternatives, is usually carried out in the morning hours when the plants are fully turgid to ensure effectiveness and ease of operation. By tailoring disbudding practices to the type of chrysanthemum, growers can achieve the desired flower number and quality to meet diverse market demands.

Disbudding in spray chrysanthemums: For spray or decorative chrysanthemums, the central flower bud (i.e. apical bud) is removed to promote the growth of lateral flowers (Fig. 6.9A). This results in a spray of 10–12 evenly sized blooms, all at a similar developmental stage and attractively arranged on the stem. Spray chrysanthemums are characterized by their smaller flowers and multiple blooms per stem (Fig. 6.9B), ideal for creating lush ornamental arrangements.

Disbudding in standard (disbud) chrysanthemums: To produce disbud or standard chrysanthemums, all the axillary buds are removed as early as possible, avoiding branching, and to focus growth on the central flower bud

Fig. 6.9. Disbudding in spray cut chrysanthemums (A), where the central flower bud (2–3 cm in size) is removed to encourage the growth of lateral flowers. A greenhouse view of spray cut chrysanthemums after disbudding (B), showing evenly developed lateral flowers arranged attractively on the stems. (Photo courtesy of Malleshaiah SharathKumar.)

(Fig. 6.10A). This technique produces large, single blooms (Fig. 6.10B), with flower diameters ranging from 10 to 15 cm or more, depending on the cultivar and flower type. This disbudding process is even more labor intensive compared with the aforementioned one.

6.6 Harvest

6.6.1 Harvesting stages

The harvesting or marketing stage of chrysanthemums depends on its type (whether standard/disbud, spray, cut chrysanthemums, or potted plants). Additional factors, such as the distance between production and sale, storage duration, and consumer preferences, further influence the decision on the optimal harvest stage.

6.6.1.1 Standard/disbud chrysanthemums

For local markets, chrysanthemums are ideally harvested when the outer petals are fully opened but not overripe or at the full anthesis stage. For distant markets, flowers are harvested at the half-open stage, when the outer two to three rows of ray florets have just begun to unfurl. This allows the flowers to continue toward anthesis during transport, reaching their peak bloom closer to the point of sale and ensuring freshness and visual appeal upon arrival.

6.6.1.2 Spray chrysanthemums

Among spray types, there are three broad flower types, each with distinct harvesting stages:

- *Spray daisy chrysanthemums*: Harvest when older flowers are fully open, but no pollen has been shed. For highest quality, three to four open flowers per spray stem are ideal, while five to six open flowers are acceptable for the second-highest grade.
- *Spray anemone or decorative types*: Harvest when older flowers are open but before the disc flowers begin to elongate, preserving flower shape and preventing premature aging.
- *Santini chrysanthemums*: Harvest when at least eight flowers are open within the top 5 cm of the umbel. The maximum flower diameter should be around 4 cm, which ensures that the flowers meet the required size and quality standards.

6.6.2 Harvesting process for cut chrysanthemums

Harvesting chrysanthemums has traditionally been a labor-intensive process, involving manual removal of plants from the soil. Workers would pull entire plants by hand, often leading to repetitive strain injuries caused by the physical

Fig. 6.10. Disbudding in standard or single-flower cut chrysanthemums (A), where the central flower bud is retained while all lateral flower buds are removed to focus growth on the central flower bud. A greenhouse view of standard or single-flower cut chrysanthemums after disbudding (B), showing evenly developed, large-sized single flowers per stem. (Photo courtesy of Malleshaiah SharathKumar.)

effort and continuous bending required to pick flowers. With over 50% of labor in chrysanthemum greenhouses dedicated to harvesting and processing, growers have increasingly turned to mechanization and automation to enhance efficiency and reduce physical strain on workers. Modern automated harvesting systems now cover the entire process, from cutting to sorting, bunching, finishing, and bundling chrysanthemums. Workers are primarily tasked with placing the mechanically cut stems onto conveyor belts and arranging the sleeved bunches into boxes.

High-tech greenhouses have introduced advanced harvesting robots (Fig. 6.11). Designed to adapt to varying bed widths, the machines cut six to seven rows of chrysanthemum stems simultaneously. Equipped with automatic shut-off features, these machines stop when the collection tray is full, allowing workers to remove the harvested stems before resuming operation. The speed of the machines can be continuously adjusted to optimize the harvesting process. The use of harvesting machines has proven to be highly efficient in high-tech greenhouses, which often require the harvesting of thousands of stems daily. These machines can process up to 5000 stems per hour, significantly reducing labor requirements. Additionally, they contribute to sustainability by chopping residual stems into small pieces and redistributing them into the soil, increasing the organic matter content.

Variations in stem readiness or flower development can result in some stems being discarded. Quality control takes place at two stages: manual inspection, where workers remove substandard stems during the conveyor belt process, and automated sorting, where advanced systems equipped with image analysis technology evaluate bloom size, color, and overall appearance to automatically discard unsuitable stems. The level of automation depends on the grower's investment in technology. High-tech greenhouses tend to use advanced sorting systems, while others employ a combination of manual and automated sorting (see later). This blend of mechanized harvesting and quality control enables consistent product standards while addressing labor challenges in large-scale production.

6.6.3 Postharvest handling in cut chrysanthemum

Once chrysanthemums are harvested, they undergo several essential postharvest steps, including checking/sorting, packing, and cooling (refer to Chapter 11). These steps are as critical as the cultivation process itself, ensuring the flowers maintain their quality and freshness. In high-tech greenhouses, postharvest processes are often highly automated, improving efficiency and ensuring consistent quality control to meet the stringent standards required by the market.

- *Cutting and initial handling*: The harvesting machine cuts the flower stems, and workers manually bundle them into groups of five and place them on the conveyor belt (Fig. 6.12). Alternatively, this bundling process can also

Fig. 6.11. Harvesting machines or robots in high-tech greenhouses. (Photo courtesy of Malleshaiah SharathKumar.)

take place during the sorting or grading stage in the packaging area, depending on the operational setup. These bundles are then placed on a conveyor belt for transport for further processing (Figs 6.12 and 6.13). The flower stems are transported, often through underground systems (Fig. 6.14), to the packaging area, where they undergo quality inspections. These checks may be conducted manually, with workers discarding substandard stems, or automatically using camera systems equipped with image analysis to grade flowers based on predefined standards. Automated sorting systems identify and remove lower-quality stems before the sleeving process.

- *Bunching and sleeving*: In the packaging area, machines cut the stems to the desired length, remove lower leaves, often yellowed or damaged, and bind the stems into bundles of five. The bundles are then sleeved automatically, ensuring consistent handling and reducing reliance on manual labor. The weight of chrysanthemum stems is a key quality factor, influenced by the number of stems per bunch. It is common practice to assess the weight of individual stems before bunching or the overall weight of a five-stem bunch as part of quality evaluation. Bunches are weighed after the sleeving stage (Fig. 6.15). Flower bunches are prepared for market by either being box-stored for export or placed in flower buckets for local retailers (Fig. 6.16).

Fig. 6.12. Machine-cut chrysanthemum stems are manually gathered by workers. Higher-quality standard stems are bundled into groups of five, while lower-quality stems are manually discarded by workers. The bundles are then placed onto a conveyor belt for transportation to the packaging area for further processing. (Photo courtesy of Malleshaiah SharathKumar.)

Fig. 6.13. Conveyor belt transporting cut chrysanthemum stems to the packaging area. (Photo courtesy of Malleshaiah SharathKumar.)

- *Packing and cooling*: After the sleeving process, chrysanthemum bunches are packed into boxes, marking a critical phase in maintaining postharvest quality. The adoption of a cold chain immediately after harvest is essential for preserving freshness and extending flower longevity. Temperature control is maintained throughout all postharvest stages, starting with precooling and continuing until the flowers reach retail. Flowers are typically cooled to temperatures below 10°C, with an optimal range of 3–5°C. Advanced cooling techniques, such as beginning with a low-pressure cooling step, involve placing boxes of cut flowers in a forced air cooler to remove field heat before storage or shipping. Cold air is drawn into the boxes through round vents at the ends, rapidly cooling the flowers. These methods store flower stems under reduced pressure and low temperatures, which helps suppress ethylene production by limiting oxygen and other volatile compounds, thereby preserving freshness and extending vase life. It is recommended to maintain a pressure of 40–60 mm Hg to prolong the vase life of many cut flowers, including chrysanthemums. By adopting these advanced cooling technologies, growers can manage large volumes of flowers while upholding high-quality standards, meeting the demands of both retailers and consumers.

6.7 Pot Chrysanthemum Cultivation Practices

Aside from cut flower, chrysanthemum is also a popular pot plant. Specialized cultivars of chrysanthemums differing in size, morphology, flower number, color, and shape are widely available as potted plants. Their production systems can vary a lot, from highly technical, automated operations to small-scale,

Fig. 6.14. An opening in the central aisle allows for underground transport of cut chrysanthemum stems to the central processing shed or packaging area via a conveyor belt. (Photo courtesy of Malleshaiah SharathKumar.)

family-run operations that sell predominantly locally or regionally. This section provides a brief overview of potted chrysanthemum cultivation practices, emphasizing that production methods differ significantly based on regional practices, cultivation traditions, varietal characteristics, and cost considerations. Pots and containers used for potted chrysanthemums are also diverse, varying in height, diameter, and volume, offering consumers multiple options. The cultivation cycle for potted chrysanthemum typically spans 80–90 days, with timing influenced by the variety and seasonal conditions and whether the growing cycle starts from rooted or unrooted cuttings.

6.7.1 Container (pot) sizes

Container sizes in potted chrysanthemum production are chosen based on seasonal sales, market demand, and plant characteristics. For spring sales, smaller

Fig. 6.15. Bundled cut flowers on a weighing scale as part of the automated sorting system. The machine sorts and creates bunches according to the set number of branches, weight, and defoliation length. (Photo courtesy of Malleshaiah SharathKumar.)

Fig. 6.16. Cut flowers bundled and placed in plastic sleeves, ready for marketing at auction centers or for retailers. (Photo courtesy of Malleshaiah SharathKumar.)

containers, such as 8–15 cm pots, with or without handles for hanging, are commonly used. In summer and fall, larger containers like 15–20 cm plastic or fiber pots and 4–8 l nursery containers are preferred (Syngenta Flowers, 2016). Additionally, there is increasing demand for large pot sizes, including 25-, 30-, and 35 cm pots, which are available in various colors and styles to suit consumer preferences and seasonal trends. Container selection also considers the type of chrysanthemums being grown. For example, smaller flowers, such as spray types, may suit smaller pots, while larger disbud or cushion types often require larger pots. The number of plants per pot, typically three to five, also influences container size. Larger pots accommodate more plants, yielding fuller and more impressive displays.

6.7.2 Substrate composition for potted chrysanthemum

The growing substrate for potted chrysanthemum production should retain enough water after irrigation, while providing sufficient drainage for proper aeration. Peat-based mixes are a popular choice among growers, while bark-based mixes, which are more porous, are also widely used but may require 20% additional nitrogen in the feed program. Regardless of the mix, it must be well drained with 12–18% air-filled pore space and maintain a pH range of 5.8–6.2. EC levels should range between 0.7 and 1.0 $dS{\cdot}m^{-1}$ to ensure nutrient availability. Lightweight mixes can cause instability, especially during handling in the production chain, making bark-based or heavier components more suitable for ensuring pot stability. Substrates can be formulated on-site or purchased from specialized suppliers, and their physical and chemical properties must be optimized to support plant growth throughout the production cycle and consumer use. Common materials include pine bark, peat, coir pith, carbonized rice husks, and charcoal. In advanced systems, soil use is minimized or eliminated, though up to 25% soil may be included in cost-sensitive setups. Depending on the availability, a typical substrate formulation may consist of 40% pine bark, 25% soil, 25% peat or manure, and 15% carbonized rice husk. Microbiological safety is crucial; substrate components must be analyzed to prevent contamination and disease. The substrate should promote a well-distributed root system by balancing macro- and micropores for proper aeration and moisture retention. During substrate preparation, chemical analysis is necessary to assess the nutrient concentration, applying 1–2 kg of 20–20–20 n–P–K fertilizer per cubic meter of substrate is generally recommended.

6.7.3 Planting

For optimal establishment and growth, planting of potted chrysanthemum cuttings into moist growing media is recommended. Planting into dry media can hinder initial growth and reduce the plant's establishment potential.

Cuttings should be planted deep enough to cover the root ball (block) and part of the stem base.

6.7.4 Critical environmental requirements for pot chrysanthemum

6.7.4.1 Light

The light requirements for potted chrysanthemums closely align with those for cut flower production, particularly in the regulation of long and short days to control flowering. Photoperiodic strategies, such as night breaks or day extensions for supplementary lighting, are commonly utilized to ensure proper flowering timing. To manage temperatures and enhance the vibrant green color of the plants, as well as to prevent flower discoloration, shading screens are often employed. However, it is essential to carefully monitor light intensity when using screens to avoid excessive reduction, which could negatively impact plant growth and quality.

6.7.4.1.1 LONG DAYS FOR VEGETATIVE GROWTH. The supply of long days for potted chrysanthemum production mirrors the strategy used in cut flower cultivation (described earlier) but is adjusted for fewer days to ensure that plant height aligns with pot size while balancing green biomass and the production of colorful flowers. When rooted cuttings are planted, long days are applied for 12–15 days. For unrooted cuttings planted directly into pots, the duration is extended to 18–21 days, depending on factors such as variety, season, and pruning (pinching) practices. In regions where natural daylight exceeds 14 hour, long-day conditions can be achieved without supplementary lighting during certain times of the year. Long-day treatments following pruning promote the differentiation of nodes and leaves inside buds, contributing to increased plant size and vigor. With the discontinuation of incandescent lamps due to their low efficiency, LED lighting has become the preferred choice for photoperiodic control. LEDs offer enhanced efficiency, sustainability, and precision, making them ideal for modern chrysanthemum production.

6.7.4.1.2 SHORT DAYS FOR FLOWERING. Short days in potted chrysanthemum production are applied using blackout screens, similar to the approach used in cut flower cultivation. These coverings are used until flowers begin to open. Typically, flower bud development begins in the sixth week after planting, with flowers starting to open around the tenth week. By the 11th to 12th week, when 50–60% of the inflorescences are open, the plants are market ready. Understanding a cultivar's sensitivity to photoperiodic induction allows growers to optimize the application of short days, reducing the number of days required and hence production costs. By carefully managing the interaction of light, temperature, and cultivar characteristics, growers can ensure optimal growth and flowering, and hence deliver high-quality potted chrysanthemums to the market.

6.7.4.2 Temperature

Optimal cultivation of potted chrysanthemums is achieved at average daytime and nighttime temperatures of approximately 25 and 18°C, respectively. However, certain well-adapted cultivars can maintain good production and quality under slightly higher temperatures, offering flexibility in varying climatic conditions.

6.7.4.3 Irrigation and nutrients

During the initial days after planting, misting the cuttings is beneficial to maintain turgidity until the roots begin actively absorbing water. This practice ensures successful establishment and promotes healthy development. In specialized production systems, irrigation is typically carried out using a drip system, where each pot or container receives water directly, a method most commonly applied in large cushion-type cultivation. This method also facilitates fertigation, delivering nutrients efficiently with the water. In more mechanized greenhouses, the ebb-and-flow method is also commonly used for watering and fertilization, providing consistent moisture and nutrient delivery. Fertilization rates vary depending on the growing medium, fertilizer type, and application frequency. Water-soluble fertilizers, controlled-release fertilizers (CRFs), or a combination of both can be used based on grower preference and system requirements. A lack of adequate fertility during the vegetative development phase negatively impacts plant size and quality. High ammonium levels can help mitigate crown budding but may also delay flowering. To support vigorous growth, liquid fertilization is recommended at planting. Immediately after planting, irrigating the cuttings with a complete N–P–K (20–20–20) fertilizer is recommended. The choice of fertilizer is influenced more by water quality than by specific chrysanthemum preferences. During the vegetative phase, EC values of 0.7–1.0 $dS \cdot m^{-1}$ are optimal for healthy development. CRFs can be incorporated into the growing medium before planting or applied as a top dressing afterward. The choice of CRFs often depends on the duration of the cultivation. Products with a release rate of 3–4 or 5–6 months are commonly selected for shorter crop cycles or more intensive growing systems. However, some growers prefer longer-release fertilizer products with durations of 5–8 months, as they provide an extended supply of nutrients for varieties with longer production cycles, regions with prolonged growing seasons, or to ensure nutrient availability at the consumer.

6.7.4.4 Growth regulators

B-Nine® (Daminozide) is a widely used growth regulator for potted chrysanthemums, typically applied at rates of 1000–5000 ppm, with 2500 ppm serving as a standard starting point (Green Leaf Plants, 2015; Syngenta Flowers, 2016). The specific rate depends on factors such as the variety's vigor, ambient temperature, and crop growth stage. A range of growth regulators is available for potted chrysanthemums, including Abide/A-Rest, Dazide/B-Nine, Bonzi/Paczol, and Concise (Whipker, 2018). Concentrations range from 2.5 to 10 ppm for spray applications, with lower rates recommended for less vigorous cultivars. Sequential applications of lower concentrations are generally more effective in producing uniformly shaped plants compared to a single high-concentration spray (Whipker, 2018). By selecting the appropriate growth regulator and concentration based on the cultivar and growth conditions, growers can achieve optimal plant size, uniformity, and quality in potted chrysanthemum production.

6.7.4.5 *Apical pruning (pinching)*

Potted chrysanthemums are cultivated in two primary forms: spray or cushion types, characterized by multiple stems or flowers per branch, and disbud types, where a single stem supports a large apical flower. Pinching is a key practice for shaping the plant, regulating growth, and optimizing flowering, ensuring an aesthetic balance between the plant and its container. These techniques cater to diverse market preferences, enhancing the versatility of chrysanthemum production.

6.7.4.5.1 PINCHING TIMING AND TECHNIQUES Modern garden pot mum varieties often do not require pinching due to advancements in breeding that encourage natural branching. In trials, no-pinch production, which reduces labor demand, has proven effective for many newer varieties. However, pinching may still be necessary in specific situations, such as when cuttings are rooted in large plugs, held too long in small plugs, or highly vigorous and tall-growing cultivars. For crops requiring pinching, the optimal time is when the plant's roots are well established, and new growth measures 2.5–3.8 cm, typically 10–14 days after planting. The top 1.2 cm of growth is pinched out. Machine pinching prior to transplanting into final containers is a labor- and cost-saving alternative. With today's improved varieties, multiple pinches are rarely necessary. Tailoring pinching practices to the variety's growth habit allows growers to optimize plant structure and flowering while minimizing labor and production costs. This adaptability supports both traditional and modern cultivation approaches, enabling the production of high-quality potted chrysanthemums that meet the diverse demands of the market.

6.7.4.5.2 SPRAY OR CUSHION TYPES Pinching is an essential practice for spray or cushion types of potted chrysanthemums. The process involves removing the apical bud, leaving four to six leaves on the plant (Fig. 6.17A and B). This is generally carried out 18–25 days after planting or 3–5 days after spacing the pots. Pinching disrupts apical dominance, encouraging lateral shoot growth and increasing the number of flowers per plant. This technique also ensures the plant's shape is well balanced, preventing the central flower from blooming prematurely, which can result in uneven flowering. For tall chrysanthemum varieties, height control can be managed by pruning after the initiation of short-day conditions. This approach halts the differentiation of new buds while inducing existing buds to flower alongside the growth of sprouts. Conversely, for shorter varieties, applying short-day treatment after pruning ensures proper flowering and maintains a balanced plant structure. This tailored approach to pinching and pruning ensures optimal growth and flowering across different varieties of potted chrysanthemums. To meet market demands for greater vibrancy, cushion types of potted chrysanthemums are cultivated in both single-color varieties per pot and diverse color combinations per pot (Fig. 6.18A and B). These options appeal to seasonal and festive consumer preferences, offering visually striking displays tailored to specific occasions and celebrations.

In the cultivation of disbud types of potted chrysanthemums, a single flower per plant develops by removing all lateral shoots and buds, allowing the central apical flower to grow to its full size. This method is traditionally associated with the ball-type chrysanthemum group, known for producing large, uniform, and visually striking blooms (Fig. 6.20B). In modern markets, there is a growing preference for potted disbuds featuring vibrant color combinations, catering to consumer demand for bold and visually appealing plants. This trend highlights the versatility and aesthetic value of disbud chrysanthemums, making them a popular choice for premium ornamental displays.

6.8 Readiness for Marketing

For potted chrysanthemums, the transition from greenhouse to market is referred to as the marketing stage rather than a harvesting stage (Fig. 6.19). At around 70 days, flower buds begin to open, signaling the approaching end of the production cycle, which concludes between 80 and 90 days. In spray-type chrysanthemums, 50–60% of the flowers fully open (Fig. 6.20A), whereas in disbud chrysanthemums, all three plants per pot reaching 50% anthesis (Fig. 6.20B) are considered ready for market. During this phase, the plants remain in their pots, with their stage of development serving as a key indicator of market readiness. Plants are carefully selected based on attributes such as health, vigor, uniform height, color, and the degree of flower opening. Unlike cut flowers, potted plants continue to grow and develop after leaving the greenhouse, provided they are handled well. This ongoing growth highlights the importance of the marketing stage in ensuring plants retain their aesthetic appeal and vitality during transport, display, and sale. After selection, pots are placed in perforated plastic sleeves to prevent moisture buildup around the inflorescence, preserving flower quality (Fig. 6.21A and B). They are then packed into cardboard or wooden boxes or loaded into trolleys to ensure safe transport. Proper handling, including maintaining an

Fig. 6.17. Potted spray chrysanthemums ready for pinching at the ten-leaf stage (A) and after pinching with seven to eight leaves (B). (Photo courtesy of Malleshaiah SharathKumar.)

Fig. 6.18. Potted spray chrysanthemums with a single-color variety per pot (A) and dual-color varieties per pot (B). (Photo courtesy of Malleshaiah SharathKumar.)

Fig. 6.19. Market readiness of potted spray (A) and disbud (B) chrysanthemums with 50–60% of the flowers open. (Photo courtesy of Malleshaiah SharathKumar.)

appropriate transport temperature of around 20°C and ensuring sufficient moisture in the substrate, is crucial for sustaining plant health and promoting further development until they reach the consumer.

Fig. 6.20. Potted spray (A) and disbud (B) chrysanthemums packed in plastic sleeves, ready for marketing. (Photo courtesy of Malleshaiah SharathKumar.)

Fig. 6.21. Potted disbud chrysanthemums packed in plastic sleeves that are ready for marketing. (Photo courtesy of Malleshaiah SharathKumar.)

6.9 Recent Developments in High-Tech Greenhouse Cultivation of Chrysanthemums

6.9.1 Hydroponic cultivation of chrysanthemums

Greenhouse cultivation of chrysanthemums for cut flowers has traditionally relied on soil and organic matter mixtures in ground-level beds. While effective for maximizing yield, this method often results in fertilizer losses, environmental issues, and challenges with salinization and soilborne pathogens or pests. Reusing the soil medium across production cycles deteriorates its physical properties, causing compaction, reduced porosity, and excessive water retention, which can harm plant growth. Hydroponic cultivation offers an alternative, eliminating soil-related challenges while maintaining similar planting, management, and flowering practices. This method also avoids the need for steaming as an energy-intensive soil disinfection practice. Dutch chrysanthemum growers are exploring the possibility of adopting closed-loop hydroponic systems, which prevent fertilizer and pesticide runoff into soil and groundwater. Research confirms that hydroponically grown chrysanthemums are comparable to soil-grown counterparts in branch structure, plant health, flower quality, and weight (Barbosa *et al.*, 2015; Neefjes, 2017; Vertify,

2024). With its potential to maintain high production standards while minimizing environmental impact, further optimization of hydroponic systems for chrysanthemum cultivation holds significant promise. If economic feasibility is achieved and waterborne diseases and pathogens are effectively managed, hydroponics could become the future of chrysanthemum cultivation.

In hydroponic cultivation using closed systems, the nutrient solution is recirculated either continuously or intermittently. There is no universal formulation or perfect recipe for these solutions, as nutrient absorption, transport, and distribution depend on factors such as variety, season, and the crop's developmental stage. Chrysanthemums have specific nutritional demands, with nitrogen (N) and potassium (K) being especially important. Ensuring adequate nitrogen availability during the first 50 days of growth is critical for optimal development. Nutrient concentrations in the solution fluctuate as plants absorb nutrients at varying rates throughout their growth phases. Regular monitoring of EC, pH, and mineral composition and replenishment of nutrients is crucial for optimal plant growth and quality. EC is a reliable method for evaluating nutrient levels, reflecting the total soluble salt content in the solution. EC level of 1 $dS \cdot m^{-1}$ corresponds to approximately 640–700 $mg \cdot l^{-1}$ of total dissolved salts. The recommended total nutrient concentration is 1000–1500 $mg \cdot l^{-1}$, equivalent to an EC range of approximately 1.5–2.2 $dS \cdot m^{-1}$. Maintaining this range ensures an appropriate osmotic pressure for efficient nutrient uptake by the roots.

6.9.2 Rooting of chrysanthemum cuttings

Transition from conventional single-layer greenhouses to high-tech multilayer vertical rooting facilities (vertical farming) presents a remarkable opportunity for producing cuttings or young plants (SharathKumar *et al.*, 2020). Utilizing multiple layers improves land-use efficiency. This space-efficient approach has been implemented in the multistory (vertical) rooting system of the company Deliflor (Deliflor Hoogveld, 2023; www.deliflorhoogveld.nl). This rooting facility is spread over 8.5 ha and can process 2.4 million chrysanthemum cuttings daily, 12 million weekly, and 500 million annually. Cuttings are sourced from the company's African production site and rooted through a fully automated process. The facility's three-layer cultivation system, supported by advanced climate control technologies, optimizes space utilization and ensures ideal rooting conditions. The initial phase of rooting takes place in the bottom two cultivation layers, which rely solely on LED lighting. The rooting phase is completed in the top layer, where daylight is combined with LED light. The 12-day rooting process is divided evenly among the three cultivation levels, each tailored with specific light intensity and humidity settings to enhance rooting efficiency. The highly mechanized facility utilizes advanced machinery to manage all processes, eliminating physical human interaction with the plants and ensuring a contamination-free environment. This innovative approach represents a significant advancement in efficiency, productivity, and space optimization for chrysanthemum propagation. After approximately 2 weeks in the rooting layers, the chrysanthemum cuttings are ready for transport and delivery to growers for planting.

Traditionally, peat blocks are used as rooting substrates for chrysanthemum cuttings. However, recently, plugs have been introduced as an alternative. Plugs significantly enhance root development, ensuring better plant establishment even with a reduced amount of substrate per plant. This advancement optimizes space and resource use while also shortening the cultivation cycle, leading to faster harvest-ready plants.

6.10 Conclusion and Future Perspectives

High-tech greenhouse cultivation of chrysanthemums continues to advance, enabling year-round production under controlled conditions and ensuring consistent yields of high-quality flowers. These systems are evolving with increasing technological integration, driving improvements in productivity and sustainability. Further optimization of key factors, such as light, temperature, RH, ventilation, and CO_2 levels, is vital for economic feasibility. The knowledge presented in this chapter is tailored to high-tech greenhouse systems and highlights practices specific to these advanced environments. Regional adaptations, accounting for local environmental conditions, infrastructure, and resource availability, will be essential to ensure the effective application of the concepts shared here.

Innovations such as multilayer rooting systems are set to further enhance cutting quality and rooting efficiency, while robotics-driven approaches, which integrate traditional practices with modern technology, will continue to evolve, improving sustainability and operational efficiency. Hydroponic cultivation is also expected to become more widely practiced with advantages in terms of environmental impact and increased resource use efficiency, although economic feasibility is still a key limitation. Furthermore, advancements in data-driven cultivation, including automated data integration, image analysis, and ultrasound technologies, are poised to streamline chrysanthemum production and trade. These technologies not only boost efficiency but also enable precise predictions of plant traits, including postharvest vase life. These technological advancements highlight the potential for transformative progress in high-tech chrysanthemum cultivation in the years ahead. Future developments will likely refine these innovative techniques and adapt them to greenhouse systems worldwide, ensuring that chrysanthemum production remains at the forefront of innovation, efficiency, and adaptability in the face of emerging challenges and opportunities.

References

A1 Group (2024) Reduce day length with blackout screens. Available at: https://a1group.nl/en/screening/horizontal-screening/blackout-screens/ (accessed 12 December 2024).

Barbosa, J.G., Barbosa, M.S., Almeida, D.B., Sá, P.G., Finger, F.L. *et al.* (2015) Production and postharvest quality of chrysanthemum flowers grown in hydroponic system under different N:K ratios. *Acta Horticulturae* 1060(42), 281–287.

Carvalho, S.M.P. (2003) Effects of growth conditions on external quality of cut chrysanthemum: Analysis and simulation. PhD Thesis, Wageningen University, The Netherlands.

Carvalho, S.M.P., van Noort, F.R., Postma, R. and Heuvelink, E. (2008) Possibilities for producing compact floricultural crops. Report/Wageningen UR Greenhouse Horticulture; no.173, Wageningen UR Greenhouse Horticulture. Available at: https://edepot.wur.nl/11685

Deliflor Hoogveld. (2023) Consistently high quality chrysanthemum cuttings are of great importance. Available at: www.deliflorhoogveld.nl (accessed 30 December 2024).

de Visser, P. (2024) CO2-gebruik moet efficiënter. WPR-1282. Wageningen, Stichting Wageningen Research, Wageningen Plant Research, Business Unit Glastuinbouw. Available at: https://www.kasalsenergiebron.nl/content/research/E21009_Eindrapport_EfficienterCO2doseren.pdf

Dutch Greenhouses (2024) Blackout screens & light deprivation screens. Available at: https://dutchgreenhouses.com/en/screening/blackout-screens (accessed 12 December 2024).

Green Leaf Plants. (2015) Pot mum cultural information [Online]. Available at: https://www.yumpu.com/en/document/read/11890131/pot-mum-cultural-information-syngenta-flowers

Higuchi, Y., Sumitomo, K., Oda, A., Shimizu, H. and Hisamatsu, T. (2012) Day light quality affects the night-break response in the short-day plant chrysanthemum, suggesting differential phytochrome-mediated regulation of flowering. *Journal of Plant Physiology* 169(18), 1789–1796. DOI: 10.1016/j.jplph.2012.07.003.

Kawata, J., Toyoda, T., Uda, M., Okimura, M., Shibata, M. *et al.* (1987) Factors controlling the flowering time of chrysanthemums. *Bulletin of the National Research Institute of Vegetables, Ornamental Plants and Tea* A1, 187–222.

Nakano, Y., Higuchi, Y., Sumitomo, K. and Hisamatsu, T. (2015) Delay of flowering by high temperature in chrysanthemum: Heat-sensitive time-of-day and heat effects on *CsFTL3* and *CsAFT* gene expression. *Journal of Horticultural Science & Biotechnology* 90, 143–149.

Neefjes, H. (2017) Hydroponically grown chrysanthemum is a real alternative. Available at: https://www.hortipoint.nl/floribusiness/hydroponically-grown-chrysanthemum-is-a-real-alternative/ (accessed 30 December 2024).

Nozaki, K. and Fukai, S. (2008) Effects of high temperature on floral development and flowering in spray chrysanthemum. *Journal of Applied Horticulture* 10, 8–14.

SharathKumar, M., Heuvelink, E. and Marcelis, L.F.M. (2020) Vertical farming: Moving from genetic to environmental modification. *Trends in Plant Sciences* 25(8), 724–727. DOI: 10.1016/j.tplants.2020.05.012.

Shibata, M. (1997) Studies on responses to temperature and photoperiod and breeding of spray type chrysanthemums with summer-to-autumn flowering. *Bulletin of the National Research Institute of Vegetables, Ornamental Plants and Tea* 12, 1–71.

Syngenta Flowers (2016). Available at: https://www.syngentaflowers.com/ams/sites/g/files/kgtney2381/files/file_field/import/field_media_link/sites/g/files/zhg721/f/potmumculturalinformation_1.pdf (accessed 22 December 2024).

Vertify (2024) Future-proof cultivation of cut flowers on water. Available at: https://www-vertify-nl.translate.goog/projecten/toekomstbestendig-snijbloemen-telen-op-water/?_x_tr_sl=auto&_x_tr_tl=en&_x_tr_hl=auto (accessed 13 December 2024).

Whipker, B. (2018). Available at: https://www.fine-americas.com/wp-content/uploads/2017/01/PGR_GUIDE_2017_LR.pdf

Open-Field Cultivation of Chrysanthemum

7

M. Ganga[1]*, S. Ganesh[1], Rajiv Kumar[2] and Malleshaiah SharathKumar[3]*

[1]Department of Floriculture & Landscape Architecture, Horticultural College & Research Institute, Tamil Nadu Agricultural University, Coimbatore - 641 003, Tamil Nadu, India.
[2]Division of Flower and Medicinal Crops, Indian Institute of Horticultural Research, Hessaraghatta Lake Post, Bengaluru, 560 089, Karnataka, India.
[3]Horticulture and Product Physiology, Department of Plant Sciences, Wageningen University and Research, Wageningen 6700 AA, The Netherlands.

Abstract

Chrysanthemums hold a prominent place in global floriculture, valued as both cut flowers and potted plants. However, their cultivation practices differ significantly across regions. Unlike in the Western world, where chrysanthemums are primarily cultivated in controlled environments for ornamental purposes, in India, they are widely grown in open fields for loose flower production. These flowers are predominantly used in making garlands for various festivals and religious celebrations, reflecting their cultural and economic importance in the Asian subcontinent, particularly in India. This chapter focuses on the cultivation of chrysanthemums in open fields for loose flower production, highlighting the traditional and modern practices that have evolved to meet the demands of garland-making and religious festivities. Furthermore, it delves into the advancements made by Indian research institutes in improving chrysanthemum varieties, addressing challenges, and optimizing cultivation techniques under India's favorable agroclimatic conditions.

Keywords

Asian flowers, chrysanthemums, flower garlands, loose flowers, short-day plant

*Corresponding authors: sharath.malleshaiah@wur.nl; sharath.malleshaiah@deliflor.nl; sharath.malleshaiah@gmail.com and ganga.m@tnau.ac.in

DOI: 10.1079/9781789248289.0007

7.1 Introduction

Chrysanthemums are highly valued for their wide variety of colors, forms, sizes, and diverse uses. Globally, the cultivation of chrysanthemums is gaining importance due to their ease of cultivation, high economic returns, and increasing market demand. They are popular in the floriculture industry both as cut flowers and potted plants. In contrast to Western countries, chrysanthemums in India are predominantly cultivated in open fields for loose flower production, while cut flowers are grown under protected conditions, such as shade net houses or polyhouses. Flowers hold a significant place in Indian culture, serving decorative, religious, social, commercial, and aesthetic purposes. Chrysanthemum loose flowers (capitula with short stalks) are commonly used in garlands and floral ornaments for party decorations and religious offerings (Figs 7.1, 7.2 and 7.3). India's agroclimatic conditions, particularly in the tropical and subtropical zones, are highly favorable for open-field chrysanthemum cultivation, especially for loose flower production.

Fig. 7.1. Chrysanthemum loose flowers prepared for sale in local markets of southern India during the early-morning hours (A). Chrysanthemum flowers are sold by weight for various uses, including special decorations and devotional arrangements at homes and temples (B). Loose flowers in colors such as white, yellow, and purple to reddish hues are the most preferred in southern India (C). Garlands made from chrysanthemums, along with other loose flowers, such as roses, tuberose, and marigold, are prepared daily for sales for religious and cultural purposes (D). (Photo courtesy of Malleshaiah SharathKumar.)

Fig. 7.2. Garland designs created by men using various colored loose chrysanthemum flowers, such as yellow and white, combined with other flowers, such as roses and tuberose, to create vibrant color combinations. Additionally, various garland patterns are finished, displayed, and ready for daily sales, catering to religious and cultural purposes. (Photo courtesy of Malleshaiah SharathKumar.)

Fig. 7.3. Various colored garlands made using chrysanthemum loose flowers are offered to the Ganesha (Vinayagar or Vinayaka) idol in temples for religious purposes. (Photo courtesy of K. Anitha.)

In tropical zones, high temperatures persist for most of the year, with minimal variation between day and night temperatures. Photoperiod plays a critical role in chrysanthemum cultivation, as it is a short-day plant requiring 13–14 hour of darkness to induce flowering and 11–12 hour of daylight for vegetative growth. During the hotter months (February to July), sunlight averages 12–13 h day^{-1}, supporting planting and vegetative growth. In the rainy and winter months, daylight decreases to 9–10 hour, triggering the transition from vegetative to generative phases, meeting the optimum photoperiodic requirement of 13–14 hour of darkness. In southern India, particularly in Tamil Nadu and Karnataka, these natural photoperiod conditions support open-field chrysanthemum cultivation, since flowering aligns with the winter season.

7.2 Characteristics of Chrysanthemum as Loose Flowers

Chrysanthemum cultivars are preferred for loose flower production if they exhibit a longer blooming season, profuse branching, high yield, and a strong ability to recover from the plucking pressure experienced during flower harvesting. Desired traits also include a pleasant fragrance, fluffy flowers with a diameter of approximately 5 cm, with no visible disc florets (Fig. 7.1A and C). Cultivars with good color retention particularly in shades of yellow, red, and white are in demand (Fig. 7.1C).

7.3 Chrysanthemum Crop Improvement in India

Chrysanthemums are unique due to their extensive range of flower and vegetative variations, as well as their adaptability to diverse agroecological conditions. The flowers display significant diversity in shape, size, and color. This genetic variability within the chrysanthemum gene pool has been instrumental in establishing the crop's current prominence (Datta and Janakiram, 2015). Such genetic diversity is critical for developing new flower shapes and forms, providing the raw material for future breeding programs. Modern chrysanthemum cultivars have been developed through a combination of interspecific crosses, selection, open pollination, hybridization, spontaneous and induced mutations, and chimera management. As a cross-pollinated crop, new varieties are frequently grown from seedlings derived through either natural or controlled pollination. Conventional breeding methods remain widely employed and play a crucial role in shaping the genetic structure and variability of the crop. One of the primary challenges in chrysanthemum breeding is the excessive length of ray florets, which can impede pollination. Additionally, self incompatibility among different varieties and species, along with the dense ray florets in double or anemone flower types, often prevents seed set. Seed production is typically observed only in single and semidouble varieties. The development of chrysanthemum varieties is driven by collaborative efforts involving farmer interest groups, scientists from state and central institutes, universities, and nurseries. Conventional breeding techniques, such as crossbreeding and mutation breeding, are the primary methods used to create new cultivars. However, classical breeding relies heavily on phenotypic selection to identify superior progenies or mutants, a process that can be labor-intensive and inefficient. This inefficiency is particularly evident for traits influenced by significant genotype × environment (G × E) interactions.

7.3.1 Indian research institutes and state universities involved in chrysanthemum improvement

Continuous efforts are being made to develop new chrysanthemum varieties with improved yield, diverse colors, and enhanced resistance to biotic and abiotic stresses. Extensive research on chrysanthemum breeding is conducted at leading institutions under the Council of Scientific and Industrial Research (CSIR), Indian Council of Agricultural Research (ICAR), and State Agricultural Universities (SAUs) across India. Prominent institutes, such as the National Botanical Research Institute (NBRI), Lucknow, under CSIR; and ICAR institutions, such as the Indian Agricultural Research Institute (IARI), New Delhi; Indian Institute of Horticultural Research (IIHR), Bengaluru; and Directorate of Floricultural Research (DFR), Pune, have made significant contributions to the development of new varieties and advanced management practices (Tejaswini *et al.*, 2021). Additionally, SAUs, such as Tamil Nadu Agricultural University (TNAU), Coimbatore; Sri Konda Laxman Telangana State Horticultural University (SKLTSHU), Hyderabad; Bidhan Chandra Krishi Vishwavidyalaya (BCKV), West Bengal; Dr YS Parmar University of Horticulture and Forestry, Solan, Himachal Pradesh; G.B. Pant University of Agriculture and Technology, Uttarakhand; Punjab Agricultural University (PAU), Punjab; and Maharana Pratap University of Agriculture and Technology (MPUAT), Rajasthan, are conducting interdisciplinary research on both fundamental and applied aspects of chrysanthemum cultivation. Many of these efforts are coordinated under the All India Coordinated Research Project (AICRP) on Floriculture, the key research activities of which include as follows:

- Germplasm enrichment and characterization
- Utilization of germplasm in breeding programs
- Development of new varieties through hybridization, mutation breeding, and other advanced techniques
- Agro-technology standardization
- Disease management strategies and
- Research on postharvest physiology to improve shelf life and quality.

These collaborative research initiatives are vital in creating innovation and ensuring sustainable chrysanthemum cultivation practices across India.

7.3.2 Introduction and selection

In southern Indian states, such as Tamil Nadu, Karnataka, and Andhra Pradesh, small-flowered chrysanthemum cultivars in white and yellow shades have traditionally been grown for loose flower production. The introduction of exhibition-type chrysanthemum cultivars in eastern India, particularly in regions such as Kolkata and Sikkim (Table 7.1), catalyzed the development of many new cultivars across various research institutes and SAUs. Despite its success, the selection process for chrysanthemum breeding has several

Table 7.1. List of introduced chrysanthemum cultivars. (From authors)

Country from which introduced	Name of the cultivar
Australia	J.S. Lioyd, Louisa Pockette, Willium Turner, and T.W. Pockette
Japan	Ajina Purple, Kenroku Kangiku, Kiku Biori, Tokyo, Otome, Zakura
France	Gloria Deo, S.L. Andre Raffaud, and Sancho
United Kingdom	Alfred Wilson, Alfred Simpson, Beatrice May, Pink Cloud, Princess Anne, and Woolman Century
United States	Casa Grande, Mountaineer, Nob Hill, Snowball, Potomac, and Peacock
Malaysia	Marigold chrysanthemum—introduced to Karnataka, India, by nurserymen, relatively less photosensitive, suitable for loose flowers
New Zealand	Gusman Red, Jane Sharp, Orange Fair Lady, Nancy Ferneaux

limitations. These include unknown parentage, the necessity of raising a large seedling population, and the inability of improved cultivars with long, tubular, or double ray florets to produce seeds. The single-plant selection method remains the most employed approach, leading to the development of many outstanding spray and loose flower cultivars (Table 7.2).

7.3.3 Hybridization

Chrysanthemum is highly heterozygous and prone to inbreeding depression and self incompatibility, making crossbreeding a powerful method for developing modern cultivars. This approach is both simple and effective, as F_1 progenies derived from parents with contrasting traits often exhibit significant phenotypic variations (Su *et al.*, 2019). Successful crossbreeding programs require careful consideration of several key factors, including the fertility of cross-combinations, qualitative analysis of target traits, and the selection of superior hybrid progenies (Table 7.3).

7.3.4 Mutation breeding

In chrysanthemum, seed set is often absent in many genotypes across diverse agroclimatic locations, despite extensive hybridization attempts. To overcome this limitation and induce genetic variability, mutation breeding has been employed as an effective strategy for developing novel and superior varieties. For detailed information on mutation breeding in chrysanthemum, readers are referred to Datta (2023).

Table 7.2. List of chrysanthemum cultivars evolved through selection from Indian Institutes and State Agriculture Universities. (From authors)

Cultivar	Characteristics
Tamil Nadu Agricultural University, Coimbatore, Tamil Nadu	
CO 1	Selection from Hosur-type that bears attractive canary yellow for a longer period; the weight of a single flower is around 2.5 g, with a yield potential of 16.7 t ha^{-1}.
CO 2	Clonal selection from the variety Jaya introduced from NBRI, Lucknow. Flowers have a rhodamine purple color with a yield potential of 20.3 t ha^{-1}.
MDU 1	Early-type cultivar that requires almost 104 days for flowering. Large and attractive flowers in sulfur-yellow color with a yield potential of 30.59 t ha^{-1}.
Indian Institute of Horticultural Research, Bengaluru, Karnataka	
Arka Rakhee	Open-pollinated seedling of Lord Doonex. Primrose yellow with red stripes on ray florets, suitable for a pot plant.
Arka Pink Star	Selection from a half-sib population of cv. Punjab Gold. Early-flowering (63 days) plant with a dwarf stature, suitable for pot culture and as a loose flowers. Flowers are semidouble with pink color, early flowering, and yield is around 135 flowers/plant.
Arka Anirudh	This is a half-sib seed selection from the variety "Lalpari." Plants are dwarf, with a semi-erect growth habit and show early flowering (62.77 days). Suitable for pot culture/bedding, flowers are semidouble, with maroon-yellow color (The Royal Horticultural Society (RHS) color: 17A, Yellow 7 Orange Group, Fan 1), having five to six rows of ray florets, and are resistant to white rust disease (*Puccinia horiana* Henn.).
Arka Dhaval	This is a half-sib seed selection from the variety "White Prolific." Plants are dwarf, with a spreading growth habit and show early flowering (69.17 days). Suitable for pot culture/bedding, flowers are semidouble, pinkish white in color (RHS color: 3D, Yellow Group, Fan 1) with three to four rows of ray florets.

Arka Manohar	This is a half-sib seed selection from the variety "Sunil." Plants are dwarf with an erect growth habit and show early flowering (66.52 days). Suitable for pot culture/bedding, with semidouble flowers, deep pinkish-purple in color (RHS color: 71B, Red-Purple Group, Fan 2) with five to six rows of ray florets.
Punjab Agricultural University, Ludhiana, Punjab	
Punjab Shingar	Floriferous variety with compact and upright growth habit with decorative-type flowers in white color that are suitable for loose flower purposes.
Punjab Mohini	Dwarf variety with a compact growth habit, does not require pinching and staking. Flowers are single spray-type, white with yellow center, and are suitable for pot culture.
Garden Beauty	Open-pollinated seedlings of variety Flirt. Maroon, spoon-type flowers, suitable for a garden plant.
Indian Agricultural Research Institute, New Delhi	
Pusa Chitraksha	Open-pollinated seedling of cv. Lalpari. Plant is bushy, spray-type, produces single flowers that are deep magenta in color. Suitable for a potted plant and for garden display.
Pusa Sona	Open-pollinated seedling of cv. Sadbhawana. Spray-type, produces single flowers that are yellow in color. Suited for pot mums.
Pusa Sundri	It is spreading-type, with vibrant yellow flowers, that creates a striking "golden carpet" effect when grown in pots. It blooms profusely, bearing around 170 flowers per plant that last for 7 weeks in pot conditions and 40 days in the field, making it ideal for pot displays and garden aesthetics.
Pusa Guldatsa	Open-pollinated seedling of cv. Lalpari produces semidouble, medium-sized (3.8 cm) flowers with orange-red ray florets and yellow discs. The plant reaches 58 cm in height and spreads 50 cm, bearing blooms that last up to 48 days under field conditions. It has low maintenance, requiring no pinching or staking, and is ideal for spray and pot culture.
Pusa Shwet	Open-pollinated seedling of cv. Lalpari. Semi-spreading growth habit. Semidouble, medium-sized flowers with white ray florets and yellow disc. Suitable for pot culture and garden display purposes.

Continued

Table 7.2. Continued

Cultivar	Characteristics
Pusa Aditya (JG 09–01)	Open-pollinated seedlings of cv. Jaya. The plants are bushy (30–35 cm branches), medium in height (55–60 cm), and have a moderate spread (45–50 cm). This spray-type is a unique, bearing star-shaped semidouble flowers. The flowers are yellow in color at the periphery with orange-red color in the center. The florets are spatulate with a distinct keel. It is suitable for cut flower and potted plant.
Dr Yashwant Singh Parmar University of Horticulture and Forestry, Solan, Himachal Pradesh	
Solan Shringar	Seedling selection of "Honey Comb." Double-type, white-colored flowers. Suitable for cut, pot, and loose flower production.
CSIR-National Botanical Research Institute, Lucknow, Uttar Pradesh	
Birbal Sahni	Open-pollinated seedling of cv. Nanako. It requires almost 121 days for flowering. Flowers are pompon-type, snow white in color, with an average yield of 32 q ha^{-1}.
Shanti	Small-flowered, white decorative-type; plants show profuse branching with long, erect stem, good for cut flower and garland purposes.
Y2K	Dwarf, bushy, and compact, round plants that are suitable for mini-culture. White anemone-type small flowers with 370 flowers per plant.
Sadbhavana	Small-flowered, red-yellow, open-disc, double, Korean-type mini chrysanthemum.
Kargil 99	Small-flowered, spoon-shaped, purplish mauve florets, leaves variegated.

Table 7.3. List of chrysanthemum cultivars evolved through hybridization from Indian Institutes and State Agriculture Universities. (From authors)

Cultivar	Parentage	Description
ICAR-Indian Institute of Horticultural Research, Bengaluru, Karnataka		
Arka Swarna	Nanako × CO.1	Superior hybrid that bears pompon-type yellow flowers with higher number of flowers per plant and a longer flowering duration. Suitable for cut flower and loose flower purposes.
Arka Ganga	Flirt × Red Gold	Bears attractive double spray-type white-color flowers with a pink tinge. Requires 127 days for flowering and produces around 143 flowers per plant. Suitable for loose flowers and cut flowers.
Arka Kirti	Angela × GP1	Dwarf and compact early-type hybrid that requires 88 days for flowering. Flowers are of semidouble spray-type that have white to pinkish tinge. Produces around 119 flowers per plant, with a yield of 168 g of flowers per plant. Flowers have a shelf life of 7.5 days. Suitable for bedding, potting, and floral decoration.
Arka Chandrika	Angela × GP1	Early-type hybrid that requires 92 days for flowering. Bears white decorative semidouble-type flowers, produces 130 flowers per plant, with a shelf life of 7 days. Suitable for loose flowers, flower arrangement, decoration, and as a cut flower.
Arka Chandrakant	Angela × GP67	Dwarf hybrid with an average plant height of 35–40 cm. Floriferous hybrid, requires almost 120–125 days to produce decorative-type white flowers. On an average, it produces 149 flowers with a yield of 272 g per plant. Flowers are used as loose flowers in garland-making and as a bedding plant in landscape gardening.

Continued

Table 7.3. Continued

Cultivar	Parentage	Description
Arka Nilima	Flirt × Valentine	Floriferous hybrid that bears purple decorative spray-type flowers produced from 116 days after transplanting (DAT). Flowers have a diameter of 4.86 cm, with an average flower weight of 1.95 g. A single plant can bear up to 144 flowers, with an average flower yield of 281 g per plant. Suitable for loose flowers and cut flowers (spray-type) purpose that has a vase life of 14 days.
Arka Ravi	Mundial × Arka Swarna	Early-type hybrid that requires 102 days for flowering. Flowers are semidouble to spray-type and peach-colored. Suitable for cut flower with a vase life of 12 days.
Arka Indira	Cross between open-pollinated seedling of Lord Doones (LD 14) and hybrid seedling of Flirt × Valentine	Hybrid with high multiplication rate by cuttings. It flowers 125 DAT. Produces double spray-types of flowers that are yellow in color at the bud stage, buttercup yellow at picking, and later fade to aureolin color. Each flower is up to 6.3 cm in diameter. The average yield of flowers per plant is 300 g. Suitable for loose flowers, cut flowers, bedding, and potting.
Arka Red Gold	Flirt × Valentine	Flowers profusely starting from 120 DAT. Flowers are decorative-type, orange to yellowish with brick-red stripes on petals; flower diameter is 5.1 cm with an average of 400 g of flowers per plant. Suitable for cut flower, bedding, potting, wreaths, and religious offerings.

Arka Ravi Kiran	Flirt × Valentine	It takes 122 days to produce stellate-type flowers that are reddish-brown in color. Produces 86 flowers per plant; flower diameter is 6.12 cm. Suitable for cut flower purpose and has a vase life of 12 days.
Arka Pankaj	Lord Doonex-14 × Flirt × Valentine-1	It flowers early and takes about 114 days to bloom. It bears pink, attractive flowers, which are decorative-type. The diameter of each flower is 4.8 cm, and each weighs 2.68 g. It is high-yielding and produces 140 flowers with a yield of 363 g per plant. Its attractive flowers on stiff stalks make them suitable for cut flower purpose. It is also ideal as loose flowers for floral decorations.
Arka Yellow Star	Flirt × Valentine-22	Flowers are yellow, semidouble, used for garland, floral decoration, and bouquet making.
CSIR-National Botanical Research Institute, Lucknow, Uttar Pradesh		
NBRI Indiana	Little Darling (orange) × Nanako (yellow)	Small-flowered, yellow, pompon-type, suitable for mini and pot culture.
NBRI Kusum	Haldighati (yellow) × Sharad Kanti (yellow)	Small-flowered, yellow, open-disc, single spray-type.
NBRI Little Darling	White Charm (white) × Jubilee (bronze)	Small-flowered, terracotta, pompon-type mini chrysanthemum.
NBRI Mini Jessie	Cameo (pink) × Jessie (purple)	Small-flowered, cineraria-type mini chrysanthemum, requires neither pinching nor staking.
Punjab Agricultural University, Ludhiana		

Continued

Table 7.3. Continued

Cultivar	Parentage	Description
Punjab Gold	Flirt × Gul-e-Sahir	Plants are dwarf, early-flowering, bearing golden-yellow, semidouble flowers of 5.0 cm diameter. Flowers are of coppery-red color at the time of opening and turn to golden-yellow at anthesis. Suitable for pot culture.
Anmol	Panchoo × Rage	Bright yellow anemone-type flowers, suitable for growing in pots.
Royal Purple	Bindiya × Rage	Purplish pink anemone-type flowers, suitable for pot culture.
Yellow Delight	White Bouquet × Gul-e-Sahir	An early-flowering variety with bright yellow pompon-type flowers, suitable for cut flower.
Winter Queen	White Bouquet × Flirt	Pink spoon-type flowers suitable for garden decoration.
Autumn Joy	White Bouquet × Flirt	It is early flowering, highly floriferous with pink decorative flowers, suitable for commercial cultivation and for garden display.

7.3.4.1 Spontaneous mutation

Bud sports, or spontaneous somatic mutations, have played a pivotal role in the evolution of many chrysanthemum varieties worldwide. Approximately 30% of chrysanthemum cultivars are estimated to have originated as sports. In India, several outstanding chrysanthemum cultivars have been developed through spontaneous mutations (Table 7.4) highlighting the significance of this natural phenomenon in cultivar improvement (Datta, 2013).

7.3.4.2 Induced mutagenesis

Extensive research has been conducted on the improvement of *Chrysanthemum morifolium* through induced mutagenesis by scientists across various countries. A wide range of physical mutagens including X-rays, gamma rays, fast neutrons, thermal neutrons, and radioactive phosphorus as well as chemical mutagens, such as ethylene imine, ethyl methane sulfonate (EMS), and colchicine, have been employed in these studies. These efforts have significantly advanced our understanding of practical applications for crop improvement using both classical and modern induced mutagenesis techniques.

Key aspects of induced mutagenesis encompass several critical steps to ensure effective application and desired outcomes (Table 7.5). Radiosensitivity analysis is essential to determine the optimal exposure levels for various plant materials, ensuring minimal damage while maximizing mutagenic efficiency. Careful selection of plant material is crucial, as it influences the mutagenesis success rate and the subsequent traits expressed in the progeny. Mutagen exposure methods, including chemical and physical agents, are tailored to the specific requirements of the crop and the desired mutations. Determining the optimal mutagen dose is vital to balance mutation induction with plant

Table 7.4. Elite varieties of chrysanthemum developed by spontaneous mutation. (From authors)

Variety	Description	Parent variety
Kasturba Gandhi	White flowers, reflexed	Mahatma Gandhi (mauve)
Sonar Bangla	Cream yellow flowers, incurved	Snowball (white)
White Cloud	White	Pink Cloud (pink)
Sharda	Yellow flowers	Sharad Shobha (white)
Queen of Tamluk	Yellow flower	Casa Grandi (white)
JS Lloyd	Light yellow, intermediate	William Turner (white)
Kasturba Gandhi	White flower	Mahatma Gandhi (Mauve)
Sonar Bangla	Yellow flower	Snowball (white)
White Cloud	White flower	Pink Cloud (pink)
R Venkataraman	Yellow flower	SS Arnold (white)
William Turner	White flower	Pink Turner (pink)

Table 7.5. Varieties developed through induced mutations. (From authors)

Variety	Breeding institute and description
Indian Institute of Horticultural Research (ICAR), Bengaluru, Karnataka	
Arka Usha Kiran	Induced mutant from cv. Arka Kirti. It requires 121 days for flowering and may produce up to 170 flowers per plant, each having a diameter of 5 cm. Flowers are semidouble with prominent disc in yellow color. Suitable for loose flowers, bedding, and pot plants. Flowers have a shelf life of 7 days.
Arka Yellow Gold	Induced mutant of cv. "Flirt." On an average, it takes 140 days to flower. It bears yellow-colored decorative flowers. The diameter of a single flower is approximately 5.26 cm, and the weight is 2.16 g. Petals have brick-red-colored stripes. It produces 113 flowers per plant. Flower yield per plant is 244 g, good for loose flower purposes. It is also suitable for cut flower (spray-type) purpose and has a vase life of 14 days.
CSIR-National Botanical Research Institute (NBRI), Lucknow, Uttar Pradesh	
Agnishika	Gamma ray-induced (1.5 and 2.0 krad). Decorative-type flowers, disc florets are not visible due to highly developed ray florets, Erythrite red-colored flowers. Suitable for garland and bedding.
Batik	Gamma ray-induced (2.0 krad) mutant of cv. Flirt, flowers are small and double spray-type, red-colored, with yellow stripes.
Navneet Yellow	Gamma ray-induced (1.5 krad) mutant of cv. Navneet that bears yellow flowers.
Asha	Gamma ray-induced (1.5 krad) mutant of cv. Hope. Large, reflexed-type, and cream-colored flowers.
NBRI Kesar	Gamma ray (7.5 Gy)-induced mutant of cv. Puja. Flowers are decorative-type with yellow-pink petals on adaxial and pink on abaxial surfaces.
NBRI Him Jyothi	Gamma ray-induced mutant of cv. NBRI Himanshu, dwarf and floriferous in nature. Suitable for pot culture and landscape gardening.
Shekhar	Gamma ray-induced mutant of cv. Su-Neel that produces decorative mauve flowers that bloom during late December to mid-February.
NBRI Pukhraj	Induced mutant of cv. NBRI Himanshu. Plant is dwarf, with yellow-colored flowers that bloom from late November to early January. Suitable for a potted plant and for landscape gardening.

Basanti	Gamma ray-induced (1.5 krad) mutant from E-13. Small, pompon-type, and yellow-colored flowers.
Himani	Gamma ray-induced (2.0 krad) mutant from E-13. Small, pompon-type, and white-colored flowers.
Kumkum	Gamma ray-induced (2 and 2.5 krad) mutant from M-71. Brown-colored flowers.
Navneet	Gamma ray-induced (2 krad) mutant from Kalyani Mauve. Small, decorative-type flowers, creamy-white in color.
Purnima	Gamma ray-induced mutant from Otome Zakura. Large, pompon-type, white-colored flowers.
Shabnam	Gamma ray-induced (1.5 krad) mutant from D-5. Decorative-type, magnolia-purple flowers. The flowers have distinct shape, with small appendage-like structure at the tip of each floret.
Sharad Har	Gamma ray-induced (1.5 krad) mutant from Sharad Mala. Flowers are yellow-colored, small, and decorative-type.
Sheela	Gamma ray-induced (2 and 2.5 krad) mutant from Himani. Flowers are small, pompon-type, and canary yellow-colored.
Subarna	Gamma ray-induced color mutant of cv. Flirt. Flowers are small, double, spray-type, and yellow in color.
Indian Agricultural Research Institute (IARI), New Delhi	
Pusa Centenary	An induced radio mutant of a var. Thai Chen Queen. Plants are vigorous-growing; flowers possess brilliant yellow incurving ray and disc florets, which are fleshy and delicate, suitable for cut flower and exhibition purpose.
Pusa Anmol	Gamma ray-induced mutant of cv. Ajay, highly floriferous and bushy variety that bears yellowish-pink flowers. Thermo- and photo-insensitive-type and produces three flushes in a year during October–November, February–March, and June–July. Suitable for loose flowers and cut flowers (spray-type).
Pusa Arunodaya	Gamma ray-induced pink-colored mutant of cv. Thai Chen Queen; medium and spreading-type that bears orange-colored flowers. Flowers are semidouble-type, with incurving ray florets that are 7–8 cm in diameter. Suitable for cut flower, pot culture, and exhibition purposes.
Pusa Kesari	Gamma ray-induced mutant from cv. Thai Chen Queen. Flowers are double-type and red in color. Suitable for both exhibition purpose and cut flower.

viability. Advanced techniques, such as combined treatments, recurrent irradiation, split doses, colchimutagenesis, and *in vitro* mutagenesis, expand the scope of mutagenesis, allowing for more precise and diverse applications. Mutation detection and frequency analysis, coupled with spectrum evaluation, provide insights into the types of mutations generated. Additionally, understanding the nature of chimerism and employing both classical and modern methods for chimera management play a pivotal role in stabilizing desired mutations and achieving uniformity in the resulting plant population.

Radiosensitivity assessment is a prerequisite for large-scale irradiation aimed at inducing mutations. For this purpose, large numbers of chrysanthemum cultivars have been exposed to gamma rays at doses of 1.5, 2.0, and 2.5 krad. These studies evaluated the effects of radiation on flower type, shape, color, chromosome number, interphase nuclear volume (INV), interphase chromatin volume (ICV), 2cDNA content, and chromosomal aberrations. Numerous studies have identified suitable radiation doses for inducing somatic mutations in chrysanthemum. Notably, Datta and his research group at NBRI determined that the lethal dose for 50% of the individuals (LD_{50}) for chrysanthemum cultivars ranges from 1.5 to 2.5 krad.

Development of chlorophyll-variegated varieties: Chlorophyll-variegated chrysanthemum varieties have been developed at CSIR-NBRI using both classical breeding and induced mutation techniques. Through conventional breeding and selection, six chlorophyll-variegated varieties were developed: DWS-2, DWS-12, DWS-15, B-16, B-17, and OO-2. Additionally, gamma ray-induced mutagenesis led to the development of two more chlorophyll-variegated varieties, Kargil 99 and Niharika (Datta, 1990).

Mutation in flower morphology: Several chrysanthemum mutants with altered flower morphology have been developed at CSIR-NBRI, Lucknow, and subsequently commercialized. Notable examples include Tulika, Shabnam, Lalima Tubular, Cosmonaut, Jhalar, Nirbhik, and Nirbhaya.

7.3.4.3 In vitro mutagenesis

In traditional approaches, *in vivo* mutagen-treated plants are grown under field conditions. However, this method is often constrained by limited land availability and funding, restricting the size of the plant population that can be treated and screened. To address these limitations, the concept of *in vitro* mutagenesis was introduced, providing opportunities to induce a larger number of mutants, including stable ones. Significant progress has been made in utilizing *in vitro* mutagenesis for chrysanthemum improvement. A notable example is the development of NaCl-tolerant chrysanthemum plants through chemical mutagenesis. A stable NaCl-tolerant variant was successfully developed using whole-plant selection and treatment with the chemical mutagen EMS (Hossain *et al.*, 2004; Hossain *et al.*, 2007; Nasri *et al.*, 2022). In addition, chrysanthemum has been regenerated from a single cell via somatic embryogenesis, a technique particularly suited for managing single-cell mutation events. The somatic embryogenesis protocol for chrysanthemum has been successfully standardized (Mandal and Datta, 2005).

Isolation of mutants through tissue culture: Tissue culture techniques are widely employed for the isolation and propagation of chrysanthemum mutants, enabling the stable regeneration of plants with desirable traits. Gamma-ray irradiation of cv. "Lilith" induced morphological changes, including chlorophyll variegation in leaves and sectorial flower color mutations (Dwivedi *et al.*, 2000). Chimeric yellow ray florets were cultured in Murashige and Skoog (MS) medium, regenerating plants that flowered true to the original color. Additionally, somatic embryogenesis was used for single-cell mutation management, demonstrating potential for developing novel ornamental cultivars (Mandal and Datta, 2005). A regeneration system for ray florets of a chrysanthemum mutant achieved 95.56% microshoot regeneration (Verma *et al.*, 2012). Maximum proliferation (100%) was observed with a combination of 6-Benzylaminopurine (BAP) (5.0 mg l^{-1}), 1-Naphthaleneacetic acid (NAA) (0.1 mg l^{-1}), and gibberellic acid (GA_3, 0.2 mg l^{-1}). The *in vitro*-raised plants produced true-to-type red flowers, confirming the stability of the induced mutation. Regeneration protocols were developed for a yellow-flowered mutant derived from cv. Lalit (white) (Kumar *et al.*, 2017b). The study achieved 92% microshoot proliferation using BAP (4.0 mg l^{-1}) and NAA (0.05 mg l^{-1}). Rooting success (86%) was obtained with half-strength MS medium supplemented with 0.5 mg l^{-1} NAA. The *in vitro*-raised plants exhibited bright yellow flowers, in stark contrast to the original, white-flowered variety.

Continued efforts are underway to develop chrysanthemum varieties suitable for the Indian market and diverse agro-climatic conditions across different regions of India. Multiple varieties in the pipeline, developed by various institutes and universities, have been registered with ICAR–NBPGR, New Delhi (Table 7.6).

7.4 Climatic and Soil Requirements

Tropical and subtropical climates are ideal for cultivating chrysanthemums in open fields. The optimal temperature range for chrysanthemum cultivation is 20–28°C during the day and 15–20°C at night, with a relative humidity of 70–80%. As chrysanthemums are obligate short-day plants, planting should be timed to ensure that flowering coincides with short-day photoperiod conditions. For vegetative growth, chrysanthemums thrive under long-day conditions with bright sunlight and temperatures between 20 and 27°C. In contrast, flower bud development and flowering require short-day conditions and cooler temperatures ranging from 10 to 27°C. These temperature ranges and light conditions are critical for achieving optimal plant performance. Chrysanthemums grow best in well-drained, sandy loam soils with a pH of 6.5–7.0 and an electrical conductivity (EC) of 1.0–1.5. Due to their shallow, fibrous root system, chrysanthemums are particularly sensitive to waterlogging, which can lead to diseases such as root rot and wilt, often exacerbated

Table 7.6. Chrysanthemum germplasm registered with ICAR-NBPGR, New Delhi for novel and unique traits. (From authors)

National Identity	INGR No.	Pedigree	Novel unique features
IC568708	10114	Selection from cv. Punjab Gold	Early and off-season flowering, suitable for pot and garden use
IC568709	10115	Selection from cv. Punjab Gold	Early and off-season flowering, suitable for pot and garden use
IC623437	17059	Half-sib progeny from cv. Pink cloud	Flower color: 77.B, Purple group, Fan 2. Stellate ray florets (Cylindrical in shape)
IC623438	17060	Half-sib progeny from cv. Red Stone	Flower color: 162.D, Gray yellow group, Fan 4. Stellate ray florets (Cylindrical in shape)
IC635436	20106	Yellow Puma × White Star	Yellow-colored flower. Double flower shape (8.36 cm diameter). Spray-type.
IC635437	20107	Yellow Puma × Shyamal	Brick-red-colored flower with bicolored florets (yellow color on floret tips). Spray-type.
IC635438	20108	Yellow Puma × Shyamal	Pink-colored flower. Flower diameter = 9.94 cm. Plant height = 100.97 cm.
IC635439	20109	Poornima × Shyamal	Dark, pink-colored flower. Spatulate (fluted) florets. Flower diameter = 7.99 cm.
IC638881	21108	Half-sib selection from the var. PAU D-1	Florets spatulate in shape. Long peduncle (8–12 cm).
IC638882	21109	Half-sib selection from the var. PAU D-1	Flowers possess mild fragrance. The plant bears cream white-colored, ligulate-type flowers.
IC638883	21110	Half-sib selection from the var. PAU A-64	It is spray chrysanthemum and suitable for pot mums and garden decoration. The plant bears attractive, yellow-colored, single-type flowers. Plant gives an appearance of leaflessness and dome shape during flowering.

by poor soil aeration. Continuous chrysanthemum cultivation can negatively impact soil health, reducing plant growth, soil microbial diversity, and altering microbial community structure. Additionally, soil nutrient levels decline with continuous cropping, increasing the incidence of soilborne diseases. Deep tillage is an effective method to improve nutrient utilization from deeper soil layers and to enhance the physical and chemical properties of the soil in monoculture cropping systems. To maintain soil fertility and ensure sustainable crop yields, the application of vermicompost combined with 50% nitrogen (N), phosphorus (P), and potassium (K) (NPK) fertilizers and a microbial consortium is highly effective not only to support plant health but also to promote a more balanced soil ecosystem.

7.5 Plant Propagation

Chrysanthemums are commercially propagated asexually through various vegetative methods, including apical stem cuttings, leaf cuttings, suckers, and micropropagation.

7.5.1 Suckers

In chrysanthemums, a sucker is a lateral branch that emerges from the main stem near ground level and runs underground for a short distance. Suckers produce adventitious roots, and the aboveground shoots can develop into new plants. Well-rooted suckers are typically transplanted directly into the field, offering a simple and effective propagation method.

7.5.2 Terminal cuttings

The most widely adopted commercial propagation method for chrysanthemums is rooting of terminal stem cuttings taken from healthy mother plants (Mukherjee, 2008). The ideal cutting length is 5–7 cm, with three to four leaves. Dipping the cuttings in indole butyric acid (IBA) at concentrations of 1500–2500 ppm promotes early and uniform rooting. These cuttings are commonly planted in sand or cocopeat beds under shaded conditions or covered with a 100 µm polythene sheet to maintain high humidity essential for rooting. Grewal *et al.* (2005) reported significantly higher rooting percentages in terminal cuttings of cv. Snowball when treated with IBA (400 ppm), with rooting success peaking in the first week of June (56.7%) compared to the first week of July (50.5%). The integrated nutrient management approach significantly enhanced rooting behavior in chrysanthemums, yielding the highest rooting percentage and a greater number of roots under a treatment of 22.5 g m^{-2} of NPK combined with vermicompost (1 kg m^{-2}) and biofertilizers (*Azotobacter*), phosphate-solubilizing bacteria (PSB), and vesicular arbuscular mycorrhiza (VAM) (Laishram *et al.*, 2016).

7.5.3 Stem segment and leaf cutting with axillary bud

Propagation by terminal stem cuttings typically produces only one plant per stem, resulting in lower multiplication rates. To address this limitation, alternatives, such as leaf bud cuttings and stem fragments with one leaf and an axillary bud, have been explored. Singh and Chettri (2013) demonstrated the effectiveness of leaf cuttings consisting of the leaf blade, petiole, and an attached axillary bud under *in vivo* conditions. This method has proven to be reliable for propagating the loose flower chrysanthemum variety CO.1, producing strong, transplantable plants. Plants of CO.1 propagated from leaf bud cuttings showed accelerated growth when treated with coconut water or Keradix-F at 20 ppm and rooted in enriched cocopeat media (Soundarya, 2018). Higher multiplication rates have been achieved using stem fragments with one leaf and an axillary bud as propagules (Cojocariu and Tanase, 2019). This technique is particularly advantageous when propagation (mother plant) material is limited, enabling the production of multiple plants from a small amount of initial material.

7.6 Field Preparation, Planting Season, and Spacing

Light and temperature are critical environmental factors influencing the growth and flowering of chrysanthemums. Both photoperiod and light intensity significantly affect their development. As short-day plants, chrysanthemums typically initiate flowering between September and December, when the photoperiod is shorter. In India, the ideal planting season is from May to July, with late-flowering varieties planted until August. This timing is strategically aligned with India's festive season, which runs from September to January. Growers often cultivate three crops annually, with the first crop planted in August, the second in September, and the third in October. Chrysanthemums require approximately 6 months to mature and produce flowers, with the first crop typically blooming by January. Planting is generally carried out at a spacing of 30 × 30 cm or 30 × 20 cm, allowing for 63,000–65,000 plants ha^{-1}. However, studies on *Chrysanthemum coronarium* (garland chrysanthemum) by Mali *et al.* (2017) and Nehra and Munendra (2019) suggest that a wider spacing of 45 × 60 cm, combined with 200 kg ha^{-1} of nitrogen, results in better outcomes, including maximum plant spread and shorter flowering duration. Prior to planting, the roots of cuttings are treated with 0.2% Ceresan or Captan to prevent diseases. Raised beds covered with black plastic mulch are commonly used to suppress weed growth and retain soil moisture (Fig. 7.4). A plant density of 12 plants m^{-2} was found to enhance overall growth and flowering parameters, as reported by Mitra and Pal (2008). Furthermore, unpinched, single-stemmed plants demonstrated greater leaf area, earlier flowering, larger and heavier flowers, longer and thicker flower stalks, and maximum shelf life, particularly in the cultivar Chandrama. In addition, Chaitra *et al.* (2018) highlighted that

Fig. 7.4. Open- field chrysanthemum crop at planting stage (A) and vegetative stage (B). (Photo courtesy of Shambu GB.)

applying 150% of the recommended dose of fertilizers in combination with mulching significantly improved flower yield, plant growth, and flowering characteristics.

7.7 Irrigation

Chrysanthemum cultivation requires frequent and thorough irrigation, particularly before the onset of the monsoon season. However, it is advisable to avoid watering during midday, when temperatures are high to prevent stress on the plants. An open drainage system is essential in the planting beds, as chrysanthemums are highly sensitive to excess water. Waterlogging, especially during the rainy season, should be strictly avoided to minimize the risk of root rot and other water-related issues. During the vegetative stage, irrigation is typically carried out twice a week during the first month and then reduced to weekly intervals, depending on soil moisture and prevailing weather conditions. Raised beds equipped with drip irrigation and mulching are effective in conserving soil moisture and controlling weed growth (Fig. 7.4). Irrigating chrysanthemums at 100% field capacity significantly enhances optimal crop performance, both in terms of flower yield and shelf life.

7.8 Crop Nutrition

Chrysanthemums are heavy feeders and require balanced fertilization to ensure optimal growth and flowering. To maintain soil health and enhance organic matter, a basal application of 8 t acre^{-1} of well-decomposed farmyard manure (FYM) should be incorporated at least 15 days prior to planting. It is essential to ensure that the FYM or organic compost is free from pathogens, nematodes, weed seeds, and other harmful organisms. The land should be prepared into appropriately sized beds to facilitate uniform nutrient distribution. Inorganic fertilizers should be applied based on soil analysis and recommended nutrient levels to optimize growth while minimizing environmental impact. A typical recommendation is 40:20:40 kg NPK acre^{-1} (Nair *et al.*, 2021). During the vegetative growth stage, nitrogen application at 200 mg kg^{-1} of soil promotes profuse branching, especially for achieving a topiary form. This rate should be reduced to 100 mg kg^{-1} during the flowering phase to support bloom development without excessive vegetative growth (Janipour *et al.*, 2017).

7.8.1 Fertigation

Fertigation significantly enhances fertilizer use efficiency and yield in chrysanthemum cultivation. Applying the full recommended dose of water-soluble fertilizers, combined with liquid biofertilizers through weekly fertigation, promotes early flowering and maximizes plant height, spread, flowering duration, and the yield of loose and cut flowers, particularly in the chrysanthemum variety "Marigold" (Sahana *et al.*, 2019). In chrysanthemum cv. MDU-1, using 75% of the recommended fertilizer dose (RDF), supplemented with 5 t ha^{-1} vermicompost and a 0.2% humic acid foliar spray, improved key growth parameters, such as plant height, leaf number, leaf area, and branching (Kumar *et al.*, 2019). This approach achieves high yields while saving 25% of fertilizers, making it a cost-effective strategy. For cost-effective fertigation, a basal application of 8 t acre^{-1} of FYM, along with 16.8 kg Urea, 70 kg Single Super Phosphate (SSP), and 12.8 kg Muriate of Potash (MOP) per acre, is recommended at planting. In Karnataka, a major chrysanthemum-growing region, the recommended fertigation rate for the variety "Marigold" is 100:150:100 kg NPK ha^{-1} (Sahana *et al.*, 2019).

7.8.2 Biofertilizers and biostimulants

The application of biofertilizers and biostimulants significantly enhances both flower quality and yield in chrysanthemums. A combination of 15 g m^{-2} each of NPK, 1 kg m^{-2} of vermicompost, and biofertilizers has been shown to increase the population of Azotobacter (10.50×10^6 cfu g^{-1}) and phosphorus-solubilizing bacteria (12.83×10^6 cfu g^{-1}), resulting in higher flower yields (Laishram *et al.*, 2013; Nasri and Ghaderi, 2015). For improved flower quantity, stalk girth, and flower diameter, a foliar application of 2%

panchagavya + 4% seaweed extract is recommended, particularly for the variety "Amalfi" (Vetrivel *et al.*, 2017). Incorporating Arka Microbial Consortium at 12.5 kg ha^{-1} further improves flower yields, while the addition of 2 kg each of Azospirillum and Phosphobacteria, mixed with FYM at a rate of 100 kg ha^{-1} at planting, provides additional benefits. Chrysanthemums grown for loose flowers respond particularly well to organic manures. Studies have shown that applications of 20 t ha^{-1} of vermicompost or 25 t ha^{-1} of goat manure significantly increase the number of flowers per plant, flower yield per ha^{-1}, and cost–benefit ratio, particularly in the cultivar "Dolly White" (Premkumar *et al.*, 2016).

7.8.3 Micronutrients

Micronutrient deficiencies can significantly hinder crop growth and yield in open-field chrysanthemum cultivation. Deficiencies in iron and copper, for instance, delay flowering and impede bud development, while boron deficiency results in irregular floret formation, necrosis, and incurved inflorescences. Regular inspection and timely application of micronutrients are, therefore, essential for optimal crop performance. Micronutrient enrichment, particularly with iron and zinc, has proven to be effective in regulating flowering and enhancing flower yield. For example, foliar application of $FeSO_4$ at 0.8% has been shown to extend the flowering duration, increase yield (Naveen Kumar *et al.*, 2009), and improve postharvest life in the variety CO.1. Likewise, a combined foliar spray of $FeSO_4$ + $ZnSO_4$ (0.5%) in the cultivar "Mayur," resulted in increased branching, earlier flower initiation, larger flower diameters, and longer flower longevity, thereby enhancing overall yield (Vanlalruati *et al.*, 2019). Additionally, a fortnightly foliar spray of $ZnSO_4$ (0.25%) + $FeSO_4$ (0.2%) + Boric acid (0.1%) is widely recommended for improved growth and higher yields in chrysanthemums (Nair *et al.*, 2021). This integrated micronutrient management approach is essential for addressing deficiencies and maximizing productivity in open-field cultivation systems.

7.9 Weeding, Interculture, and Mulching

Timely weeding is essential in chrysanthemum cultivation to prevent growth constraints and yield loss. Manual weeding and hoeing, typically performed 8–10 times a year, help loosen the soil, improving porosity and aeration. Additionally, sugarcane trash mulching at 5 t ha^{-1} has been found to promote vigorous growth and enhance flower quality in *Chrysanthemum* cv. "Ratlam Selection" (Vamaja *et al.*, 2021). However, the labor-intensive nature, high costs, and time requirements of manual weeding make it impractical for large-scale operations. Herbicides offer an economical and efficient alternative for weed management, significantly enhancing flower yield. Butachlor applied at 1.0–1.5 kg ha^{-1}, in combination with hand weeding, has proven highly effective in reducing weed competition and boosting yields (Kaur *et al.*, 2014; Kumar *et al.*, 2017a). Similarly, pendimethalin, applied at 0.64–1.0 kg ha^{-1} followed by hand weeding 40 DAT, has shown excellent weed control and promoted growth in chrysanthemums (Sharma *et al.*, 2014). For optimal weed management, a pre-plant application of pendimethalin at 1.0 kg ha^{-1}, followed by hand weeding at 30 and 60 days after planting, has been recommended. This approach not only minimizes weed growth but also improves flower productivity (Bala, 2017; Kumar *et al.*, 2017a).

7.10 Crop Management Practices

To enhance plant growth and vigor while maintaining flower quality, various crop management practices, such as pinching, disbudding, and de-suckering, are widely employed in chrysanthemum cultivation.

7.10.1 Pinching

Pinching involves the removal of the growing tips of the plant to reduce its height and encourage the growth of vegetative laterals, promoting axillary branching. This practice is commonly applied to crops propagated through suckers and terminal cuttings. Pinching is typically performed manually using the thumb and forefinger (Fig. 7.5). The first pinching is carried out when plants reach a height of 15–20 cm and have three to four pairs of leaves. A second pinching may be required if plants exhibit straggly or lean growth. Two types of pinching are performed based on the extent of removal—*soft pinching:* involves removing the soft tip of the shoot along with two to three open leaves. *Hard pinching:* involves removing a longer portion of the shoot, up to the hardened part.

Pinching offers several benefits, including delaying flowering, which allows shoots to mature physiologically and support better flower development (Ehsanullah *et al.*, 2021). It also checks apical dominance, redirecting the plant's energy toward producing more branches and flowers. Pinching at 25 DAT has been shown to result in early bud initiation and increased flower yield per plant, plot, and hectare (Kour, 2009; Thumar *et al.*, 2020). Double-pinch plants demonstrated significant advantages, including maximum plant spread, a higher number of suckers per plant, and an improved crop growth rate (Mounika *et al.*, 2019). Additionally, pinching performed at 30 DAT or 4–6 weeks after transplanting was found to be the most effective in promoting better overall growth, flowering, and yield (Kedar *et al.*, 2021; Nagdeve *et al.*, 2021; Yogita *et al.*, 2021). However, early pinched plants showed greater plant spread and stem girth, while non-pinched plants were consistently taller (Dorajeerao and Mokash, 2020). Non-pinched plants recorded earlier bud initiation, fully opened flowers, and better flower quality (Salve *et al.*, 2016; Jindal *et al.*, 2018). Later pinching at 50 DAT extended the flowering span (Jindal *et al.*, 2018).

Fig. 7.5. Pinching of terminal shoot with thumb and forefinger. (Photo courtesy of Malleshaiah SharathKumar.)

7.10.2 De-suckering

During the vegetative growth phase, new suckers develop at the base of the plants. De-suckering involves the periodic removal of these suckers to maintain a healthy plant canopy and promote vigorous shoots that will ultimately produce flowers.

7.10.3 Staking

Staking is essential for preventing lodging, providing vertical support, and maintaining the plants' proper shape and bloom. Bamboo sticks are commonly used for staking, ensuring plants remain erect and well-structured.

7.11 Growth Regulators for Crop Management

Growth regulators play a pivotal role in enhancing the growth, flowering, and overall performance of chrysanthemum plants. Numerous studies have highlighted their effects on various growth parameters and yields. GA_3 at 200 ppm significantly increased the number of flowers per plant (148.73), while lower concentrations of 50 ppm promoted earlier flower formation. A combination of GA_3 (150 ppm) and cycocel (CCC, 3000 ppm) enhanced plant spread, stem diameter, number of branches, days to first flower bud appearance, flower size, weight, and the total number of flowers per plant (Singh *et al.*, 2018). At a concentration of 300 ppm, GA_3 maximized plant height, spread, number of branches, suckers, leaves, leaf area, plant fresh and dry weight, number of flowers, flower size, persistence, and promoted early flowering (Alhajhoj, 2017).

Benzyl adenine (BA) at 200 ppm, applied as a foliar spray, enhanced the keeping quality of *Chrysanthemum* cv. "Punjab Shyamli," improving postharvest performance (Singh and Bala, 2018). CCC at 3000 ppm significantly increased the number of flowers per plant (Dorajeerao and Mokashi, 2012). Additionally, 1.5 l nitroxin + 25 mg l^{-1} paclobutrazol improved flower production, fresh and dry weight of shoots and flowers, and boosted chlorophyll, carotenoids, soluble carbohydrates, and protein content (Nasri and Ghaderi, 2015). In conclusion, growth regulators, such as GA_3, BA, CCC, MH, paclobutrazol, and biofertilizers, such as nitroxin, have proven to be effective tools for optimizing chrysanthemum growth and performance. Their strategic application can enhance flowering, growth parameters, and postharvest quality while reducing reliance on synthetic fertilizers.

7.12 Harvest Indices, Yield and Postharvest Handling

For loose flower purposes, flowers are ready for harvest when 2–3 rows of ray florets are perpendicular to the flower stalk (Fig. 7.6). Fully or semi-opened flowers are hand-picked with the peduncle at 3- to 4 day intervals. Chrysanthemum plants start yielding flowers after 3–4 months of transplanting, depending on the variety. The average yield of loose flowers is 20–25 t ha^{-1}. Loose chrysanthemums are packed in bamboo baskets or gunny bags for marketing. Harvested chrysanthemum loose flowers are transported to local markets using bamboo baskets or plastic crates (Fig. 7.7), which typically hold 10–15 kg of flowers. Additionally, gunny bags are commonly used, with a capacity of 30–35 kg. These gunny bags help retain moisture, keeping the flowers cool during long-distance transport.

Fig. 7.6. Open-field chrysanthemum crop in full bloom for harvesting; filed view (A) and close-up of flowers that are ready for harvest (B). (Photo courtesy of Shambu GB.)

7.13 Physiological Disorders in Chrysanthemum Cultivation

Open-field chrysanthemum cultivation is susceptible to several physiological disorders that can significantly impact flower quality and yield. These include floral abnormalities, such as quilling of florets, crown bud, bleaching, blindness, and chrysanthemum phyllody.

Fig. 7.7. Harvested chrysanthemum loose flowers ready for transport to local markets using plastic crates (A); the same plastic crates were also used in the local markets to display the flowers for sale (B). (Photo courtesy of Malleshaiah SharathKumar.)

Quilling of florets: This results in twisted, cup-shaped petals due to boron deficiency and low night temperatures (<15°C). This can be addressed by applying borax (3 g l^{-1}) during vegetative growth, bud initiation, and flowering stages. Night temperatures below 15°C during flower development can cause pink discoloration on white petals. Similarly, low light intensity and lower night temperatures during flower development can result in quilling of florets, where petals twist and become cup-shaped, while peripheral florets lose turgidity.

Crown bud: This is a common issue in chrysanthemums, characterized by severely retarded bud development and glossy, enlarged involucral bracts. It often occurs when plants experience long days following a few short days. Proper planting schedules can mitigate this disorder.

Photoperiod and nutrient deficiencies: Prolonged photoperiods combined with low night temperatures can delay or inhibit floral induction. Nutrient deficiencies, particularly copper and manganese, can exacerbate this. Applying manganese sulfate (3 g l^{-1}) and copper sulfate (1 g l^{-1}) during vegetative, bud initiation, and flowering stages can control these abnormalities.

Bleaching: Bleaching of petals and bract formation on the receptacle may occur during floret formation under long days and high temperatures.

Blindness: Blindness occurs when flower buds fail to develop, often due to low night temperatures and short days during the bud formation stage. Symptoms include a rosette growth of leaves, failure of central petals to develop, and unopened flowers. Proper regulation of light and temperature can reduce its occurrence.

Chrysanthemum phyllody: This disorder involves the transformation of floral organs into leafy structures, leading to stunted and dwarfed plants. Affected plants display reddish older leaves, clustering stems with shortened internodes (witch's broom effect), and deformed or absent flowers. The disorder alters flower parts into vegetative structures during early development.

7.13.1 Management strategies

Physiological abnormalities in chrysanthemums are often associated with unfavorable environmental conditions, including inappropriate photoperiods, temperature extremes, and nutrient deficiencies. Proactive management practices can mitigate these issues, ensuring optimal flower quality and improved yields. Some key strategies include:

- *Maintaining optimal light and temperature conditions*: Providing consistent and favorable light and temperature levels during critical growth stages to support normal development.
- *Timely nutrient management*: Ensuring an adequate supply of essential nutrients such as boron, copper, and manganese through targeted foliar applications.
- *Aligning planting schedules*: Scheduling planting to coincide with favorable photoperiods and temperature conditions to minimize stress and prevent abnormalities.

7.14 Conclusion and Future Prospects

Open-field cultivation of chrysanthemums plays a vital role in the floriculture industry, particularly in India and other Asian countries, where these flowers are widely used for garlands, religious ceremonies, and cultural festivities. Open-field cultivation remains cost-effective and well suited for large-scale production under favorable agroclimatic conditions. However, challenges such as disease susceptibility, climate variability, and fluctuating market demand continue to impact growers. Future advancements should focus on breeding disease-resistant and high-yielding varieties to enhance productivity and adaptability to changing climatic conditions. The adoption of sustainable farming practices, including integrated pest and disease management and water-efficient irrigation systems, could help to improve environmental sustainability while reducing input costs.

Additionally, the integration of advanced digital agriculture tools, such as remote sensing, AI-driven disease detection, and precision irrigation, will help optimize chrysanthemum crop production by enabling real-time monitoring and early intervention. Strengthening postharvest handling, cold storage, and efficient market linkages will further enhance price realization and reduce postharvest losses, making the supply chain more resilient. With these advancements, open-field chrysanthemum farming can remain a sustainable, profitable, and competitive segment of the global floriculture industry, catering to both domestic and international markets.

References

Alhajhoj, M.R. (2017) Effects of foliar application of plant growth regulators on growth and flowering characteristics of chrysanthemum cv. Paintball. *Pakistan Journal of Life and Social Sciences* 15(2), 114–119. Available at: http://www.pjlss.edu.pk/pdf_files/2017_2/114-119.pdf

Bala, M. (2017) Weed management effect on vegetative growth and flowering parameters of chrysanthemum. *Indian Journal of Weed Science* 49, 303. DOI: 10.5958/0974-8164.2017.00081.8.

Chaitra, G.S., Seetharamu, G.K., Kumar, R., Munikrishnappa, P.M. and Shivanna, M. (2018) Effect of different levels of macronutrients (NPK) and mulching on growth, quality and yield of chrysanthemum (*Dendranthema grandiflora*) cv. Marigold. *Indian Journal of Agricultural Sciences* 88(6), 871–876. DOI: 10.56093/ijas.v88i6.80591.

Cojocariu, A. and Tanase, C. (2019) Development and testing a new technology for production of chrysanthemums planting material (*Chrysanthemum* s.l.). *Journal of Plant Development* 26, 93–107. DOI: 10.33628/jpd.2019.26.1.93.

Datta, S.K. (1990) Induction and analysis of somatic mutations in garden chrysanthemum. *Advances in Horticultural Science* 1, 241–254.

Datta, S.K. (2013) *Chrysanthemum morifolium* Ramat-a unique genetic material for breeding. *Science and Culture* 79(7–8), 307–313.

Datta, S.K. and Janakiram, T. (2015) Breeding and genetic diversity in *Chrysanthemum morifolium* in India: A review. *The Indian Journal of Agricultural Sciences* 85(11), 1379–1395. DOI: 10.56093/ijas.v85i11.53372.

Datta, S.K. (ed.) (2023) Chrysanthemum. In: *Role of Mutation Breeding in Floriculture Industry*. Springer, Singapore. DOI: 10.1007/978-981-99-5675-3_8.

Dorajeerao, A.V.D. and Mokashi, A.N. (2012) Yield and quality parameters of garland chrysanthemum (*Chrysanthemum coronarium* L.) as influenced by growth regulators/chemicals. *Indian Journal of Plant Sciences* 1(1), 16–21.

Dorajeerao, A.V.D. and Mokash, A.N. (2020) Floricultural crop management. *Indian Journal of Plant Sciences* 41(1), 9–15. Available at: http://14.139.158.107:8079/varieties/chrysanth_variety.htm

Dwivedi, A.K., Banerji, B.K., Chakrabarty, D., Mandal, A.K.A. and Datta, S.K. (2000) Gamma ray-induced new flower colour chimera and its management through tissue culture. *Indian Journal of Agricultural Sciences* 70(12), 853–855.

Ehsanullah, M., Tarapder, S.A., Maukeeb, A.R.M., Khan, A.U. and Khan, A.U. (2021) Effect of pinching on growth and quality flower production of chrysanthemum (*Chrysanthemum indicum* L.). *Journal of Multidisciplinary Applied Natural Science* 1(2), 62–68. DOI: 10.47352/jmans.v1i2.15.

Grewal, H.S., Kumar, R. and Chauhan, R. (2005) Effect of IBA and NAA on rooting in chrysanthemum (*Dendranthema grandiflora* Tzevlev) terminal cuttings. *Journal of Ornamental Horticulture* 8(3), 230–232.

Hossain, Z., Mandal, A.K.A., Shukla, R. and Datta, S.K. (2004) NaCl stress—its chromotoxic effects and antioxidant behavior in roots of *Chrysanthemum morifolium* Ramat. *Plant Science* 166(1), 215–220. DOI: 10.1016/j.plantsci.2003.09.009.

Hossain, Z., Mandal, A.K.A., Datta, S.K. and Biswas, A.K. (2007) Development of NaCl-tolerant line in *Chrysanthemum morifolium* Ramat. Through shoot organogenesis of selected callus line. *Journal of Biotechnology* 129(4), 658–667. DOI: 10.1016/j.jbiotec.2007.02.020.

Janipour, B., Khandan-Mirkohi, A., Jamali, F. and Khalighi, A. (2017) Influence of nitrogen nutrition in topiary of Chrysanthemum (*Chrysanthemum grandiflorum*). *Iranian Journal of Horticultural Science* 4(8), 237–247.

Jindal, M., Thumar, B.V. and Hallur, V. (2018) Effect of planting time and pinching on flowering and flower quality of Chrysanthemum CV. Ratlam Selection. *Journal of Pharmacognosy and Phytochemistry* 7(4), 390–393. Available at: https://www.phytojournal.com/archives/2018/vol7issue4/PartG/7-3-622-617.pdf

Kaur, R., Bala, M. and Kaur, T. (2014) Chemical weed management in Chrysanthemum. *Indian Journal of Weed Science* 46(4), 396–398.

Kedar, D.P., Panchbhai, D.M. and Chatse, D.B. (2021) Influence of pinching on growth, flowering and yield of different flower crops. *International Journal of Current Microbiology and Applied Sciences* 10(3), 1854–1859. DOI: 10.20546/ijcmas.2021.1003.233.

Kour, R. (2009) Flowering production as affected by spacing and pinching in chrysanthemum cv. Flirt. *International Journal of Agricultural Sciences* 5(2), 588–589.

Kumar, A., Kumar, M., Ghosh, S., Tewari, T. and Bhardwa, S.B. (2017a) Effect of weed management practices in chrysanthemum (*Dendranthema grandiflora* T.) under Tarai conditions of Uttarakhand. *Journal of Crop and Weed* 13(2), 203–207.

Kumar, G., Sindhu, S.S., Kumar, S. and Vanlalruati, V. (2017b) *In vitro* isolation, regeneration, and purification of yellow mutant in Chrysanthemum (*Chrysanthemum morifolium*) cv. Lalit through ray floret regeneration. *Indian Journal of Agricultural Sciences* 87(7), 958–963. DOI: 10.56093/ijas.v87i7.71959.

Kumar, S., Sreedar, C., Sanjeev, K., Muralidharan, A. and Elakkuvan, S. (2019) Studies on the effect of integrated nutrient management on the growth parameters of Chrysanthemum cv. MDU-1. *Plant Archives* 19(2), 2743–2746.

Laishram, N., Dhiman, S.R., Gupta, Y.C., Bhardwaj, S.K. and Singh, A. (2013) Microbial dynamics and physico-chemical properties of soil in the rhizosphere of chrysanthemum (*Dendranthema grandiflora*) as influenced by integrated nutrient management. *Indian Journal of Agricultural Sciences* 83(4), 447–455. Available at: https://epubs.icar.org.in/index.php/IJAgS/article/view/28856/13053

Laishram, N., Dhiman, S.R., Gupta, Y.C., Singh, A. and Bhardwaj, S.K. (2016) Quality cutting production of chrysanthemum (*Dendranthema grandiflora*) as influenced by integrated nutrient management of the plants during production phase. *Indian Journal of Agricultural Sciences* 86(11), 1449–1454. DOI: 10.56093/ijas.v86i11.62926.

Mali, G.L., Moond, S.K., Mishra, A., Sharma, M.K. and Singh, B. (2017) Effect of planting geometry and nitrogen on growth, flowering, and yield of chrysanthemum (*Chrysanthemum coronarium* L.). *Progressive Horticulture* 49(1), 75–78.

Mandal, A.K.A. and Datta, S.K. (2005) Direct somatic embryogenesis and plant regeneration from ray florets of chrysanthemum. *Biologia Plantarum* 49(1), 29–33. DOI: 10.1007/s10535-005-0033-6.

Mitra, M. (Sarkar) and Pal, P. (2008) Performance of *Chrysanthemum* × *morifolium* Ramat. cv. 'Chandrama' grown at different levels of planting density and stem maintained per plant. *Natural Product Radiance* 7(2), 146–149.

Mounika, C.H., Suseela, T., Dorajeerao, A.V.D., Subbaramamma, P. and Sujatha, R.V. (2019) Effect of pinching and growth regulators on vegetative and physiological parameters. *International Journal of Current Microbiology and Applied Sciences* 8(8), 2031–2042. DOI: 10.20546/ijcmas.2019.808.237.

Mukherjee, D. (2008) *Speciality Cut Flowers Production Technologies*. Naya Udyog, Kolkata, India.

Nagdeve, N.S., Khobragade, H.M., Thakare, A.A., Gajbhiye, R.P. and Mandhare, K.S. (2021) Effect of plant spacing and pinching on growth and flower yield of annual chrysanthemum. *International Journal of Chemical Studies* 9(1), 491–495. DOI: 10.22271/chemi.2021.v9.i1g.11279.

Nair, S.A., Smitha, G.R., Safeena, S.A. and Aswath, C. (2021) *Annual Report 2021*. ICAR-Indian Institute of Horticultural Research (ICAR-IIHR), Bengaluru, India, pp. 1–48.

Nasri, F. and Ghaderi, N. (2015) *Journal of Science and Technology of Greenhouse Culture* 5(20), 111–124.

Nasri, F., Zakizadeh, H., Vafaee, Y. and Mozafari, A.A. (2022) *In vitro* mutagenesis of *Chrysanthemum morifolium* cultivars using ethylmethanesulphonate (EMS) and mutation assessment by ISSR and IRAP markers. *Plant Cell, Tissue and Organ Culture* 149, 657–673. DOI: 10.1007/s11240-021-02163-7.

Naveen Kumar, P., Misra, R.L., Dhiman, S.R., Ganga, M. and Kameswari, L. (2009) Effect of micronutrient sprays on growth and flowering of chrysanthemum. *Indian Journal of Agricultural Sciences* 79, 426–428.

Nehra, M.K. and Munendra, S. (2019) The effect of nitrogen and potash fertilizer and spacing on plant growth parameters. *Annals of Horticulture* 12(1), 74–79.

Premkumar, A., Punetha, P., Bohra, M., Upadhyay, S. and Nautiyal, B.P. (2016) Response of chrysanthemum cultivar Dolly White to different sources and combinations of organic manures under the mid hill regions of Uttarakhand. *International Journal of Agriculture Sciences* 8(59), 3294–3297.

Sahana, A.S., Biradar, M.S., Hiremath, S.M. and Somanagouda, G. (2019) Influence of source and levels of fertilizers for fertigation in chrysanthemum variety Marigold. *Journal of Farm Sciences* 32(4), 497–500.

Salve, D.M., Panchbhai, D.M., Badge, S. and Satar, V. (2016) Growth and flower yield of chrysanthemum as influenced by varieties and pinching. *Plant Archives*

16(2), 826–828. Available at: https://www.plantarchives.org/PDF/162/826-828.pdf

Sharma, G., Shrivastava, A., Dhakre, D.S. and Singh, D.P. (2014) Effect of weed management practices in chrysanthemum (*Dendranthema grandiflora* T.) under Chhattisgarh plains agro-climatic condition. *International Journal of Bio-Resource and Stress Management* 5(3), 400–403. DOI: 10.5958/0976-4038.2014.00587.9.

Singh, P. and Chettri, R. (2013) A new propagation method for rapid multiplication of chrysanthemum under *in vivo* conditions. *International Journal of Conservation Science* 4, 95–100.

Singh, T. and Bala, M. (2018) Effect of foliar spray of benzyl adenine, gibberellic acid and putrescine on post-harvest keeping quality of chrysanthemum. *Agricultural Research Journal* 55(2), 386. DOI: 10.5958/2395-146X.2018.00073.X.

Singh, J., Nigam, R., Singh, R., Kumar, A. and Kumar, A. (2018) Effect of gibberellic acid and cycocel on growth, flowering and yield of chrysanthemum (Dendranthema grandiflora Ramat) cv. Birbal Sahni. *Journal of Pharmacognosy and Phytochemistry* 7(SP1), 2753–2758. Available at: https://www.phytojournal.com/archives/2018/vol7issue1S/PartAO/SP-7-1-776.pdf

Soundarya, S.M. (2018) Studies on the effect of media and growth regulators on different planting materials for rapid multiplication of chrysanthemum. MSc thesis, Tamil Nadu Agricultural University, Coimbatore, India.

Su, J., Jiang, J., Zhang, F., Liu, Y., Ding, L. *et al.* (2019) Current achievements and future prospects in the genetic breeding of chrysanthemum: A review. *Horticulture Research* 6, 109. DOI: 10.1038/s41438-019-0193-8.

Tejaswini, P., Bharathi, T.U. and Nair, S.A. (2021) *Arka Flower Varieties and Production Technology*. Director ICAR-Indian Institute of Horticulture Research, Bengaluru. e-book. Available at: https://iihr.res.in/arka-ebook/52

Thumar, B.V., Chovatiya, N.V., Butani, A.M., Bhalu, V.B. and Jindal, M. (2020) Effect of planting time and pinching on flowering and yield of chrysanthemum (*Dendranthema grandiflora*) cv. Ratlam selection. *Journal of Pharmacognosy and Phytochemistry* 9(2), 414–416. Available at: https://www.phytojournal.com/archives/2020/vol9issue2/PartH/9-1-543-847.pdf

Vamaja, S.M., Bhatt, D.S., Bhatt, S.T., Chawla, S.L., Patel, M.A. *et al.* (2021) Effect of mulching on growth, flowering and yield of chrysanthemum (*Chrysanthemum morifolium* Ramat.) cv. Ratlam selection. *Journal of Crop and Weed* 17(3), 233–236. DOI: 10.22271/09746315.2021.v17.i3.1515.

Vanlalruati, V., Anand, P., Kumar, G. and Tiwari, A.K. (2019) Effect of saline stress on growth and biochemical indices of chrysanthemum (*Chrysanthemum morifolium*) germplasm. *Indian Journal of Agricultural Sciences* 89(1), 41–45. DOI: 10.56093/ijas.v89i1.86102.

Verma, A.K., Prasad, K., Singh, S. and Kumar, S. (2012) *In vitro* isolation of red coloured mutant from chimeric ray florets of chrysanthemum induced by gamma-ray. *Indian Journal of Horticulture* 69(4), 562–567. Available at: https://journal.iahs.org.in/index.php/ijh/article/view/2043

Vetrivel, T., Jawaharlal, M., Arulmozhiyan, R. and Kannan, M. (2017) Effect of biostimulants on growth and yield of chrysanthemum (*Dendranthema grandiflora* Tzvelev.) var. Amalfi under protected cultivation. *Ecology, Environment and Conservation* 23(4), 2297–2301.

Yogita, J., Sahu, T.L. and Sushila, D. (2021) Effect of pinching on flowering and quality parameters of chrysanthemum varieties. *The Pharma Innovation Journal* 10(10), 1579–1581. Available at: https://www.thepharmajournal.com/archives/2021/vol10issue10/PartW/10-10-100-674.pdf

8

Crop Models for Chrysanthemum

Oliver Körner[1]* and Ep Heuvelink[2]*

[1]Leibniz-Institute of Vegetable and Ornamental Crops, Großbeeren, Germany
[2]Horticulture and Product Physiology, Department of Plant Sciences, Wageningen University and Research, Wageningen 6700 AA, The Netherlands

Abstract

Models of growth and quality in chrysanthemum pot plant and cut flower production can be used as input to IT-based decision support systems (DSSs) for planning and online monitoring, and model-based predictions can be directly input to climate controllers. There is a large variety of models that can be used for these purposes. Mathematical models range from simple regression models, or black box big-data algorithms, to explanatory highly parameterized crop physiology models. Each model type has its advantages and disadvantages: data-based models can be used as reliable predictors within the scope they are trained for. Explanatory models can be used to understand processes and find reasons for cultivation problems, when used as input to a planning or monitoring tool. We explain the differences between model families and guide through a process of creating simple models from scratch. However, we mainly focus on and explain the setup of process-based (explanatory) modeling. The crop itself is central, represented in crop growth models and models for external plant quality. We show how photo-assimilates and quality are related, for example, through leaf area formation (external quality) that has a direct feedback on photosynthesis (growth) and vice versa. Biochemical leaf photosynthesis models are introduced, and the scale-up to the crop level is explained. We discuss how climate variables that influence external quality variables, such as stem length, flower number, and so on, are incorporated in models. Eventually, some possible applications of model implementation into IT-based DSSs and climate controllers are introduced and discussed.

Keywords

AI, decision support systems, explanatory models, photosynthesis-driven models, plant quality models, regression models

*Corresponding authors: ep.heuvelink@wur.nl; koerner@igzev.de

DOI: 10.1079/9781789248289.0008

8.1 Introduction

Greenhouse production of chrysanthemum (*Chrysanthemum morifolium* ssp.) needs cultivar-specific knowledge on development and growth responses to the greenhouse climate. How can crop models improve control and monitoring systems for decision support in chrysanthemum? Are current chrysanthemum crop models suitable, or what are the obstacles, and what is still needed to employ chrysanthemum crop models in horticultural production with reliable decision support systems (DSSs) or model-based climate control?

Chrysanthemum, as a short-day plant, in the Northern Hemisphere, is mainly cultivated in climate-controlled greenhouses, with options for day length control with both blackout screens and supplementary lighting. Crop growth and product quality depend on production system and environmental conditions including the number of long days before applying short days. With perfect timing of this control scheduling, the chrysanthemum crop is able to attain a certain mass/weight and leaf area index (LAI) before the photo-sensitive period for flower induction starts. In that short-day period, day length (i.e. period with illumination in 24 hour) must be reduced to the cultivar-specific maximum period. Although different aspects of flowers are a major quality aspect in chrysanthemum, plant architecture is as important (Carvalho and Heuvelink, 2001). To achieve high-quality produce in commercial practice, large amounts of resources in greenhouse climate control for heat and electricity are used (Körner and Van Straten, 2008). Energy consumption strongly varies throughout the year in Northern Europe, for example, in the Netherlands. For a 12-week cultivation, energy consumption was seven times larger when planting in January compared to planting in May was reported (Körner and Challa, 2004a). Various methods for energy saving in chrysanthemum production with modified climate control regimes have been developed (Körner and Challa, 2003b; Körner *et al.*, 2006), in addition to new greenhouse technology (Raaphorst *et al.*, 2020). However, in all aspects, energy-saving measures must go hand-in-hand with plant quality.

In both cut chrysanthemum and potted varieties, compactness (next to many other attributes) is a major quality concern (Carvalho and Heuvelink, 2001), which, besides the use of chemical growth retardants (Langton, 1998), which experiences reduced societal acceptance, can be achieved through directed temperature control (difference between the mean day and night temperatures (DIF), DROP) (Cuijpers and Vogelezang, 1992; Cockshull *et al.*, 1995; Hansen *et al.*, 1996), light quality (Lund *et al.*, 2008), and a combination of both (Carvalho *et al.*, 2002). In the last decades, dynamic climate regimes were designed for energy saving in greenhouses that also obey the quality demand of chrysanthemum production (Körner and Challa, 2003b, 2004a, b; Aaslyng *et al.*, 2005). However, to cope with this multifactor complexity in chrysanthemum production with photoperiodic aspects, temperature sensitivity, crop growth driven by photosynthesis and also disease and quality monitoring, models could help in both decision support and climate control. In decision support, models could be used for planning and online monitoring (Körner, 2019), while in climate control, model-based predictions are useful as direct input to the climate controllers (Janka *et al.*, 2016; Körner *et al.*, 2021).

With the increasing complexity of production, models become increasingly important (Gurban and Andreescu, 2018). In fact, while the abovementioned complexity solely describes the plant responses to climate variables, the main target in commercial chrysanthemum production is high yield of high product quality at low usage of resources, eventually yielding the best return on investment (ROI) for the grower. To reduce energy consumption as a major cost, and reduce resource consumption in general, greenhouse heating and the use of chemicals should be reduced (Pearson *et al.*, 1995; Khattak and Pearson, 1997) while achieving high-quality crops by intelligent climate regimes (Iddio *et al.*, 2020). For that, however, a combination of well-performing crop physiological models, greenhouse physical models, and sensors is necessary (Aiello *et al.*, 2018; Watson *et al.*, 2018; Körner, 2019). With the combination of models and sensors, optimal climate control can yield optimized growth conditions when using cultivar- and location-specific constraints in the control algorithm (Rytter *et al.*, 2012).

Mathematical models have been developed for a wide range of purposes and with an even wider range of prediction quality. The simplest models describe an isolated process in a defined setup, while complex models incorporate various processes, system dynamics, and probabilities. In fact, while all models are simplifications of the reality, one needs to be sure about the types of models and their purposes; for example, while a trained black box model used in artificial intelligence in greenhouses (e.g. Zou *et al.*, 2017; Moon *et al.*, 2018; Taki *et al.*, 2018) can be used to detect the plant disease white rust (*Puccinia horiana*) from an image of a chrysanthemum leaf, the same model can never tell the reasons behind such a disease, that is, what process had led to it and what needs to be adjusted to avoid the problems another time. On the other hand, a model that consists of a set of explanatory mathematical equations to predict the risk to get that certain plant disease, and as such can help to control the climate to avoid it, using time courses of climate and microclimate data as input, cannot be used to determine whether this disease is present in the crop.

In the following, we will elaborate on the different types of models and their applicability in chrysanthemum production in greenhouses. We will show the differences in model types, model applications, and their applicability in simulation studies or real-time implications; how different models can be designed, the problem and complexity of parameterization, and pros and cons of model types in that regard; and how to incorporate mathematical models in DSSs for cultivation planning or monitoring, or for application in greenhouse climate control. We will answer the introductory question of what model work is still needed for reliable DSSs or model-based climate control in greenhouse-grown chrysanthemum.

8.2 Model types

Models are schematic representations of systems, and a system is part of reality, defined by its boundaries (Wit, 1969); they are simplified systems from reality (De Wit and Van Keulen, 1987). In general, a distinction can be made between descriptive and explanatory models. Descriptive models usually consist of one or more mathematical equations, which are derived from measured datasets, while explanatory models consist of quantitative descriptions of mechanisms and processes, and are based on scientific theory and hypotheses (Penning de Vries *et al.*, 1989; Lee, 2002). As such, descriptive models in their basic form are only applicable to the situations they were parameterized for and cannot be transferred to other circumstances without new parameterization (Marcelis *et al.*, 1998). While descriptive models offer the possibility to consider subprocesses, a causal relationship is usually not present.

Models can also be categorized using different criteria, such as their base construction or their behavior (as summarized in Table 8.1). In behavioral terms, models can either be static or dynamic. The often-used pseudo-dynamic models in crop growth and greenhouse climate simulations are an intermediate type of model system, which are based on static models incorporated in a dynamic simulation environment (see "*Dynamic Modelling*"). In construction type, roughly two families of models can be distinguished, that is, data-based models and deterministic models, while each of them has subcategories. The borders of model classification are, however, not strict. While a vast amount of literature has been published to distinguish and classify model families, the systematics remain complex and blurry; for example, a descriptive model can have either a deterministic or a data-based origin (not indicated in Table 8.1).

Static models describe the status of a system at a fixed time, while dynamic models are used in simulations and are able to predict the change of the system in time. An example of a simple static regression model is the allometric individual leaf area estimation from measured leaf length and width, as it is often carried out as input to further model applications for crop transpiration, photosynthesis, or simply light interception (Schwarz and Kläring, 2001). Examples of chrysanthemum leaf allometry are presented by Fanourakis *et al.* (2021). A further enlargement of this kind of model structure was suggested by incorporating the plant's interaction with its environment, including a fitted temperature sum as additional variable (Yu and Körner, 2020). Without the causal relationship of this additional input variable, however, the nature of the model as a static regression model type remains unchanged.

Data-based models originating from big data are trained black box models. Complex machine learning algorithms, such as neural networks (Fitz-Rodríguez and Giacomelli, 2009; Miranda *et al.*, 2017; Taki *et al.*, 2018), genetic algorithms, or nonlinear autoregressive exogenous models (Wang *et al.*, 2017), are used to fit large datasets. Depending on the amount and variation of the input data, big-data-based models can rather well predict or simulate reality. However, the reasoning of a model decision is completely

Table 8.1. Overview of model types according to main headline classification (combinations of classifications are possible, e.g. black box models have often been highly predictive, while explanatory models are usually non-predictive). Adapted from author.

Model classification							
Construction				Behavior		Application	
Data-based		Deterministic		Static	Dynamic	Predictive	Explanatory
Mathematical data fitting and big-data machine learning		Mathematical description of individual processes		Describing static behavior of a system	Describing changes on variates over time	Aim to accurately estimate the statistical probability that a future event will occur	Identifying individual factors and processes that are associated with the outcome
Regression	Black box	Descriptive	Explanatory				
Simple data fit of experimental results	Complex machine learning systems with large amount of data	Describes the system with its relationship to its environment	Establishing causal relationships between the various data and processes				

hidden in the self-learning algorithm; therefore, the terminology "black box." Contrary to deterministic models, black box models show only the main result; deeper analyses and consideration of the subprocesses are not possible. To really understand the processes and the system's behavior, deterministic–explanatory models are the choice to make. However, there are good reasons for using one or the other model family. Each type has its pros and cons; for example, while a black box model is well suited to detect facts in complex data structures such as images and is easy to handle (Behmann *et al.*, 2015), deterministic–explanatory models can analyze all subprocesses (Marcelis *et al.*, 1998), while, as a drawback, they are often highly over-parameterized (Metselaar, 1999). The larger a deterministic–explanatory model becomes, the higher the risk of mismatch and model instability, that is, the prediction quality decreases.

Fan *et al.* (2015) demonstrate a novel knowledge-and-data-driven modeling (KDDM) approach for simulating tomato plant growth that consists of two sub-models. One sub-model is derived from available domain knowledge, including known relationships from physiological mechanistic models; the other is constructed solely from data without using any domain knowledge. This could also be a very valuable approach for crop growth modeling in chrysanthemum, as the case study by Fan *et al.* (2015) confirms that the KDDM approach inherits advantages from both the knowledge-driven and data-driven approaches. Several ways of combining black box and explanatory modeling into KDDM to make use of the best of both worlds are presented by Zhang *et al.* (2022).

In the last decades, another model family has been developed, the functional structural plant models (FSPMs; e.g. Buck-Sorlin *et al.*, 2011), which have been thoroughly discussed in journal special issues (Evers *et al.*, 2018; Louarn and Song, 2020). These models generate a virtual plant structure and can be applied to study geometric aspects, such as plant architecture interactions, in a cropping system in chrysanthemum (De Visser *et al.*, 2006). FSPMs have also been combined with deterministic crop growth sub-models such as a photosynthesis model. These are able to simulate and predict plant phenotypic plasticity in development and growth in chrysanthemum (Kang *et al.*, 2012).

In the following, the creation and usage of different models are explained with leaf area modeling as an example. The simplest model is a mathematical equation with only one output and one input variable with a fixed proportionality; that is,

$$a = xb \tag{8.1}$$

where x would be the fitted parameter to the variable b in order to predict a. This type of regression model can have different behaviors, depending on the context. One step further is using two input variables c and b describing one output a, which can have different forms, for example:

$$a = cb \quad or \quad a = xcb \quad or \quad a = xcb^2 \tag{8.2}$$

In chrysanthemum, an example of such a model would be the allometric calculation of a leaf area (A_l) from leaf length (L_l) and leaf width (W_l):

$$A_l = L_l W_l \tag{8.3}$$

as this relation is usually too simple to calculate A_l, extensions to this model are needed, for example, adding fitting variables, as it was recently done for chrysanthemum leaves (Fanourakis *et al.*, 2021). Many options of empirical data-fits were tested, leading to a set of different equations, for example:

$$A_l = a + bL_l W_l \tag{8.4}$$

with a and b being fitted parameters. This equation had the best fit, with R^2 values of 0.84 or 0.88 in calibration and validation, respectively, when tested with six or nine chrysanthemum cultivars (Fanourakis *et al.*, 2021). The simplicity of this model is also its strength. Only two fitted parameters and two measured variables are needed to calculate A_l with an accuracy of 84 or 88%. This static model can serve the purpose of calculating leaf area when only a ruler is available. However, no information on dynamics or processes leading to a prediction of A_l is present in this model. This model, in its present state, could be classified as a static regression model. Its usability is limited to static leaf area estimation.

Information on the processes involved could change this model to an explanatory type and give the user information on the processes leading to a certain leaf size (or its deviation from it). In that case, a causal relationship between a, b, and A_l would be input to the model, and could evolve into this chosen form (for explaining purposes, an arbitrary form is used in the present case):

$$A_l = \varphi a + \psi bL_l W_l \tag{8.5}$$

where φ and ψ could be variables related to physiological processes in the leaf, for example, a causal relationship between the microclimate around the leaves and processes as leaf photosynthesis P_l, stomatal conductance g_s, abscisic acid content in the leaves $[ABA]_l$, soil water capacity ψ_v, and so on. In that case, φ and ψ could be summarized as

$$\varphi = f_\varphi(P_l, g_s, [ABA]_l, \psi_v) \tag{8.6}$$

$$\psi = f_\psi(P_l, g_s, [ABA]_l, \psi_v) \tag{8.7}$$

where f denotes the general term for a mathematical function; f can, in that regard, take any type of complexity of an (explanatory) model. In fact, the variables P_l, g_s, $[ABA]_l$, and ψ_v, which all have a proven effect on leaf growth and expansion, could have any form of a variable, being a result of a calculation with sub-models themselves (all types possible) or a measured/predicted entity. Also interrelations are possible, with, for example, g_s as input to the P_l

model (and vice versa) (Körner *et al.*, 2007a), ψ_v as influence on g_s (Liu *et al.*, 2009), or ψ_v influencing $[ABA]_l$ (Tardieu *et al.*, 2010).

The next step is applying the model in a dynamic process. Dynamic modeling (described in more detail in the next section) is used in simulations. The modeler is observing the change of one or several variables in time, for example, leaf area development from leaf unfolding to senescence for each time step of, for example, hours. For dynamic modeling, the numeric value of variables can change at each time step. This dynamic behavior can in fact also be achieved using static models by creating a pseudo-dynamic model.

When the simulation time step is low enough, a pseudo-dynamic model works surprisingly well. With larger simulation time steps, however, the results can be rather unstable and change from extremely high to extremely low values for each simulation time step, resulting in the risk for oscillation and instability. A better option is to build the dynamics directly into the models. Then, mathematical algorithms are used to solve the models' dynamic behavior in their form as differential equations, for example:

$$k\frac{A_1}{\mathrm{d}t}=[m(t)-A_1]\,\mathrm{d}t \tag{8.8}$$

where m is $\varphi a + \psi b\, L_l W_l$; and we set the size of the leaf as finite, denoted by k. The general terms of models and their interpretation in relation to leaf area simulation example are summarized in Table 8.2.

8.3 Dynamic Modelling

In crop growth models, the processes can be sorted into sub-modules that together form a flow chart (Fig. 8.1). Only the light intercepted by the green leaf area leads to photosynthesis (production of assimilates). The photosynthesis products (photo-assimilates) are used to form biomass (dry mass; assimilate pool), and part of the photosynthetic products are used in respiration (maintenance respiration, and growth respiration) as a result of the conversion from primary photosynthetic products to structural dry matter. Finally, dry mass is distributed over the different plant organs. An increase in leaf dry weight results in an increase in leaf area (calculated by multiplying by SLA = specific leaf area), which results in the new leaf area for simulation in the next time step (e.g. minutes, hours, or 1 day).

These processes are highly dynamic and interactive. The photosynthesis-driven crop growth (cf. Chapter 8.4, this volume) is a classical dynamic model (Challa, 1990), and the processes are often incorporated in yield simulators (Vanthoor, 2011; Körner and Hansen, 2012).

In simulation modeling, mathematical models can be roughly categorized into static and dynamic models. However, due to the complexity of many explanatory models, static models are often used in dynamic simulations. Then, static models become pseudo-dynamic by simply defining a loop where output variables in step x become input variables in step $x + 1$. By this procedure,

Table 8.2. Examples of model equations related to classification. Adapted from author.

	Data-based—Regression (Static)	Deterministic–Explanatory (Static)	Deterministic–Explanatory (Dynamic)
General term	$h = a + bxy$	$\mathrm{h} = \mathrm{u\ a} + \mathrm{n\ b}\ [\mathrm{x\ y}]$	$\dot{h} = \dot{u}\dot{a} + n\,\mathrm{b}\ [\mathrm{x\ y}] - \text{-h*}$ $\frac{\mathrm{h}}{\mathrm{d}t} = [\{ua + nb\ [xy]\}(t) - h]\ \mathrm{d}t$
Specified for leaf area simulation	$A_l = a + b\ L_l\ W_l$	$A_l = \varphi a + \psi b\ Ll\ \mathrm{W}l$	$k\frac{A_1}{\mathrm{dt}} = [m(t) - A_1]\ \mathrm{dt}$

*The point above the variables denotes for a matrix notation.

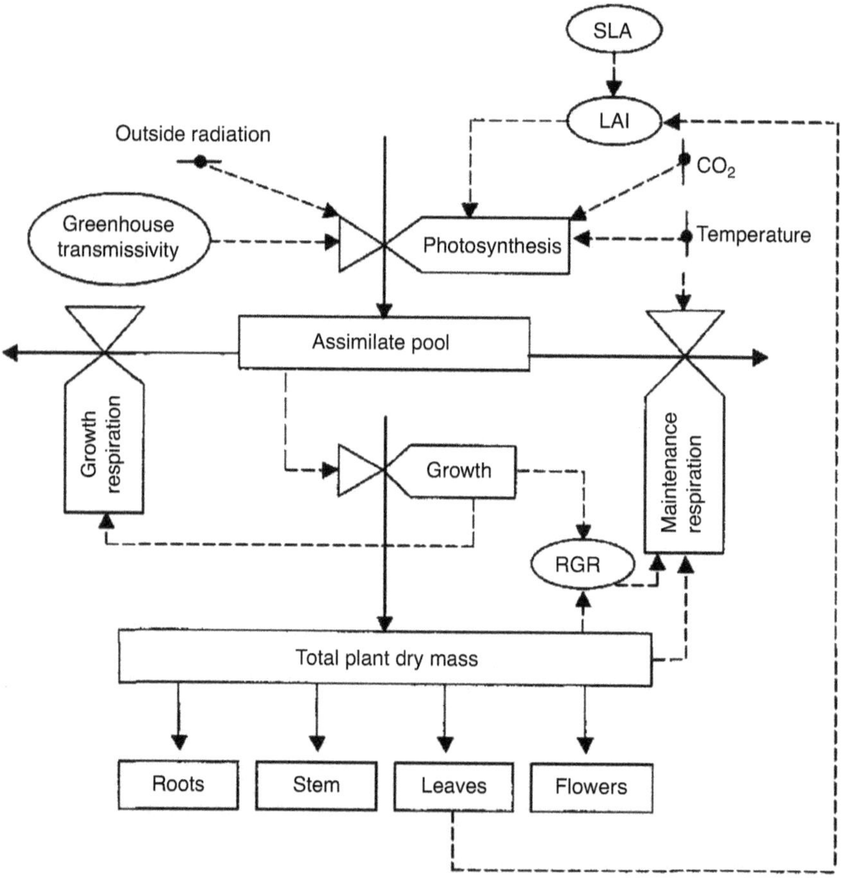

Fig. 8.1. Schematic presentation of a photosynthesis-driven crop growth model, with abbreviations: specific leaf area (SLA), leaf area index (LAI), relative growth rate (RGR). (Photo courtesy of Carvalho, 2003.)

with increasing number of simulation time steps, the error of static models in dynamic simulations can be reduced. While in horticultural crop modeling (as is the case with chrysanthemum), dynamic changes of plant physiological responses are known, until today, many growth models in crop simulation modeling rely on static equations of, for example, photosynthesis and stomatal conductance. However, incorporating real dynamics in simulation modeling could be the better option (depending on aim).

A case scenario is the artefact of diurnal hysteresis in the course of the daily photosynthesis rate. Often, photosynthesis in the morning hours reaches a higher assimilation rate with the same light level as compared to the afternoon (Fig. 8.2). According to Levy *et al.* (2004) there are several hypotheses for the mechanisms of diurnal hysteresis: (1) circadian rhythm, (2) diurnal changes in nutrient and CO_2 availability, (3) decrease in photosynthesis due to photoinhibition after long exposure to high light intensities, and (4) an increase in photorespiration in the afternoon. Dynamic models would enable to incorporate this afternoon decrease, either with process-based equations (when the actual reason is known and quantified) or by empirical functions. In static models, this "error" could be covered with an additional correction factor based on time of day; while the actual physiological processes that led to that decrease are not part of the static model and any dynamics (i.e., changes to the process) influenced by the complex interplay of the climate variables on the photosynthetic response would not be predicted by the model. A difference in actual and predicted photosynthesis would be the result.

8.4 Photosynthesis-Driven Growth Models for Chrysanthemum

A crop growth model as "user" of a photosynthesis model consists of various sub-models, with photosynthesis as its most important part. Crop growth is

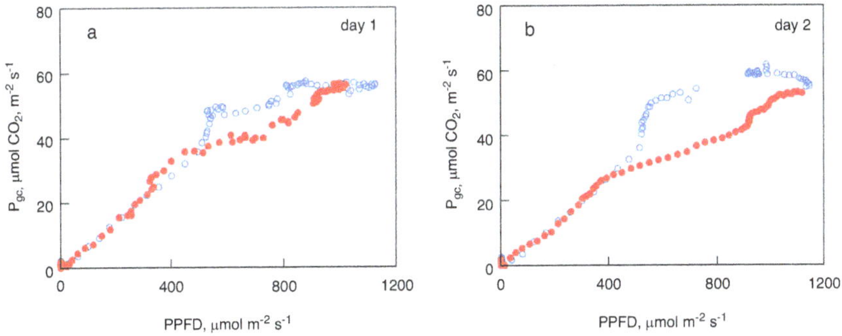

Fig. 8.2. Measured crop photosynthesis (P_{gc}) in cut chrysanthemum during 2 days with hysteresis: morning (blue) and afternoon (red); data are based on measurements. (Adapted from Körner *et al.*, 2007b)

usually modelled as the total increase of dry matter (W_{dc}), which is widely used in crop growth modeling and was already adopted in the first version of a photosynthesis-driven crop growth model for cut chrysanthemum, that is, CHRYSIMv1.0 (Lee, 2002).

$$\frac{dW_{dc}}{dt} = C_f(P_{gc} - R_m) \tag{8.9}$$

with crop gross photosynthesis ($P_{gc,}$ µmol CO_2 m^{-2} s^{-1}), maintenance respiration ($R_{m,}$ µmol CO_2 m^{-2} s^{-1}), and an efficiency factor for the conversion of assimilates to dry mass (C_f). In many models, C_f is constant at about 0.7, while in reality, it integrates conversion efficiencies of all plant organs, which are dynamic. Next to photosynthesis, R_m is the overall ruling entity in crop growth models. R_m is the critical process restricting crop growth with increasing weight of the crop. R_m depends on the existing biomass at each time point t, according to the full set of differential equations for W_{dc}. A simple model form that describes R_m well is

$$R'_m = m\,Q_R^{\,p\,(T_c - T_b)} \tag{8.10}$$

where m is the respiration coefficient at base temperature T_b, Q_R is the Q_{10} factor for respiration, which is calculated according to Arrhenius (Körner *et al.*, 2002), p is the factor related to the Q_R, and T_c is crop temperature. The full equation would then be

$$\frac{dW_{dc}}{dt} = C_f(P_{gc} - R'_m W_{dc}) \tag{8.11}$$

However, this equation does not fit in all circumstances for chrysanthemum at all times, as it underestimates growth at low growth rates, due to which an empirical correction was introduced to the CHRYSIMv1.0 model, based on an early observation in tomato (Heuvelink, 1996).

$$\frac{DW_{dc}}{dt} = C_f\left(P_{gc} - R'_m W_{dc}(1 - e^{-\beta g_r})\right) \tag{8.12}$$

with a regression parameter β and relative growth rate (RGR) (g_r).

The core of any explanatory crop growth model (including chrysanthemum), however, is the photosynthesis sub-model, calculating net and gross leaf photosynthesis (P_{nl}, P_{gl}; µmol CO_2 m^{-2} s^{-1}) that eventually are upscaled to the crop level in order to reach net and gross crop photosynthesis (P_{nc}, P_{gc}; µmol CO_2 m^{-2} s^{-1}). The difference between net and gross photosynthesis is dark respiration. In fact, many model types exist for photosynthesis (cf. Table 8.1) that can serve various purposes in chrysanthemum cultivation; for example, Ozturk *et al.* (2012) developed an empirical leaf photosynthesis model suitable for simulating leaf assimilation at fast-changing light and temperatures.

In this section, we focus on explanatory photosynthesis model types as sub-model for crop growth simulations. Photosynthesis is defined here as CO_2 uptake (photo-assimilation) of a leaf, including all involved subprocesses. This, as such, underlines the complexity of defining a model and its boundaries.

The basic process underlying crop growth is photosynthesis, the biochemical process transforming light energy to chemical energy, using CO_2 as the building material of the plant, and having O_2 as a waste product. The latter process is highly temperature-dependent. In simple terms, the three variables—net absorbed light (R_{na}), intercellular CO_2 concentration (C_i, μmol mol^{-1}), and leaf temperature (T_L K)—are the driving entities for net photosynthesis of a leaf (i.e. P_{nl}), which is usually given as the rate of moles CO_2 fixation in time (i.e. μmol [CO_2] m^{-2} s^{-1}). Fig. 8.3 illustrates the effect of temperature, CO_2, and light on cut chrysanthemum P_{gc} in a two-dimensional way (Körner *et al.*, 2009). Models simulating photosynthesis in spray-cut chrysanthemum (i.e. multiheaded) can neglect a possible negative feedback effect caused by low sink demand (Carvalho *et al.*, 2006).

To model the influence of light and CO_2, various photosynthesis models have been developed that could be used as sub-models in any C3-crop growth model, including chrysanthemum. A widely used biochemical photosynthesis model (Farquhar *et al.*, 1980) (i.e. Farquhar–von Caemmerer–Berry (FvCB) model) is used in its original form, with adaptations (Spitters *et al.*, 1989; Gijzen, 1992), or recent updates, as summarized by Yin *et al.* (2021). The FvCB model calculates assimilation A of the minimum of three entities of limitation to CO_2 assimilation, with RuBisCo-limited CO_2 assimilation rate (A_c, μmol CO_2 m^{-2} s^{-1}), the electron (e^-) transport-limited rate (A_j, μmol CO_2 m^{-2} s^{-1}), and the triose phosphate utilization limited rate (A_p, μmol CO_2 m^{-2} s^{-1}).

$$A = \min(A_c, A_j, A_p) \tag{8.13}$$

This model family is largely based on fitted parameters from CO_2 response curves (Sharkey *et al.*, 2007). For greenhouse crops, equations of this rather complex biochemical process-based model compilation were used and adopted to conditions in crop stands, enabling a straightforward upscaling from leaf to crop level. For that, the equations were partly restructured and partly simplified, and set in a light–response curve perspective for greenhouse crops, using a negative exponential light–response curve (Spitters *et al.*, 1986; Gijzen and Goudriaan, 1989; Gijzen, 1992; Goudriaan and Van Laar, 1994). The simple form of this equation, as used by Lee (2002) for leaf photosynthesis in cut chrysanthemum, is

$$P_g = P_{g,\max}\left(1 - e^{\frac{-\varepsilon R_{na}}{P_{g,\max}}}\right) \tag{8.14}$$

with gross photosynthesis (P_g, μmol CO_2 m^{-2} s^{-1}), maximum gross photosynthesis $P_{g,max}$ (μmol CO_2 m^{-2} s^{-1}), chemical light use efficiency (ε, mol CO_2

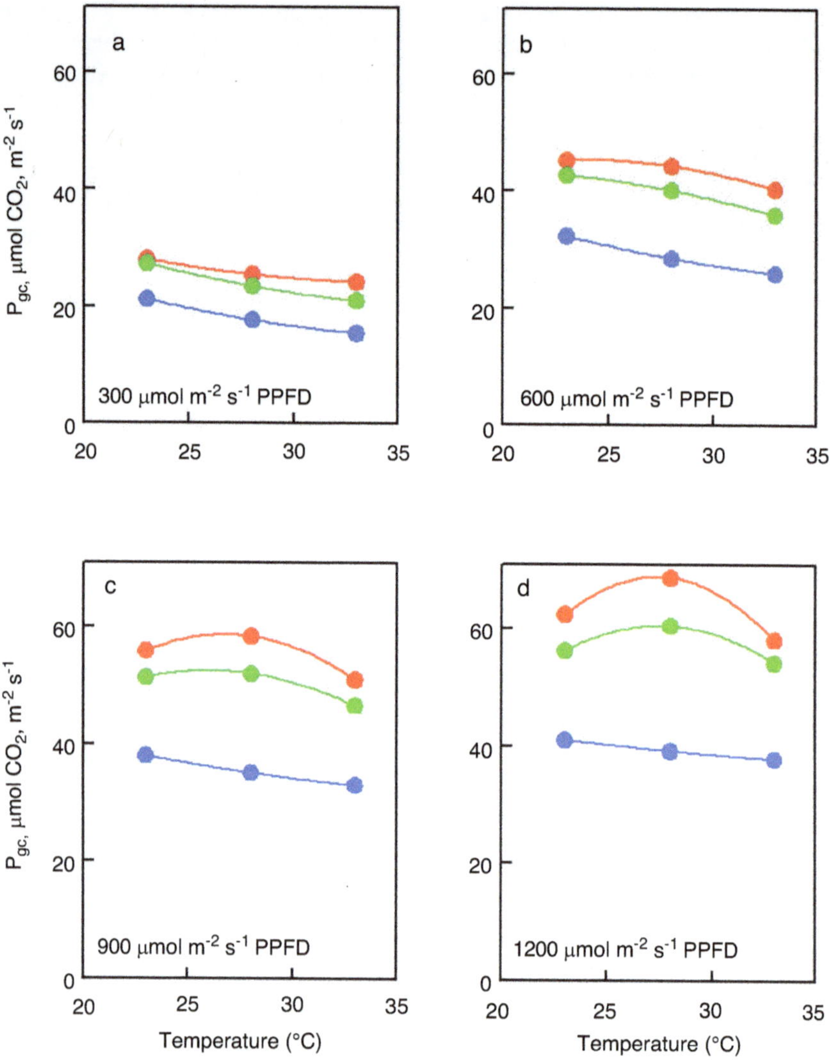

Fig. 8.3. Measured and fitted crop gross photosynthesis at different temperatures, three CO_2 levels (400 blue; 700 green; 1000 red, μmol mol^{-1}), and four radiation levels (photosynthetic photon flux density, PPFD) in cut chrysanthemum. (Reproduced Körner *et al.*, 2009, with permission from Taylor and Francis.)

mol^{-1} photons), and net absorbed photosynthetic active radiation (PAR) (R_{na}, μmol m^{-2} s^{-1}). P_n is then simply calculated from P_g and mitochondrial dark respiration R_d, μmol CO_2 m^{-2} s^{-1})

$$P_n = P_g - R_d \tag{8.15}$$

To solve this equation, the two variables, $P_{g,max}$ and ε, need to be determined. Both are dynamic functions of primary leaf temperature and leaf internal CO_2 concentration (C_i). For chrysanthemum, the set of equations to calculate $P_{g,max}$ and ε are presented in a model description (Körner and Challa, 2004b).

Crop photosynthesis is the sum of the photosynthesis of all leaves in the canopy. If all leaves would contribute with the same strength, a simple calculation of the individual leaf areas of all leaves in vertical direction *z* and horizontal direction *x* in the canopy and photosynthesis would be enough.

$$P_{nc} = P_{nl} A_{l}{}_{x_1 z_1}^{x_n z_n} \tag{8.16}$$

However, when upscaling from leaf to crop level, the absorption of diffuse and direct light in crop models is usually calculated separately. Absorption of a single (direct) light beam is calculated from path length through the canopy, leaf area traversed, average projection of leaves into the direction of the beam, reflection, and scattering for shaded and sunlit leaves (Goudriaan and Van Laar, 1994). Crop gross assimilation is then separately calculated from absorbed total light fluxes for shaded and sunlit leaves ($R_{na,shd}$ and $R_{na,sun}$, respectively) to yield shaded and sunlit photosynthesis ($P_{gc,shd}$ and $P_{gc,sun}$, respectively). P_{gc} is then the sum of those.

$$P_{gc} = P_{gc,sun} + P_{(gc,shd)} \tag{8.17}$$

While differences in leaf photosynthesis at different vertical positions have been observed for both cut chrysanthemum (Körner *et al.*, 2009) and potted chrysanthemum (Weerakkody and Suriyagoda, 2015), the larger the crop height/area ratio, the less accurate a two-dimensional crop model becomes. A crop canopy as cut chrysanthemum with its distinct rectangular shape can be seen either as a block with a rectangular three-dimensional grid with width, depth, and height of the block, or as an infinite canopy. In the latter case (two-dimensional), only the light beam in the vertical depths of a canopy is needed to estimate crop photosynthesis, and the Gaussian integration procedure can be applied (Goudriaan and Van Laar, 1994). A chrysanthemum crop, however, is, in most cases, restricted in space. Light beams from the side influence P_{gc} estimation and ignoring these would underestimate the real crop photosynthesis of a chrysanthemum stand (Körner *et al.*, 2007b). The calculation of a block-shaped canopy demands a three-dimensional model, where each leaf is set in the three-dimensional space in *x*, *y*, and *z* directions, using maximum leaf gross photosynthesis and leaf chemical light use efficiency ($P_{gl,max}$; ε_l) as inputs, for a grid with *n* points in each direction (Kläring and Körner, 2020).

$$P_{gc,sun} = \int_{z_1,y_1,x_1}^{z_n,y_n,x_n} \left[P_{gl,sun} \begin{Bmatrix} x_1 \ldots x_n \\ y_1 \ldots y_n \\ z_1 \ldots z_n \end{Bmatrix} \right] = \mathrm{P}_{gl,max} \left(1 - \frac{-\varepsilon_l \left[R_{n,a,sun} \begin{Bmatrix} x_1 \ldots x_n \\ y_1 \ldots y_n \\ z_1 \ldots z_n \end{Bmatrix} \right]}{\mathrm{P}_{gl,max}} \right) \tag{8.18}$$

$$P_{gc,sun} = \int_{z_1,y_1,x_1}^{z_n,y_n,x_n} \left[P_{gl,sun} \begin{Bmatrix} x_1 \dots x_n \\ y_1 \dots y_n \\ z_1 \dots z_n \end{Bmatrix} \right] = P_{gl,max} \left(1 - \frac{-\varepsilon_l \left[R_{n,a,sun} \begin{Bmatrix} x_1 \dots x_n \\ y_1 \dots y_n \\ z_1 \dots z_n \end{Bmatrix} \right]}{P_{gl,max}} \right) \tag{8.19}$$

This model is, however, still built as a semi-dynamic model. In a dynamic model, the fourth dimension (i.e. time) would need to be included. In the approach described earlier, variation between leaves in the canopy and the physiological changes of leaves with age influencing photosynthetic capacity (Bielczynski *et al.*, 2017) is missing. This could be incorporated by a leaf-age dependent sub-model as part of the crop-canopy model. In existing models, photosynthetic capacity of leaves of different ages and positions in the canopy is mostly assumed to be uniform, while it is known that different leaf anatomical properties change with age and affect CO_2 assimilation (Berghuijs *et al.*, 2015). Incorporating time dynamics in both leaf expansion (as worked out earlier, section Model types) and leaf aging processes is the next step in chrysanthemum crop growth modelling.

An explanatory photosynthesis model "ideally" includes all known processes involved in CO_2 fixation of a leaf, down to the cell and gene level. Today, however, most explanatory photosynthesis models that are used stop at the cell level, sometimes even higher. The more detailed processes are then modeled with fixed parameters, often fitted to laboratory measurements (Yu *et al.*, 2020). This again shows the difficulty in classification of a model as provided in Table 8.1. In fact, no fully explanatory model exists in crop science. At the lower aggregation level, there are always black-box approaches, and the base to all models is still data fitting in the lower model boundary.

The upper boundary for a P_{nl} model is defined by the climatic conditions the leaf is facing, the so-called microclimate with its boundary layer (i.e. a thin still layer of humid and O_2-rich air around the leaf). For a well-performing assimilation process in the chloroplasts, C_i is needed. Hence, the pathway for CO_2 migrating from the air through the boundary layer and the leaf stomata to the intercellular space in the leaves is of utmost importance, when modeling photosynthesis. For that, all leaf photosynthesis models have a stomata conductance subroutine (g_s; i.e. the conductivity for CO_2 molecules through the leaf-openings in µmol $[CO_2]$ m^{-2} s^{-1}). Many models also use a term for the boundary layer conductance (g_b; µmol $[CO_2]$ m^{-2} s^{-1}). Well-organized models combine the interactions of g_s, g_b, and P_{nl}, as was done in chrysanthemum (Janka *et al.*, 2016). However, g_s and g_b are highly dynamic variables, and interact with R_n and a fourth not-yet-introduced climate variable, relative humidity (RH), in models often defined as air vapor pressure deficit (VPD_a, kPa). For g_s model calculations, however, the water vapor difference between the boundary layer and the intercellular space, that is, leaf-to-air vapor pressure deficit (VPD_l, kPa), is used (Körner *et al.*, 2021). The effect of g_s on P_{gc} is illustrated in Fig. 8.4.

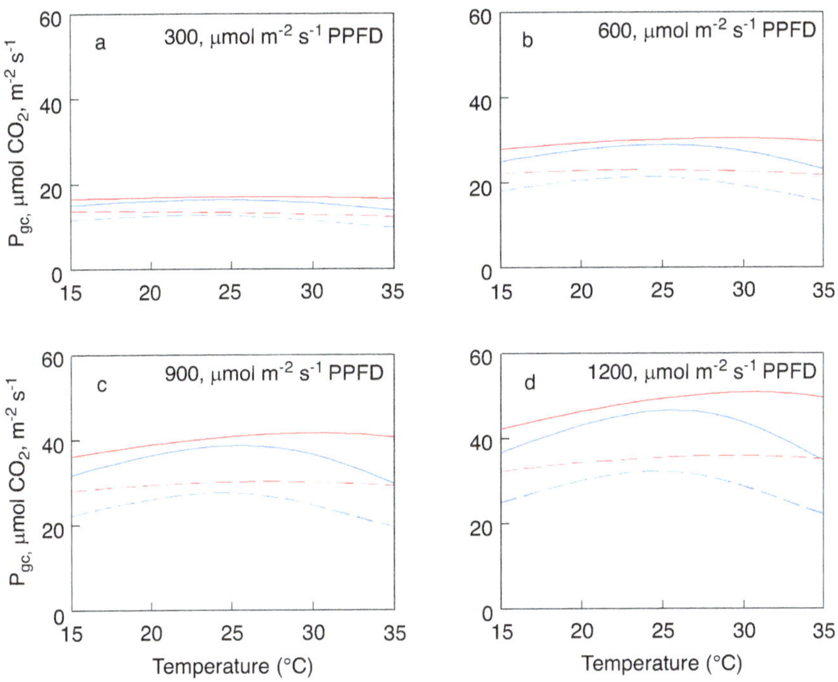

Fig. 8.4. Simulated crop gross photosynthesis over a range of temperatures at 400 (dashed line) or 1000 (solid line) μmol mol^{-1} CO_2 at four different light levels. Simulations were conducted with either a fixed stomatal conductance for H_2O of 500 mmol m^{-2} s^{-1} (red) or a stomata conductance calculated with a model (blue) implemented in the same crop photosynthesis model in cut chrysanthemum (Photo courtesy of Körner, 2003.)

A simple empirical, but widely applied model for g_s is the Ball–Woodrow–Berry (BWB) model (Ball *et al.*, 1987).

$$g_s = g_0 + m_s P_{nl} \frac{h_s}{(C_s/p_a)} \tag{8.20}$$

where C_s (μbar; air CO_2 partial pressure) and h_s (leaf surface RH fraction) are the driving variables; g_0 is the leaf minimum (residual) g_s, where P_{nl} reaches 0, while m_s is an empirical coefficient for the sensitivity of g_s to P_{nl}. The model has been adapted and its accuracy improved by several scientists (Gutschick and Simonneau, 2002; Liu *et al.*, 2009; Körner *et al.*, 2021), whereas earlier g_s was often set constant (Gijzen, 1992; Heuvelink, 1996; Lee, 2002). The effect of employing a g_s model as such is shown in Fig. 8.4; the relations of g_s and P_{gc} as function of temperature at different CO_2 and humidity levels are illustrated in Fig. 8.5.

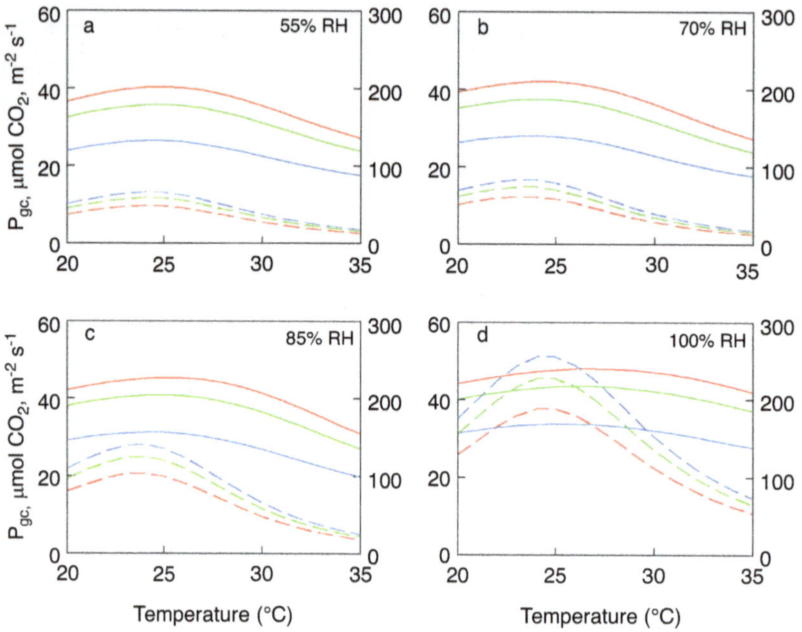

Fig. 8.5. Simulated crop gross photosynthesis over a range of temperatures, CO_2 levels 400 (blue), 700 (green), or 1000 (red) μmol mol^{-1} CO_2 at four different relative humidity levels (RH) in steps of 15%.

In fact, temperature has a strong effect on photosynthesis. High temperatures accelerate all processes, especially respiration, which leads to stress. The critical temperature limit of the thermo-tolerance of photosystem II in chrysanthemum leaves was estimated to be 38°C (Janka *et al.*, 2012).

The first two variables that need to be determined as inputs for a leaf photosynthesis model are the basic resources, that is, R_n and C_a. Without either of these two, growth will not be possible. Too low R_n or C_i (as calculated from C_a) for a longer time will result in plant stress and eventually cell death will occur. In the case of a lack of C_i, the competition of O_2 and CO_2 at the enzyme Rubisco will yield additional photorespiration and thus a reduction of earlier fixed carbon.

The importance of choosing the correct model compilation can differ according to the use-case the model is employed. Often the difference of a high- or low-quality model becomes obvious in the extremes. While under normal circumstances of climate, and so on, the model predictions could be similar, differences become evident with low or high rates of the process; an example is given with simulating P_{gc} in cut chrysanthemum with different choice on g_s sub-models and its fit to measured data (Fig. 8.6).

8.5 External Crop Quality Models

In agriculture, in general, the focus in crop modeling has traditionally been on biomass production. Only a limited number of models exist for horticultural

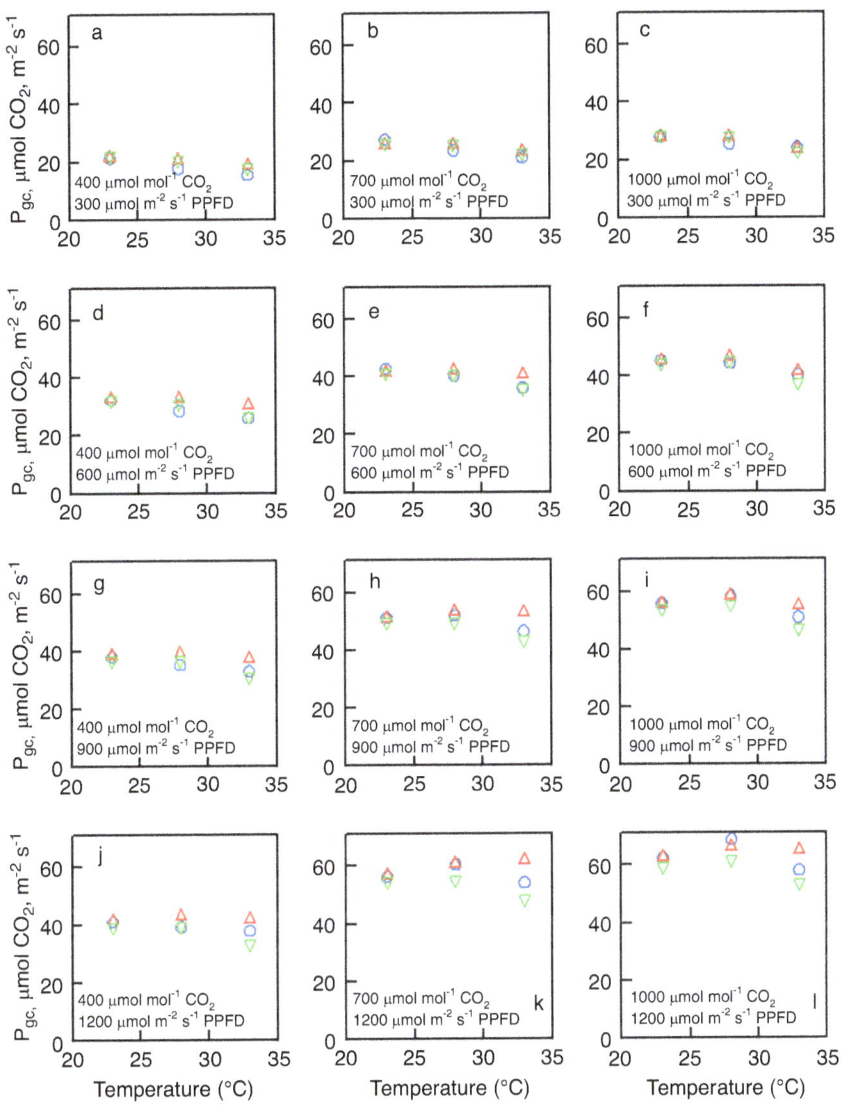

Fig. 8.6. Measured (○) and simulated (green and red triangles) crop gross photosynthesis at three different temperatures with a variation in irradiation and CO_2 concentration (increasing from *a* to *l*) with two versions of a stomatal model implemented in the same crop photosynthesis model (Photo courtesy of Körner, 2003.)

product quality (Heuvelink *et al.*, 2004). Product quality can be defined as the degree to which a product meets the expectations of the user (Meiselman and MacFie, 1996). Most quality models for ornamental plants focus on external (visual) quality (e.g. internode length and plant height, number of leaves and flowers, flower size). In addition, this paragraph is limited to external product quality, and an important and hardly modelled topic, that is, prediction of vase life based on pre- and postharvest conditions, is omitted. Especially for modeling, visual quality of ornamental crops, such as chrysanthemum, plant architecture is essential. The amount of photosynthates calculated by a crop growth model can be used as input for a stem-length prediction model as indicated by Carvalho and Heuvelink (2004). Stem length as key crop quality parameter is determined by temperature, DIF (Moe, 1989), and long-day period. These environmental factors influence internode appearance rate (IAR) and internode elongation rate (IER), and the minimum required photo-assimilates are modeled as discussed in Section 8.4.

Both development and growth determine plant architecture. As for many crops, descriptive or empirical models for chrysanthemum have been developed, including plant height (Pearson *et al.*, 1995), internode elongation (Karlsson and Heins, 1994; Schouten *et al.*, 2002), total plant dry weight (Lee *et al.*, 2003), leaf area (Lee and Heuvelink, 2003), time to flowering (Adams *et al.*, 1998), flower development (Hiden and Larsen, 1994), and flower diameter (Nothnagl *et al.*, 2004). Combined light and temperature cycles influence elongation by affecting the circadian clock. The complexity of the interaction between clock, entrainment signals, and circadian gating makes static elongation models of limited value for predicting growth in a crop production environment (Bours, 2014). Carvalho *et al.* (2003) conducted a case study to show that the use of a dynamic model that incorporates some quality aspects is extremely valuable in terms of decision support, as several options are possible to obtain a given quality, and these options are very much dependent on the planting week. In their study, a photosynthesis-driven crop growth model for predicting total dry mass in cut chrysanthemum (Lee *et al.*, 2002) was validated and further used to address some practical questions concerning the choice for the duration of the long-day period and planting density. Carvalho and Heuvelink (2004) selected and estimated model parameters based on greenhouse experiments conducted in different seasons for building modules for plant height, number of flowers per plant, and individual flower size. Increased assimilate availability by higher daily incident PAR (4.2–18.3 mol m^{-2} d^{-1}), higher CO_2 concentration (345–623 μmol mol^{-1}), and lower plant density (80–32 plants m^{-2}) hardly influenced stem length (maximum increase of 10%). Nevertheless, the duration of the long-day period (0–21 days) and temperature (16–28°C) were required inputs when predicting stem length, since these strongly enhanced the number of internodes and the average internode length. Number of flowers, excluding flower buds (F_n), could be predicted (Carvalho and Heuvelink, 2004); $r^2 = 0.84$; Fig. 8.7) based on the total shoot fresh mass per plant (M_{ft}):

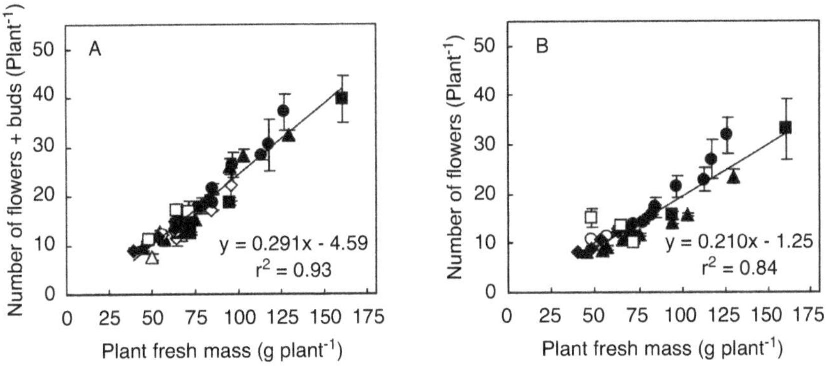

Fig. 8.7. Relationship between total number of flowers per plant, including flower buds and total plant fresh mass (A), and number of flowers per plant, excluding flower buds (B) in eight experiments indicated by different symbols (Carvalho and Heuvelink, 2004). (Reproduced from Carvalho and Heuvelink, 2004, with permission from International Society for Horticultural Science (ISHS).)

$$F_n = 0.210\,M_{ft} - 1.25 \tag{8.21}$$

In contrast, individual flower size of the fully open flowers did not respond to the assimilate availability, showing a rather constant value: 0.21 g flower^{-1}; 32 cm^2 flower^{-1}. However, when the incident PAR during the short-day period was under a threshold value (7 mol m^{-2} d^{-1}; i.e. autumn crops), individual flower size decreased linearly with decreasing PAR, and higher temperatures resulted in smaller flowers (Carvalho and Heuvelink, 2004).

The external quality model of Carvalho and Heuvelink (2004) is explanatory and built up from individual modules, such that it is relatively easy to adapt to other cultivars. Several model parameters are certainly cultivar-specific. These parameter values may be determined from a limited number of experiments. For example, considering that the relationship between M_{ft} and F_n is linear,

$$F_n = 0.210\,M_{ft} - 1.25 \tag{8.22}$$

only two extreme conditions, resulting in a low and in a high M_{ft}, are needed for each new cultivar to determine this relationship. The model could include cultivar correction factors, which are set to 1 for "Reagan Improved" (the cultivar used by Carvalho and Heuvelink, 2004) and adapted for other cultivars by calculating the ratio between a parameter for a given cultivar and for "Reagan Improved." The advantage of this approach is the limited number of experiments that it will require to adjust the modules to other cultivars.

As mentioned earlier, FSPMs create a virtual plant structure and hence are excellent tools for prediction of external chrysanthemum quality (including visualization) as influenced by environmental factors and crop management

(Kang *et al.*, 2012); Fig. 8.8). The basipetal flowering sequence in chrysanthemum is described and modelled by Kang *et al.* (2012) as follows: stacked leaves appear in the side shoots and start to expand in the vegetative stage, whereas the associated internodes remain dormant until the flowering stage. Branches of higher phytomer ranks elongate earlier than those of lower ranks (earlier appearance), which leads to the basipetal flowering sequence. The time lag between the appearance and elongation of a branch, called the "growth delay," needs to be quantified in order to determine when the side-shoot internodes and flowers become significant sinks. As higher-ranking branches show less growth delay, a position-dependent growth delay function is defined, in which the coordinate (phytomer rank) of each phytomer and branch is specified. The coordinate (x) of a first-order branch is the phytomer rank of this branch on the main stem, and the maximum number of leaves in this branch is $T_1(x)$. The coordinate of a second-order branch is (x, y), with y being the phytomer rank

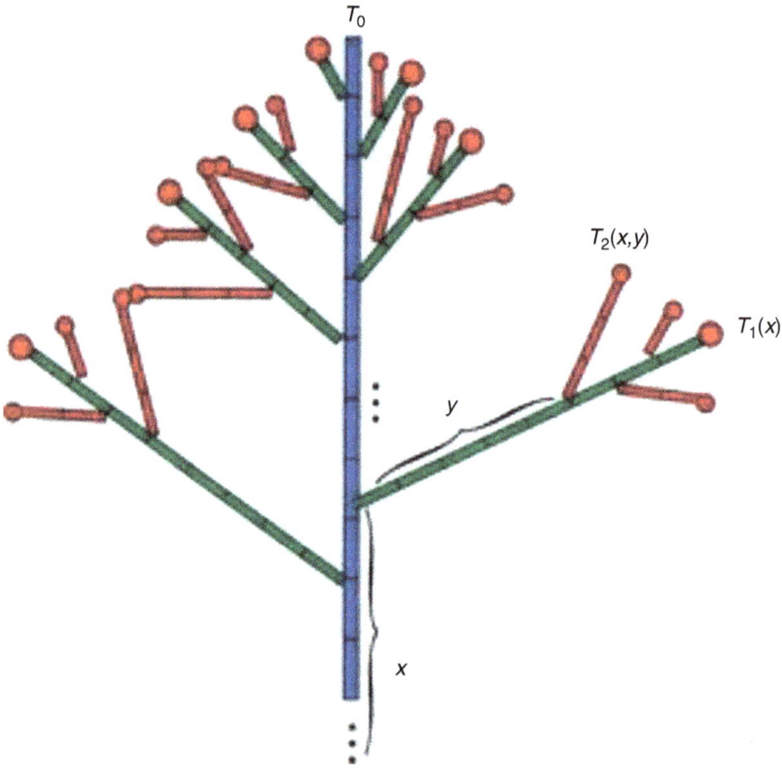

Fig. 8.8. Illustration of the aboveground topological structure of a chrysanthemum plant with first-order (green) and second-order (red) branches (Adapted from Kang *et al.*, 2012.)

of this branch on its mother branch and x being the coordinate of its mother branch (Fig. 8.8); the maximum number of leaves in this branch is $T_2(x, y)$.

The growth delays (in cycles) of a first-order branch, $d_0(x)$, and a second-order branch, $d_0(x, y)$, are defined in a recursive manner (Fig. 8.8) (Kang *et al.*, 2012):

$$\begin{aligned} &d_0(x)=g_1.c_1^{(T_0-x)}.(T_0-x)\ x\leq T_0, g_1\geq 0, c_1\leq 1 \\ &d_0(x,y)=d_0(x)+g_2.c_2^{T_1(x)-y}.(T_1(x)-y) \\ &y\leq T_1(x), g_2\geq 0,\ c_2\geq 1, \end{aligned} \tag{8.23}$$

where g_1 and g_2 are the initial delay intensities for the first- and second-order branches, respectively; c_1 and c_2 are two coefficients describing the non-linear increase in delay intensity along the mother axis. The lower phytomeric ranks (x and y) give greater delays; second-order branches (x, y) have longer growth delays than their mother branch (x). When both c_1 and c_2 are equal to one, the growth delay becomes a linear function of branch coordinate (x, y). Under this condition, if $g = 2$, branches start to grow one by one, as proposed by De Visser *et al.* (2006).

This concept is used in FSPM and can, for example, be employed in light interception studies. Kang *et al.* (2012) calibrated their FSPM using 8 out of 16 different environmental regimes where chrysanthemum plants were grown in climate chambers: four different temperatures (15, 18, 21, and 24°C) were combined with four different light intensities (40, 51, 65, and 100%, where 100% is 340 μmol m^{-2} s^{-1}). Periodic measurements included plant height, flower number, and major organ dry mass (main and side-shoot stems, main and side-shoot leaves, and flowers). The model was calibrated on eight treatments. It was capable of simulating multiple plant phenotypes (flower number, organ biomass, plant height) with visual output (Fig. 8.9). Furthermore, it predicted well the phenotypes of the other eight treatments (validation) through parameter interpolation. This model could potentially serve to bridge models of different scales and to link energy input to crop output in greenhouse chrysanthemum production.

8.6 Model-Based Decision Support Systems

Process-based crop simulation models are dynamic computational tools that simulate the development and growth of a crop in relation to environmental conditions (e.g. air temperature, soil water, evaporative demand, and atmospheric CO_2 concentration) and management practices (e.g. sowing date, N fertilizer application, and crop residue) (Muller and Martre, 2019). These model compilations are useful as inputs to model-based DSSs or to climate controllers (cf. next section *Model-based greenhouse climate control*). DSSs can be grouped into two main types, that is, real-time monitoring tools and planning tools (Körner, 2019). According to that theory, the tools are either used

Fig. 8.9. Virtual chrysanthemum architectures under conditions (from left to right) of 15°C, 40% light (T15L40); 15°C, 100% light (T15L100); 24°C, 40% light (T24L40); and 24°C, 100% light (T24L100). Photographs of real plants are shown on the right side for reference (100% light=340 μmol $m^{-2}s^{-1}$) (Kang *et al.*, 2012). (Reproduced from Kang *et al.*, 2012, with permission from John Wiley and Sons.)

before cultivation starts (planning tools) or during cultivation (monitoring tools). Both are model-based DSSs and can rely on the same framework of mathematical models (Körner and Holst, 2017).

A DSS system for monitoring and control options is suggested for chrysanthemum greenhouse production (Fig. 8.10). The system's core is a chrysanthemum crop growth model, a possible combination of the models for crop growth, crop development, and crop quality explained earlier. A model-based DSS relies primarily on its model structure and the prediction quality of the models involved. For this purpose, the nature of the model classification (cf. Table 8.1) is of less importance. Each DSS needs to be designed for its purpose, while different purposes need different model types.

A classical model-based DSS would start with a descriptive model (shown as model structure in Fig. 8.10), which can be filled with any type of unparametrized sub-models; for example, models for stomatal conductance g_s, leaf size A_l, leaf temperature T_l, and leaf photosynthesis P_l. Model parameterization is based on measured data of either underlying processes of g_s, A_l, and T_l, when adhering to the introduced example (explanatory models), or measurements of g_s, A_l, and T_l combined with environmental variables (data-based models). The various parameterized sub-models will eventually be compiled to simulation models of g_s, A_l, or T_l (bullet point 5 in Fig. 8.10). Also a model containing all three model predictors (i.e. g_s, A_l, or T_l) is a possibility. While the sub-models are calibrated in a stand-alone mode, a calibration of the full model, including sub-model interactions and commonly used parameters, is a good idea; for example, there are interactions between g_s and T_l, A_l, and T_l, and thus a highly interactive system exists in a model including all three predictors. In principle, after calibration, the model would be ready to use.

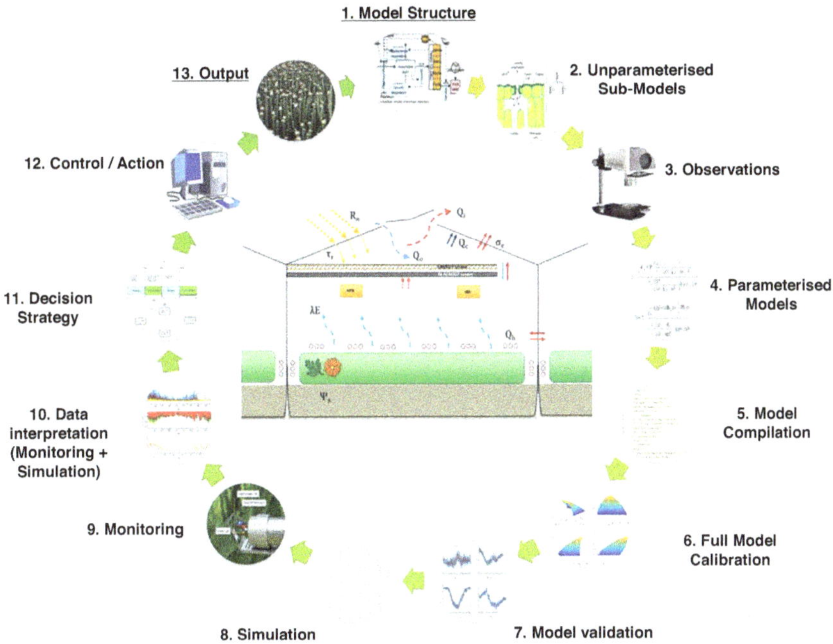

Fig. 8.10. Scheme of a system-development procedure for a sensor- and model-based real-time monitoring system for decision support.
(Photo courtesy of Oliver Körner.)

However, its predictive quality should be tested and confirmed in a validation study, using an independent set of data from various sources; for example, the effects of soil water potential Ψ_v (cf. Chapter 8.2, this volume) on growth and quality of cut chrysanthemum were modelled for greenhouse irrigation and climate management (Lin *et al.*, 2011). If the chrysanthemum g_s model involves Ψ_v (Liu *et al.*, 2009), the influence of Ψ_v needs to be validated with a range of different soil types, other than the soil type the model was calibrated upon. After successful validation, that is, a statistically satisfying agreement between measured and simulated data from the independent dataset, the model is ready for implementation in the DSS monitor or soft-sensor framework (Körner, 2019).

In the second part of the "wheel" in Fig. 8.10, the model will be combined with data from technical measurement devices, and measured and model-simulated data are analyzed in combination (Yu *et al.*, 2020) for chrysanthemum production analyses on, for example, early stress leading to production losses. At that point, the decision strategy is employed to determine the type of action.

A simple example: The DSS soft-sensor system (De Koning, 2006; Körner, 2019), as sketched in Fig. 8.10, detects a difference between measured and simulated T_l of 3.5 K. At the same time, soil monitoring reveals a lower than usual Ψ_v, which in fact triggers a low g_s (as influencing factor to T_l) to increase T_l when solar radiation is high. The reason for that is a disturbed xylem water flux due to a too low Ψ_v, which reduces g_s and consequently transpiration of the leaf. The missing energy exchange at the leaf surface induces an increase in leaf temperature, as incoming solar energy is accumulating in the chrysanthemum leaf. Wilting and crop quality deterioration would be consequences when no action follows, thus creating a risk of economic losses for the chrysanthemum producer.

The model system can eventually be used in scenario analyses finding the best suitable solution in terms of cultivation possibilities and economics. Sure, the direct solution is increasing Ψ_v, while unfolding the shade screens would probably be faster and more effective as an immediate solution to avoid crop damage. The latter action has either been done manually or could be implemented in a greenhouse climate controller (see section *Model-based greenhouse climate control*).

In that, the soft-sensor DSS could (1) reveal in an early stage when a problem in a cultivation exists, (2) support the analysis of the underlying processes, and (3) find solutions with regard to the best decision.

8.7 Model-Based Greenhouse Climate Control

To evaluate the energy costs of achieving high-quality short, compact chrysanthemum plants, different set point strategies could be compared by a model-based simulator (Körner *et al.*, 2004). To find the optimum of maximized energy saving, reductions of chemical biocide and crop yield and quality, modern greenhouses, new materials, and new technology are merged with advanced climate control concepts, as suggested by (Körner, 2003), that is, based on the method of optimal greenhouse climate control (Gal *et al.*, 1984; Van Henten, 1994; Tap, 2000). In this concept, greenhouse climate controllers aim at optimal crop growth and performance with low energy consumption, making independent decisions based on optimal climate control algorithms (Rytter *et al.*, 2012; Miranda *et al.*, 2017). This method is only applied in few cases in commercial practice.

Although climate computers are a common good in most greenhouses today, forming the basis for computerized greenhouse climate control (Körner, 2003), despite the growing usage of AI also in horticulture, with, for example, developments as the autonomous greenhouse challenge (autonomo usgreenhouses.com), the acceptance of a grower-independent greenhouse climate control in commercial practice is still low. In particular, producers of ornamental crops as chrysanthemum are risk-avoiding, as the smallest cultivation error can lead to deviations from the top quality, meaning a drastic drop in market price for the produce. Besides reliable greenhouse technology, well-validated models are a basic need for a wide acceptance of an optimal climate control strategy (Seginer, 1993). Another shortcoming is that often the systems do not contain all the necessary models for complete security in cultivation as, for example, disease risk prediction.

As discussed earlier, models of different types have been developed for many greenhouse crops, including chrysanthemum. While the major aim of greenhouse climate control is the optimization of resource consumption and production targets, mathematical models are tools to achieve that goal. In chrysanthemum, several processes are sensitive to multiple climate factors, to a different extent. This makes chrysanthemum cultivation highly complex and demands not only models but also reassurance technologies. The combination of mathematical models with data information from physical sensors in the crop, the so-called soft-sensors (De Koning, 2006; Körner, 2019), as part of a high-tech monitoring system, is used by most participants in the aforementioned autonomous greenhouse challenge. The step from measured and simulated data to an action is the control step (cf. Fig. 8.10). This can either be done manually by a person using the DSS as help or automatically through a climate computer. In the latter case, measured and model-calculated data are used as input to the control algorithm. Based on these, the algorithm determines set points and sends these to the actuators. The feedback of the actuators' state is then used in the next control time step. In the more advanced algorithms of model-based predictive climate control (Rytter *et al.*, 2012) (predicted), future states of, for example, greenhouse climate or crop behavior are used as input to the decision matrix. This method needs well-performing models of both greenhouse physics and crop growth and physiology, as well as their interactions.

In theory, when cultivating in modern climate-controlled greenhouses, keeping the greenhouse climate to the cultivar-specific climate set points will result in the desired crop yield, plant quality, and scheduling. An experienced grower will be able to "read" the crop and adjust the climate controller when needed. A fully automated climate control is possible, but due to the risks of crop damage for the grower, it is not fully exploited. However, energy costs for heating and supplementary lighting are, despite the increasing technological development of greenhouses, still a major factor in chrysanthemum production and solutions for finding an optimum in production are desired.

Earlier, dynamic model-based climate regimes, based on average temperature control over a certain time period, and modeled plant responses were designed and successfully tested in chrysanthemum production for both potted plants (Ottosen *et al.*, 2005) and cut chrysanthemum (Körner and Challa, 2003a; Körner *et al.*, 2004). The concepts were successfully implemented in a climate controller (Körner and Challa, 2004b; Aaslyng *et al.*, 2005), and substantial energy savings could be achieved. The plant physiological background of this type of climate control is the fact that chrysanthemum plant development is reacting to a mean temperature within certain boundaries in time and temperature. Leaf unfolding and flower development rate, for example, are, within a limited temperature range, responding to 24 hour mean temperature (Karlsson *et al.*, 1989; Moe and Heins, 1990), and temperature integration (Hurd and Graves, 1984) can be applied. In this way, mean temperature can be attained at its desired level while heating is shifted to periods of lower costs (Lacroix and Kok, 1999), for example, temperature integration with heating under energy screens during winter nights can reduce energy consumption (Körner *et al.*, 2004). Temperature integration implies a crop response to temperature sum rather than a precise temperature regime, and this can only occur in a temperature range where the rates of plant processes, for example, growth rate and flower development rate, show a linear relationship with temperature.

The difference between a free dynamic model-based climate control and standard regime was evaluated by Körner and Van Straten (2008). Model-based dynamic control allows the greenhouse climate to follow the natural dynamics within certain limits (Fig. 8.11), while the climate set points for the actuators as vent openings, screens, and heating valves are controlled regularly and adjusted in each control time step (commonly each 5–10 min). With these measures, energy saving of more than 30% with 39% increased crop production weight could be achieved compared to a standard rigid climate control (Körner, 2003). Thus, production efficiency could be strongly improved by a model-based climate control when growers accept dynamics in climate control (compared to the commonly applied rigid regimes).

8.8 Conclusion and Future Prospects

In this chapter, the purpose of crop models in chrysanthemum cultivation was discussed next to general model thinking theory. It is evident that modeling chrysanthemum physiology and its interaction with the environment is not a clear, straight issue. Depending on the need and purpose of the respective model, whether it is used as a prediction of a present semi-static state as leaf area or image analysis, various model types exist. While data-based models, as simple regression or more advanced black box data-trained models, can be used as reliable predictors within the scope they are trained for, the processes leading from data-input to data-output are hidden. Explanatory process-based models, on the other hand, can be used to understand the processes between input and output, while their complexity and often over-parameterization are the major obstacles. For the two main model output targets in chrysanthemum, as discussed in this chapter, that is, crop growth (increase of dry matter) and external quality, all model types could theoretically be used. For the two main application families that employ mathematical models, that is, DSSs or real-time climate controllers, the purpose is key for the choice of modeling. While photosynthesis-based growth models are still being improved, including more detailed molecular processes and sub-models for, for example, stomatal conductance, modeling external crop quality mainly relies on early

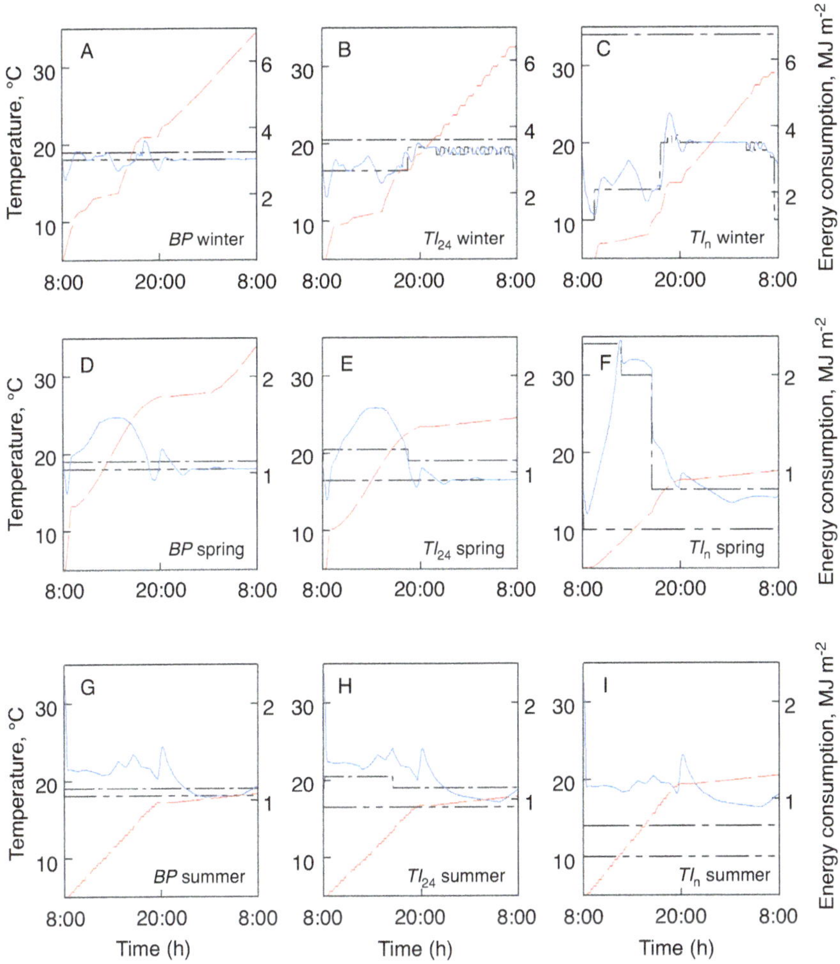

Fig. 8.11. Model-simulated greenhouse climate set points for heating and ventilation (black lines), temperature (blue), and cumulative energy consumption (red). Simulations over a 24-h period in winter (A–C), spring (D–E), and summer (G–I). Simulations with common rigid temperature control (BP), 24-h temperature integration (Körner *et al*., 2004) (TI24) or temperature integration with extended dynamics (TIn) (Photo courtesy of Körner and Challa, 2003a). Data taken, figure modified.

work of the last decades. Although FSPM has been successfully introduced to model external chrysanthemum quality, and potential use cases as cropping structures or supplementary light applications are present, chrysanthemum models have rarely found their way to commercial practice. The large variety of chrysanthemum cultivars with significant differences in plant architecture, such as leaf size, internode length, flower number, flower size, and so on, are probably the main obstacles in commercial user acceptance of chrysanthemum models. In addition, here, explanatory models would be superior, as in theory each cultivar only needs a parameter–value list as exchangeable item. For that, however, experimental measurements are needed. A parameter reduction process, resulting in explanatory models containing the essential parameters, would help in reducing that demand. In addition, a standardized measurement procedure to attain those parameter values is probably the way forward. With that, coded and compiled models (e.g. in exchangeable and standard *dll*-format) could easily be used in commercial practice for DSSs or in climate controllers. This would improve chrysanthemum cultivation by optimizing resource consumption, yield, product quality, and economic benefit for the grower.

References

Aaslyng, J.M., Ehler, N. and Jakobsen, L. (2005) Climate control software integration with a greenhouse environmental control computer. *Environmental Modelling & Software* 20(5), 521–527. DOI: 10.1016/j.envsoft.2003.11.015.

Adams, S.R., Pearson, S. and Hadley, P. (1998) The effect of temperature on inflorescence initiation and subsequent development in chrysanthemum cv. Snowdown (*Chrysanthemum* x *morifolium* Ramat.). *Scientia Horticulturae* 77, 59–72. DOI: 10.1016/S0304-4238(98)00163-0.

Aiello, G., Giovino, I., Vallone, M., Catania, P. and Argento, A. (2018) A decision support system based on multisensor data fusion for sustainable greenhouse management. *Journal of Cleaner Production* 172, 4057–4065. DOI: 10.1016/j.jclepro.2017.02.197.

Ball, J.T., Woodrow, I.E. and Berry, J.A. (1987) A model predicting stomatal conductance and its contribution to the control of photosynthesis under different environmental conditions. In: Biggins, J. (ed.) *Progress in Photosynthesis Research*. Martinus Nijhoff, pp. 221–224. DOI: 10.1007/978-94-017-0519-6_48.

Behmann, J., Mahlein, A.K., Rumpf, T., Römer, C. and Plümer, L. (2015) A review of advanced machine learning methods for the detection of biotic stress in precision crop protection. *Precision Agriculture* 16(3), 239–260. DOI: 10.1007/s11119-014-9372-7.

Berghuijs, H.N.C., Xinyou., Y., Ho, Q.T., Van der Putten, P.E.L., Verboven, P. *et al.* (2015) Modelling the relationship between CO_2 assimilation and leaf anatomical properties in tomato leaves. *Plant Science* 238, 297–311. DOI: 10.1016/j.plantsci.2015.06.022.

Bielczynski, L.W., Lacki, M.K., Hoefnagels, I., Gambin, A. and Croce, R. (2017) Leaf and plant age affects photosynthetic performance and photoprotective capacity. *Plant Physiology* 175(4), 1634–1648. DOI: 10.1104/pp.17.00904.

Bours, R. (2014) Antiphase light and temperature cycles disrupt rhythmic plant growth: The *Arabidopsis* jetlag. Wageningen, The Netherlands: Wageningen University. Available at: https://edepot.wur.nl/304356

Buck-Sorlin, G., De Visser, P.H.B., Henke, M., Sarlikioti, V., Ven der Heijden, G. *et al.* (2011) Towards a functional–structural plant model of cut-rose: Simulation of light environment, light absorption, photosynthesis and interference with the plant structure. *Annals of Botany* 108(6), 1121–1134. DOI: 10.1093/aob/mcr190.

Carvalho, S.M.P. (2003). Wageningen, The Netherlands: Wageningen University. Available at: https://edepot.wur.nl/193310

Carvalho, S. and Heuvelink, E. (2004) Modelling external quality of cut chrysanthemum: Achievements and limitations. *Acta Horticulturae* 654, 654. DOI: 10.17660/ActaHortic.2004.654.33.

Carvalho, S.M.P. and Heuvelink, E. (2001) Influence of greenhouse climate and plant density on external quality of chrysanthemum (*Dendranthema grandiflorum* (Ramat.) Kitamura): First steps towards a quality model. *Journal of Horticultural Science and Biotechnology* 76(3), 249–258. DOI: 10.1080/14620316.2001.11511359.

Carvalho, S., Dhonukshe, V.B. and Heuvelink, E. (2003) Interactive effects of duration of long-day period and plant density on external quality of cut chrysanthemum. *Acta Horticulturae* 624, 335–342. DOI: 10.17660/ActaHortic.2003.624.46.

Carvalho, S.M.P., Heuvelink, E., Cascais, R. and Van Kooten, O. (2002) Effect of day and night temperature on internode and stem length in cut chrysanthemum: Is everything explained by DIF? *Annals of Botany* 90(1), 111–118. DOI: 10.1093/aob/mcf154.

Carvalho, S.M.P., Heuvelink, E., Harbinson, J. and Van Kooten, O. (2006) Role of sink–source relationships in chrysanthemum flower size and total biomass production. *Physiologia Plantarum* 128(2), 263–273. DOI: 10.1111/j.1399-3054.2006.00733.x.

Challa, H. (1990) Crop growth models for greenhouse climate control. In: Rabbinge, R., Goudriaan, J. and Van Keulen, H. (eds) *Theoretical Production Ecology: Reflections and Prospects*. Pudoc, Wageningen, The Netherlands, pp. 125–145. Available at: https://edepot.wur.nl/171939

Cockshull, K.E., Langton, F.A. and Cave, C.R.J. (1995) Differential effects of different DIF treatments on chrysanthemum and poinsettia. *Acta Horticulturae* 378, 15–25. DOI: 10.17660/ActaHortic.1995.378.1.

Cuijpers, L.H.M. and Vogelezang, J.V.M. (1992) DIF and temperature drop for short day pot plants. *Acta Horticulturae* 327, 25–32. DOI: 10.1016/0304-4238(95)00783-P.

De Koning, A.N.M. (2006) Models and sensors in greenhouse control. *Acta Horticulturae* 718, 175–182. DOI: 10.17660/ActaHortic.2006.718.19.

De Visser, P.B., Heijden, G.V.D., Marcelis, L.F.M., Carvalho, S. and Heuvelink, E. (2006) A functional-structural model of chrysanthemum for prediction of ornamental quality. *Acta Horticulturae* 718, 59–66. DOI: 10.17660/ActaHortic.2006.718.5.

De Wit, C.T. and Van Keulen, H. (1987) Modelling production of field crops and its requirements. *Geoderma* 40(3–4), 253–265. DOI: 10.1016/0016-7061(87)90036-X.

Evers, J.B., Letort, V., Renton, M. and Kang, M. (2018) Computational botany: Advancing plant science through functional–structural plant modelling. *Annals of Botany* 121(5), 767–772. DOI: 10.1093/aob/mcy050.

Fan, X.-R., Kang, M.-Z., Heuvelink, E., de Reffye, P. and Hu, B.-G. (2015) A knowledge-and-data-driven modeling approach for simulating plant growth: A case study on tomato growth. *Ecological Modelling* 312, 363–373. DOI: 10.1016/j.ecolmodel.2015.06.006.

Fanourakis, D., Kazakos, F. and Nektarios, P.A. (2021) Allometric individual leaf area estimation in chrysanthemum. *Agronomy* 11(4), 795. DOI: 10.3390/agronomy11040795.

Farquhar, G.D., Von Caemmerer, S. and Berry, J.A. (1980) A biochemical model of photosynthetic CO_2 assimilation in leaves of C3 species. *Planta* 149(1), 78–90. DOI: 10.1007/BF00386231.

Fitz-Rodríguez, E. and Giacomelli, G.A. (2009) Yield prediction and growth mode characterization of greenhouse tomatoes with neural networks and fuzzy logic. *Transactions of the ASABE* 52(6), 2115–2128. DOI: 10.13031/2013.29200.

Gal, S., Angel, A. and Seginer, I. (1984) Optimal control of greenhouse climate: Methodology. *European Journal of Operations Research* 17, 45–56. DOI: 10.1016/0377-2217(84)90007-9.

Gijzen, H. (1992) Simulation of photosynthesis and dry matter production of greenhouse crops. Wageningen, The Netherlands, CABO-DLO. Available at: https://edepot.wur.nl/216772

Gijzen, H. and Goudriaan, J. (1989) A flexible and explanatory model of light distribution and photosynthesis in row crops. *Agricultural and Forest Meteorology* 48(1–2), 1–20. DOI: 10.1016/0168-1923(89)90004-X.

Goudriaan, J. and Van Laar, H.H. (1994) *Modelling Potential Crop Growth Processes*. Kluwer, Dordrecht, The Netherlands. DOI: 10.1007/978-94-011-0750-1.

Gurban, E.H. and Andreescu, G.-D. (2018) Greenhouse environment monitoring and control: State of the art and current trends. *Environmental Engineering and Management Journal* 17(2), 399–416. DOI: 10.30638/eemj.2018.041.

Gutschick, V. and Simonneau, T. (2002) Modelling stomatal conductance of field-grown sunflower under varying soil water content and leaf environment: Comparison of three models of stomatal response to leaf environment and coupling with an abscisic acid-based model of stomatal response to soil drying. *Plant Cell and Environment* 25(11), 1423–1434. DOI: 10.1046/j.1365-3040.2002.00937.x.

Hansen, H.T., Hendriks, L., Ueber, E. and Andersen, A.S. (1996) Effect of a low temperature period (drop) during different periods on morphogenesis of dendranthma 'Surf. *Fuchsia 'Beacon', Verbena 'Karmonrosa', and Pelargonium 'Pulsar Red'. Gartenbauwissenschaft* 61(4), 188–196.

Heuvelink, E. (1996) Tomato growth and yield: Quantitative analysis and synthesis. Department of Horticulture. Wageningen Agricultural University, Wageningen, The Netherlands. Available at: https://library.wur.nl/WebQuery/wurpubs/33806

Heuvelink, E., Tijskens, P. and Kang, M.Z. (2004) Modelling product quality in horticulture: An overview. *Acta Horticulturae* 654, 19–30. DOI: 10.17660/ActaHortic.2004.654.1.

Hiden, C. and Larsen, R. (1994) Predicting flower development in greenhouse grown chrysanthemum. *Scientia Horticulturae* 58, 123–138. DOI: 10.1016/0304-4238(94)90133-3.

Hurd, R.G. and Graves, C.J. (1984) The influence of different temperature patterns having the same integral on the earliness and yield of tomatoes. *Acta Horticulturae* 148, 547–554. DOI: 10.17660/ActaHortic.1984.148.69.

Iddio, E., Wang, L., Thomas, Y. and McMorrow, G. (2020) Energy efficient operation and modeling for greenhouses: A literature review. *Renewable and Sustainable Energy Reviews* 117(3), 109480. DOI: 10.1016/j.rser.2019.109480.

Janka, E., Ottosen, C.O. and Körner, O. (2012) Log-logistic model analysis of optimal and supraoptimal temperature effect on photosystem II using chlorophyll a fluorescence in chrysanthemum (*Dendranthema grandiflora*). *Acta Horticulturae* 957, 297–302. DOI: 10.17660/ActaHortic.2012.957.34.

Janka, E., Körner, O., Rosenqvist, E. and Ottosen, C.O. (2016) A coupled model of leaf photosynthesis, stomatal conductance, and leaf energy balance for chrysanthemum (*Dendranthema grandiflora*). *Computer and Electronics in Agriculture* 123, 264–274. DOI: 10.1016/j.compag.2016.02.022.

Kang, M., Heuvelink, E., Carvalho, S.M.P. and de Reffye, P. (2012) A virtual plant that responds to the environment like a real one: The case for chrysanthemum. *New Phytologist* 195(2), 384–395. DOI: 10.1111/j.1469-8137.2012.04177.x.

Karlsson, M.G. and Heins, R.D. (1994) A model of chrysanthemum stem elongation. *Journal of the American Society for Horticultural Science* 119(3), 403–407. DOI: 10.21273/JASHS.119.3.403.

Karlsson, M.G., Heins, R.D., Erwin, J.E., Berghage, R.D., Carlsson, W.H. *et al.* (1989) Temperature and photosynthetic photon flux influence chrysanthemum shoot development and flower. *Journal of the American Society of Horticultural Science* 114(1), 158–163. DOI: 10.21273/JASHS.114.1.158.

Khattak, A.M. and Pearson, S. (1997) The effect of light quality and temperature on the growth and development of chrysanthemum cvs. *Bright Golden Anne and Snowdon. Acta Horticulturae* 435, 113–121. DOI: 10.17660/ActaHortic.1997.435.10.

Kläring, H.P. and Körner, O. (2020) Design of a real-time gas-exchange measurement system for crop stands in environmental scenarios. *Agronomy* 10(737), 1–22. DOI: 10.3390/agronomy10050737.

Körner, O. (2003) Crop based climate regimes for energy saving in greenhouse cultivation. Wageningen University, Wageningen, The Netherlands. Available at: https://edepot.wur.nl/27517

Körner, O. and Challa, H. (2003a) Design for an improved temperature integration concept in greenhouse cultivation. *Computers and Electronics in Agriculture* 39(1), 39–59. DOI: 10.1016/S0168-1699(03)00006-1.

Körner, O. and Challa, H. (2003b) Temperature integration and DIF in cut chrysanthemum. *Journal of Horticultural Science and Biotechnology* 78(3), 335–342. DOI: 10.1080/14620316.2003.11511628.

Körner, O. and Challa, H. (2004a) Energy saving climate control regime for greenhouse cut chrysanthemum. *Acta Horticulturae* 633, 489–496. DOI: 10.17660/ActaHortic.2004.633.61.

Körner, O. and Challa, H. (2004b) Temperature integration and process-based humidity control in chrysanthemum. *Computers and Electronics in Agriculture* 43(1), 1–21. DOI: 10.1016/j.compag.2003.08.003.

Körner, O. and Van Straten, G. (2008) Decision support for dynamic greenhouse climate control strategies. *Computers and Electronics in Agriculture* 60(1), 18–30. DOI: 10.1016/j.compag.2007.05.005.

Körner, O. and Hansen, J.B. (2012) An on-line tool for optimising greenhouse crop production. *Acta Horticulturae* 957(1), 147–154. DOI: 10.17660/ActaHortic.2012.957.16.

Körner, O. and Holst, N. (2017) An open source greenhouse modelling platform. *Acta Horticulturae* 1154, 241–248. DOI: 10.17660/ActaHortic.2017.1154.32.

Körner, O., Challa, H. and Van Ooteghem, R.J.C. (2002) Modeling temperature effects on crop photosynthesis at high radiation in a solar greenhouse. *Acta Horticulturae* 593, 137–144. DOI: 10.17660/ActaHortic.2002.593.17.

Körner, O., Bakker, M.J. and Heuvelink, E. (2004) Daily temperature integration: A simulation study to quantify energy consumption. *Biosystems Engineering* 87(3), 333–343. DOI: 10.1016/j.biosystemseng.2003.11.003.

Körner, O., Andreassen, A.U. and Aaslyng, J.M. (2006) Simulating dynamic control of supplementary lighting. *Acta Horticulturae* 711(1), 151–156. DOI: 10.17660/ActaHortic.2006.711.17.

Körner, O., Aaslyng, J.M., Andreassen, A.U. and Holst, N. (2007a) Microclimate prediction for dynamic greenhouse climate control. *HortScience* 42(2), 272–279. DOI: 10.21273/HORTSCI.42.2.272.

Körner, O., Van't Ooster, A. and Hulsbos, M. (2007b) Design and performance of a measuring system for CO_2 exchange of a greenhouse crop at different light levels. *Biosystems Engineering* 97(2), 219–228. DOI: 10.1016/j.biosystemseng.2007.02.018.

Körner, O., Heuvelink, E. and Niu, Q. (2009) Quantification of temperature, CO_2, and light effects on crop photosynthesis as a basis for model-based greenhouse climate control. *Journal of Horticultural Science & Biotechnology* 84(2), 233–239. DOI: 10.1080/14620316.2009.11512510.

Körner, O. (2019) Models, sensors and decision support systems in greenhouse cultivation. In: Marcelis, L.F.M. and Heuvelink, E. (eds) *Achieving Sustainable Greenhouse Cultivation*. Burleigh Dodds, pp. 379–412. DOI: 10.1201/9780429266744.

Körner, O., Fanourakis, D., Hwang, M.C., Hyldgaard, B., Tsaniklidis, G. *et al.* (2021) Incorporating cultivar-specific stomatal traits into stomatal conductance models improves the estimation of evapotranspiration enhancing greenhouse climate management. *Biosystems Engineering* 208(2), 131–151. DOI: 10.1016/j.biosystemseng.2021.05.010.

Lacroix, R. and Kok, R. (1999) Simulation-based control of enclosed ecosystems—A case study: Determination of greenhouse heating setpoints. *Canadian Agricultural Engineering* 41(3), 175–183.

Langton, F.A. (1998) Regulation of stem extension by temperature. In: Cockshull, K.E., Gray, D., Seymour, G.B. and Thomas, B. (eds) *Genetic and Environmental Manipulation of Horticultural Crops*. CAB International, Wallingford, UK, pp. 191–203.

Lee, J.H. (2002) Analysis and simulation of growth and yield of cut chrysanthemum. Wageningen University, Wageningen, The Netherlands. Available at: https://edepot.wur.nl/192250

Lee, J.H. and Heuvelink, E. (2003) Simulation of leaf area development based on dry matter partitioning and specific leaf area for cut chrysanthemum. *Annals of Botany* 91(3), 319–327. DOI: 10.1093/aob/mcg015.

Lee, J.H., Heuvelink, E. and Challa, H. (2002) A simulation study on the interactive effects of radiation and plant density on growth of cut chrysanthemum. *Acta Horticulturae* 593, 151–157. DOI: 10.17660/ActaHortic.2002.593.19.

Lee, J.H., Goudriaan, J. and Challa, H. (2003) Using the expolinear growth equation for modelling crop growth in year-round cut chrysanthemum. *Annals of Botany* 92(5), 697–708. DOI: 10.1093/aob/mcg195.

Levy, O., Dubinsky, Z., Schneider, K., Zakai, D. and Gorbunov, M.Y. (2004) Diurnal hysteresis in coral photosynthesis. *Marine Ecology Progress Series* 268, 105–117. DOI: 10.3354/meps268105.

Lin, L., Li, W., Shao, J., Luo, W., Dai, J. *et al.* (2011) Modelling the effects of soil water potential on growth and quality of cut chrysanthemum (*Chrysanthemum morifolium*). *Scientia Horticulturae* 130(1), 275–288. DOI: 10.1016/j.scienta.2011.06.008.

Liu, F., Andersen, M.N. and Jensen, C.R. (2009) Capability of the 'ball-berry' model for predicting stomatal conductance and water use efficiency of potato leaves under different irrigation regimes. *Scientia Horticulturae* 122(3), 346–354. DOI: 10.1016/j.scienta.2009.05.026.

Louarn, G. and Song, Y. (2020) Two decades of functional-structural plant modelling: Now addressing fundamental questions in systems biology and predictive ecology. *Annals of Botany* 126(4), 501–509. DOI: 10.1093/aob/mcaa143.

Lund, J.B., Körner, O., Aaslyng, J.M. and Blom, T.J. (2008) Stem elongation of chrysanthemum in response to end-of-day light treatments and daylength. *Acta Horticulturae* 766, 109–114. DOI: 10.17660/ActaHortic.2008.766.12.

Marcelis, L.F.M., Heuvelink, E. and Goudriaan, J. (1998) Modelling biomass production and yield of horticultural crops: A review. *Scientia Horticulturae* 74, 83–111. DOI: 10.1016/S0304-4238(98)00083-1.

Meiselman, H.L. and MacFie, H.J.H. (1996) *Food Choice, Acceptance and Consumption.* Blackie, London, UK.

Metselaar, K. (1999) Auditing predictive models: A case study in crop growth. Wageningen Agricultural University, Wageningen, The Netherlands. Available at: https://edepot.wur.nl/196301

Miranda, L., Lara, B., Rocksch, T., Dannehl, D. and Schmidt, U. (2017) Neural networks for simulation of transpiration and photosynthesis in a semi-closed greenhouse. *Acta Horticuturae* 1170, 193–200. DOI: 10.17660/ActaHortic.2017.1170.22.

Moe, R. (1989) Differansen mellom dag- og nattemperatur kontroller plantehoyden i veksthuskulturene. *Gartneryrket* 7, 26–28.

Moe, R. and Heins, R.D. (1990) Control of plant morphogenesis and flowering by light quality and temperature. *Acta Horticulturae* 272, 81–89. DOI: 10.17660/ActaHortic.1990.272.11.

Moon, T.W., Jung, D.H., Chang, S.H. and Son, J.E. (2018) Estimation of greenhouse CO_2 concentration via an artificial neural network that uses environmental factors. *Horticulture, Environment, and Biotechnology* 59(1), 45–50. DOI: 10.1007/s13580-018-0015-1.

Muller, B. and Martre, P. (2019) Plant and crop simulation models: Powerful tools to link physiology, genetics, and phenomics. *Journal of Experimental Botany* 70(9), 2339–2344. DOI: 10.1093/jxb/erz175.

Nothnagl, M., Kosiba, A. and Larsen, R. (2004) Predicting the effect of irradiance and temperature on the flower diameter of greenhouse grown chrysanthemum. *Scientia Horticulturae* 99, 319–329. DOI: 10.1016/S0304-4238(03)00096-7.

Ottosen, C.O., Rosenqvist, E., Aaslyng, J.M. and Jakobsen, L. (2005) Dynamic control in combination with average temperature control saves energy in ornamentals. *Acta Horticulturae* 691, 133–139. DOI: 10.17660/ActaHortic.2005.691.14.

Ozturk, I., Holst, N. and Ottosen, C.O. (2012) Simulation of leaf photosynthesis of C3 plants under fluctuating light and different temperatures. *Acta Physiologiae Plantarum* 34(6), 2319–2329. DOI: 10.1007/s11738-012-1033-8.

Pearson, S., Hadley, P. and Wheldon, A.E. (1995) A model of the effect of day and night temperature on the height of chrysanthemum. *Acta Horticulturae* 378, 71–80. DOI: 10.17660/ActaHortic.1995.378.9.

Penning de Vries, F.W.T., Jansen, D.M., Ten Berge, H.F.M. and Bakema, A. (1989) Simulation of Ecophysiological Processes of Growth in Several Annual Crops. Pudoc, Wageningen. Available at: https://edepot.wur.nl/108856

Raaphorst, M., De Gelder, A., Helmus, L., van der Stoep, R. and De Veld, P. (2020) *Met LED Naar Een Perfecte Chrysant Teelt, Energiezuinige Demonstratieteelten Bij Delphy Improvement Centre. Wageningen UR, Wageningen.* Available at: https://doi.org/10.18174/537104

Rytter, M., Sørensen, J.C., Sørensen, B.N. and Körner, O. (2012) Advanced model-based greenhouse climate control using multi-objective optimization. *Acta Horticulturae* 957, 29–35. DOI: 10.17660/ActaHortic.2012.957.2.

Schouten, R.E., Carvalho, S.M.P., Heuvelink, E. and Van Kooten, O. (2002) Modelling of temperature-controlled internode elongation applied to chrysanthemum. *Annals of Botany* 90, 353–359. DOI: 10.1093/aob/mcf196.

Schwarz, D. and Kläring, H.P. (2001) Allometry to estimate leaf area of tomato. *Journal of Plant Nutrition* 24(8), 1291–1309. DOI: 10.1081/PLN-100106982.

Seginer, I. (1993) Crop models in greenhouse climate control. *Acta Horticulturae* 328, 79–98. DOI: 10.17660/ActaHortic.1993.328.5.

Sharkey, T.D., Bernacchi, C.J., Farquhar, G.D. and Singsaas, E.L. (2007) Fitting photosynthetic carbon dioxide response curves for C3 leaves. *Plant, Cell & Environment* 30(9), 1035–1040. DOI: 10.1111/j.1365-3040.2007.01710.x.

Spitters, C.J.T., Toussiant, H.A.J.M. and Goudriaan, J. (1986) Separating the diffuse and direct component of global radiation and its implication for modeling canopy photosynthesis. *Part I. Components of Incoming Radiation. Agricultural and Forest Meteorology* 38(1–3), 217–229. DOI: 10.1016/0168-1923(86)90060-2.

Spitters, C.J.T., Van Keulen, H. and Van Kraalingen, D.W.G. (1989) A simple and universal crop growth simulator: SUCROS87. In: Rabbinge, R., Ward, S.A. and Van Laar, H.H. (eds) *Simulation and System Management in Crop Protection*. Pudoc Wageningen, Wageningen, pp. 147–181. Available at: https://edepot.wur.nl/171923

Taki, M., Mehdizadeh, S.A., Rohani, A., Rahnama, M. and Rahmati-Joneidabad, M. (2018) Applied machine learning in greenhouse simulation; new application and analysis. *Information Processing in Agriculture* 5(2), 253–268. DOI: 10.1016/j.inpa.2018.01.003.

Tap, R.F. (2000) Economic-based optimal control of greenhouse tomato crop production. Wageningen University, Wageningen. Available at: http://library.wur.nl/WebQuery/wurpubs/fulltext/195235

Tardieu, F., Parent, B. and Simonneau, T. (2010) Control of leaf growth by abscisic acid: Hydraulic or non-hydraulic processes? *Plant, Cell & Environment* 33(4), 636–647. DOI: 10.1111/j.1365-3040.2009.02091.x.

Van Henten, E.J. (1994) Greenhouse climate management: An optimal control approach. Wageningen Agricultural University, Wageningen, The Netherlands. Available at: https://edepot.wur.nl/205106

Vanthoor, B. (2011) A model-based greenhouse design method. Wageningen University, Wageningen, The Netherlands. Available at: https://edepot.wur.nl/170301

Wang, H., Sánchez-Molina, J.A., Li, M., Berenguel, M., Yang, X.T. *et al.* (2017) Leaf area index estimation for a greenhouse transpiration model using external climate conditions based on genetics algorithms, back-propagation neural networks and nonlinear autoregressive exogenous models. *Agricultural Water Management* 183(82), 107–115. DOI: 10.1016/j.agwat.2016.11.021.

Watson, R.T., Boudreau, M.-C. and Van Iersel, M.W. (2018) Simulation of greenhouse energy use: An application of energy informatics. *Energy Informatics* 1(1), 1–14. DOI: 10.1007/s42162-018-0005-7.

Weerakkody, W.A.P. and Suriyagoda, L.D.P. (2015) Estimation of leaf and canopy photosynthesis of pot chrysanthemum and its implication on intensive canopy management. *Scientia Horticulturae* 192, 237–243. DOI: 10.1016/j.scienta.2015.05.028.

Wit, D. (1969) Dynamic concepts in biology, prediction and measurement of photosynthetic productivity. In: *Proceedings of the IBP/PP Technical Meeting*, pp. 17–23.

Yin, X., Busch, F.A., Struik, P.C. and Sharkey, T.D. (2021) Evolution of a biochemical model of steady-state photosynthesis. *Plant, Cell & Environment* 44(9), 2811–2837. DOI: 10.1111/pce.14070.

Yu, W. and Körner, O. (2020) Effect of temperature and CO_2 concentration on leaf expansion in a tomato canopy. *Acta Horticulturae* 1296, 509–516. DOI: 10.17660/ActaHortic.2020.1296.66.

Yu, W., Körner, O. and Schmidt, U. (2020) Crop photosynthetic performance monitoring based on a combined system of measured and modelled chloroplast electron transport rate in greenhouse tomato. *Frontiers in Plant Science* 11(1038), 1–15. DOI: 10.3389/fpls.2020.01038.

Zhang, N., Zhou, X., Kang, M., Hu, B.G., Heuvelink, E. *et al.* (2022) Machine learning versus crop growth models: An ally, not a rival. *AoB PLANTS* 15(2), 1–7. DOI: 10.1093/aobpla/plac061.

Zou, W., Yao, F., Zhang, B., He, C. and Guan, Z. (2017) Verification and predicting temperature and humidity in a solar greenhouse based on convex bidirectional extreme learning machine algorithm. *Neurocomputing* 249, 72–85. DOI: 10.1016/j.neucom.2017.03.023.

9

Chrysanthemum Pests and Their Management

Malleshaiah SharathKumar[1,2]*, N. Aswathanarayana Reddy[3], Marcella Bovio[4] and Lotte Caarls[4]

[1]Horticulture and Product Physiology, Department of Plant Sciences, Wageningen University and Research, Wageningen 6700 AA, The Netherlands
[2]Deliflor Chrysanten B.V., Korte Kruisweg 163, 2676 BS Maasdijk, The Netherlands
[3]College of Horticulture, University of Horticultural Sciences, Kolar 563 103, Bagalkot, India
[4]Plant Breeding, Department of Plant Sciences, Wageningen University and Research, Wageningen 6700 AA, The Netherlands

---

Abstract

This chapter provides an up-to-date overview of both fundamental and applied research related to the biology, feeding behavior, symptoms of damage, and integrated pest management (IPM) strategies for controlling major and minor pests affecting chrysanthemums in greenhouse and open-field cultivation. Emphasis is placed on the principles of IPM, the use of biological control methods, and the potential of insect-resistant varieties to effectively manage key pests in the production of chrysanthemum. Additionally, the chapter explores advanced pest management strategies, innovative tools, and the concept of host plant resistance to insects. It covers modern phenotyping methods, plant defense mechanisms, specialized metabolites, and induced defenses, offering comprehensive insights into maintaining the health and productivity of chrysanthemum plants while addressing the challenges posed by various pest species.

Keywords

Biological control, chrysanthemum, greenhouse cultivation, insect pests, integrated pest management, open-field cultivation, resistant varieties

---

*Corresponding author: sharath.malleshaiah@wur.nl; sharath.malleshaiah@deliflor.nl; sharath.malleshaiah@gmail.com

DOI: 10.1079/9781789248289.0009

9.1 Introduction

Chrysanthemums are a popular flower crop, offering a diverse spectrum of colors and shapes throughout the year. Unfortunately, their production is persistently threatened by infestations from various insect pests throughout their life cycle. Major insect pests targeting chrysanthemums include aphids, thrips, whiteflies, serpentine leaf miners, bud borers, and spider mites. Some pests directly cause damage through feeding, whereas others indirectly cause damage by transmitting several viral diseases. All these pests contribute to significant economic losses in chrysanthemum cultivation. Consequently, understanding pest identification, symptoms, damage, and control strategies is imperative for ensuring the production of high-quality crops. The adverse effects of pesticide use on human health, beneficial insects, and the environment are becoming increasingly apparent, leading to rising public opposition to their usage. For instance, the European Union aims to reduce pesticide use by 50% by 2030 (European Union, 2018). Consequently, there is an urgent need for alternative solutions among breeders and growers of chrysanthemums. Successful pest management involves integrating various strategies such as exclusion, scouting, and appropriate cultural and sanitation practices, which may involve the use of chemicals and bioagents. This chapter provides an overview of major pests affecting chrysanthemums, with specific attention to considerations for greenhouse and open-field cultivation. Emphasis is placed on the principles of integrated pest management (IPM) and the utilization of biological control methods, as well as the concept of host plant resistance to insects in chrysanthemum. The insect pest complex in chrysanthemums comprises various species that affect the plants at different growth stages.

A total of 14 insect species have been identified, each contributing uniquely to pest dynamics (Sreedhar *et al.*, 2020). Depending on the type of cultivation and location, pest infestations vary in terms of economic damage and crop loss, particularly affecting the quality. Key sucking pests such as thrips (*Frankliniella occidentalis* and other thrips species), aphids (*Macrosiphoniella sanborni* and *Myzus persicae*), red spider mite (*Tetranychus urticae*), and mealybugs (*Maconellicoccus hirsutus*) persist from the seedling to flower maturity stages, causing significant sap loss and potential virus transmission. Additionally, leaf miner (*Liriomyza trifolii*), chrysanthemum bud borer (*Helicoverpa armigera*), which damages leaves, young shoots, and flower buds, and leaf caterpillar (*Spodoptera litura*), known for feeding on foliage, are major pests. Other pests, including the hairy caterpillar, semilooper, looper, grasshoppers, and ash weevil, also contribute to the pest complex but are less economically damaging. For a detailed overview of the pest complex affecting chrysanthemums, including species, impact, and severity status, refer to Table 9.1.

9.2 Integrated Pest Management in Chrysanthemum Cultivation: General Tips and Considerations

Pest control measures exert different selective pressures on pest populations depending on whether they are applied individually or as part of an IPM

Table 9.1. Pest complex on chrysanthemum: overview of pests, damaging stages, and severity status. (From authors)

S. no.	Common name	Scientific name	Family and order	Damaging stage	Severity status
1	Aphids	Green peach aphid (*Myzus persicae*), pea aphid (*Acyrthosiphon pisum*), cotton aphid (*Aphis gossypii*), chrysanthemum aphid (*Macrosiphoniella sanborni*)	*Aphididae, Hemiptera*	Nymph and adult	Major
2	Thrips	*Frankliniella occidentalis, Thrips tabaci, Thrips setosus, Microcephalothrips abdominalis, Thrips nigropilosus, Thrips parvispinus*	*Thripidae, Thysanoptera*	Nymph and adult	Major
3	Red spider mite	*Tetranychus urticae*	*Tetranychidae, Trombidiformes*	Nymph and adult	Major
4	Leaf miner	*Liriomyza trifolii, Chromatomyia syngenesiae*	*Agromyzidae, Diptera*	Larva	Major
5	Lygus species	*Lygus rugulipennis, Lygus lineolaris, Lygus elisus, Lygus shulli, Lygus hesperus*	*Miridae, Hemiptera*	Nymph and adult	Major
6	Bud borer	*Helicoverpa armigera*	*Noctuidae, Lepidoptera*	Larva	Major
7	Leaf caterpillar	*Spodoptera litura*	*Noctuidae, Lepidoptera*	Larva	Major
8	Mealybugs	*Maconellicoccus hirsutus*	*Pseudococcidae, Hemiptera*	Nymph and adult	Major
9	Bihar hairy caterpillar	*Spilosoma obliqua*	*Arctiidae, Lepidoptera*	Larvae	Minor
10	Painted lady butterfly	*Vanessa cardui*	*Nymphalidae, Arthropoda*	Larvae	Minor
11	Semilooper	*Plusia orichalcea*	*Noctuidae, Lepidoptera*	Larvae	Minor
12	Looper	*Trichoplusia ni*	*Noctuidae, Lepidoptera*	Larvae	Minor
13	Grasshopper	*Chrotogonus sp*	*Acrididae, Orthoptera*	Nymph and adult	Minor
14	Ash weevil	*Myllocerus subfasciatus*	*Curculionidae, Coleoptera*	Grub and adult	Minor

strategy (Fig. 9.1). When used individually, control measures, particularly chemical pesticides, can exert strong directional selection, leading to rapid development of resistance in pest populations. However, when these measures are combined within an IPM framework, which includes a variety of strategies such as biological control, mechanical methods, cultural practices, and the judicious use of pesticides, the selective pressures are more varied. This integrated approach can include fluctuating and diversifying selection pressures, which reduce the likelihood of pests developing resistance. IPM strategies prioritize sustainable methods, resorting to chemical controls only when necessary, thereby promoting environmental sustainability and delaying the evolution of resistance in pest populations.

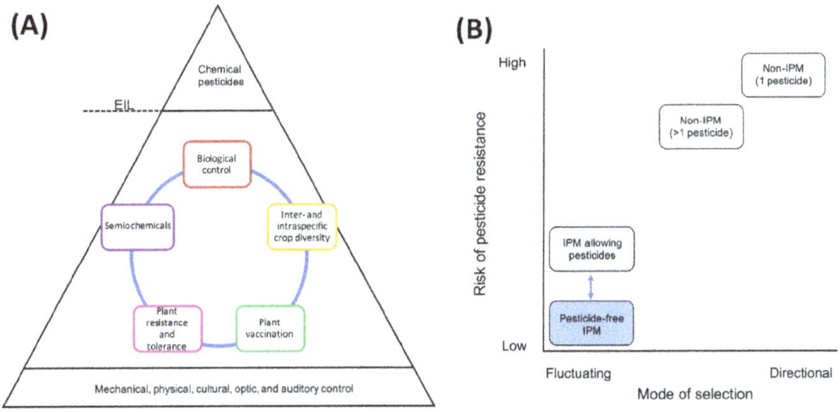

Fig. 9.1. Pest control measures in integrated pest management (IPM). (A) IPM pyramid: The base of the pyramid represents sustainable preventive and curative control methods, such as mechanical and physical actions, which are broad and foundational. The midsection includes ecologically based methods, emphasizing the importance of nonchemical interventions. The smaller top section represents chemical pesticide control, used only when the economic injury level (EIL) is reached. This pyramid model emphasizes prioritizing sustainable practices over chemical interventions. (B) The mode of selection of pest control measures to avoid the risk of pesticide resistance: The conceptual illustration demonstrates how various pest management strategies, including IPM and non-IPM approaches, affect the evolution of pesticide resistance in pests. Sustainable IPM methods, such as biological control and resistance breeding, create fluctuating or diversifying selection pressures, which can help prevent resistance. In contrast, relying solely on pesticides, especially outside of an IPM framework, applies strong directional selection, increasing the likelihood of resistance. By incorporating diverse methods such as crop rotation and intercropping within IPM, the strategy mitigates resistance risk by promoting genetic diversity. The illustration also depicts a sustainability spectrum, from conventional methods (white) to more sustainable practices (blue). Minimizing pesticide use and integrating it with other IPM measures enhances sustainability, balancing effective pest control with reduced environmental impact and resistance risk. (Photo courtesy of Karlsson Green *et al.*, 2020, CC BY 4.0.)

For effective pest control in chrysanthemum cultivation, employing IPM is key. This strategy combines various methods to control pests in a way that is environmentally sustainable and economically viable. PAMS, which stands for **P**revention, **A**voidance, **M**onitoring, and **S**uppression, is a classification system developed for the control of pests. Here are some general tips and considerations for implementing IPM in chrysanthemum cultivation:

9.2.1 Prevention (before the pest has established or appeared)

- **Inspection upon arrival and use of certified pest-free plant material:** Upon the arrival of chrysanthemum plant materials (rooted or unrooted cuttings), rigorous inspection and the use of certified pest-free material are essential to minimize the risk of introducing pests into protected chrysanthemum cultivation. The initial inspection is crucial in both greenhouse and open-field cultivation, planting material should be carefully examined for any signs of pest infestation. If an infestation is detected or pest-related symptoms are observed, removing infested material or damaged leaves is recommended to prevent the spread of pests. Managing pest infestations at an early crop stage is significantly more effective than at later growth stages, where dense foliage can obstruct proper treatment application and pest control efforts.
- **Utilize resistant varieties and optimizing plant spacing:** Selecting or breeding chrysanthemum varieties that are naturally resistant or less susceptible to major pests can significantly enhance the plant's natural defenses against infestations. Additionally, ensuring adequate plant spacing is crucial for promoting good air circulation and reducing humidity levels around the foliage. These conditions create a less favorable environment for pests, thereby minimizing infestation risks and supporting a healthier, more productive growing system.
- **Exclusion:** Exclusion is a fundamental tactic in greenhouse IPM programs, primarily relying on physical barriers to prevent insect pests from infiltrating cultivation areas, particularly in greenhouse environments. Utilizing protective screens and nets allows growers to effectively shield crops from common pests such as thrips, aphids, leaf miners, and whiteflies. To ensure complete exclusion, it is essential to use insect-proof screens with mesh sizes smaller than 200 μm, effectively blocking even the smallest pests from accessing the plants or cultivation area. Proper implementation of exclusion techniques reduces pest pressure, minimizes the reliance on chemical control, and promotes a sustainable, pest-free cultivation environment.
- **Preseason or pre-crop cleanup:** Preseason or pre-crop cleanup and sanitation are crucial for establishing a pest-free environment before introducing a new crop into greenhouse or open-field cultivation. It is essential to thoroughly remove all plant debris and weeds, as these can harbor pests from previous crops. Many pests also reside on other crops or broadleaf weeds, making it imperative to control vegetation around the greenhouse

or open-field chrysanthemum cultivation. To minimize risk, avoid planting other crops adjacent to chrysanthemum cultivation areas and actively manage broadleaf weed growth along the perimeter. Implementing a fallow period of 2–4 weeks can significantly reduce the pest load, disrupting pest life cycles and ensuring a cleaner growing environment for the next crop.

- **Sanitation:** Sanitation involves removing infested materials and potential infestation sources, followed by thorough disinfection of surfaces. Key sanitation practices include using clean planting materials to prevent initial infestations, ensuring proper ventilation to reduce humidity and discourage pest populations, and utilizing clean or sterile soils, tools, and equipment to prevent pest and disease carryover. Maintaining a clean, mowed area around the greenhouse helps minimize pest invasions from outdoor weeds, while removing trash and plant debris eliminates pest breeding grounds. Additionally, thorough cleaning of the greenhouse after each production cycle, combined with soil sanitation techniques such as solarization or steam sterilization, significantly lowers pest populations before planting, creating a more controlled and pest-free environment for the next crop.
- **Ultraviolet (UV) radiation-absorbing sheets:** UV radiation-absorbing sheets are gaining recognition as a pivotal component within IPM programs for crop protection. These sheets leverage the unique visual behaviors of insects, which rely on their compound eyes, specifically tuned to detect UV wavelengths, for crucial activities such as orientation, navigation, feeding, and mating. These spectrally modified sheets contain a UV-absorbing additive mixed into the raw materials, effectively blocking most UV wavelengths below 370–380 nm while allowing the transmission of photosynthetically active radiation (400–700 nm). The strategic deployment of UV-blocking materials in greenhouse cladding disrupts insect UV perception, significantly reducing the infiltration of pests such as whiteflies, aphids, thrips, and leaf miners from external sources into the protected crop area. Studies have demonstrated that UV-absorbing plastic roofs are particularly effective in deterring thrips, leading to a substantial reduction in their presence and the damage they cause (Nguyen *et al.*, 2009).
- **Implementing climate control and employee training in greenhouses:** Climate control systems play a crucial role in regulating temperature and humidity within greenhouses, creating an environment that is less conducive to pest proliferation. By maintaining optimal climate conditions, growers can minimize the risk of pest outbreaks and enhance overall plant health. Proper climate management helps suppress pest populations by reducing high humidity levels, which are often favorable for pests such as thrips, aphids, and fungal pathogens. In addition to climate regulation, regular training sessions for nursery, farm, and greenhouse workers are essential. These sessions should focus on pest identification, emphasizing the importance of hygiene and sanitation to prevent pest spread. Educating employees about pest life cycles, detection methods, and proper hygiene practices equips them with the necessary skills to identify the early signs of

infestation and apply preventive measures efficiently. Well-trained personnel can contribute significantly to early pest management interventions, reducing reliance on chemical control methods and promoting a more sustainable greenhouse environment.

9.2.2 Avoidance (if there are or have been occurrences of the pest)

- **Immediate isolation of infested plants and implement buffer zones:** As soon as pests are detected, affected plants should be promptly removed or isolated to prevent the spread of pests to healthy plants. This action helps to contain the infestation and significantly reduces the overall pest load, thereby protecting the remainder of the crop. Additionally, establishing buffer zones around the affected areas can effectively prevent the migration of pests to unaffected parts of the nursery, open field, or greenhouse. These buffer zones serve as physical barriers that impede or halt the movement of pests, providing an additional layer of protection and further minimizing the risk of a widespread infestation.
- **Using trap crops and intercropping in open-field cultivation:** Planting trap crops that attract pests more than the main crop is an effective pest management strategy, particularly in open-field chrysanthemum cultivation. These crops, planted around or among chrysanthemums, divert pests away from valuable flowers, concentrating infestations in a controlled area for easier management or eradication. Trap crops such as marigold (*Tagetes* sp.), basil (*Ocimum basilicum*), and allium species have been effective against aphids (Ben-Issa *et al.*, 2017). Additionally, white mustard (*Sinapis alba*) has been found significantly more attractive than chrysanthemums to both adult and nymph *Lygus rugulipennis* (Woelke *et al.*, 2023). Additionally, intercropping, the practice of planting different species together, can enhance pest control in open-field cultivation. Introducing trap crop species that naturally repel pests or mask the presence of chrysanthemums can confuse pests, significantly reducing the likelihood of attacks on the main crop. Intercropping not only aids in pest management but also promotes a more sustainable and resilient cultivation system.
- **Balanced use of fertilizer:** A balanced fertilization regimen is essential for maintaining healthy crop growth while reducing pest susceptibility. Excessive nitrogen application promotes lush growth, making plants more attractive to aphids and other pests by providing an abundant food source (de Haan and Buitenhuis, 2022). Conversely, adequate potassium strengthens plant resistance against pests. To prevent pest outbreaks, fertilizers should be applied judiciously, following recommended rates based on soil tests and crop requirements. Proper nutrient management helps maintain optimal plant health while minimizing conditions favorable for pest infestations.
- **Regular cleaning and maintenance:** Keeping the growing area clear of debris and fallen leaves is crucial, as these can serve as breeding grounds for

pests. Regular cleaning helps eliminate potential pest habitats, reducing the risk of infestations. This includes removing plant residues and thoroughly cleaning the cultivation area to maintain a tidy environment that is less conducive to pest proliferation.

9.2.3 Monitoring (watch out for pests or pest progression)

Monitoring in IPM extends beyond mere detection; it is integral to the evaluation and refinement of pest management strategies. This ongoing cycle of assessment and adjustment ensures that approaches adapt to evolving pest dynamics and remain effective over time. For instance, managing flying insects such as whiteflies or thrips involves strategically placing yellow sticky traps near the crops. These traps not only aid in monitoring pest populations but also help in their reduction. Regular checks and adjustments of these traps are necessary to respond to changing infestation levels effectively. Larger pests, such as caterpillars, require regular visual inspections as part of the monitoring process. These inspections help in spotting infestations early, allowing for timely interventions that can prevent significant damage to the crops. The most important monitoring tactics in IPM involve several key strategies, such as:

- **Scouting and early detection:** Scouting and early detection are crucial for effectively managing pests in chrysanthemum cultivation, particularly within protected environments such as greenhouses (Fig. 9.2). Regular monitoring involves systematic inspections of the plants at predefined intervals throughout the growing season, especially under conditions favorable for pests. This process includes examining all parts of the plant—from the soil surface to the upper foliage, paying close attention to the undersides of leaves where many pests typically gather. To aid in monitoring, various tools such as sticky traps and pheromone traps are used. Yellow sticky cards, generally sized 4″ × 12″ or 8″ × 12″, are extensively utilized to monitor populations of flying pests such as whiteflies, aphids, thrips, and leaf miners (Fig. 9.2). For specifically targeting thrips, blue-colored sticky traps prove effective. These traps should be strategically placed in a grid pattern throughout the greenhouse, approximately 1–2 cards per 100 m^2, or more for mass trapping objectives. Positioned about 4″–6″ above the plant canopy, these traps should be adjusted higher as plants grow. For effective management, it is important to map out the location of each sticky card within the greenhouse and check them regularly, ideally during each biweekly scouting visit. During inspections, the number of pests on each card is recorded on a field data sheet. Sticky cards should be replaced once 60–70% of their surface area is covered with trapped insects. This method of early detection through regular and structured monitoring allows for timely and targeted pest management interventions, preventing significant damage to crops and ensuring effective pest control.

- **Accurate identification:** Identifying the correct species of pest is crucial for choosing the most effective management strategy. Misidentification can result in inappropriate treatment approaches, potentially leading to ineffective control and unnecessary pesticide use. Accurate identification aids in the application of targeted control measures, which are more effective and less harmful to the environment.
- **Consistent inspection practices and develop a recordkeeping system:** It is important to inspect the entire greenhouse in a uniform and consistent pattern to ensure no areas are overlooked. Because pests do not distribute themselves evenly across the crop, thorough scouting is necessary to obtain an accurate estimate of the pest population. When crops are young, it is crucial to examine all leaves, as early-stage detection is key to preventing widespread infestation. Additionally, implement a detailed recordkeeping program that logs pest occurrences, population levels, and the effectiveness of different control measures. This historical data serves as a valuable decision-making tool to anticipate pest outbreaks and determine the most effective time for interventions.

9.2.4 Suppression (remove the pest)

9.2.4.1 Cultural methods

- **Manipulate the environment or habitat alteration:** Using mulches helps suppress weed growth, which can serve as alternative hosts for pests, thereby creating a habitat less conducive to pest infestations. This practice not only controls weeds but also modifies environmental conditions to reduce pest proliferation and enhance overall crop health. Additionally, minimizing potential pest habitats near production areas by managing plant residues, reducing excess moisture, and clearing field margins of debris and weeds helps eliminate breeding and hiding spots for pests, further reducing infestation risks.
- **Maintain healthy plants by pinching and pruning:** Ensuring proper watering, fertilization, and spacing promotes healthy plant growth, making them less susceptible to pest attacks. Promptly removing and destroying infested plants prevents the spread of pests to healthy crops. Pinching or pruning off damaged plant parts, flowers, and spotted leaves (those with insect larvae or egg deposits) effectively reduces pest populations. To prevent further infestation, removed plant debris should be immediately placed in a covered container before disposal. This practice limits pest spread and enhances IPM efficiency.

9.2.4.2 Biological methods

Biological control in pest management utilizes living agents such as beneficial insects, mites, bacteria, fungi, and nematodes to suppress pest populations.

Fig. 9.2. Scouting (A) and crop monitoring using yellow or blue sticky cards (B and C) involve the systematic observation of crops throughout the growth cycle in greenhouse environments. This practice enables timely and targeted pest management interventions, ensuring effective control of pests such as whiteflies, thrips, leaf miners, and winged aphids. (Photo courtesy of Malleshaiah SharathKumar.)

Although not 100% effective on their own and often requiring initial effort and adjustments in production methods, these agents significantly reduce the reliance on chemical pesticides. However, they may be more susceptible to pesticides than the pests themselves and typically act slowly, requiring early introduction and, in some cases, multiple releases of biological agents. The success of biological control relies on integrating other IPM measures, such as maintaining sanitation, employing proper cultural practices, and selecting resistant cultivars. The most used biological control agents are detailed in Table 9.2. For large-scale and efficient distribution of natural enemies or biological agents, advanced systems such as the monorail- or pipe-rail-based release system (full-automatic system) are utilized (Fig. 9.3). The distribution system features a fully automatic blower designed for large surfaces and full-field distribution in the high-tech cultivation of chrysanthemums. Its special fan, combined with the revolving carrying case with two adjustable dosing and mixing pots, ensures the uniform release of carrier material containing natural enemies. The airflow generated by the fan guarantees even distribution over distances of 10–12 m, facilitating the effective spread of natural enemies across extensive chrysanthemum cultivation areas in high-tech greenhouses. It offers higher efficiency compared with manual application, with no negative impact on the quality of natural enemies.

9.2.4.3 Chemical methods

IPM programs aim to reduce, rather than eliminate, pesticide use. They focus on minimizing the amount and frequency of pesticide applications and encourage the use of less toxic alternatives. While older broad-spectrum pesticides have been effective, their toxicity has raised concerns, leading to the development and registration of safer chemicals for greenhouse use. However, sustainable pest suppression requires combining multiple nonchemical tactics with careful, targeted pesticide applications. Pesticides should be used judiciously when pest populations exceed manageable levels, starting with the least toxic options and applying them precisely where needed. Proper timing and dosage are crucial to avoid resistance and ensure effectiveness. The use of certain chemicals against certain pests is discussed in the respective sections of pest control measures.

- **Targeted pesticide use:** While aiming to keep pesticide use to a minimum, select pesticides that are least harmful to beneficial insects and the environment when necessary. Prioritize targeted applications, such as soil drenches, over broad-spectrum sprays to minimize disruption to natural predators.
- **Strategic application:** Employ pesticides as a last resort when other measures fail, focusing on the least persistent and least toxic options to avoid long-term environmental impact. Select the right chemical for the specific pest and apply it correctly to minimize environmental impact and avoid harming nontarget organisms. Rotate pesticides to prevent the development of resistance.

Fig. 9.3. The monorail- or pipe-rail-based release system (full-automatic system) used for the release of biological agents. (Photo courtesy of Malleshaiah SharathKumar.)

9.3 Alternative or Environmentally Friendly Products for Pest Management

Alternative pest management products are essential for sustainable crop protection measures. Horticultural oils, which act as insecticides, miticides, and fungicides, are safe for plants and mammals but require careful application to avoid phytotoxicity. These oils can contain petroleum, paraffinic, mineral, or neem as active ingredients. Horticultural or insecticidal soaps, made from potassium salts of fatty acids, disrupt the osmotic potential of pest cells, causing them to burst, and are effective against the pupal and nymph stages of pests. Insect growth regulators (IGRs), such as diflubenzuron and tebufenozide, target developing insects, causing death or sterility, and are only effective against immature stages of pests. These are best used as part of a rotation strategy with products affecting the adult stage. Botanical pesticides, derived from plants, offer effective pest control with quick decomposition and low environmental impact. Neem-based products, containing azadirachtin, interfere with the normal molting process of insects and are safe for plants, mammals, and the environment. Pyrethrins, derived from chrysanthemums, provide effective pest control by interrupting nerve impulses in insects. These alternative products, integrated into pest management strategies, reduce the need for conventional pesticides and support environmentally friendly practices.

9.4 Aphids (*Hemiptera:Aphididae*)

Chrysanthemums are susceptible to infestation by various aphid species, including the green peach aphid (*Myzus persicae* Sulzer), pea aphid (*Acyrthosiphon pisum* Harris), melon or cotton aphid (*Aphis gossypii* Glover), and chrysanthemum aphid (*Macrosiphoniella sanborni* Gillette), all of which share similar biology and feeding behaviors. Aphids are typically small insects, measuring between 1.4 and 2.2 mm in length, and depending on the species, they are known for their shiny, dark reddish-brown to blackish-brown coloration, often referred to as "blackflies" due to their appearance (Frank, 2016; Baker, 2019).

9.4.1 Biology, feeding behavior, and symptoms of damage

During warmer months, most female aphids produce live young, called nymphs, through vivipary. However, they can also undergo sexual reproduction in response to environmental changes such as day length and temperature. An average female produces between 50 and 200 nymphs during her lifetime, with female nymphs beginning to reproduce within 7–10 days, asexually producing several offspring per day. Consequently, a single aphid on a plant can lead to a population of hundreds within a short period. Aphids exist in both winged and wingless forms, depending on population density and host plant conditions. Winged aphids are particularly problematic due to their ability to disperse throughout cropped areas.

Aphids possess small, soft bodies with piercing-sucking mouthparts, which they use to insert into the phloem tissue of plants and suck cell sap. They tend to aggregate around terminal buds and feed on new growth (Fig. 9.4). Aphid feeding causes significant damage to flowers, leading to malformed, withered, and prematurely dried blooms, resulting in 20–40% crop losses (Zahedi, 1999; Emam, 2016). In addition to direct feeding damage, aphids transmit various plant viruses such as chrysanthemum virus B and chrysanthemum vein mottle virus (Blackman and Eastop, 1984; Jaskiewicz *et al.*, 2001). Severe infestations result in the excretion of honeydew onto stems, leaves, and flowers, serving as a substrate for sooty mold fungus, which impedes photosynthesis and diminishes the ornamental value of the crop (Frank, 2016).

9.4.2 Management

Monitoring and trapping: Aphids can enter greenhouses through incoming planting materials, vents, clothing of personnel, and tools. Maintaining high sanitation standards helps prevent pest transfer. Regular monitoring by inspecting the undersides of leaves for shiny honeydew patches aids in early identification. Scouting should focus on new growth and leaf undersides. The presence of winged aphids on yellow sticky cards indicates an extensive population, necessitating an aggressive weekly scouting program from spring through late autumn, particularly for genotypes highly susceptible to aphids.

Pest control sprays: Aphids, being soft-bodied pests, are effectively managed with horticultural soaps and oils. Sprays of plant-based extracts such as pongamia or neem oil at 3 ml l^{-1}, or 4% pongamia seed kernel extract at 10-to-15-day intervals, have proven efficient in controlling aphid infestations (Reddy, 2010). Oils should be mixed with soap at 0.5 ml l^{-1} of water for proper emulsification. Chemical control includes spraying dimethoate 30 EC (1.6 ml l^{-1}), oxydemeton methyl 25 EC (2 ml l^{-1}), or acephate 75 SP (1 g l^{-1}) at the onset of infestation, with repeat sprays after ten days, if necessary. Rotating pesticides is crucial to prevent resistance. Key volatiles such as cis-4-thujanol have shown efficacy as repellents against chrysanthemum aphids and can be used in natural insecticide formulations.

Biological control: Biological control agents are commercially utilized to manage aphids, such as *Verticillium lecanii* (1×108 cfu g^{-1}) at 3 g l^{-1} (Askary *et al.*, 1998), or the release of 1-day-old *Chrysoperla carnea* larvae at a 50:1 ratio (Scopes, 1969). Parasitoids such as *Aphidius* spp. and predators such as coccinellids, syrphids, and chrysopids are effective against aphids. During the action of natural enemies, insecticide application should be avoided. Breeding resistant cultivars with higher concentrations of aphid-repelling bioactive volatile organic compounds (VOCs) offers an eco-friendly approach to aphid control in chrysanthemums (Zhong *et al.*, 2022). Utilizing the aromatic wild species *Chrysanthemum aromaticum* as breeding material can also create aphid-resistant chrysanthemums.

9.5 Thrips (*Thysanoptera*: *Thripidae*)

Thrips are notorious pests and present a year-round challenge for chrysanthemum growers globally. The primary species attacking chrysanthemums include the western flower thrips (WFT, *Frankliniella occidentalis* Pergande), *Thrips tabaci* Lindeman, and composite thrips (*Microcephalothrips abdominalis* Crawford). Additionally, *Thrips setosus* Moulton and *Thrips nigropilosus* Uzel can also cause significant damage to chrysanthemums. Over the past two decades, *Thrips parvispinus* Karny has markedly expanded its geographical distribution (Thorat *et al.*, 2022). Recently, this species has been reported in Spain and the Netherlands, with changes in agricultural practices and climate potentially exacerbating its impact in Europe. Thrips pose a major problem due to their rapid life cycle, high reproductive rate, and the cryptic behavior of larvae and adults, which enables them to evade contact with insecticides. Moreover, in recent years, thrips have developed resistance to several insecticides (Morse and Hoddle, 2006). These polyphagous invaders have a broad host range, infesting over 500 species across 65 plant families in various field and greenhouse crops (Yudin *et al.*, 1986; Reddy, 2010; Mouden *et al.*, 2017).

9.5.1 Biology, feeding behavior, and symptoms of damage

Thrips, minute insects ranging from 0.2 to 3.0 mm in size, undergo a life cycle comprising six distinct stages: egg, two nymphal stages, two pupal stages, and an

Table 9.2. Pests and their commercially available natural enemies for chrysanthemum integrated pest management. (From authors)

Pests	Natural enemy	Type	Stage affected	References
Aphids	*Aphidius colemani*	Parasitoid	Nymph and adult	Scopes (1969); Askary *et al.* (1998); Cloyd (2019)
	Aphidius ervi	Parasitoid	Nymph and adult	
	Aphelinus abdominalis	Parasitoid	Nymph and adult	
	Aphidius matricariae	Parasitoid	Nymph and adult	
	Aphidoletes aphidimyza	Predator	Larval	
	Chrysoperla carnea	Predator	Larval	
	Hippodamia convergens	Predator	Adult	
	Adalia bipunctata	Predator	Adult	
	Verticillium lecanii	Pathogen	All stages	
Thrips	*Orius tristicolor*	Predator	Nymph and adult	Waterhouse and Norris (1989); Messelink and de Kogel (2013)
	Chrysoperla carnea	Predator	Larval	
	Amblyseius montdorensis	Predatory mite	Nymph and adult	
	Amblyseius limonicus	Predatory mite	Nymph and adult	
	Ceranisus menes	Parasitoid	Larval	
	Ceranisus americensis	Parasitoid	Larval	
	Haplothrips spp.	Predatory thrips	Nymph and adult	
	Geolaelaps aculeifer	Soil predator	Nymph and adult	
	Amblyseius spp.	Predatory mite	Nymph and adult	
	Neoseiulus spp.	Predatory mite	Nymph and adult	

Red spider mite	*Phytoseiulus persimilis*	Predatory mite	Nymph and adult	Scopes and Biggerstaff (1973); Wardlow (1986); Bartel and Bennison (2021)
	Amblyseius andersoni	Predatory mite	Nymph and adult	
	Neoseiulus californicus	Predatory mite	Nymph and adult	
	Neoseiulus cucumeris	Predatory mite	Nymph and adult	
	Amblyseius swirskii	Predatory mite	Nymph and adult	
	Transeius montdorensis	Predatory mite	Nymph and adult	
	Feltiella acarisuga	Predatory midge	Larval	
Serpentine leaf miner	*Hemiptarsenus varicornis*	Parasitoid	Larval	Mujica *et al.* (2016)
	Chrysocharis parksi	Parasitoid	Larval	
	Diglyphus isaea	Parasitoid	Larval	
	Diglyphus crassinervis	Parasitoid	Larval	
	Neochrysocharis formosa	Parasitoid	Larval	
	Steinernema feltiae	Entomopathogenic nematode	Larval	
Lygus species	*Geocoris* spp.	Predator	Nymph and adult	Ferguson *et al.* (2012)
	Nabis spp.	Predator	Nymph and adult	
	Orius tristicolor	Predator	Nymph and adult	
	Anaphes iole	Parasitoid	Egg	
	Peristenus spp.	Parasitoid	Nymph	
	Orius insidiosus	Predator	Nymph	

Continued

Table 9.2. Continued

Pests	Natural enemy	Type	Stage affected	References
Bud borer	*Trichogramma chilonis*	Parasitoid	Egg	van den Berg *et al.* (1988); Liu *et al.* (2016); Raiz *et al.* (2021)
	Chrysoperla carnea	Predator	Larval	
	Apanteles sp.	Parasitoid		
	Meteorus pulchricornis	Parasitoid		
	Habrobracon hebetor	Parasitoid		
	Cotesia kazak	Parasitoid		
	Campoletis chlorideae	Parasitoid		
	Diadegma spp.	Parasitoid		
	Trichogramma pretiosum	Parasitoid		
	Trichogrammatoidea annulata	Parasitoid		
Leaf caterpillar	*Spodoptera nuclear polyhedrosis virus*	Pathogen		
	Trichogramma chilonis	Parasitoid	Egg	
	Chrysoperla carnea	Predator	Larval	
Mealybugs	*Cryptolaemus montrouzieri*	Predator		Kairo *et al.* (2000); Mani and Krishnamoorthy (2001, 2008);Hoy *et al.* (2023)
	Anagyrus kamali	Parasitoid		
	Achrysopophagus sp.	Parasitoid		

Fig. 9.4. Colony of aphids feeding on chrysanthemum leaves and flower buds (A) and a close-up of aphids feeding on a flower bud (B). (Photo courtesy of Malleshaiah SharathKumar and Saskia Sip.)

adult stage (Lewis, 1997; Fig. 9.5). Notably, several thrips species reproduce parthenogenetically, allowing them to reproduce without fertilization, producing males through arrhenotokous reproduction and females through thelytokous reproduction. The duration of their life cycle varies depending on temperature, ranging from 9 to over 40 days. The life expectancy and cycle length of adult thrips are also temperature-dependent, with an optimal range between 25 and 30°C, influenced by food quality. Their reproduction rate increases most rapidly at around 30°C. Additionally, thrips can survive during winter months in greenhouses if adequate prevention and control measures are not in place.

WFT begin laying eggs approximately 3 days after emerging (Waterhouse and Norris, 1989). Using their saw-like ovipositor, females partially or fully insert eggs into incisions made in the parenchyma tissue of leaves and flowers. The eggs are tiny, opaque, kidney-shaped, and typically scattered throughout flowers, although females may occasionally arrange them in rows alongside or beneath leaf veins (Lewis, 1973). Both larvae and adults are active feeding stages. The first instar nymphs are slender, wingless, and pale yellow, measuring between 0.2 and 1 mm in length, and begin feeding immediately upon hatching. The second instar larvae are also wingless and exhibit distinct differences in antennal shape compared with the first instar. During the nonfeeding stages, prepupae and pupae are generally inactive and have a glass-like appearance. Adult thrips have slender, dorsoventrally flattened bodies ranging from 1.5 to 2 mm in length (Fig. 9.5A and B), with two fringed wings (Heming,

1985). Adult females vary in color from pale yellow to dark brown and measure approximately 1.5 mm in length, while adult males are yellow and smaller, averaging around 1 mm in length (Mound, 2005; Ghosh *et al.*, 2021).

Thrips, both adults and nymphs, possess piercing-sucking mouthparts that puncture plant tissues and extract sap from young, developing leaves by lacerating the epidermal and mesophyll cells of leaves and flowers (Mound, 2005). Their feeding damage is evident in affected leaves, which develop silvery and discolored spots (known as silver damage), crinkling and upward curling along the margins, elongated petioles, stunted growth, reduced leaf size, and diminished flower yield (Fig. 9.6). On flowering plants, thrips target the sap at the bases of open flowers or deep within flower buds. Infested flower buds often fail to open and show silvery streaks or mottled distortions on their petals, leading to discoloration, withering, and drying of the flowers (Nair, 1975). In advanced stages of infestation, flower buds and flowers become unmarketable and prematurely fall.

Thrips primarily feed on foliage, young unopened leaves (buds), and flowers, with flowers being preferred due to their high nutritional value (De Jager *et al.*, 1993; Kirk, 1997; Mound and Kibby, 1998; Reitz, 2002). They cause two main types of damage: growth damage and silver damage (De Jager *et al.*, 1993; Kos *et al.*, 2014). Growth damage occurs when thrips feed on developing tissues, affecting the expanding primordial cells of young chrysanthemum leaves and petals. This results in distorted growth and deformed flower buds that fail to open during subsequent development. Silver damage occurs when thrips feed on mature chrysanthemum leaf cells, introducing air into the cells and causing the affected tissue to appear silvery (Fig. 9.6B and C). Although this visual damage may not significantly reduce chrysanthemum production, it severely impacts the aesthetic quality of the flowers, often rendering them less marketable or even unsellable.

Beyond the direct damage caused by feeding, thrips also act as vectors for viral diseases such as *Tomato spotted wilt virus* (TSWV; Fig. 9.7), which can result in total crop failure (German *et al.*, 1992; Maris *et al.*, 2004). *Frankliniella occidentalis* is one of the primary vectors of tospoviruses (He *et al.*, 2020; Fig. 9.7). Young larvae acquire the virus, where it replicates and migrates from the midgut to the salivary glands (Moritz *et al.*, 2004). The adults that emerge from these infected larvae transmit the virus while probing and feeding, although they do not pass the virus to their eggs (Wijkamp *et al.*, 1996).

9.5.2 Management

Thrips can cause significant damage to chrysanthemum crops, but their populations can be managed effectively with the right strategies. A comprehensive approach involving constant monitoring, biological control, and preventive measures is essential. The following key steps can help control thrips and maintain healthy chrysanthemum plants:

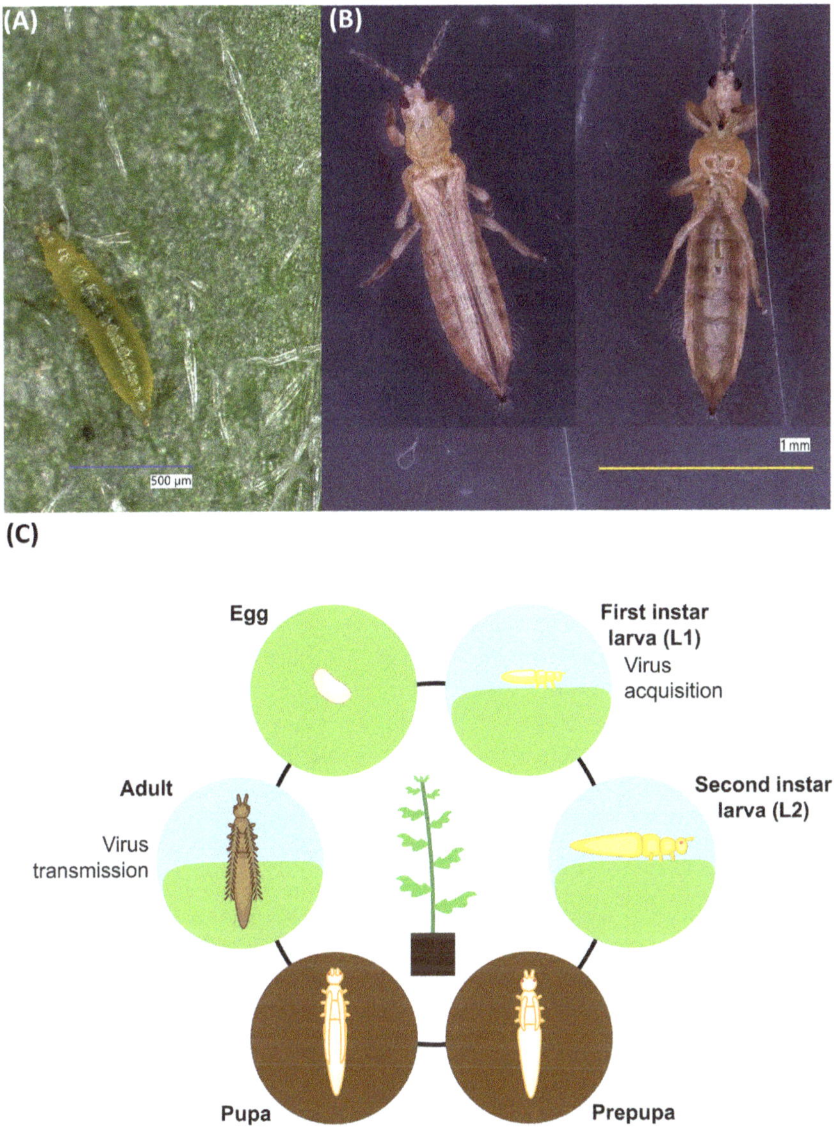

Fig. 9.5. *Frankliniella occidentalis* — adult thrips (1.5–2 mm, slender and dorsoventrally flattened), showing second-instar larva (A) and adult female (B). The life cycle of thrips (C) begins when adults lay eggs inside plant tissue. The first instar larvae hatch and develop into second instar larvae, both of which feed on the plant. The nonfeeding pupal stages typically occur in the soil. Adults emerge from the pupae and return to the plant to feed and reproduce. Adults can transmit plant viruses after acquiring them as L1 larvae. Plant resistance can affect any stage of the life cycle of thrips. (Photo courtesy of Marcella Bovio.)

Selection of pest-free planting material: Historically, chrysanthemum growers propagated new plants using older ones, inadvertently introducing thrips to new crops. To mitigate this issue, it is crucial to use younger mother plants instead of older ones to generate new propagating material. For chrysanthemum cutting propagators sending vegetative material to other greenhouses, it is essential to break the emergence and transport cycle by consistently starting with clean, thrips-free material. Selection of thrips-resistant cultivars is also advised.

Monitoring and trapping: Regular inspection of chrysanthemum plants for the signs of thrips infestation is essential. Symptoms include discolored, deformed leaves, buds, or leaves displaying a silvery appearance. To effectively monitor the plants, gently shake or tap the flowers and shoots over a sheet of white paper for closer inspection. For trapping adult thrips, consider using flowering chrysanthemum plants as "trap" plants during the vegetative stage (Buitenhuis *et al.*, 2007). Flowering chrysanthemums or gerberas are particularly effective as trap plants in potted chrysanthemum cultivation before flowering. The strategic placement of trap plants is crucial for attracting migrating thrips.

Thrips commonly invade commercial greenhouses via infested nursery transplants or through open vents. Therefore, using insect netting in the vents as a preventive management strategy is essential. Trap plants should be strategically positioned near vents, doorways, or between flowering and nonflowering crops for a few days to intercept incoming thrips. Promptly removing flowering plants from the greenhouse or implementing a wider separation between flowering and nonflowering plants can significantly reduce thrips outbreaks, ensuring healthier and more resilient crops. In open-field cultivation, controlling surrounding weeds can reduce the density of thrips (Matsuda and Ichihara, 2022).

Using an insect screen to exclude thrips is an effective and straightforward solution. However, selecting the appropriate screen requires careful consideration of mesh design, functionality, and properties. The maximum recommended hole size for excluding WFT is 192 μm. Considering the unique size and body geometry of thrips, maximum height at the abdomen (0.243 mm) and thorax (0.262 mm), and maximum width at the wider part of the abdomen (0.315 mm) and thorax (0.296 mm), modern screens are designed with consistent hole sizes of 0.15×0.15 mm. These screens are more effective than standard insect screens, creating the ability to keep thrips out of the greenhouse while maintaining adequate ventilation. They are considered 99% effective for the precise exclusion of thrips from greenhouses (Svensson, 2025).

Soil and crop management: Thrips spend two stages of their life cycle in the soil, providing an opportunity for targeted control. Effective methods include soil sterilization and the application of granular systemic pesticides to eliminate thrips residing in the soil. Another preventive measure is to dip rooted cuttings in imidacloprid during planting (0.3 ml l^{-1} for 20 min), which

Fig. 9.6. Thrips feeding with piercing and sucking mouthparts results in the removal of cell contents, causing white/silvery feeding scars and malformation on leaves (A, B, and C) or translucent spots and streaking on flower petals (D, E, and F), collectively referred to as "silver" damage on leaves and flowers. (Photo courtesy of Malleshaiah SharathKumar.)

can help curb thrips multiplication. To further mitigate thrips infestation, consider installing yellow or blue sticky traps (approximately 15 traps per hectare) positioned about 10 cm above the crop. These traps are effective in capturing adult thrips. Additionally, certain cultural practices can have a significant impact on WFT populations. For example, the application of higher levels of nitrogen fertilizers has been shown to influence thrips populations (Davies *et al.*, 2005; Chau and Heinz, 2006).

Fig. 9.7. *Tomato spotted wilt virus* (TSWV), a tospovirus, is transmitted in greenhouses by western flower thrips (*Frankliniella occidentalis*), causing symptoms such as blotching chlorosis in circular and angular patterns, necrosis, or angular necrosis of leaves, along with plant stunting or yellowing (A), and rings, brown necrotic spotting, and graying of chrysanthemum leaves (B) (Photo courtesy of Malleshaiah SharathKumar.)

Thrips oviposition varies depending on plant nutrition and the effect of fertilizer, with more eggs being laid on leaves from the high-fertilizer treatment compared with the low-fertilizer treatment. When given the choice between leaves from plants grown with high or low fertilizer, thrips preferred to oviposit on leaves from the high-fertilizer-treatment plants (de Haan and Buitenhuis, 2022). Manipulating plant nutrition can affect oviposition preference in thrips, and cultivars differ greatly in susceptibility to thrips. Therefore, reducing fertilizer applications could lead to cost savings for growers, directly through a reduction in input costs and indirectly by reducing losses caused by the high population of thrips.

Pest control sprays: If infestation becomes severe, consider using horticultural oil sprays or insecticidal soaps. Thoroughly spray the underside of leaves and terminal buds, as these are common hiding spots for thrips. It is important to note that spraying may not eliminate all thrips due to their concealed locations and the development of insecticide resistance. To manage thrips populations, spray insecticides such as Ethofenprox 10EC (1 ml l^{-1}), Fipronil 5SC (1 ml l^{-1}), or Spinetoram 11.7 SC (0.5 ml l^{-1}) in rotation at 15-day intervals. Frequent use of broad-spectrum synthetic insecticides can lead to the development of insecticide resistance in thrips populations, so they should be used judiciously.

Biological agents and beneficial insects: Modern IPM strategies emphasize the introduction of natural predators, such as predatory bugs of the genus *Orius* and predatory mites from the genera *Amblyseius* and *Neoseiulus*, to maintain a thrips-free environment for chrysanthemum cultivation. Early implementation of biological control measures, along with providing high-quality nutrition, supports the development of a robust population of the predatory bug *Orius*. Among beneficial insects, the minute pirate bug, *Orius tristicolor* White, plays a pivotal role as an effective thrips predator. Additionally, other beneficial bugs from the genera *Geocoris* and *Nabis*, as well as lacewing larvae of *Chrysoperla carnea* Stephens, contribute to effective thrips control (Waterhouse and Norris, 1989).

Predatory mites such as *Transieus* and *Amblyseius* species, along with predatory thrips such as *Haplothrips*, are also commonly utilized. However, their effectiveness in controlling thrips populations is typically limited, except when applied within a controlled greenhouse IPM program. Soil predators, in combination with foliage predators, can be helpful in reducing thrips numbers by decreasing the number of WFT emerging from soils. For example, *Geolaelaps aculeifer* significantly reduced the number of adult thrips on the foliage of chrysanthemums by about 40% (Manners *et al.*, 2013). Therefore, soil predators may assist in managing WFT during moderate pest pressure, when there are relatively high numbers of thrips pupating in the soil media.

In recent years, biological agents such as *Amblyseius montdorensis*, *Amblyseius limonicus*, *Ceranisus menes*, and *Ceranisus americensis* have gained increasing prominence for controlling WFT (Loomans *et al.*, 2006; Messelink and de Kogel, 2013). These agents represent a promising approach to enhancing thrips management in horticultural settings. By integrating these biological agents and beneficial insects into IPM strategies, chrysanthemum growers can effectively manage thrips populations while minimizing the use of chemical pesticides and promoting a sustainable and environmentally friendly approach to pest control.

9.6 Red Spider Mite (*Trombidiformes*: *Tetranychidae*)

Spider mites, highly polyphagous herbivores, are significant global agricultural pests responsible for substantial economic losses (Puspitarini *et al.*, 2021). They persistently threaten a broad spectrum of ornamental plants, including chrysanthemum, dahlia, rose, gerbera, zinnia, and marigold. This cosmopolitan species of spider mite is particularly prevalent in greenhouse environments. There are many species of spider mites, but one of the most important is the two-spotted spider mite (*Tetranychus urticae*), which attacks a wide range of plants.

9.6.1 Biology, feeding behavior, and symptoms of damage

Two-spotted spider mites are small (approximately 0.5 mm) and are typically found on the underside of leaves. They are yellow-brown to dark green with two distinctive dark spots on both sides of their bodies. Spider mites lay round and translucent eggs, typically on the underside of older leaves. Young mites (nymphs) and summer adults are green with two black patches on their backs. In the autumn, the females turn brick-red prior to overwintering. Development from egg to adult can occur within 10–21 days, depending on temperature—the higher the temperature, the shorter the development period. An adult female spider mite can produce between 100 and 200 eggs in her lifetime, which hatch into tiny larvae within days. These larvae quickly progress through multiple nymph stages before maturing into adults. In hot weather, the development from egg to egg-laying female can occur in as little as 6 days. Notably, eggs and resting stages demonstrate greater resistance to pesticides compared with actively feeding stages. Due to their minuscule size and preference for feeding on the underside of leaves, spider mites often remain unnoticed until their population surges. These pests thrive in hot and arid conditions, making them particularly troublesome in protected environments.

Damage caused by spider mites can often be mistaken for that caused by leafhoppers or thrips (Bartel and Bennison, 2021). Typically, the white spots resulting from leafhopper damage are larger and more irregular compared with those from spider mites, and cast skins are often visible. Thrips damage is distinguishable by the presence of small black fecal spots on the affected white areas. Both adult and nymph spider mites primarily feed by scraping and consuming oozing sap from epidermal tissues, including chlorophyll pigment. Their feeding damage is not immediately visible on standard chrysanthemum foliage, which may result in infestations going unnoticed until flower buds begin to show color. At this stage, mites tend to concentrate around the buds, where their feeding can cause significant damage, including webbing on opening flowers. As the infestation progresses, spider mites spread across all parts of the plant, often forming colonies covered by white-silken webs. Both adults and nymphs create yellow feeding scars and cause discoloration on young leaves. Severely affected leaves turn reddish-brown and bronze, eventually withering and drying out. Spider mite infestations can also negatively impact flower development, reducing both the aesthetic quality and market value of chrysanthemums.

9.6.2 Management

A combination of cultural, biological, and conventional chemical control measures provides the best approach for controlling mite infestations as part of an IPM program.

Soil and crop management: Implementing a variety of cultural control strategies can help minimize potential pest threats to crops and reduce the need for frequent interventions. Start with clean propagating material and avoid using cuttings from infested mother plants. Inspect all incoming plant shipments in a quarantine area and either reject or isolate any infested batches. Clean up plant debris and apply a recommended biocide between crop cycles. Eliminate weeds in both cropped and noncropped areas, as they can serve as alternative hosts for pests. For example, Salix and willowherb are particularly vulnerable to attacks from the two-spotted spider mite. Dispose of crop debris, unmarketable plants, and weeds promptly and carefully. Use sealed bags or covered containers placed as far from production areas as possible before disposal.

Monitoring and trapping: Monitoring and trapping are essential components of effective crop management. Regular, thorough inspections, particularly focusing on the undersides of leaves where pests often reside, are crucial. These inspections should be performed by individuals trained to identify the early signs of mite damage and distinguish between different mite species. Since mites do not fly, sticky traps are ineffective for monitoring mite outbreaks. Instead, it is recommended to use a hand lens with at least 10 × magnification to accurately detect the presence of mite damage, mites, or eggs on the plants. This approach ensures that any infestations are caught early, allowing for timely intervention.

Pest control sprays: Pest control sprays are essential for managing mite infestations effectively. It is crucial to apply pesticides thoroughly to the undersides of plant foliage, where mites typically gather, before the flower buds open. In open-field conditions, you can use Dicofol 18.5 EC at a rate of 2.5 ml or wettable sulfur 50 WP at 3 g l^{-1}. Under polyhouse (greenhouse) conditions, a targeted application of acaricides is recommended based on the specific needs, including options such as Abamectin 1.9 EC at 0.5 ml l^{-1}, Dicofol 18.5 EC at 2.5 ml l^{-1}, or Fenazaquin 10 EC at 1 ml l^{-1}. To prevent resistance, rotate these acaricides with plant-based products such as pongamia oil or neem oil at 8–10 ml l^{-1}, or neem soap at 10g l^{-1}. If the infestation is severe, removing and destroying heavily infected leaves or plants might be necessary, along with repeating the pesticide application every 5–7 days as needed.

Biological control: A variety of commercially available predators are integral to biological control programs, targeting pest mites and other pests such as whitefly and thrips. The choice of predatory mite species depends on environmental conditions, specific pest mites, and the need for control of additional pests. *Phytoseiulus persimilis* is the most used predatory mite for managing the two-spotted spider mite. As a specialist predator coevolved with *Tetranychus* species, *P. persimilis* is highly effective against this pest. Adult *P. persimilis* are slightly larger than pest mites, orange-red, shiny, and highly active, consuming both mobile stages and eggs. It is recommended to release approximately ten predators per 200 plants throughout the chrysanthemum crop's lifespan, increasing to ten predators per ten plants in infested areas (Scopes and Biggerstaff, 1973; Wardlow, 1986). Optimal effectiveness is achieved at temperatures between 15 and 25°C and relative humidity levels above 75% (Bartel and Bennison, 2021).

Several other predatory mites can supplement *P. persimilis* or target other pest mite species. *Amblyseius andersoni* is a generalist predator notable for its cold tolerance and omnivorous diet, feeding on various mite species, thrips larvae, pollen, fungal spores, and honeydew. It is less susceptible to some plant protection products. *Neoseiulus californicus* is effective against spider mites and tarsonemid mites and feeds on pollen. This species tolerates high temperatures and low humidity better than *P. persimilis* and is less susceptible to certain plant protection products.

Neoseiulus cucumeris is primarily used for thrips control, but it can also prey on spider mites and tarsonemid mites. *Amblyseius swirskii, A. limonicus,* and *Transeius montdorensis* are recommended for thrips and whitefly control but will also feed on pollen and some mite species. *T. montdorensis* is particularly effective in reducing spider mites, eriophyid mites, and tarsonemid mites. *Feltiella acarisuga*, a predatory midge, is useful for "hot spot" treatments of spider mites. Supplied as pupae in cartons, they should be placed in shady areas near infestations. Optimum conditions are 20–27°C and 80% relative humidity. The adult midges emerge and locate spider mite colonies to lay their eggs, dispersing more easily than predatory mites.

By integrating these biological control agents into pest management programs, growers can effectively manage spider mite populations while minimizing the use of chemical pesticides and promoting sustainable, environmentally friendly practices (Bartel and Bennison, 2021).

9.7 Leaf Miner (*Diptera*:*Agromyzidae*)

The serpentine leaf miner (*Liriomyza trifolii* Burgess) and the chrysanthemum leaf miner (*Chromatomyia syngenesiae*) are significant pests affecting chrysanthemums and other members of the *Asteraceae* family, such as aster, gerbera, and dahlia (Minkenberg and van Lenteren, 1986; Gratwick, 1992).

9.7.1 Biology, feeding behavior, and symptoms of damage

Adult flies of leaf miners are tiny, measuring 1.3–2.3 mm in length and 0.5–0.75 mm in width. They are pale yellow and metallic with plum-red eyes and a yellow spot on the scutellum. Females are larger than males and can lay up to 300 eggs over 17–21 days, depending on host and temperature (Charlton and Allen, 1981; De Jong and Rademaker, 1991). eggs are laid singly in the leaf epidermis; they are white and translucent when freshly laid and turn opaque as they develop. eggs hatch in about 2–3 days. The larvae are small, apodal, and orange-yellowish in color. Pupation occurs in the larval mines, soil, or occasionally on the leaf surface. The pupal stage, lasting 5–12 days, does not cause feeding damage (Nair, 1975; Reddy, 2010).

Adult chrysanthemum leaf miners are small, black to gray flies with yellow markings. Females puncture leaves to feed on plant sap and lay eggs within the leaf tissues. The primary damaging phase is the larval stage, during which the larvae live in and nourish themselves from the leaf tissue. They create characteristic tunnels, or mines, in the foliage of chrysanthemums and other host plants within the *Asteraceae* family. The larvae feed by tunnelling between the upper and lower epidermal layers of the leaf, creating serpentine mines (Fig. 9.8). These mines appear as silvery-white or transparent trails on the leaves, indicating the larval feeding path (Dudek, 2016). Severe infestations can result in extensive mining, leading to a significant reduction in photosynthetic activity, which impedes plant growth and overall health. Affected leaves often become chlorotic, and heavy infestations can cause premature leaf drop, reducing the aesthetic and commercial value of chrysanthemum crops. The damaged leaf cells around the mines die, forming brown patches, and the plants appear stippled or speckled (Reddy, 2010).

9.7.2 Management

Monitoring and trapping: Proactive measures, including greenhouse sanitation and regular monitoring, are essential for the early detection and control of leaf miner infestations. Collecting and disposing of severely affected plants, eggs, and larvae can prevent pest population buildup. Installing yellow sticky traps effectively monitors adult fly populations.

Pest control sprays: Products containing Abamectin are highly effective against leaf miners and relatively less harmful to beneficial insects. In open-field conditions with high infestations, removing infected leaves and applying a combination of chemical treatments, such as triazophos (0.15%) mixed with neem oil (7.5 ml l^{-1}) or Abamectin 1.9 EC (0.4 ml l^{-1}), followed by deltamethrin 2.8 EC (1 ml l^{-1}) at 10-day intervals, can be effective. In protected conditions, it is advisable to avoid frequent synthetic pesticide spraying. At most, one application of deltamethrin 2.8 EC (1 ml l^{-1}), Cypermethrin 25 EC (0.5 ml l^{-1}), or Triazophos 40 EC (2 ml l^{-1}) may be administered. Avoiding repeated use of the same insecticide is crucial to prevent resistance buildup (Reddy, 2010).

Additionally, dipping chrysanthemum cuttings in 3% neem oil for 2 hour and foliar spraying with neem botanical at 0.4% can reduce pupae and adults.

Biological control: The serpentine leaf miner has several natural enemies, including larval parasitoids such as *Hemiptarsenus varicornis*, *Chrysocharis parksi*, *Diglyphus isaea*, *Diglyphus crassinervis*, and *Neochrysocharis formosa* (all Eulophidae). Entomopathogenic nematodes such as *Steinernema feltiae* are also used to control the serpentine leaf miner (Mujica *et al.*, 2016). In open-field conditions, applying neem cake at 250 kg ha^{-1} to furrows or raised beds before planting can deter pupation of pests in the soil. Installing yellow sticky traps effectively monitors fly populations. Spraying neem seed powder extract at 4%, neem soap at 1%, or azadirachtin (10,000 ppm) at 1 ml l^{-1} at 15-day intervals after planting, with repeated applications every ten days, is recommended for control.

9.8 Lygus Species (*Hemiptera*:*Miridae*)

The European tarnished plant bug (*Lygus rugulipennis*) is a pest found throughout Europe (Woelke *et al.*, 2023). It feeds on over 400 crop plants, ranging from peach trees to wheat to lettuce. *L. rugulipennis* is a significant pest in Dutch chrysanthemum nurseries, causing substantial crop losses. Other species, such as *Lygus lineolaris*, are widespread pests in North America, with a broad appetite, feeding on over half of all commercially grown crop plants (Dixon and Fasulo 2021). This bug can be found from northern Canada to southern Mexico. Additionally, other species such as *Lygus elisus*, *L. shulli*, and *L. hesperus* are also reported.

9.8.1 Biology, feeding behavior, and symptoms of damage

Lygus bugs begin their life cycle by laying eggs in young plant tissue, which incubate for 7–12 days. A single female can lay an average of 96 eggs, with some laying up to 140. Nymphs emerge from the eggs, resembling aphids but distinguished by their harder exoskeletons and absence of aphid cornicles. They progress through five nymphal stages, growing from about 1 to 3–4 mm in length (Capinera, 2001). Upon reaching maturity, nymphs transform into adult lygus bugs, which are greenish to brownish with reddish-brown wing markings. Adults are capable of flight and are generally larger in females. Overwintered adults differ in color and form from new generation adults and seek sheltered areas such as leaf litter and bark. They emerge in April and early May to feed, mate, and lay eggs.

Lygus bugs are notorious for their destructive feeding habits, which involve puncturing plant tissues with their piercing mouthparts and feeding on the plant's sap. The damage results from both the physical injury caused by the bug's feeding mechanism and the plant's reaction to the saliva introduced during feeding. Plants attacked by lygus bugs exhibit symptoms such as discoloration, deformation of shoots and stems, curling of leaves, and lesions on plant tissues. These insects prefer young, developing plant tissue, making these areas especially vulnerable. Consequently, the damage often includes the development of multiple shoots, distortion of flowers, and the shedding of floral buds (Fig. 9.9).

The consequences of lygus bug feeding vary widely, from the destruction of the growing point in young seedlings to the deformation of leaves, which may appear crinkled and riddled with holes. The bugs introduce salivary enzymes during feeding, causing chemical damage to plant cells. The injury caused by the bugs can inhibit growth due to substances induced by the insects, while their saliva may contain plant growth-promoting substances that stimulate surrounding tissue growth.

Infested plants are noticeably shorter, up to 15 cm less in height, compared with noninfested plants of the same age. This damage occurs when one or a few *L. rugulipennis* adults puncture the plant's apical meristem, leading to the splitting of the growing point and inhibiting its growth. Additionally, *L. rugulipennis* damages flowers, causing spots on petals, irregular growth, and indentations, rendering the flowers unmarketable. In greenhouse ornamental production, *L. rugulipennis* primarily becomes problematic when pesticide use is reduced, particularly in cases where biological control methods are employed. This pest can cause significant damage to crops such as chrysanthemum and gerbera and can disrupt biocontrol programs designed to manage other pests.

9.8.2 Management

Monitoring and trapping: Effective monitoring and trapping methods are crucial for early detection and management of lygus bug infestations in greenhouse crops (Haye *et al.*, 2006). Yellow sticky cards, commonly used for monitoring other pests, are effective in attracting lygus bugs and should be regularly inspected to assess activity and population levels. Routine examination of young flowers and leaves provides valuable insights into their presence, allowing for timely intervention. Installing insect barriers over greenhouse vents, doorways, and other openings significantly reduces the likelihood of lygus bugs entering the greenhouse; a mesh screen with an appropriately large mesh size is sufficient for excluding these relatively large insects while maintaining proper airflow. Additionally, water traps equipped with white LED lights have been shown to catch 20–30 times more *L. rugulipennis* than sex pheromone-based traps, making them an effective trapping method in greenhouses (van Tol *et al.*, 2022).

Soil and crop management: Effective soil and crop management is crucial for controlling lygus bugs. Weed control in the immediate vicinity of the greenhouse is essential to prevent the development of large bug populations. Mowing or brush cutting in spring can force resident bug populations in weeds to migrate into the greenhouse, exacerbating the problem. Therefore, it is best to manage weeds during autumn to discourage overwintering. Creating a wide, well-mowed strip around the greenhouse is advisable. Additionally,

Fig. 9.8. Leaf miner infestation on the front (A) and rear (B) sides of a chrysanthemum leaf. A characteristic symptom of leaf miner damage is the serpentine mining pattern within the leaf tissue, caused by larvae feeding inside the leaf. Adult females puncture leaves to feed on plant sap and lay eggs within the leaf tissues. The larvae live inside the leaf and nourish themselves by consuming plant tissue. (Photo courtesy of Malleshaiah SharathKumar.)

managing weeds during winter is essential. When adult lygus bugs emerge, they will migrate to chrysanthemums if weeds senesce or are removed. Mowing or discarding weeds or alternative hosts while lygus bugs are in the nymphal stage can be an effective control measure.

Trap crops: Utilizing trap crops is an effective strategy for managing *L. rugulipennis*, particularly given the limited availability of specific biological controls (Hennekam *et al.*, 2012). One approach involves planting flowering hosts near chrysanthemum plants to attract lygus bugs during the vegetative stage, but this requires careful management. Vacuuming lygus bugs from trap crops can help reduce their numbers, although this method may inadvertently remove beneficial arthropods as well. Studies have found that *Sinapis alba*, especially in its vegetative stage, is highly attractive to both nymphs and adults of *L. rugulipennis* compared with chrysanthemums (Woelke *et al.*, 2023). *Medicago sativa* has also shown potential as a successful trap crop for *Lygus* spp. (Sevacherian and Stern, 1974; Swezey *et al.*, 2013). *Matricaria recutita* (chamomile) could also be an alternative as a trap plant for *Lygus rugulipennis*. This strategy delayed the population buildup of nymphs on strawberry plants, effectively diverting the pest and protecting the crops (Easterbrook and Tooley, 1999). Sunflower has been identified as an effective trap crop for *L. rugulipennis*. The headspace of sunflowers attracted a higher number of adult *L. rugulipennis*, making it a valuable tool in managing this pest (Ondiaka *et al.*, 2016).

Pest control sprays: Effective chemicals for controlling *Lygus* bugs include novaluron, a growth regulator affecting immature stages, and sulfoxaflor, which interacts with nicotinic acetylcholine receptors. Acephate, a standard commercial insecticide, has also shown efficacy. Although neonicotinoids such as imidacloprid and thiamethoxam were historically used, resistance has developed. To reduce resistance risk, it is essential to rotate these chemicals and employ varied modes of action, prioritizing novel insecticides targeting unique metabolic processes or physiological systems.

Biological control: *Lygus* bug populations can be regulated by various naturally occurring predators such as *Geocoris* spp. (big-eyed bugs), *Nabis* spp. (damsel bugs), *Orius tristicolor* (minute pirate bugs), and several spider species. The commercially available egg parasitoid *Anaphes iole* targets *Lygus* bug eggs, while *Peristenus* wasps lay their eggs inside *Lygus* nymphs, with the developing larvae consuming the nymphs from within. *Orius insidiosus* is also effective, particularly in greenhouse settings, where they attack small *Lygus* nymphs. *Beauveria bassiana*, a fungal control agent, manages *Lygus* bug populations and is registered for use against various greenhouse pests in Ontario (Ferguson *et al.*, 2012).

9.9 Bud Borer (*Lepidoptera*:*Noctuidae*)

The bud borer, *Helicoverpa armigera* (Hübner), is a major pest of many economically important crops in Asian open-field cultivation, including chrysanthemum, cotton, pigeonpea, chickpea, sunflower, tomato, sorghum, millet, okra, and corn (Manjunath *et al.*, 1989; Sharma, 2001). *Helicoverpa armigera* is highly adaptable and polyphagous, with a preference for nitrogen-rich, harvestable parts of valuable crops. At present, *Helicoverpa armigera* is distributed across most of Oceania, Asia, Africa, and southern Europe and has recently spread to South America (Raiz *et al.*, 2021).

9.9.1 Biology, feeding behavior, and symptoms of damage

Helicoverpa armigera has a short generation time of approximately 30 days in the field. Its life cycle is characterized by destructive feeding behavior and distinct symptoms on host plants. eggs are laid singly on young terminal leaves and unopened or partially opened flower buds. Upon hatching, larvae feed on tender foliage, leaving circular holes on leaves. As plants mature, larvae migrate to flower buds and flowers, feeding on the inner contents. In warm weather, larvae feed for 2–3 weeks before pupating in the soil. Each female can lay up to 1000 eggs in her lifespan (Rajapakse and Walter, 2007; Raiz *et al.*, 2021).

Fig. 9.9. Lygus feeding damage on chrysanthemum leaves (A) and flowers (B). (Photo courtesy of Ferguson *et al.*, 2012.)

The biology and behavior of *H. armigera* complicate pest management. Females can disperse 5–10 km during oviposition and undertake long-distance migrations over 100 km, assisted by wind currents, covering distances up to 2000 km. Under favorable conditions, eggs hatch in 3 days. With a wide host range and a generation time as short as 30 days, *H. armigera* can produce up to 12 generations per year, making it a significant pest management challenge.

H. armigera has shown resistance to synthetic pesticides and transgenic crops producing *Bacillus thuringiensis* toxins (Bt crops), particularly in fourth and fifth instar larvae. Accurate identification is crucial, especially in the The European and Mediterranean Plant Protection Organization (EPPO) region, where similar species such as *H. zea* and *H. punctigera* may be present. Differentiating *H. armigera* may require the examination of male genitalia or molecular tests such as DNA barcoding. Reports of *H. armigera* in light traps during summer in northern EU countries indicate its potential to move to higher latitudes. Facultative diapause allows populations to persist under harsh conditions. Its migratory behavior and international trade can transport eggs and larvae, posing a significant phytosanitary risk.

Detecting larvae can be challenging as they are often hidden within plant organs, such as flowers. Round holes and damaged flower buds contaminated with larval excreta are clear signs of infestation, leading to secondary infections and flower rot. Infested buds may fail to open, resulting in deformed flowers. Bud borer infestation typically peaks between summer and the start of the rainy season. Boreholes may be visible on the plant's surface, but dissecting plant organs may be necessary to identify larvae. These larvae prefer the reproductive parts of the host plant, with the lower surface of leaves being a commonly favored oviposition site.

9.9.2 Management

Helicoverpa armigera is a challenging pest due to its polyphagous nature, larvae residing within host plants, and resistance to pesticides, requiring a comprehensive approach. Various control measures, including synthetic insecticides, phytopesticides, microbial pesticides, macro-biocontrol agents, and genetically modified crops (e.g. Bt cotton), have been trialed or proposed. Effective management necessitates an IPM approach that combines biological, chemical, and physical control measures for optimal efficacy (Raiz *et al.*, 2021).

Monitoring and trapping: The adoption of sex-pheromone traps has proven superior to light traps for monitoring adult moth populations. These traps aid in identifying and counting moths and disrupt their mating behaviors. Furthermore, they are valuable for detecting the early spring emergence of adult moths, facilitating timely control measures and distribution assessments (Reddy and Manjunatha, 2000; Baker *et al.*, 2011). Bucket traps with pheromone bait are highly efficient in capturing *Helicoverpa* moths, especially for mass trapping of adult moths. Approximately 30–50 pheromone traps per hectare have led to significant captures (Guerrero *et al.*, 2014; Shah *et al.*, 2017). Traps should be placed at 1.5–1.8 m above the ground (Aheer *et al.*, 2009).

Summer trap crops: Another valuable approach is the use of summer trap crops. These crops divert *H. armigera* away from economically important plants while simultaneously fostering populations of natural enemies that prey on the pest. Sorghum, for instance, has been utilized as a trap crop, leading to increased parasitism rates by *Trichogramma chilonis*, a natural enemy of *H.*

armigera. To ensure their effectiveness, trap crops should be destroyed before *H. armigera* larvae enter the pupal stage, preventing further pest development (Virk *et al.*, 2004).

Soil and crop management: Many standard crop management strategies commonly employed to enhance overall crop health are also applicable to *H. armigera* control. These practices include proper tillage, irrigation, and the timely removal of crop residues. Crop rotation can be particularly effective in preventing the buildup of pest populations. Periodic mechanical collection and destruction of damaged flower buds, flowers, and mature larvae have also been effective, reducing bud borer incidence to less than 2%.

Pest control sprays: Sprays of endosulfan (0.07%) or methyl parathion (0.05%) applied at the appearance of eggs on buds and tender foliage can effectively control borer damage. Integrating these treatments with other chemical control options such as nuclear polyhedrosis virus of *H. armigera* (HaNPV) with UV protectants, neem oil, neem seed kernal extract (NSKE), and *Bacillus thuringiensis* var. *kurstaki* for early instar larvae, and deltamethrin, Indoxacarb 14.5 SC, and Thiodicarb 75 WP for grown-up larvae, provides a comprehensive approach to managing *Helicoverpa armigera* while minimizing crop damage and losses.

Biological control: Natural enemies play a crucial role in managing *Helicoverpa armigera*, with various egg, larval, and pupal parasitoids and predators identified (van den Berg *et al.*, 1988). These include species from the families *Braconidae* (*Apanteles* sp., *Meteorus pulchricornis*, *Habrobracon hebetor*, *Cotesia kazak*), *Ichneumonidae* (*Campoletis chlorideae*, *Diadegma* spp.), *Trichogrammatidae* (*Trichogramma pretiosum*, *Trichogrammatoidea annulata*, and *Trichogramma chilonis*), *Scelionidae*, and *Tachinidae*. Predators include Chrysopidae, Nabidae, Anthocoridae (e.g. *Orius* spp. and *Geocoris* spp.), Miridae, Coleoptera, and Araneae (Raiz *et al.*, 2021). Successful outcomes have been observed in specific crops through the augmentation of natural enemies and the implementation of complex biocontrol systems involving multiple predators and parasitoids. For example, in cotton crops in China, *Trichogramma chilonis* has proven effective as the primary egg parasitoid (Liu *et al.*, 2016).

While natural enemies provide significant benefits, they may not be sufficient to entirely control *H. armigera* populations, especially in areas susceptible to migratory invasions. Several biocontrol methods have been explored. Researchers have investigated the potential of the NPV for biocontrol in citrus trees (Moore *et al.*, 2004) and its applicability in cotton and chickpea (Jeyarani *et al.*, 2010). Additionally, *Bacillus thuringiensis* has demonstrated significant reductions in European *H. armigera* populations (Avilla *et al.*, 2005). Mohsin *et al.* (2015) reported five species of predatory spiders (*Lycosa terrestris*, *Plexippus paykulli*, *Thomisus projectus*, *Cyclosa insulana*, and *Pardosa birmanica*) that were found to feed on *H. armigera* larvae. Notably, a new virus with potential for biocontrol, *Heliothis virescens ascovirus 3i* (HvAV-3i), was recently identified (Chen *et al.*, 2020). Biological control methods, such as the release of natural enemies such as *Trichogramma chilonis* and *Chrysoperla carnea*, have successfully managed *H. armigera* populations, with releases timed to coincide with flowering periods.

9.10 Leaf Caterpillar (*Noctuidae*: *Lepidoptera*)

The leaf caterpillar *Spodoptera litura* is a pest with widespread distribution across tropical and subtropical regions worldwide. The leaves of commercially grown *Chrysanthemum morifolium* at the seedling stage are particularly susceptible to infestation by the larvae of *S. litura*, commonly known as the tobacco cutworm (TCW), especially in open fields. TCW is a major pest that feeds on plant leaves in tropical and subtropical areas of Asia, negatively affecting the growth of *C. morifolium* and reducing its ornamental value. This pest poses a threat to a diverse range of host plants, including chrysanthemums, tobacco, castor, cruciferous crops, tomato, and gladiolus.

9.10.1 Biology, feeding behavior, and symptoms of damage

The leaf caterpillar breeds continuously throughout the year, following a life cycle that begins with the laying of eggs, which are arranged in clusters and covered with brown hairs. These eggs typically hatch within 3–5 days. The caterpillar undergoes six larval stages, achieving full development in approximately 15–30 days. Subsequently, it enters the pupal stage, lasting 7–15 days. The entire life cycle can be completed within a broad range of 32–80 days, facilitating the occurrence of multiple overlapping generations within a single year.

The adult leaf caterpillar exhibits brown forewings adorned with wavy white markings, while its hind wings are predominantly white with a distinctive brown patch along the wing's margin. The eggs, laid in clusters on foliage during dusk, are covered with brown body hairs. Early instar larvae are light green with black heads or spots and tend to display gregarious behavior. As they mature, the larvae become pale greenish with dark markings and may continue to feed gregariously during their early stages.

The leaf caterpillar exhibits distinct feeding behaviors and causes noticeable damage to host plants. During their early instar stages, the larvae feed in groups on the ventral surface of leaves. However, as they progress through their development, they scatter and feed voraciously, often consuming entire leaves during dusk, resulting in a skeletonized appearance. To avoid exposure during the daytime, these larvae seek refuge in the soil or under crop debris. In severe infestations, they may not only defoliate the plant but also target tender twigs and even floral parts.

9.10.2 Management

Monitoring and trapping: The presence of newly hatched larvae can be detected by the "scratch" marks they make on the leaf surface. Older larvae are nocturnal feeders and are usually found in the soil around the base of plants during the day. They chew large areas of the leaf and can strip a crop of its leaves at high population densities. To manage *Spodoptera litura* effectively, growing castor plants along field borders and irrigation channels as trap crops can be beneficial. Flooding the field can force hibernating larvae out of their hiding places, and regularly collecting and destroying egg masses, gregarious larvae, and fully grown caterpillars can reduce the pest population.

Set up plastic funnel pheromone traps with Spodolure at a rate of 15 ha^{-1} to attract male moths. Additionally, installing and operating light traps at a rate of 2–3 ha^{-1} during evening hours (between 6:30 and 11:30 PM) can attract adult moths. Erecting bird perches encourages insectivorous birds to prey on feeding larval populations present on the crop canopy.

Poison bait can be prepared by mixing 10 kg of rice bran or wheat bran with 2 kg of jaggery. Allow the mixture to ferment for 12 hour after adding 2–3 l of water in the morning. In the evening of the next day, introduce 250 g of Thiodicarb formulation. Broadcast this bait between crop rows or over the bed. Caterpillars are attracted to the fermenting jaggery, ingest the bait, and perish. Repeat baiting as necessary.

Biological control: Utilize *Spodoptera* NPV by spraying at a rate of 250 LE ha^{-1}. Enhance its effectiveness by incorporating 1% jaggery or 2.5 kg crude sugar and 0.1% teepol during evening applications. Implement six releases of *Trichogramma chilonis* at a rate of 50,000 ha^{-1} $week^{-1}$, coordinating the releases with the flowering period. Release *Chrysoperla carnea* at a rate of 50,000 eggs or grubs ha^{-1} on a weekly basis, starting from 30 days after planting.

Chrysanthemums employs an effective indirect defense strategy against herbivores by releasing herbivore-induced plant volatiles (HIPVs). These HIPVs act as chemical signals that attract natural enemies of herbivores, such as predatory and parasitic insects, crucially protecting chrysanthemums from insect damage. The complex blend of sesquiterpenes and monoterpenes serves as the chemical cue for this defense strategy. The biosynthesis of these volatiles is facilitated by the upregulation of terpene synthase genes following herbivory, enhancing the plant's ability to attract beneficial insects and reduce herbivore pressure (Xu *et al.*, 2021). These new insights could be instrumental in aiding the prevention and control of TCW larvae in *C. morifolium*. Understanding the role of HIPVs and the associated upregulation of terpene synthase genes (*CmjTPS*) in indirect defense mechanisms and exploring the function of *CmjTPS* genes in *C. morifolium* can provide valuable information for developing more effective pest management strategies, leveraging the plant's natural defense systems to mitigate herbivore damage.

Pest control sprays: Consider chemical intervention when necessary. Options include spraying *Bacillus thuringiensis* at 2 g l^{-1}, Indoxacarb 14.5 SC at 0.75 ml l^{-1}, Spinosad at 0.75 ml l^{-1}, Thiodicarb 75 WP at 0.75 g l^{-1}, or Chlorpyriphos 20 EC at 2.5 ml l^{-1}. Choose the appropriate chemical based on the severity of the infestation and adhere to recommended application rates. New molecules such as chlorantraniliprole, spinosad, and emamectin benzoate have shown promising results against *S. litura* (Gadhiya *et al.*, 2014).

9.11 Mealybugs (*Pseudococcidae*:*Hemiptera*)

The pink hibiscus mealybug (PHM, *Maconellicoccus hirsutus* Green) poses a significant threat to tropical and subtropical regions of Asia, Africa, and Australia, with its accidental introduction into the Caribbean and the Americas raising serious concerns (Watson, 2022). The risk of further dissemination, including its potential arrival in the continental United States and Europe, remains a significant concern. Originally prevalent in regions such as central and northern Africa, India, Pakistan, northern Australia, and southeastern Asia (Kairo *et al.*, 2000), the PHM has recently infiltrated tropical areas in the western hemisphere. This highly polyphagous pest, native to southern Asia, affects various economically important crops in Europe (Bragard *et al.*, 2022). It targets over 200 plant species, including ornamental crops such as chrysanthemums, citrus, coffee, anthurium, bougainvillea, croton, heliconia, ixora, and hibiscus (Williams, 1996; Kairo *et al.*, 2000; Watson, 2022). Efforts are underway to standardize its common name as "pink hibiscus mealybug (PHM)," despite its wide range of host plants. Its rapid spread is facilitated by the transport of infested ornamentals, underscoring the need for vigilant monitoring and control measures to curb its proliferation (Kairo *et al.*, 2000).

9.11.1 Biology, feeding behavior, and symptoms of damage

The reproductive cycle of the PHM is rapid, with females perishing shortly after laying their eggs. Freshly laid eggs are initially orange but turn pink before hatching and are usually found in egg sacs. First instar nymphs, also known as crawlers, disperse by walking and through wind dispersal, covering significant distances in search of suitable host plants. Under optimal development temperatures around 29°C up to 15 generations can be produced annually, leading to the potential for extensive pest populations (Chong *et al.*, 2008).

Adult PHM measure about 3 mm in length and are pink, covered with a waxy secretion. Females are typically obscured by a white mealy wax with short waxy filaments, while crushed adults release pink body fluids. In contrast, adult males are smaller, reddish-brown, and possess one pair of wings along with two long waxy "tails." The adult female has a body length of approximately 3 mm and is wingless, covered in white, flocculent wax. It features two short, inconspicuous caudal filaments and lacks lateral wax filaments. The female's body and fluids are reddish, and she secretes a cotton-like egg mass, irregular in shape, containing 300–600 pink eggs. First instar nymphs, or pink crawlers, emerge from these eggs. In tropical climates, a single generation takes about 30 days to complete. Adult males, smaller than females, are reddish-brown and possess one pair of wings, along with two long wax caudal filaments. Males, unlike females, have nonfunctional mouthparts and do not feed, living only a few days. Unmated females release a sex pheromone to attract males for mating.

Mealybugs feed on plant tissues, injecting toxic saliva that causes leaf distortion and curling, leading to stunted growth and twisted shoots (Williams, 1996; Kairo *et al.*, 2000; Watson, 2022). They excrete honeydew, which promotes the growth of sooty mold. Severe infestations can kill plants, with damage severity depending on the host plant's health. Young or weak plants are more susceptible. The saliva injected by *Maconellicoccus hirsutus* during feeding likely contains a phytotoxic substance that impacts host plants differently depending on their tolerance (Williams, 1996). Tolerant species experience stunted growth at the growing points and stem axils, leading to reduced internode extension and leaf expansion. Highly susceptible plants may show severe crumpling of leaves following brief feeding by crawlers, and significant infestations can lead to total defoliation and plant death. As the plant deteriorates, mealybugs move from the dying tips to healthier tissues, spreading the infestation from shoot tips downwards to the trunk.

9.11.2 Management

Management of the PHM involves a multifaceted approach combining biological and chemical strategies. Native natural enemies are expected to provide some control, and classical biological control methods are being explored (Kairo *et al.*, 2000; Mani and Krishnamoorthy, 2001, 2008; Hoy *et al.*, 2023). The ladybug *Cryptolaemus montrouzieri*, known for feeding on mealybugs, can be an asset under high mealybug populations, although it may not be sufficient for lower population levels. Care should be taken not to destroy *Cryptolaemus* larvae, which resemble mealybugs but are crucial predators. *C. montrouzieri* is unable to survive the cold of winter in sufficient numbers to be effective. Therefore, several effective parasitoids identified in Asia and other regions, such as *Anagyrus kamali*, offer promising control options. *Anagyrus kamali* has shown exceptional potential in suppressing PHM populations. Mealybugs parasitized by *A. kamali* produce distinct "mummies" with exit holes for adult wasps.

The effectiveness of pesticide sprays is generally limited due to the mealybug's habit of hiding in crevices and its protective waxy covering. The careful selection of pesticides is essential to avoid harming the natural enemies of the mealybug, which play a critical role in long-term population control. The first instar stage of the mealybug is most susceptible to pesticides (Persad and Khan, 2000); however, all pesticides tested were highly toxic to the key biological control agent, *Anagyrus kamali*.

In cases of severe infestations, systemic pesticides offer effective control of mealybugs, often requiring only a single application. In contrast, spray treatments typically necessitate two or three repetitions to achieve satisfactory results. Additionally, inorganic oil emulsion sprays have been found effective (Jalaluddin and Sadakathulla, 1998). Abrine, an alkaloid from *Abrus precatorius* seeds, could significantly impact mealybug population density (Anitha *et al.*, 1999). Successful mealybug control depends on the meticulous and comprehensive application of IPM strategies. These strategies must combine biological control methods with careful pesticide use to manage mealybug populations effectively.

9.12 Host Plant Resistance to Insects in Chrysanthemum

Host plant resistance to insects is a strategy employed by plant breeders to enhance plants' ability to withstand or deter insect pests. This form of resistance is a crucial component of IPM as it reduces reliance on chemical pesticides, thereby promoting a more sustainable approach to crop protection. Plants can exhibit resistance to insects through several mechanisms:

Antibiosis: This occurs when the plant exerts mechanisms that reduce survival, reproduction, or pest proliferation. It can involve the production of compounds that are toxic or otherwise detrimental to the insect and excretion of glue-like substances that immobilize and trap insects.

Antixenosis (nonpreference): This form of resistance makes the plant less attractive or more difficult for insects to locate, recognize, or feed on. It can involve physical traits such as tough leaves or trichomes (hair-like structures) or chemical cues that deter insects.

Tolerance: This refers to the plant's ability to withstand damage from insect feeding and still produce an acceptable yield. Tolerant plants can recover better from the damage or may sustain less damage due to their growth characteristics.

Developing and utilizing host plant resistance often involves breeding programs where plants are selectively bred to enhance these resistant traits. It can also involve genetic engineering to introduce new traits into the plant genome. To combat insect pests on chrysanthemums, a viable strategy is the development of insect-resistant varieties through breeding for host plant resistance. This approach, when combined with the use of biological control agents, can significantly reduce pest populations in greenhouses, minimize insecticide use, and mitigate damage caused by pests. Such resistance traits can be identified in both cultivated chrysanthemum varieties and their wild relatives.

Evaluation of chrysanthemum germplasm for resistance against various insects has been conducted, revealing some cultivated varieties that are less susceptible or resistant to several species. Notable genetic variation in susceptibility to WFT has been found in cultivated chrysanthemum (de Jager *et al.*, 1995; de Kogel *et al.*, 1998; Ohta, 2002; Kos *et al.*, 2014). Similarly, significant differences in susceptibility to aphids, including *Myzus persicae* (Wyatt, 1969), *Aphis gossypii* (Storer and Van Emden, 1995; Bethke *et al.*, 1998), and *Macrosiphoniella sanborni* (Reddy and Janakiram, 2004; He *et al.*, 2011; Fu *et al.*, 2018; Surani *et al.*, 2022), have been documented.

Differences in susceptibility to the leaf miner (*Liriomyza trifolii*) were also observed (de Jong and Van De Vrie, 1987; Kos *et al.*, 2014). Furthermore,

Spodoptera exigua exhibited preferences for specific chrysanthemum cultivars (Yoshida and Parrella, 1991). Interestingly, some cultivars show resistance to multiple pests; for instance, one cultivar resistant to thrips was also the least susceptible to leaf miners (de Jong and Van De Vrie, 1987). Additionally, resistance overlaps among WFT, leaf miners, and two-spotted spider mites were identified (Kos *et al.*, 2014).

Wild relatives within the chrysanthemum genus offer valuable sources of insect resistance that can potentially be introgressed into cultivated varieties. For example, *Chrysanthemum makinoi* and *C. aromaticum* have been found resistant to the chrysanthemum aphid, *M. sanborni* (Cheng *et al.*, 2011; Zhong *et al.*, 2022). Resistance to insects is also present in the closely related mugwort genus (*Artemisia*), such as resistance to aphids in *Artemisia vulgaris* (Deng *et al.*, 2010). These resistance traits can be incorporated into cultivated chrysanthemum through intergeneric and interspecific hybridization followed by backcrossing (Su *et al.*, 2019). Successful hybridization between cultivated chrysanthemum (*Chrysanthemum* × *morifolium*) and species from genera such as *Ajania*, *Artemisia*, and other *Chrysanthemum* species has been reported (Deng *et al.*, 2010; Cheng *et al.*, 2011; Deng *et al.*, 2011). The importance of leveraging genetic diversity within and related to chrysanthemum for developing pest-resistant cultivars, which can lead to more sustainable and effective pest management strategies in horticulture.

9.12.1 Phenotyping methods

Insect resistance in cultivated and wild plants can be assessed using various assays, generally categorized into choice and no-choice assays. In choice assays, insects are given the freedom to move between different plants, allowing them to select plants for feeding or oviposition. This method helps assess nonpreference or repellence. In contrast, no-choice assays restrict insects to a specific plant or plant part using sleeves, clip-on cages, or Petri dishes. Parameters such as insect-inflicted plant damage, the number of insects per plant, or the effects on insects (e.g. mortality, reproduction, development time, adult emergence, and population development) are measured. By evaluating how insects perform on specific plants, antibiosis-type resistance, which affects insect survival, reproduction, or development, can be identified.

Traditionally, phenotyping for insect resistance involves large screenings that are slow and labor-intensive, requiring plants to be infested and individually scored for plant damage or insect counts. The identification of new sources of resistance may be expedited through novel high-throughput screening methods developed in recent years (Goggin *et al.*, 2015). These advanced methods may employ video tracking to observe insect behavior on plants, followed by software analyses to identify resistant plants (Kloth *et al.*, 2015). Additionally, quantification of insects on plants can be accelerated using AI-based algorithms coupled with image acquisition software (Devi *et al.*, 2023). By leveraging these modern phenotyping techniques, researchers can enhance the efficiency and accuracy of screening for insect resistance in plants, ultimately contributing to more effective and sustainable pest management strategies.

9.12.2 Plant defense mechanisms

Plants have evolved a range of defense mechanisms to protect themselves from insect herbivory. Many of these defenses are constitutive, meaning they are always present. The first physical barriers that insects encounter when feeding on plant tissue include a thick surface wax, cuticle, thorns, cell walls, and trichomes. These physical structures can impede insect feeding and movement. Additionally, some trichomes have glands that produce and store toxic or adhesive compounds, which can kill or immobilize insects (Stavrinides and Skirvin, 2003). In chrysanthemum, two types of trichomes are found: T-shaped and glandular trichomes (Guo *et al.*, 2020). The density of T-shaped trichomes has been shown to contribute to aphid resistance (Cheng *et al.*, 2011), while glandular trichomes can trap aphids by secreting toxins on the plant surface (Ramirez *et al.*, 2012). However, the role of trichomes in resistance against thrips has not yet been demonstrated (Chen *et al.*, 2020b). Understanding the diverse defense mechanisms in plants, including physical barriers and chemical deterrents, is crucial for developing strategies to enhance plant resistance to pests. By leveraging these natural defenses, it is possible to reduce reliance on chemical pesticides and promote sustainable agricultural practices.

9.12.3 Metabolites

Plants constitutively produce many different classes of metabolites and specialized metabolites are likely involved in defense against insects in chrysanthemum. For example, the concentration of chlorogenic acid and isobutyl amide has been shown to correlate with resistance against thrips in cultivated chrysanthemum (Tsao *et al.*, 2005; Leiss *et al.*, 2009). Additionally, wild relatives of chrysanthemum may contain insecticidal metabolites. The wild relative *Tanacetum cinerariifolium* is traditionally ground to produce insecticides due to its production and accumulation of insecticidal pyrethrins in glandular trichomes on disc ray florets (Yang *et al.*, 2012; Lybrand *et al.*, 2020). Moreover, numerous sesquiterpenoids and monoterpenoids have been identified in *Chrysanthemum indicum* and *C. makinoi* (Jiang *et al.*, 2021). These compounds are also produced in the related *Artemisia* genus, with studies examining the essential oils of *A. feddei*, *A. keiskeana*, and *A. vulgaris* and their insecticidal properties (Kim, 1997; Deng *et al.*, 2010; Suleimen *et al.*, 2019). Some volatile compounds also play a role in plant–insect interactions by increasing the attraction of natural enemies to pest insects. These interactions underline the potential of using plant metabolites in developing effective pest management strategies, reducing reliance on synthetic insecticides and fostering sustainable agricultural practices.

9.12.4 Induced defenses

In addition to constitutive defenses, plants can activate or enhance their physical or chemical defenses upon herbivory, a phenomenon known as induced defense. This type of defense is primarily modulated by the phytohormones jasmonic acid (JA), salicylic acid (SA), and ethylene (Broekgaarden *et al.*, 2015; Erb and Reymond, 2019). The activation of induced defenses leads to the production of defense proteins, such as antioxidant enzymes, and enzymes that generate specialized metabolites against insects (Xia *et al.*, 2023). Induced defenses can make plants less susceptible to insect attacks. For example, in response to aphid infestation, chrysanthemums increase sugar metabolism to synthesize other defensive compounds (Xia *et al.*, 2023). Additionally, treatment with JA has been shown to reduce thrips-induced silver damage in chrysanthemums (Chen *et al.*, 2020a). Following infestation by larvae of the TCW, chrysanthemums became more attractive to two natural enemies of the pest (Xu *et al.*, 2021). These examples underscore the importance of induced defenses in enhancing plant resistance to herbivores, thereby providing a valuable tool for IPM strategies.

9.13 Conclusion and Future Prospects

Effective pest management in chrysanthemum cultivation requires an integrated approach that combines cultural, chemical, and biological methods to achieve sustainable and efficient control. Biological control agents, including parasitoids, predators, and entomopathogenic nematodes, play a vital role in reducing pest populations while minimizing environmental impact. The strategic deployment of these natural enemies, coupled with the use of resistant cultivars and optimized agricultural practices, enhances the resilience of chrysanthemum crops against pest infestations.

Future advancements in chrysanthemum pest management will focus on improving biological control techniques, developing environmentally friendly biopesticides, and integrating genetic resistance into breeding programs. Modern high-throughput phenotyping methods, such as video tracking and AI-based algorithms, are revolutionizing the identification of pest-resistant plants, accelerating breeding processes, and enhancing crop resilience. Furthermore, gaining deeper insights into plant defense mechanisms such as physical barriers like trichomes and chemical deterrents produced by specialized metabolites will strengthen the foundation for sustainable pest management practices.

Induced plant defenses, regulated by phytohormones such as JA, SA, and ethylene, can enhance resistance to herbivores by activating defense proteins and metabolites. These defense mechanisms can be strategically triggered to improve pest management outcomes. By integrating advanced phenotyping technologies, leveraging natural plant defense mechanisms, and employing modern biocontrol techniques, chrysanthemum growers can develop more effective, sustainable, and environmentally friendly pest management strategies. Continued research and innovation in IPM will be crucial to addressing evolving pest challenges and ensuring long-term sustainability in chrysanthemum production (Dixon and Fasulo, 2021).

References

Aheer, G.M., Ali, A. and Akram, M. (2009) Effect of weather factors on populations of *Helicoverpa armigera* moths at cotton-based agro-ecological sites. *Entomological Research* 29, 36–42.

Anitha, B., Arivalagan, M., Sundari, M.S.N. and Durairaj, G. (1999) Effect of alkaloid abrine, isolated from *Abrus precatorius* Linn. seeds on mealy bug, *Maconellicoccus hirsutus* green. *Indian Journal of Experimental Biology* 37(4), 415–417.

Askary, H., Carriere, Y., Belanger, R.R. and Brodeur, J. (1998) Pathogenicity of the fungus *Verticillium lecanii* to aphids and powdery mildew. *Biocontrol Science and Technology* 8, 23–32.

Avilla, C., Vargas-Osuna, E., Gonzalez-Cabrera, J., Ferre, J. and Gonzalez-Zamora, J.E. (2005) Toxicity of several δ-endotoxin of *Bacillus thuringiensis* against *Helicoverpa armigera* (Lepidoptera: Noctuidae) from Spain. *Journal of Invertebrate Pathology* 90, 51–54.

Baker, J. (2019) Chrysanthemum aphid, PDIC factsheets, North Carolina state extension publications. Available at: https://content.ces.ncsu.edu/chrysanthemum-aphid# (accessed 28 May 2024).

Baker, G.H., Tann, C.R. and Fitt, G.P. (2011) A tale of two trapping methods: *Helicoverpa* spp. (Lepidoptera, Noctuidae) in pheromone and light traps in Australian cotton production systems. *Bulletin of Entomological Research* 101, 9–23.

Bartel, E. and Bennison, J. (2021) Spider mites: Two-spotted spider mite (*Tetranychus urticae*). *ADAS Horticulture*. Available at: https://horticulture.ahdb.org.uk/knowledge-library/spider-mites

Ben-Issa, R., Gomez, L. and Gautier, H. (2017) Companion plants for aphid pest management. *Insects* 8(4), 112. DOI: 10.3390/insects8040112.

Bethke, J.A., Redak, R.A. and Schuch, U.K. (1998) Melon aphid performance on *Chrysanthemum* as mediated by cultivar, and differential levels of fertilization and irrigation. *Entomologia Experimentalis et Applicata* 88(1), 41–47.

Blackman, R.L. and Eastop, V.F. (1984) *Aphids on World Crops*. John Wiley and Sons, Chichester, UK, pp. 47–49.

Bragard, C., Baptista, P., Chatzivassiliou, E., Di Serio, F., Gonthier, P. *et al.* (2022) Scientific opinion on the pest categorisation of *Maconellicoccus hirsutus*. *EFSA Journal* 20(1), e07024. DOI: 10.2903/j.efsa.2022.7024.

Broekgaarden, C., Caarls, L., Vos, I.A., Pieterse, C.M.J. and Van Wees, S.C.M. (2015) Ethylene: Traffic controller on hormonal crossroads to defense. *Plant Physiology* 169, 2371–2379. DOI: 10.1104/pp.15.01020.

Buitenhuis, R., Shipp, J.L., Jandricic, S., Murphy, G. and Short, M. (2007) Effectiveness of insecticide-treated and non-treated trap plants for the management of *Frankliniella occidentalis* (Thysanoptera: Thripidae) in greenhouse ornamentals. *Pest Management Science* 63, 910–917.

Capinera, J.L. (2001) *Handbook of Vegetable Pests*. Academic Press, San Diego, California.

Charlton, C.A. and Allen, W.W. (1981) The biology of *Liriomyza trifolii* on beans and chrysanthemums. In: Schuster, D.J. (ed.) *Proceedings of the IFAS–Industry*

Conference on Biological Control of Liriomyza Leafminers, University of Florida, Institute of Food and Agricultural Sciences (IFAS), Lake Buena Vista, pp. 42–49.

Chau, A. and Heinz, K.M. (2006) Manipulating fertilization: A management tactic against *Frankliniella occidentalis* on potted chrysanthemum. *Entomologia Experimentalis et Applicata* 120(3), 201–209. DOI: 10.1111/j.1570-7458.2006.00441.x.

Chen, G., Liu, H., Mo, B.C., Hu, J., Liu, S.Q. *et al.* (2020) Growth and development of *Helicoverpa armigera* (Lepidoptera: Noctuidae) larvae infected by Heliothis virescens ascovirus 3i (HvAV-3i). *Frontiers in Physiology* 11, 93. DOI: 10.3389/fphys.2020.00093.

Chen, G., Kim, H.K., Klinkhamer, P.G. and Escobar-Bravo, R. (2020a) Site-dependent induction of jasmonic acid-associated chemical defenses against western flower thrips in *Chrysanthemum*. *Planta* 251, 1–14.

Chen, G., Klinkhamer, P.G. and Escobar-Bravo, R. (2020b) Constitutive and inducible resistance to thrips do not correlate with differences in trichome density or enzymatic-related defenses in *Chrysanthemum*. *Journal of Chemical Ecology* 46, 1105–1116.

Cheng, X., Chen, S., Chen, F., Deng, Y., Fang, W. *et al.* (2011) Creating novel chrysanthemum germplasm via interspecific hybridization and backcrossing. *Euphytica* 177, 45–53.

Chong, J.H., Roda, A.L. and Mannion, C.M. (2008) Life history of the mealybug, *Maconellicoccus hirsutus* (Hemiptera: Pseudococcidae). *Environmental Entomology* 37, 323–332.

Cloyd, R. (2019) Aphid management in greenhouse production systems, Kansas State University, March. Available at: https://bookstore.ksre.ksu.edu/pubs/MF3442.pdf (accessed 28 May 2024).

Davies, F.T., Jr., He, C.J., Chau, A., Spiers, J.D. and Heinz, K.M. (2005) Fertiliser application affects susceptibility of chrysanthemum to western flower thrips- abundance and influence on plant growth, photosynthesis, and stomatal conductance. *Journal of Horticultural Science & Biotechnology* 80, 403–412.

de Haan, J. and Buitenhuis, R. (2022, February) Less is more: Reducing fertilizer application adds to chrysanthemum IPM strategy. AFE Newsletter. American Floral Endowment. Available at: https://endowment.org/wp-content/uploads/2024/03/Thrips-and-Fertilizer-Feb-2022.pdf (accessed 21 November 2023).

de Jager, C., Butôt, R., Klinkhamer, P., De Jong, T., Wolff, K. *et al.* (1995) Genetic variation in Chrysanthemum for resistance to *Frankliniella occidentalis*. *Entomologia Experimentalis et Applicata* 77(3), 277–287.

De Jager, C.M., Butôt, R.P.T., Jong, T.J., Klinkhamer, P.G.L. and Meijden, E. (1993) Population growth and survival of western flower thrips *Frankliniella occidentalis* Pergande (Thysanoptera: Thripidae) on different chrysanthemum cultivars: Two methods for measuring resistance. *Journal of Applied Entomology* 115(1–5), 519–525. DOI: 10.1111/j.1439-0418.1993.tb00422.x.

de Jong, J. and Van De Vrie, M. (1987) Components of resistance to *Liriomyza trifolii* in *Chrysanthemum morifolium* and *Chrysanthemum pacificum*. *Euphytica* 36, 719–724.

De Jong, J. and Rademaker, W. (1991) Life history studies of the leafminer *Liriomyza trifolii* on susceptible and resistant cultivars of *Dendranthema grandiflora*. *Euphytica* 56(1), 47–53. DOI: 10.1007/BF00041743.

de Kogel, W.J., van der Hoek, M., Dik, M.T., van Dijken, F.R. and Mollema, C. (1998) Variation in performance of western flower thrips populations on a susceptible and a partially resistant *Chrysanthemum* cultivar. *Euphytica* 103, 181–186.

Deng, Y., Chen, S., Lu, A., Chen, F., Tang, F. *et al.* (2010) Production and characterisation of the intergeneric hybrids between *Dendranthema morifolium* and *Artemisia vulgaris* exhibiting enhanced resistance to chrysanthemum aphid (*Macrosiphoniella sanbourni*). *Planta* 231(3), 693–703. DOI: 10.1007/s00425-009-1081-5.

Deng, Y., Chen, S., Chen, F., Cheng, X. and Zhang, F. (2011) The embryo rescue derived intergeneric hybrid between chrysanthemum and *Ajania przewalskii* shows enhanced cold tolerance. *Plant Cell Reports* 30(12), 2177–2186. DOI: 10.1007/s00299-011-1123-x.

Devi, M.G., Rustia, D.J.A., Braat, L., Swinkels, K., Espinosa, F.F. *et al.* (2023) Eggsplorer: A rapid plant–insect resistance determination tool using an automated whitefly egg quantification algorithm. *Plant Methods* 19(1), 49.

Dudek, T. (2016) Scout garden mums for insect pests to avoid surprises. MSU Extension. Available at: https://www.greenhousemag.com/news/scout-garden-mums-for-insect-pests-to-avoid-surprises/ (accessed 17 October 2022).

Dixon, W. and Fasulo, T.R. (2021). *Lygus lineolaris* (Palisot de Beauvois), tarnished plant bug (Hemiptera: Miridae). Featured Creatures, University of Florida IFAS Extension. Available at: https://entnemdept.ufl.edu/creatures/

Easterbrook, M.A. and Tooley, J.A. (1999) Assessment of trap plants to regulate numbers of the European tarnished plant bug, *Lygus rugulipennis*, on late-season strawberries. *Entomologia Experimentalis et Applicata* 92, 119–125. DOI: 10.1046/j.1570-7458.1999.00531.x.

Emam, A.S. (2016) Biological control of the chrysanthemum aphid, *Macrosiphoniella sanborni* (Gillete) by release of *Coccinella septempunctata* L. on chrysanthemum plants. *Journal of Plant Protection and Pathology* 7(5), 291–295. DOI: 10.21608/jppp.2016.50549.

Erb, M. and Reymond, P. (2019) Molecular interactions between plants and insect herbivores. *Annual Review of Plant Biology* 70, 527–557. DOI: 10.1146/annurev-arplant-050718-095910.

European Union. (2018) L132. *Official Journal of the European Union* 61, 31–44.

Ferguson, G., Murphy, G. and Shipp, L. (2012) Managing the lygus bug in greenhouse crops, ISSN 1198-712X. Available at: https://www.ontario.ca/page/managing-lygus-bug-greenhouse-crops (accessed 17 November 2022).

Frank, S. (2016) A Refresher on aphids, GrowerTalks. Available at: https://www.growertalks.com/Article/?articleid=22063 (accessed 28 May 2024).

Fu, X., Su, J., Yu, K., Cai, Y., Zhang, F. *et al.* (2018) Genetic variation and association mapping of aphid (*Macrosiphoniella sanbourni*) resistance in chrysanthemum (*Chrysanthemum morifolium* Ramat.). *Euphytica* 214, 1–9.

Gadhiya, H.A., Borad, P.K. and Bhut, J.B. (2014) Effectiveness of synthetic insecticides against *Helicoverpa armigera* (Hubner) Hardwick and *Spodoptera litura* (Fabricius) infesting groundnut. *The Bioscan* 9(1), 23–26.

German, T.L., Ullman, D.E. and Moyer, J.W. (1992) Tospoviruses: Diagnosis, molecular biology, phylogeny and vector relationships. *Annual Review of Phytopathology* 30, 315–348.

Ghosh, A., Jangra, S., Dietzgen, R.G. and Yeh, W.-B. (2021) Frontiers approaches to the diagnosis of thrips (*Thysanoptera*): How effective are the molecular and electronic detection platforms? *Insects* 12(10), 920.

Goggin, F.L., Lorence, A. and Topp, C.N. (2015) Applying high-throughput phenotyping to plant–insect interactions: Picturing more resistant crops. *Current Opinion in Insect Science* 9, 69–76.

Gratwick, M. (ed.) (1992) Leaf miners of chrysanthemum. In: *Crop Pests in the UK: Collected Edition of MAFF Leaflets*. Chapman & Hall, London, pp. 268–273.

Guerrero, S., Brambila, J. and Meagher, R.L. (2014) Efficacies of four pheromone-baited traps in capturing male *Helicoverpa* (Lepidoptera: Noctuidae) moths in Northern Florida. *Florida Entomologist* 97, 1671–1678.

Guo, Y., Zhang, T., Zhong, J., Ba, T., Xu, T. *et al.* (2020) Identification of the volatile compounds and observation of the glandular trichomes in *Opisthopappus taihangensis* and four species of Chrysanthemum. *Plants* 9(7), 855.

Haye, T., Kuhlmann, U., Goulet, H. and Mason, P.G. (2006) Controlling *Lygus* plant bugs (Heteroptera: Miridae) with European *Peristenus relictus* (Hymenoptera: Braconidae) in Canada – risky or not? *Bulletin of Entomological Research* 96, 187–196. DOI: 10.1079/BER2005414.

He, J., Chen, F., Chen, S., Lv, G., Deng, Y. *et al.* (2011) *Chrysanthemum* leaf epidermal surface morphology and antioxidant and defense enzyme activity in response to aphid infestation. *Journal of Plant Physiology* 168(7), 687–693.

Heming, B.S. (1985) Thrips (Thysanoptera) in Alberta. *Agriculture and Forestry Bulletin* 8(2), 19–24.

Hennekam, M., van Schaik, J. and van den Meiracker, R. (2012, July) *Schadelijke wantsen onder glas* [Harmful bugs under glass] (Report, pp. 1–31, in Dutch). Entocare C.V, Wageningen. Available at: https://www.glastuinbouwnederland.nl/content/glastuinbouwnederland/docs/themas/Plantgezondheid/02_onderzoeksrapporten/Eindrapport-14432.pdf

He, Z., Guo, J.-F., Reitz, S.R., Lei, Z.-R. and Wu, S.-Y. (2020) A global invasion by the thrip, *Frankliniella occidentalis*: Current virus vector status and its management. *Insect Science* 27, 626–645. DOI: 10.1111/1744-7917.12721.

Hoy, M.A., Hamon, A. and Nguyen, R. (2023) Pink hibiscus mealybug, *Maconellicoccus hirsutus* (Green). Publication #EENY-029. Entomology and Nematology Department, UF/IFAS Extension, Gainesville, Florida. DOI: 10.32473/edis-in156-2002. (accessed November 2024).

Jalaluddin, S.M. and Sadakathulla, S. (1998) Effect of atso oil emulsion sprays in the control of mealy bug *Maconellicoccus hirsutus* infesting guava fruits. *Entomon* 23(2), 151–152.

Jaskiewicz, B., Gantner, M. and Cisek, K. (2001) Chrysanthemum aphid. *Ochrona Roslin* 45(1), 35–36.

Jeyarani, S., Sathiah, N. and Karuppuchamy, P. (2010) Field efficacy of *Helicoverpa armigera* nucleo polyhedron virus isolates against *H. armigera* (Hübner) (Lepidoptera: Noctuidae) on cotton and Chickpea. *Plant Protection Science* 46, 116–122.

Jiang, Y., Zhang, W., Chen, X., Wang, W., Köllner, T.G. *et al.* (2021) Diversity and biosynthesis of volatile terpenoid secondary metabolites in the *Chrysanthemum genus*. *Critical Reviews in Plant Sciences* 40(5), 422–445.

Kairo, M.E.K., Pollard, G.V., Peterkin, D.D. and Lopez, V.F. (2000) Biological control of the hibiscus mealybug, *Maconellicoccus hirsutus* (Green) (Hemiptera: Pseudococcidae) in the Caribbean. *Integrated Pest Management Reviews* 5, 241–254. DOI: 10.1023/A:1012997619132.

Karlsson Green, K., Stenberg, J.A. and Lankinen, Å. (2020) Making sense of Integrated Pest Management (IPM) in the light of evolution. *Evolutionary Applications* 13, 1791–1805. DOI: 10.1111/eva.13067.

Kim, J.-H. (1997) Variation of Monoterpenoids in *Artemisia feddei* and *Artemisia scoparia*. *Journal of Plant Biology* 40, 267–274.

Kirk, W.D.J. (1997) Feeding. In: Lewis, T. (ed.) *Thrips as Crop Pests. CAB International.* Wallingford, UK, pp. 119–174.

Kloth, K.J., Ten Broeke, C.J., Thoen, M.P., Hanhart-van den Brink, M., Wiegers, G.L. *et al.* (2015) High-throughput phenotyping of plant resistance to aphids by automated video tracking. *Plant Methods* 11(1), 1–14.

Kos, S.P., Klinkhamer, P.G. and Leiss, K.A. (2014) Cross-resistance of *Chrysanthemum* to western flower thrips, celery leafminer, and two-spotted spider mite. *Entomologia Experimentalis et Applicata* 151(3), 198–208.

Leiss, K.A., Maltese, F., Choi, Y.H., Verpoorte, R. and Klinkhamer, P.G. (2009) Identification of chlorogenic acid as a resistance factor for thrips in *Chrysanthemum*. *Plant Physiology* 150(3), 1567–1575. DOI: 10.1104/pp.109.138131.

Lewis, T. (1973) *Thrips: Their Biology, Ecology and Economic Importance*. Academic Press, London.

Lewis, T. (1997) *Thrips as Crop Pests*. CAB International, Wallingford, UK.

Liu, B., Yang, L., Yang, F., Wang, Q., Yang, Y.Z. *et al.* (2016) Landscape diversity enhances parasitism of cotton bollworm (*Helicoverpa armigera*) eggs by *Trichogramma chilonis* in cotton. *Biological Control* 93, 15–23.

Loomans, A.J.M., Tolsma, J., Fransen, J.J. and van Lenteren, J.C. (2006) Releases of parasitoids (Ceranisus sp.) as biological control agents of western flower thrips in experimental glasshouses. *Bulletin of Insectology* 59(2), 85–97.

Lybrand, D.B., Xu, H., Last, R.L. and Pichersky, E. (2020) How plants synthesize pyrethrins: Safe and biodegradable insecticides. *Trends in Plant Science* 25(12), 1240–1251. DOI: 10.1016/j.tplants.2020.06.012.

Mani, M. and Krishnamoorthy, A. (2001) Suppression of *Maconellicoccus hirsutus* (Green) on guava. *Insect Environment* 6(4), 152.

Mani, M. and Krishnamoorthy, A. (2008) Field efficacy of Australian ladybird beetle *Cryptolaemus montrouzieri* Mulsant in the suppression of *Maconellicoccus hirsutus* (Green) on sapota. *Journal of Biological Control* 22(2), 471–473.

Manjunath, T.M., Bhatnagar, V.S., Pawar, C.S. and Sithanantham, S. (1989) Economic importance of Heliothis spp. in India and an assessment of their natural enemies and host plants. In: *Proceedings of the Workshop on Biological Control of Heliothis: Increasing the effectiveness of natural enemies*, pp. 197–228.

Manners, A.G., Dembowski, B.R. and Healey, M.A. (2013) Biological control of thrips in cut flowers. *Australian Journal of Entomology* 52, 246–258. DOI: 10.1111/aen.12020.

Maris, P.C., Joosten, N.N., Goldbach, R.W. and Peters, D. (2004) Tomato spotted wilt virus infection improves host suitability for its vector *Frankliniella occidentalis*. *Phytopathology* 94(7), 706–711. DOI: 10.1094/PHYTO.2004.94.7.706.

Matsuda, K. and Ichihara, M. (2022) Species composition and seasonal abundance of thrips (Thysanoptera: Thripidae) on open field chrysanthemum (Asteraceae) and surrounding weeds in Shizuoka Prefecture, Central Japan. *Japan Agricultural Research Quarterly (JARQ)* 56(1), 49–58. DOI: 10.6090/jarq.56.49.

Messelink, G.J. and de Kogel, W.J. (2013) *Een systeembenadering voor onderzoek aan trips-bestrijding in de sierteelt onder glas: Een visiedocument vanuit onderzoek en praktijk.* Wageningen UR Glastuinbouw, Bleiswijk, The Netherlands.

Minkenberg, O.P.J.M. and van Lenteren, J.C. (1986) The leafminers Liriomyza bryoniae and L. trifolii (Diptera: Agromyzidae), their parasites and host plants: A review. *Agricultural University of Wageningen Papers* 86(2), 1–50.

Mohsin, S.B., Li, Y., Tang, L., Maqsood, I., Ma, T. *et al.* (2015) Predatory efficacy of cotton-inhabiting spiders on *Bemisia tabaci*, *Amrasca devastans*, *Thrips tabaci* and *Helicoverpa armigera* in laboratory conditions. *Journal of Northeast Agricultural University* 22(3), 48–53. DOI: 10.1016/S1006-8104(16)30006-X.

Moore, S.D., Pittaway, T., Bouwer, G. and Fourie, J.G. (2004) Evaluation of *Helicoverpa armigera* nucleo polyhedron virus (HearNPV) for control of *Helicoverpa armigera* (Lepidoptera: Noctuidae) on citrus in South Africa. *Biocontrol Science and Technology* 14, 239–250.

Moritz, G., Kumm, S. and Mound, L. (2004) Tospovirus transmission depends on thrips ontogeny. *Virus Research* 100(1), 143–149. DOI: 10.1016/j.virusres.2003.12.022.

Morse, J.G. and Hoddle, M.S. (2006) Invasion biology of thrips. *Annual Review of Entomology* 51, 67–89. DOI: 10.1146/annurev.ento.51.110104.151044.

Mouden, S., Sarmiento, K.F., Klinkhamer, P.G. and Leiss, K.A. (2017) Integrated pest management in western flower Thrips: Past, present and future. *Pest Management Science* 73(5), 813–822. DOI: 10.1002/ps.4531.

Mound, L.A. (2005) THYSANOPTERA: Diversity and interactions. *Annual Review of Entomology* 50(1), 247–269. DOI: 10.1146/annurev.ento.49.061802.123318.

Mound, L.A. and Kibby, G. (1998) *Thysanoptera: An Identification Guide*, 2nd ed. CAB International, Wallingford, UK.

Mujica, N., Khadioli, N., Le Ru, B., Ong'amo, G. and Carhuapoma, P. (2016) American serpentine leafminer, *Liriomyza trifolii*. In: Kroschel, J., Mujica, N., Carhuapoma, P. and Sporleder, M. (eds) *Pest Distribution and Risk Atlas for Africa: Potential Global and Regional Distribution and Abundance of Agricultural and Horticultural Pests and Associated Biocontrol Agents Under Current and Future Climates*. International Potato Center (CIP), La Molina, Peru, pp. 139–153. DOI: 10.4160/9789290604761-11.

Nair, M.R.G.K. (1975). Insects and Mites of Crops in India. Indian Council of Agricultural Research (ICAR), New Delhi. pp. 404.

Nguyen, T.H.N., Borgemeister, J. and Poehling, H.M. (2009) Manipulation of ultraviolet light affects immigration behavior of *Ceratothripoides claratris* (Thysanoptera: Thripidae). *Journal of Economic Entomology* 102(4), 1559–1566.

Ohta, I. (2002) Host plant resistance in Japanese chrysanthemums against *Frankliniella occidentalis* (Thysanoptera: Thripidae) during the non-flowering stage. *Applied Entomology and Zoology* 37(2), 271–277.

Ondiaka, S., Migiro, L., Rur, M., Birgersson, G., Porcel, M. *et al.* (2016) Sunflower as a trap crop for the European tarnished plant bug (*Lygus rugulipennis*). *Journal of Applied Entomology* 140, 453–461. DOI: 10.1111/jen.12273.

Persad, A. and Khan, A. (2000) The effect of five insecticides on *Maconellicoccus hirsutus* (Green) (Homoptera: Pseudococcidae) and its natural enemies *Anagyrus kamali* Moursi (Hymenoptera: Encyrtidae), and *Cryptolaemus montrouzieri* Mulsant and *Scymnus coccivora* Aiyar (Coleoptera: Coccinellidae). *International Pest Control* 42(5), 170–173.

Puspitarini, R.D., Fernando, I., Rachmawati, R., Hadi, M.S. and Rizali, A. (2021) Host plant variability affects the development and reproduction of *Tetranychus urticae*. *International Journal of Acarology* 47(5), 381–386. DOI: 10.1080/01647954.2021.1915377.

Riaz, A., Khan, M. A., Malik, M. F. and Ullah, F. (2021). Functional response and predation rate of *Chrysoperla carnea* on aphids under laboratory conditions. *Journal of Insect Science*, 21(3), 1–7.

Rajapakse, C.N.K. and Walter, G.H. (2007) Polyphagy and primary host plants: Oviposition preference versus larval performance in the lepidopteran pest *Helicoverpa armigera*. *Arthropod-Plant Interactions* 1, 17–26. DOI: 10.1007/s11829-007-9003-6.

Ramirez, A.M., Stoopen, G., Menzel, T.R., Gols, R., Bouwmeester, H.J. *et al.* (2012) Bidirectional secretions from glandular trichomes of pyrethrum enable immunization of seedlings. *The Plant Cell* 24(10), 4252–4265.

Reddy, P.P. (2010) *Plant Protection in Horticulture*. Scientific Publishers, Jodhpur, India.

Reddy, G.V.P. and Manjunatha, M. (2000) Laboratory and field studies on the integrated pest management of *Helicoverpa armigera* (Hübner) in cotton, based on pheromone trap catch threshold level. *Journal of Applied Entomology* 124(5–6), 213–221.

Reddy, P. and Janakiram, T. (2004) Germplasm evaluation of chrysanthemum for resistance to aphid, *Macrosiphoniella sanbornii* (Gillette). *Journal of Ornamental Horticulture* 7(3–4), 27–31.

Reitz, S.R. (2002) Seasonal and within-plant distribution of Frankliniella thrips (Thysanoptera: Thripidae) in north Florida tomatoes. *Florida Entomologist* 85(4), 431–439. DOI: 10.1653/0015-4040(2002)085[0431:SAWPDO]2.0.CO;2.

Scopes, N.E.A. (1969) The potential of chrysopa carnea as a biological control agent of myzus persicae on glasshouse chrysanthemums. *Annals of Applied Biology* 64(3), 433–439. DOI: 10.1111/j.1744-7348.1969.tb02892.x.

Scopes, N.E.A. and Biggerstaff, S.M. (1973) Progress towards integrated pest control on year round chrysanthemum. In: *Proceedings of the Seventh British Insecticide and Fungicide Conference*, November 19–22, pp. 227–234.

Sevacherian, V. and Stern, V.M. (1974) Host plant preferences of lygus bugs in alfalfa-interplanted cotton fields. *Environmental Entomology* 3, 761–766.

Shah, K.D., Jhala, R.C. and Dhandge, S.R. (2017) Standardization of pheromone traps for the mass trapping of *Helicoverpa armigera* (Hübner) Hardwick in tomato. *Current Agriculture Research Journal* 5, 45–49.

Sharma, H.C. (2001) Cotton bollworm/legume pod borer, *Helicoverpa armigera* (Hübner) (Noctuidae: Lepidoptera): biology and management. In: *Crop Protection Compendium*. International Crops Research Institute for the Semi-Arid Tropics (ICRISAT), Hyderabad, India.

Sreedhar, M., Vasudha, A. and Khudus, S. (2020) Insect-pests complex studies on chrysanthemum in Pantnagar region. *Journal of Entomology and Zoology Studies* 8, 1644–1646.

Stavrinides, M. and Skirvin, D. (2003) The effect of *Chrysanthemum* leaf trichome density and prey spatial distribution on predation of *Tetranychus urticae* (Acari: Tetranychidae) by *Phytoseiulus persimilis* (Acari: Phytoseiidae). *Bulletin of Entomological Research* 93(4), 343–350.

Storer, J. and Van Emden, H. (1995) Antibiosis and antixenosis of *Chrysanthemum* cultivars to the aphid *Aphis gossypii*. *Entomologia Experimentalis et Applicata* 77(3), 307–314.

Su, J., Jiang, J., Zhang, F., Liu, Y., Ding, L. *et al.* (2019) Current achievements and future prospects in the genetic breeding of *Chrysanthemum*: A review. *Horticulture Research* 6, 109. DOI: 10.1038/s41438-019-0193-8.

Suleimen, E., Sisengalieva, G., Adilkhanova, A., Dudkin, R., Gorovoi, P. *et al.* (2019) Composition and biological activity of essential oil from *Artemisia keiskeana*. *Chemistry of Natural Compounds* 55, 154–156.

Surani, P.M., Shukla, A., Chawla, S. and Trivedi, S. (2022) Screening of different *Chrysanthemum* cultivars against aphid, *Macrosiphoniella sanborni* (Gillette). *Progressive Horticulture* 54(1), 88–91.

Svensson, L. (2025) An insect screen that offers effective thrip control without compromising ventilation. Available at: https://www.ludvigsvensson.com (accessed 25 February 2025).

Swezey, S.L., Nieto, D.J., Hagler, J.R., Pickett, C.H., Bryer, J.A. *et al.* (2013) Dispersion, distribution, and movement of *Lygus* spp. (Hemiptera: Miridae) in trap-cropped organic strawberries. *Environmental Entomology* 42, 770–778.

Thorat, S., Sisodiya, D. and Gangwar, R. (2022) Invasive thrips, *Thrips parvispinus* (Karny) an invasive threat: A review. *Environment and Ecology* 40(4A), 2170–2175.

Tsao, R., Marvin, C.H., Broadbent, A.B., Friesen, M., Allen, W.R. *et al.* (2005) Evidence for an isobutylamide associated with host-plant resistance to western flower thrips, *Frankliniella occidentalis*, in *Chrysanthemum*. *Journal of Chemical Ecology* 31, 103–110.

van den Berg, H., Waage, J.K. and Cock, M.J.W. (1988) Natural Enemies of *Helicoverpa armigera* in Africa: A Review. CAB International and Institute of Biological Control, Ascot, UK.

van Tol, R.W., Diaz Rodriguez, C.M., de Bruin, A., Yang, D., Taparia, T. *et al.* (2022) Visual attraction of the European tarnished plant bug *Lygus rugulipennis* (Hemiptera: Miridae) to a water trap with LED light in *Chrysanthemum* greenhouses and olfactory attraction to novel compounds in Y-tube tests. *Pest Management Science* 78(6), 2523–2533. DOI: 10.1002/ps.6881.

Virk, J.S., Brar, K.S. and Sohi, A.S. (2004) Role of trap crops in increasing parasitation efficiency of *Trichogramma chilonis* Ishii in cotton. *Journal of Biological Control* 18, 61–64.

Watson, G.W. (2022). *Maconellicoccus hirsutus* (pink hibiscus mealybug). In: CABI Invasive Species Compendium. CAB International, Wallingford, UK. Available at: https://www.cabi.org/isc

Wardlow, L.R. (1986) Adapting integrated pest control to work for ornamentals. *Grower* 6, 26–29.

Waterhouse, D.F. and Norris, K.R. (eds) (1989) *Frankliniella occidentalis* (Pergande) and Thrips tabaci Lindeman. In: *Biological Control: Pacific Prospects, Supplement 1*. Inkata Press, Melbourne, Victoria, pp. 25–35.

Wijkamp, I., Goldbach, R. and Peters, D. (1996) Propagation of tomato spotted wilt virus in *Frankliniella occidentalis* does neither result in pathological effects nor in transovarial passage of the virus. *Entomologia Experimentalis et Applicata* 81, 285–292. DOI: 10.1046/j.1570-7458.1996.00098.x.

Williams, D.J. (1996) A brief account of the hibiscus mealybug *Maconellicoccus hirsutus* (Hemiptera: Pseudococcidae), a pest of agriculture and horticulture, with descriptions of two related species from southern Asia. *Bulletin of Entomological Research* 86, 617–628.

Woelke, J.B., Bouw, M., Cusumano, A. and Messelink, G.J. (2023) *Lygus rugulipennis* on *Chrysanthemum*: Supplemental prey effects and an evaluation of trap plants. *Journal of Applied Entomology* 147, 157–166. DOI: 10.1111/jen.13093.

Wyatt, I. (1969) Factors affecting aphid infestation of *Chrysanthemum*. *Annals of Applied Biology* 63(2), 331–337.

Xia, C., Xue, W., Li, Z., Shi, J., Yu, G. *et al.* (2023) Presenting the secrets: Exploring endogenous defense mechanisms in *Chrysanthemum* against aphids. *Horticulturae* 9(8), 937.

Xu, M., Jiang, Y., Chen, S., Chen, F. and Chen, F. (2021) Herbivory-induced emission of volatile terpenes in *Chrysanthemum morifolium* functions as an indirect defense against *Spodoptera litura* larvae by attracting natural enemies. *Journal of Agricultural and Food Chemistry* 69, 9743–9753. DOI: 10.1021/acs.jafc.1c02637.

Yang, T., Stoopen, G., Wiegers, G., Mao, J., Wang, C. *et al.* (2012) Pyrethrins protect pyrethrum leaves against attack by western flower thrips, *Frankliniella occidentalis*. *Journal of Chemical Ecology* 38(3), 370–377. DOI: 10.1007/s10886-012-0097-7.

Yoshida, H. and Parrella, M. (1991) *Chrysanthemum* cultivar preferences exhibited by *Spodoptera exigua* (Lepidoptera: Noctuidae). *Environmental Entomology* 20(1), 160–165.

Yudin, L.S., Cho, J.J. and Mitchell, W.C. (1986) Host range of western flower thrips, *Frankliniella occidentalis* (Thysanoptera: Thripidae), with special reference to *Leucaena glauca*. *Environmental Entomology* 15(6), 1292–1295. DOI: 10.1093/ee/15.6.1292.

Zahedi, K. (1999) *Summer Crops and Ornamental Plants Pests and Control in Iran*. Iran University Press, Tehran, Iran.

Zhong, J., Guo, Y., Shi, H., Liang, Y., Guo, Z. *et al.* (2022) Volatiles mediated an eco-friendly aphid control strategy of *Chrysanthemum* genus. *Industrial Crops and Products* 180, 114734. DOI: 10.1016/j.indcrop.2022.114734.

10

Chrysanthemum Diseases and Their Management

Margery L. Daughtrey*

Long Island Horticultural Research and Extension Center, College of Agriculture and Life Sciences, Cornell University, 3059 Sound Ave., Riverhead, NY 11901, USA

Abstract

Chrysanthemum (*Chrysanthemum morifolium*), a member of the *Asteraceae* family, stands as one of the world's most beloved flowers. However, its widespread cultivation and exchange of vegetative propagation materials have inadvertently facilitated the spread of devastating diseases caused by various fungal pathogens such as Ascochyta, Septoria, and Verticillium. These diseases pose significant threats to the chrysanthemum industry, often resulting in severe symptoms and substantial economic losses. Beyond fungal pathogens, chrysanthemums are susceptible to a spectrum of diseases caused by oomycetes, bacteria, viruses, viroids, and nematodes, alongside abiotic stresses that impact both production and product quality. This chapter aims to provide a comprehensive overview of the primary diseases encountered during chrysanthemum production, focusing on their sources, spread, and life cycles. Detailed descriptions of symptoms, diagnostic techniques, and integrated disease management practices are also included. The diseases discussed herein play a critical role in the production of cut, pot, and garden mums, underlining the importance of proactive disease management strategies in ensuring chrysanthemum health and productivity.

Keywords

Botrytis cinerea, chrysanthemum, fungal and bacterial diseases, *Fusarium oxysporum*, phytoplasmas, viroids

*Corresponding author: mld9@cornell.edu

DOI: 10.1079/9781789248289.0010

Fig. 10.1. Chrysanthemums affected by *Phytophthora chrysanthemi* in an extensive landscape planting (A). Close-up of root and stem rot caused by *Phytophthora chrysanthemi* (B). (Photo courtesy of Francesca Peduto Hand.)

10.1 Introduction

Centuries of cultivation have resulted in the development of hundreds of chrysanthemum cultivars, which generally grow vigorously and without obvious weaknesses. The exceptions will be covered here, as we discuss the major problems caused by oomycetes, fungi, bacteria, viruses, viroids, and nematodes on chrysanthemum, as well as some of the abiotic stresses that may cause symptoms of ill health. Global trade has increased the spread of pathogens, so growers of chrysanthemums may at any time encounter problems outside of their previous experience. This chapter is intended to forewarn chrysanthemum producers of diseases that they should be on guard against and to remind the industry of the seriousness of diseases against which today's clean stock production systems and integrated pest management procedures are essential protection.

10.2 Fungal and Fungal-Like Diseases

10.2.1 *Phytophthora* root, stem, and crown rot

Chrysanthemum crops are not often plagued by *Phytophthora* but *P. drechsleri* and *P. chrysanthemi* were noted when Pennsylvania isolates 1975–2019 were analyzed (Lin *et al.*, 2017; Molnar *et al.*, 2020). *Phytophthora chrysanthemi*, first described from Japan (Naher *et al.*, 2011) on cut flowers and in ebb–flood hydroponic production, appears to be an emerging disease (Fig. 10.1). It has been reported from nursery plants or cut flowers in Germany, Croatia, and Ohio (Götz *et al.*, 2017; Werres *et al.*, 2017). There are also records of *P. cryptogea* and *P. nicotianae* on chrysanthemum (Farr and Rossman, 2019).

Symptoms and diagnosis: *Phytophthora* spp. can cause root rot, stunting, and off-color to necrotic foliage (Fig. 10.1). Both *P. cryptogea* and *P. nicotianae* infect roots, stems, and leaves, while *P. drechsleri* is primarily a root rot pathogen; all have a broad host range (Farr and Rossman, 2019). *P. chrysanthemi* causes root rot and stem blackening (Naher *et al.*, 2011). Identification of the better-known *Phytophthora* pathogens may be made through polymerase chain reaction (PCR), sequencing, or morphological examination. Identification of *P. chrysanthemi* may be hindered because the pathogen grows slowly and may degenerate in culture within a few weeks (Werres *et al.*, 2017). Using pimaricin, ampicillin, rifampicin, and pentachloronitrobenzene (PARP) (a selective medium) will help with initial isolation. Additionally, another method to detect the presence of *P. chrysanthemi* in soil involves using leaves as bait (Randall-Schadel, 2016). Morphological examination and sequencing may be necessary to identify *P. chrysanthemi*.

Source, spread, and life history: Sources for the cosmopolitan *Phytophthora drechsleri, P. cryptogea*, and *P. nicotianae* include other ornamentals grown in the same facilities as well as chrysanthemum cuttings. For *P. chrysanthemi*, the only known host is *Chrysanthemum*, so cuttings are the likely source. As oomycetes, *Phytophthora* pathogens will easily be moved by splashing or spread within puddles or by water flow. *Phytophthora chrysanthemi* is a high-temperature-tolerant species, tolerating up to 32°C (Werres *et al.*, 2017), similar to *P. nicotianae*.

Management: *Phytophthora* disease management centers on the exclusion of the pathogen. Growing mixes with good drainage, outdoor beds crowned to shed water, irrigation timed for early in the day will all help. Fungicides effective against *Pythium* are effective against *Phytophthora* as well. See Table 10.1 for a list of some active ingredients that may be used against *Phytophthora* root and crown rot (Daughtrey *et al.*, 2023).

10.2.2 Pythium root rot

Container production is often plagued by root rot caused by oomycetes (water molds) in the genera *Pythium, Phytopythium*, and *Globisporangium*. Pythium root rot pathogens from chrysanthemum include the *Pythium* species *P. aristosporum, P. carolinianum*, and *P. dissotocum*; the *Phytopythium* species *P. helicoides* and *P. oedochilum*; and the *Globisporangium* species *G. buismaniae, G. debaryanum, G. megalacanthum, G. mamillatum, G. rostratum, G. sylvaticum*, and *G. ultimum* (Trolinger *et al.*, 2018; Farr and Rossman, 2019). In garden mum cultivation in the USA, the species *P. aphanidermatum* is the most common.

Symptoms and diagnosis: Plants with Pythium root rot show stunting, wilting (Fig. 10.2), interveinal chlorosis, and leaf scorch. The entire root system may be involved, and disease may progress into the stem base (Fig. 10.2). Infected stems are brown to black and water soaked. Disease is likely in areas where soil is compacted and thus drains slowly. Initially, foliage takes on a grayish cast. Roots show grayish-brown discoloration and are softened. Because the pathogen rots the root cortex, only the central, xylem core remains when a plant is pulled from its pot. When temperatures are cooler

Table 10.1. Chemical and biological management tools for *Phytophthora* diseases. Author's own table.

FRAC[a] Groups	Some examples of active ingredients
4	Mefenoxam
11	Fluoxastrobin, pyraclostrobin, trifloxystrobin
14	Etridiazole
21	Cyazofamid
28	Propamocarb
P 07	Fosetyl-Al, mono- and di-potassium salts of phosphorous acid, phosphite
40	Dimethomorph, mandipropamid
43	Fluopicolide
BM 02	*Bacillus amyloliquefaciens, B. subtilis*
49	Oxathiapiprolin
M 03 + M 05	Mancozeb + copper hydroxide
BM 02	*Gliocladium catenulatum, Pseudomonas chlororaphis, Streptomyces* sp. K61, *Streptomyces lydicus, Trichoderma asperellum, T. gamsii*

[a]FRAC stands for Fungicide Resistance Action Committee. FRAC Groups indicate modes of action; see https://www.frac.info. M indicates chemical multi-site inhibitors; 'BM' indicates biologicals having multiple modes of action, thus not prone to development of resistance in the target pest population.

(10–16°C), nutrient deficiency symptoms may be the only clue that *Pythium* is present, whereas with warmer temperatures (27–32°C), plants are apt to show more dramatic symptoms (Trolinger *et al.*, 2018). Roots may be examined with a microscope for oospores or surface sterilized and cultured on agar media. Diagnosis is best done by careful inspection of roots and culturing of the pathogen; test strips that use serology to test for *Pythium* are not helpful as there are many *Pythium* spp. in soil that are not pathogens.

Source, spread, and life history: Field soil may be a source of *Pythium* and related species. Rooted cuttings may also arrive infected. Movement of water and soil by irrigation or rainfall may spread these pathogens. The microorganisms causing Pythium root rot are water molds, or oomycetes, and thus not fungi. The life cycles of *Pythium*, *Globisporangium*, and *Phytopythium* are closely tied to water. Sporangia, produced asexually, release single-celled motile zoospores that infect roots when soil is flooded. These water molds also form oospores (sexual spores) and asexual chlamydospores that survive for long periods in plant debris.

Management: It is important to remove crop debris and disinfect growing surfaces in the greenhouse to avoid Pythium root rot. Cut flowers in ground

Fig. 10.2. Root and stem rot caused by *Pythium aphanidermatum*, showing a close-up of black stem discoloration (A) and wilting plants due to root infection (B). (Photo courtesy of Margery Daughtrey.)

beds are less susceptible to Pythium root rot than container-grown plants. Overfertilization or poorly drained soil or artificial growing media provide an environment conducive to Pythium root rot. Because these pathogens are so common, growers often make preventive fungicide drenches on highly susceptible crops such as chrysanthemum. Fungicides that protect against *Phytophthora* are not necessarily effective against *Pythium*. Active ingredients with effectiveness against *Pythium* and its close relatives include cyazofamid, etridiazole, fosetyl-Al, mefenoxam, phosphites, and propamocarb (Daughtrey *et al.*, 2023). Growers should rotate among effective materials and avoid repeated use of one active ingredient with a single-site mode of action, that is, materials from one FRAC (Fungicide Resistance Action Committee) group. Resistance to mefenoxam has been well documented for greenhouse populations of *Pythium*, so rotating applications with active ingredients from other FRAC Groups is necessary. Treatments are appropriate on rooted cuttings a week after sticking and at transplanting (Trolinger *et al.*, 2018). Repeated treatments at recommended intervals may be needed to ensure root health. Biological controls used preventively are frequently employed to keep Pythium root rot problems in check. See Table 10.2 for a list of some of the FRAC Groups and active ingredients that may be used against Pythium root rot (Daughtrey *et al.*, 2023).

10.2.3 Fungal leaf and flower blights

The major leaf and flower blights on chrysanthemum include Alternaria ray speck and leaf spot (*Teretispora leucanthemi*, syn. *Alternaria chrysanthemi*, and other *Alternaria* spp.); Ascochyta ray blight (*Stagonosporopsis chrysanthemi*, syn.

Table 10.2. Chemical and biological management tools for Pythium root rot. Author's own table.

FRAC Group[a]	Some examples of active ingredients
4	Mefenoxam
14	Etriadiazole
21	Cyazofamid
28	Propamocarb
P 07	Fosetyl-Al, mono- and di-potassium salts of phosphorus acid, phosphite
43	Fluopicolide
BM 02	*Bacillus amyloliquefaciens, B. subtilis*
BM 02	*Bacillus pumilus, Gliocladium catenulatum, Pseudomonas chlororaphis, Streptomyces* sp. K61, *Streptomyces lydicus, Trichoderma asperellum, T. gamsii, T. harzianum*

[a]FRAC' stands for Fungicide Resistance Action Committee. FRAC Groups indicate modes of action; see https://www.frac.info. 'BM' indicates biologicals having multiple modes of action, thus not prone to development of resistance in the target pest population.

Mycosphaerella ligulicola and *Ascochyta chrysanthemi*); Botrytis blight (*Botrytis cinerea*); Itersonilia petal blight (*Itersonilia perplexans*); and Stemphylium ray speck (*Stemphylium lycopersici*, primarily). These diseases are not major production problems today because of sanitation practices and fungicide use; they are most often encountered on crops grown outdoors.

Symptoms and diagnosis: On leaves or flower petals, spots often begin as light yellow or tan flecks and may progress to blight (Fig. 10.3). Holding symptomatic tissue in a plastic bag for 24–48 hours with a piece of damp paper toweling allows sporulation to develop that aids identification.

- Stemphylium ray speck is an exception: surface-sterilize small pieces of spotted tissue and place on V8 juice agar at 25°C (77°F) to identify *S. lycopersici* (Nishi *et al.*, 2009).
- Ascochyta ray blight infects leaves, stems, buds, flowers, and even roots. Flowers are often attacked on one side so that petals do not develop symmetrically and petals may fall off easily if flowers are shaken (Horst and Nelson, 1997). Spore cases (pycnidia) form on colonized tissue.
- Botrytis blight leaf lesions are generally large, irregular, brown and exhibit a zonate pattern; the fungus also produces characteristic "gray mold" sporulation. The major damage is to flowers, either pre- or postharvest, causing a brown petal blight (Fig. 10.3).
- Itersonilia petal blight affects only flowers (Horst and Nelson, 1997).
- Stemphylium ray speck and Alternaria ray speck appear very similar to one another, so sporulation should be examined to distinguish them.

Fig. 10.3. Flower blight caused by *Botrytis cinerea* (A, B). Purple-bordered leaf spots caused by *Cercospora chrysanthemi* on chrysanthemum (C). (Photo courtesy of (A) Elise Lobdell, (B) Malleshaiah SharathKumar, and (C) Margery Daughtrey.)

Source, Spread and life history: These pathogens are spread by wind, splashing water, or both. In all cases, cuttings may introduce the pathogens, and infested plant debris allows the fungi to persist.

- Ascochyta ray blight infections begin from ascospores released from overwintered perithecia. Pycnidia develop best at 27°C (80°F) under moist conditions (Horst and Nelson, 1997). The fungus causing Ascochyta ray blight is limited to chrysanthemum and pyrethrum as hosts (Vaghefi *et al.*, 2016).
- Botrytis blight may be initiated when the plant surface is wet for 5 hours (Jewett and Jarvis, 2001). Latent infections initiated during production may result in postharvest disease. Botrytis blight is favored by conditions allowing condensation on plant surfaces and temperatures of 10–21°C (Horst and Nelson, 1997).
- For Itersonilia petal blight, a wetness period of 8 hours is needed for infection. A cool-weather pathogen, *I. perplexans* has an 18°C optimum for infection. Lesions develop fastest at 10–15°C, and high humidity fosters larger lesions (Trolinger *et al.*, 2018). The fungus can infect members of the Apiaceae and also members of the *Asteraceae* (McGovern *et al.*, 2006; Farr and Rossman, 2019).
- *Stemphylium lycopersici* has a wide host range (Farr and Rossman, 2019); it grows over a range of 10–30°C, and optimally at 25°C on V8 agar (Nishi *et al.*, 2009).

Management: To minimize leaf and flower blights, inspect cuttings and rogue out those with leaf spots. Regularly remove unharvestable flower heads. Avoid overhead irrigation or water early in the day to allow foliage to dry before nightfall, and do not overcrowd plants. Fungicides and/or biocontrols may be used to protect plants if a low level of symptoms is detected. Leaf and flower diseases often occur in combination, and thus materials with broad-spectrum activity are beneficial (Singh and Milne, 1974).

- For Alternaria blight, some materials in FRAC Groups 2, 3, 9, 11, 12, 19, 44, M1, M3, and M5, as well as cold-pressed neem, *Swinglea glutinosa*, and some *Streptomyces* and *Gliocladium* biocontrols are options for management (Daughtrey *et al.*, 2023).
- For Ascochyta ray blight, propiconazole (FRAC Group 3) may be used, as well as materials in FRAC Groups 1, 11, M3, and M5 and some biological controls (Daughtrey *et al.*, 2023).
- With chemical control of Botrytis blight, pathogen resistance can be a particular problem, as resistance has developed to benzimidazole and dicarboximide fungicides (LaMondia and Douglas, 1997; Yourman and Jeffers, 1999). Rotation of applications among different FRAC Groups (Table 10.3) is recommended to slow the development of resistance to the newer systemic chemicals with a single-site mode of action (Daughtrey *et al.*, 2023).
- For Itersonilia petal blight, myclobutanil and propiconazole (FRAC Group 3) and potassium bicarbonate (FRAC Group NC) are effective and superior to azoxystrobin (FRAC Group 11) treatment (McGovern *et al.*, 2006).
- Stemphylium ray blight is likely to be responsive to FRAC Groups 3 and 7, based on experiments on similar diseases in vegetables. It was reduced by chlorothalonil (FRAC Group M5) along with other flower blights in a New Zealand field study (Singh and Milne, 1974).

Table 10.3. Chemical and biological management tools for Botrytis blight. Author's own table.

FRAC[a] Groups	Some examples of active ingredients
1	Thiophanate-methyl (note widespread resistance to Group 1)
2	Iprodione (note widespread resistance to Group 2)
7	Boscalid, isofetamid
9	Cyprodinil
11	Azoxystrobin, fluoxastrobin, pyraclostrobin, trifloxystrobin
12	Fludioxonil
17	Fenhexamid
19	Polyoxin D, zinc salt
M 01	Copper hydroxide, copper octanoate, basic copper sulfate, copper sulfate pentahydrate
M 03	Mancozeb
M 05	Chlorothalonil
BM 01	BLAD
BM 02	*Gliocladium catenulatum (Clonostachys rosea), neem, cold-pressed, Pseudomonas chlororaphis, Streptomyces* sp. K61*, Streptomyces lydicus, Ulocladium oudemansii Bacillus amyloliquefaciens, B. subtilis Bacillus amyloliquefaciens, B. subtilis*
P4	Laminarin

[a]'FRAC' stands for Fungicide Resistance Action Committee. FRAC Groups indicate modes of action; see https://www.frac.info. M indicates chemical multi-site inhibitors; 'BM' indicates biologicals having multiple modes of action, thus not prone to development of resistance in the target pest population.

10.2.4 Fungal leaf spots

Cercospora leaf spot caused by *Cercospora chrysanthemi* and Septoria leaf spot caused by any of six species of *Septoria* are the most likely leaf spot diseases on chrysanthemum. These two diseases could be confused with some of the flower blights, particularly Alternaria, Ascochyta, and Botrytis blight, as these three diseases also have large leaf spots. The most common Septoria pathogen is designated *S. chrysanthemella*, which is a species complex comprising several cryptic species that need further study. *Septoria chrysanthemella* has been reported from all over the world (Farr and Rossman, 2019). Septoria and Cercospora leaf spots are seen only rarely now in temperate climates.

Symptoms and diagnosis: Cercospora leaf spot causes round to irregular brown lesions that turn gray and may show a raised purple border (Fig. 10.4). The lesions may coalesce. Septoria leaf spot caused by *S. chrysanthemella* is characterized by round-to-oval, dark brown to black leaf spots and defoliation; spots are 5–10 mm in diameter and may have a thin chlorotic

Fig. 10.4. Fusarium wilt of chrysanthemum, caused by *Fusarium oxysporum* f. sp. *chrysanthemi*. Wilting, chlorosis, and necrosis (A). Close-up of purple discoloration in the vascular system (B). (Photo courtesy of Margery Daughtrey.)

Table 10.4. Chemical and biological management tools for Septoria and Cercospora leaf spots. Author's own table.

FRAC[a] Groups	Some examples of active ingredients
1	Thiophanate-methyl
2	Iprodione
3	Propiconazole
9	Cyprodinil
11	Azoxystrobin, trifloxystrobin
12	Fludioxonil
BM 02	*Bacillus amyloliquefaciens, B. subtilis*
M 01	Copper hydroxide
M 03	Mancozeb
M 05	Chlorothalonil
P 04	Laminarin
P 05	*Reynoutria sachalinensis* extract

[a]'FRAC' stands for Fungicide Resistance Action Committee. FRAC Groups indicate modes of action; see https://www.frac.info. M indicates chemical multi-site inhibitors; 'BM' indicates biologicals having multiple modes of action, thus not prone to development of resistance in the target pest population.

halo (Waddell and Weber, 1963; Punithalingam and Wheeler, 1965). Flower quality and yield can be reduced (Alfieri, 1968). The pathogens are identified to particular genus by microscopy, observing the conidiophores (*Cercospora*) or pycnidia (*Septoria*) on the surface of leaf lesions and the characteristic sporulation. Speciation may be done by molecular sequencing.

Source, spread, and life history: Septoria leaf spot inoculum is introduced via cuttings but also might originate from other weeds or crops in the *Asteraceae* or *Dipsacaceae*. Cercospora leaf spot inoculum originates from cuttings or other members of *Asteraceae* (Farr and Rossman, 2019). Both pathogens are spread by splashing, wind, and handling. Conidiophores of *C. chrysanthemi* develop on both sides of leaf spots and produce narrow, elongated, hyaline, multiseptate spores under humid conditions. *S. chrysanthemella* forms tiny brown to black pycnidia, usually on the upper surface of leaf lesions. The conidia are hyaline and filiform, 36–65 µm long and 1.5–2.5 µm wide, with four to nine septations (Punithalingam and Wheeler, 1965). The fungus survives for 4 months in leaf debris (Alfieri, 1968).

Management: Inspect cuttings for symptoms, discard symptomatic cuttings, and begin a protective fungicide program (Table 10.4) (Daughtrey *et al.*, 2023). Avoid using overhead irrigation or water early in the day so that foliage dries before nightfall.

10.2.5 Fusarium wilt

Vascular wilt caused by *Fusarium oxysporum* f. sp. *chrysanthemi* continues to be a problem. The pathogen has been found on Marguerite daisy, osteospermum, and gerbera as well as *Rudbeckia fulgida* (Matic *et al.*, 2018). *F. oxysporum* f. sp. *tracheiphilum*, primarily a pathogen of soybean and cowpea, was also recorded as a source of wilt in chrysanthemum in North Carolina (Farr and Rossman, 2019). At present, *Fusarium oxysporum* is considered a complex of organisms not easily separated into species; there is no formal acceptance of the names of these formae speciales.

Symptoms and diagnosis: Fusarium wilt can kill plants of many chrysanthemum cultivars, while other cultivars may show no symptoms. Persistent symptoms of water stress will develop in susceptible plants. Symptoms likely include stunting, wilting, chlorosis, and leaf scorch (Fig. 10.4). The progression of wilt is systemic via vascular tissue; thus, symptoms are in some cases one-sided, affecting one side of a plant with leaf epinasty, yellowing, and eventually branch wilt. When wilting is seen, inspect the stem base for insect chewing injury and the roots for root rot. Laboratory diagnosis is needed to distinguish Fusarium wilt from Verticillium wilt or systemic bacterial diseases. To prevent such diseases, mother stock should be repeatedly culture-indexed and tested for vascular pathogens such as Fusarium wilt. This has been done by culturing the base of cuttings in an agar broth to inspect for contamination with bacterial or fungal pathogens (culture-indexing) (Horst and Nelson, 1997); today serological testing and PCR of mother stock are also used (Copes, 2018).

Source, spread, and life history: Fusarium wilt is introduced primarily via latently infected cuttings, but sometimes previous outbreaks may have left inoculum on-site. There is minimal chance of plant-to-plant spread of inoculum during production, but diseased plants should be removed once detected. The disease has a long development time that is temperature dependent. Sporulation may form aboveground on infected stems. The disease often starts when cuttings are taken from an infected mother plant, which may be growing under cooler conditions than optimum for disease development. Symptoms appear primarily during summer months when high temperatures favor the pathogen. Wilting is common at 32°C (Gardiner *et al.*, 1987). Two spore types are distinguished: single-celled microconidia and multicelled spores in a crescent-moon shape (macroconidia). These are produced loosely on the hyphae, within the xylem vessels. Resting spores (chlamydospores) are also produced and linger in the plant debris present in soil.

Management: Specialist propagators aim to sell disease-free cuttings. Growers should start with clean stock in soil or media free from the pathogen. An integrated program combining high calcium, fungicide, and fertilization with a nitrate rather than an ammonium nitrogen source was effective for Fusarium wilt management (Engelhard and Woltz, 1973). In climates without a winter freeze, Fusarium wilt may return season after season, and soil disinfestation will be needed. Ground beds should be treated between crop cycles to eliminate inoculum of *F. oxysporum*. Methyl bromide + chloropicrin fumigation was once standard practice; now it has been replaced by alternative fumigants, anaerobic soil disinfestation, solarization, or steam (Trolinger *et al.*, 2018). Treatment with fungicides alone has not been sufficient for the control of Fusarium wilt but is helpful when combined with other practices. Fungicides highly effective against *Fusarium* include azoxystrobin (FRAC Group 11) and fludioxonil (FRAC Group 12); materials such as iprodione (FRAC Group 2), thiophanate-methyl + iprodione (FRAC Groups 1 and 2), or triticonazole (FRAC Group 3) may be rotated with these highly effective active ingredients (Trolinger *et al.*, 2018).

Cultivars vary greatly in susceptibility to Fusarium wilt. Strains of *Trichoderma viride*, *Aspergillus ochraceus*, *T. harzianum*, *Trichoderma* spp., and *Streptomyces griseoviridis* have been shown to compete against *F. oxysporum* in research studies (Trolinger *et al.*, 2018), indicating a future role for biological control. Dazomet fumigation followed by application of bio-organic fertilizer with a fungal antagonist (*Paenibacillus polymyxa*) suppressed Fusarium wilt and improved plant quality in a trial of a more ecological approach to Fusarium wilt management (Zhao *et al.*, 2016). Botanical extracts of *Mentha arvensis*, *Tagetes patula*, and *Datura stramonium* gave 50–70% Fusarium wilt reduction; *Trichoderma harzianum* combined with botanical extracts was not as effective as *T. harzianum* alone (Singh and Kumar, 2011).

10.2.6 Rhizoctonia root rot, stem rot, and web blight

Stem rot is the most common symptom caused by the fungus *Rhizoctonia solani*, which has a large host range. Under highly humid conditions, a web blight (aerial blight) may also be caused by *R. solani* or binucleate *Rhizoctonia* spp. (Fig. 10.5).

Symptoms and diagnosis: The stem base is commonly attacked, resulting in tan to brown cankers (areas of stem tissue in which the cambium has been killed). Cankers may girdle the stem partially or completely. Plants may be stunted, and foliage on entirely girdled stems will wilt and die. In very humid conditions, 80–90% RH or more, web blights may develop (Copes and Scherm, 2005; Bonkowski, 2021) (Fig. 10.5). Also, in the interior of the plant canopy, extensive leaf death occurs. Microscopic examination of the affected stem and leaves is often sufficient for identification, as the fungus grows exposed on the surface of its host. Culturing from canker margins will also allow identification. Not all visible webbing indicates *R. solani*: spiders, fungus gnats, and European pepper moth (*Duponchelia fovealis*) larvae also create webs. On infected areas, the sticky, cobweb-like superficial mycelium may cause soil particles to cling in a characteristic way that aids in diagnosis (Horst and Nelson, 1997).

Fig. 10.5. Rhizoctonia web blight (aerial blight) on chrysanthemum. Symptoms in field-grown plants (A). Browning, dead foliage, and light-colored fungal hyphae in the plant interior on container-grown chrysanthemum (B). Close-up of spiderweb-like, light brown *Rhizoctonia* hyphae and necrotic leaves, and leaf spot caused by *Rhizoctonia* sp. (C and D). (Photo courtesy of Margery Daughtrey.)

Source, spread, and life history: Cuttings contaminated with *Rhizoctonia solani* may initiate a problem in production. The widespread presence of *R. solani* in nature means that contamination can also easily occur on-site. Movement of soil particles or infested plant debris by wind or splashing water or through human activity helps to disseminate *R. solani*. This fungus is relatively unique in that it persists largely as mycelium without any asexually produced spores that could be air- or waterborne. Basidiospores are produced, but rarely, and their role in epidemiology of Rhizoctonia diseases of chrysanthemum is not well understood. *R. solani* invades directly by growth of the mycelium rather than via spores. It is considered a "soilborne" disease because the fungal contamination comes from the movement of soil particles containing the fungus. Warm soils, 21–32°C, favor disease development (Horst and Nelson, 1997). High humidity at the soil surface is conducive to fungal growth. Web blight spreads most rapidly at 24–30°C, and symptoms can develop suddenly (Bonkowski, 2021). Allowing the soil or growing medium surface to dry between irrigation turns helps to control Rhizoctonia diseases. Long-lasting knots of fungal tissue called sclerotia may form in soil or on plant tissue or on the surface of pots and flats. Movement of sclerotia inoculum may carry over disease from one cultivation to another if containers are reused without thorough disinfection.

Management: Check cuttings carefully on receipt. Good sanitation practices are an important defense against Rhizoctonia stem rot. Soilless growing media are expected to be free from *R. solani*, but they will harbor the fungus once it becomes contaminated. Crops should be grown on raised benches or surrounded by ground cloth to avoid contact with field soil; hose ends in the greenhouse should be hung up rather than allowed to drop to the ground. Bagged soilless media should be stored under cover to keep it safe from blowing soil. If pots and flats cannot be discarded after a disease outbreak, clean off organic debris and then use a disinfectant treatment before reuse of containers. Regular fungicide applications may be needed to suppress *Rhizoctonia* under some circumstances. Biological controls added to the growing medium at the start of the crop reduce *Rhizoctonia* growth through various mechanisms of competition, direct parasitism, and antibiotic production. See Table 10.5 for a list of some of the FRAC Groups and active ingredients that may be used against Rhizoctonia root and stem rot, and Table 10.6 for a list of FRAC Codes for Rhizoctonia web blight disease management (Daughtrey *et al.*, 2023).

Table 10.5. Chemical and biological management tools for Rhizoctonia root and stem rot management. Author's own table.

FRAC Group[a]	Some examples of active ingredients
1	Thiophanate-methyl
2	Iprodione
3	Triflumizole
7	Flutolanil
11	Azoxystrobin, trifloxystrobin, fluoxastrobin, pyraclostrobin, trifloxystrobin
12	Fludioxonil
19	Polyoxin D zinc salt
14	PCNB
BM 02	*B. pumilus, Pseudomonas chlororaphis, Streptomyces* sp. K61, *Streptomyces lydicus, Trichoderma asperellum, T. gamsii, T. harzianum Bacillus amyloliquefaciens, B. subtilis*

[a]'FRAC' stands for Fungicide Resistance Action Committee. FRAC Groups indicate modes of action; see https://www.frac.info. 'BM' indicates biologicals having multiple modes of action, thus not prone to development of resistance in the target pest population.

Table 10.6. Chemical and biological management tools for Rhizoctonia web blight disease management. Author's own table.

FRAC Group[a]	Some examples of active ingredients
1	Thiophanate-methyl
2	Iprodione
3	Propiconazole
7	Flutolanil
9	Cyprodinil
11	Azoxystrobin, pyraclostrobin
12	Fludioxonil
19	Polyoxin D zinc salt
M 01	Copper octanoate
M 03	Mancozeb
M 05	Chlorothalonil
BM 02	*Gliocladium catenulatum* (*Clonostachys rosea*), *Pseudomonas chlororaphis*

[a]'FRAC' stands for Fungicide Resistance Action Committee. FRAC Groups indicate modes of action; see https://www.frac.info. M indicates chemical multi-site inhibitors; 'BM' indicates biologicals having multiple modes of action, thus not prone to development of resistance in the target pest population. 'NC' indicates "Not Classified".

10.2.7 Powdery mildew

Powdery mildew is not often a serious problem in chrysanthemum. *Golovinomyces cichoracearum* (syn. *Erysiphe cichoracearum*) is the main powdery mildew fungus reported from *Chrysanthemum* or *Dendranthema* spp. *G. cichoracearum* is a cosmopolitan species with a wide host range in the *Asteraceae*, *Cucurbitaceae*, and *Cannabaceae* (Farr and Rossman, 2019). Also, *G. orontii* has been reported from Italy, *E. pisi* var. *pisi* from Canada, and *G. chrysanthemi* from Montenegro (Farr and Rossman, 2019) and North America (Pscheidt and Ocamb, 2022). Powdery mildew is recognized by the superficial growth of whitish fungus on the leaves, stems, or petals, forming circular patches called colonies or coatings (Fig. 10.6). Spores (conidia) are produced on exposed conidiophores, and are dispersed by air movement. Spacing plants to improve air circulation helps in powdery mildew management. If using fungicides, materials with a single mode of action should be alternated strategically to avoid accelerating resistance development. Contact action materials such as horticultural or neem oils or potassium bicarbonate may be alternated with systemic fungicides such as strobilurins (FRAC Group 11) and demethylation inhibitors (DMIs, FRAC Group 3) in a management program. Biological controls having *Bacillus*, *Streptomyces*, or *Trichoderma* spp. as active ingredients will also reduce powdery mildew.

10.2.8 Rusts

Chrysanthemum has two major rust diseases: chrysanthemum brown rust (caused by *Puccinia chrysanthemi*) and chrysanthemum white rust (caused by *Puccinia horiana*). White rust is the more dangerous of the two because it is a quarantine pest subject to federal regulation in the USA and many other countries (USDA APHIS, 2021). Following its introduction to the UK in 1963, white rust has become so well established in Western Europe that many countries have now abandoned statutory measures against it (CABI, 2021). Regulations continue for the USA and Canada, where the disease is eradicated when detected.

Symptoms and diagnosis: When either of the two rusts infects leaves, the upper surfaces show small pale spots that may be slightly dimpled. Symptoms of the two rusts are, however, strikingly different on the leaf undersurface. The spore pustules (uredinia) of brown rust are filled with chocolate-brown masses of spores (Fig. 10.7). The spore pustules (telia) of chrysanthemum white rust, most apparent on young leaves and shoots, are pinkish or buff initially and whiten with time (Fig. 10.7). Both rusts form concentric rings of pustules. The two rusts can co-occur on the same leaf (Fig. 10.7D). Diagnosis is generally made by examination of the spore pustules on the leaf undersurface; accuracy is improved by using a microscope. When darkly pigmented natural antagonists of white rust (hyperparasites) are present, their sporulation may be mistaken for brown rust with the naked eye. PCR assays can be used for identification of white rust (Demers *et al.*, 2015).

Source, spread, and life history: The source of rusts is chrysanthemum cuttings, plants, or cut flowers—perennials in the landscape nearby may also supply inoculum. The dark urediniospores of *P. chrysanthemi* can be spread by splashing or air currents, whereas the spread of *P. horiana* is over shorter

Fig. 10.6. Symptoms of chrysanthemum powdery mildew after intensive disease development on the entire plant (A). Mycelial growth on a mature leaf (B). Microscopic view of mycelium on the surface of an infected chrysanthemum leaf (C). (Photo courtesy of Malleshaiah SharathKumar.)

Fig. 10.7. Chrysanthemum white rust (*Puccinia horiana*), lower leaf surface showing white telia (A, Photo courtesy of Lynn Hyatt). Two- celled chrysanthemum white rust teliospores as seen through a compound microscope (B, Photo courtesy of Lynn Hyatt). Chrysanthemum brown rust (*Puccinia chrysanthemi*), young pustules (uredinia) on lower leaf surfaces (C, Photo courtesy of Lynn Hyatt). Mixed infection of brown rust and white rust (D, Photo courtesy of Lynn Hyatt.) Further developed chrysanthemum brown rust, showing spore pustules forming in concentric rings (E, Photo courtesy of Lynn Hyatt). Brown teliospores (and three urediniospores) of chrysanthemum brown rust (F, Photo courtesy of Margery Daughtrey).

distances and only under wet, cloudy conditions that protect the basidiospores from drying. Both chrysanthemum rusts have an abbreviated life cycle compared to other common rust diseases. They are thus termed microcyclic and have no alternate hosts.

For *P. chrysanthemi* (brown rust), only urediniospores are produced during the growing season, and only on chrysanthemum. This repeating spore stage allows buildup of disease on plants not protected with fungicides. In the fall, two-celled teliospores form within the same pustules and may assist in overwintering. In greenhouses or milder climates, urediniospores survive throughout the year. Ideal temperatures for brown rust urediniospore germination are 16–21°C. More than 8 hours exposure at a temperature higher than 38°C will prevent infections (Horst and Nelson, 1997). Brown rust has chrysanthemum as its only host (Newcombe, 2003).

For *P. horiana*, the white rust, the life cycle is simplified further: only teliospores and basidiospores are produced in spore cushions on the leaf undersurface. Teliospores germinate over the broad temperature range of 4–23°C to produce basidiospores under very highly humid or wet (96–100% RH) conditions. Basidiospore formation takes only 3 hours at 17°C (CABI, 2021). The basidiospores, windblown to new sites, germinate when there is a film of moisture on the plant. Basidiospores establish new infections in just a few hours: only 5 hours

Table 10.7. Chemical and biological management tools for rust disease management. Author's own table.

FRAC Group[a]	Some examples of active ingredients
1	Thiophanate-methyl
3	Propiconazole, myclobutanil, triflumizole
7	Flutolanil
11	Azoxystrobin, fluoxastrobin, pyraclostrobin, trifloxystrobin
BM 02	*Bacillus amyloliquefaciens, B. subtilis*
M 01	Copper sulfate pentahydrate
M2	Sulfur
M 03	Mancozeb
M 05	Chlorothalonil
NC	Mineral oil, neem oil

[a]'FRAC' stands for Fungicide Resistance Action Committee. FRAC Groups indicate modes of action; see https://www.frac.info. M indicates chemical multi-site inhibitors; 'BM' indicates biologicals having multiple modes of action, thus not prone to development of resistance in the target pest population. 'NC' indicates "Not Classified".

of leaf wetness allow inoculum to be produced and new infections to occur. Fortunately, spread of white rust is limited by the delicacy of its basidiospores: they cannot tolerate less than 90% RH (Horst and Nelson, 1997). Wind-driven rain could conceivably move basidiospores over about 0.8 km (Zandvoort, 1968); shorter distances are more likely. The next generation of basidiospores may be released from an infected plant after 7–10 days unless temperatures above 30°C delay the process. The incubation time may be extended to 8 weeks if there is a high-temperature delay (CABI, 2021). Under moist, disease-conducive conditions, symptoms typically appear 5–14 days after infection. Chrysanthemum white rust is known to affect only species of *Chrysanthemum*, *Dendranthema*, *Leucanthemella*, and *Nipponanthemum* (USDA APHIS, 2021).

Management: For brown rust, irrigate carefully to avoid long periods of leaf wetness, scout for symptoms, and apply fungicides, if needed (see Table 10.7). The difficulty of chemical control of white rust was the original reason for the national quarantines placed on the pathogen. Dickens (1990) found that propiconazole (FRAC Group 3) halted disease development completely even when applied 4–8 days postinoculation; it was subsequently used in statutory eradication programs. More recently, several fungicide active ingredients were shown to suppress basidiospore germination in *in vitro* tests (Palmer *et al.*, 2015). Combination products showed the same low EC_{50} value as their respective strobilurin components. In the USA, myclobutanil (FRAC Group 3) is currently the fungicide most used for white rust eradication; other FRAC Group 3 (DMI) with curative properties are sometimes used as a substitute (USDA APHIS, 2021). See Table 10.7 for a listing of FRAC Groups of materials used for preventive treatments against rusts (Daughtrey *et al.*, 2023).

10.3 Bacterial Diseases

10.3.1 Bacterial leaf spot—*Pseudomonas cichorii*

Bacterial leaf spot of chrysanthemum is caused by *Pseudomonas cichorii*, a bacterium with an extensive host range. However, in chrysanthemum, only certain cultivars are highly susceptible depending upon weather conditions.

Symptoms and diagnosis: On leaves, black, irregularly shaped, water-soaked lesions appear, which dry to a lighter brown color (Fig. 10.8). Lesions run along the edge of the leaf or appear in the center of the leaf blade. Necrosis may extend into stem tissue from infected leaves and buds may be blighted, mimicking Ascochyta ray blight. To diagnose bacterial leaf spot, the margins of the lesions may be cultured on agar media and the bacterial isolates sequenced.

Source, spread, and life history: Inoculum of *P. cichorii* may come either from cuttings or other host plant species nearby. Outdoors, the bacteria are spread by wind-driven rain during warm-weather storms. Spread also happens through splashing from overhead irrigation. Infection may be latent and not detected until the environment is conducive. Warm temperatures and leaf wetness allow epidemic spread on leaves (Jones *et al.*, 1984b), with an optimum of 28°C. Higher dose of fertilization can increase disease (Jones *et al.*, 1984a). Bacteria enter leaves through natural openings or wounds, including those made by leaf miners (Matteoni and Broadbent, 1988). Overwintering within crop debris is likely in areas that are frost free. Alternative hosts of *P. cichorii* include other ornamentals as well as vegetables, especially lettuce. Symptoms vary with host plants, indicating that there may be different virulence mechanisms involved (Hikichi *et al.*, 2013; Osdaghi, 2020).

Management: Use of resistant cultivars is a wise practice; learn from disease outbreaks which cultivars to avoid. Reducing leaf wetness periods is important. Chrysanthemums grown outdoors should be irrigated at the soil line. Overhead irrigation must be done early in the day to avoid having wet foliage overnight. Remove diseased plants promptly. Chemicals are only marginally helpful against bacterial disease. Copper, or copper + mancozeb combination, may suppress bacterial growth, but copper resistance is a frequent problem because of previous overuse of copper-containing fungicides. Copper fungicides may be alternated with biocontrols that curb bacterial disease, such as those containing *Bacillus amyloliquefaciens* or *B. subtilis*. Treatments should be made at least every 7 days. Some quaternary ammonium compounds have also been demonstrated to reduce bacterial disease when used preventively.

10.3.2 Bacterial stem rot and wilt

Chrysanthemums are susceptible to soft rot and wilt diseases that were once attributed to bacteria named *Erwinia chrysanthemi* and *Erwinia carotovora*. The names of the pathogens have been changed to *Dickeya chrysanthemi* and *Pectobacterium carotovorum*, respectively (Samson *et al.*, 2005). Much of the research on these diseases is listed under the older names.

Fig. 10.8. Black leaf lesions on chrysanthemum caused by *Pseudomonas cichorii*, the pathogen responsible for bacterial leaf spot. (Photo courtesy of Margery Daughtrey.)

Symptoms and diagnosis: Internal discoloration of the vascular system and browning of stem pith may occur with *Pectobacterium* or *Dickeya* (Trolinger *et al.*, 2018). Brown to black lesions, initially water soaked, appear on leaves and stems of plants infected with *D. chrysanthemi*, which wilt and die (Fig. 10.9) (Vegh *et al.*, 2014). External symptoms from *P. carotovorum* infection are similar. Systemic bacterial and fungal diseases may be confused with one another without a laboratory diagnosis. The soft rot bacteria cause an abrupt decline in plant health when high humidity and high temperature favor rapid disease development. Fusarium wilt also progresses under warm conditions, but symptoms develop more gradually, and often on one side of a plant. Verticillium wilt also causes a more gradual decline in plant health. Isolation and sequencing of the bacterial pathogens in a diagnostic laboratory will allow precise identification of the problem.

Source, spread, and life history: Latently infected cuttings are a common source of the pathogens. The production facility may also be a source of inoculum from debris of hosts. Local spread is by irrigation splash, worker handling, or insects. With either pathogen, there may be latent systemic infection in stock plants; symptoms may appear subsequently in cuttings or rooted plants, especially under high-temperature conditions above 27°C (Trolinger

Fig. 10.9. Blackening of stems and foliage due to infection by *Dickeya chrysanthemi*. (Photo courtesy of Robert L. Wick.)

et al., 2018). Wounding provides an opportunity for bacteria to enter the plant. Bacterial soft rot pathogens have wide host ranges. Of the two, *Dickeya chrysanthemi* has a more restricted host range, but it is known to infect globe artichoke, common chicory, sunflower, tomato, guayule, and philodendron.

Management: Careful sanitation practices are needed to reduce bacterial inoculum, especially in propagation areas. Infected plants should be discarded by carrying bags for disposal to the affected plants (on the bench or in the field), and workers should wash hands after handling diseased plants. The use of disinfectants on tools, impermeable greenhouse floors, and work surfaces between crops will be beneficial. Copper is the main chemical against *Dickeya* and *Pectobacterium*, but resistant isolates are found in horticultural operations. Some quaternary ammonium compounds are beneficial antibacterial treatments. The biocontrols *Bacillus subtilis* and *Bacillus amyloliquefaciens* also provide some benefit.

10.3.3 Crown gall

Crown gall disease, caused by *Agrobacterium tumefaciens* (syn. *Rhizobium radiobacter*), affects chrysanthemum as well as many other herbaceous and woody ornamental hosts all over the world (CABI, 2019c). The nomenclature of *Agrobacterium* continues to change. Although merged with *Rhizobium* (Young *et al.*, 2001), more recent work has shown that these genera, as well as additional ones, do not differ sufficiently to merit separation (Weisberg *et al.*, 2020). For historical consistency, pathogenic isolates will be referred to here as *A. tumefaciens*. Both *A. tumefaciens* and *R. skierniewicense* sp. nov. have been found associated with galls on chrysanthemum (Pulawska *et al.*, 2012).

Symptoms and diagnosis: Galls may appear on roots and foliage or at the root crown (Fig. 10.10), appearing at the site of wounds. They are more or less round, smooth, or with a nubby texture, dirty white to tan in color, and variable in size. Chrysanthemums may be stunted as well as galled. Leaf galls generally form on plants that are systemically infected, after wounding (Fig. 10.10B) (Moore and Putnam, 2021). The pathogen may be most easily isolated from fresh, new galls on semi-selective media and then sequenced to determine identity. Isolates may be tested by PCR, identifying the presence of the Ti (tumor-inducing) plasmid.

Source, spread, and life history: *Agrobacterium tumefaciens* is found in nature but is more likely to be brought in on cuttings. Incoming material should be checked for galls. Local spread is by handling or by soil movement via wind or water splash. Bacteria can be transmitted in ebb–flood irrigation on a greenhouse bench or during cutting dips. Bacteria enter through wounds, most often cutting wounds. *A. tumefaciens* contains the Ti plasmid that bears the genes that code for crown gall formation. Galls may form in just a few weeks on herbaceous plants when temperature is above 21°C (Moore

Fig. 10.10. Crown gall caused by *Agrobacterium tumefaciens* on chrysanthemum. An unusually extensive galled area on a chrysanthemum stem (A). Small gall formed on a leaf wound in a systemically infected chrysanthemum (B). (Photo courtesy of Margery Daughtrey.)

and Putnam, 2021). The infection becomes systemic in chrysanthemum, with the bacteria inhabiting the xylem. Thus, wounds made aboveground on a systemically infected plant may also develop galls (Putnam, 2015). Bacteria can likely persist in plant debris for 2 years in field soil (Putnam, 2015).

Management: The pathogen should be excluded from all growing operations by using stock plants indexed to eliminate *A. tumefaciens*. Cuttings from systemically infected stock plants are a source. Contaminated areas will need to be steam-sterilized or solarized to eliminate the bacteria; fumigation has been ineffective in the past but in some cases, this may have been due to increased recolonization of soil by *A. tumefaciens* after methyl bromide fumigation through the creation of a biological vacuum (Yakabe *et al.*, 2010). Avoid growing chrysanthemums in heavy, wet soils. To reduce *A. tumefaciens* infections, plant beds when the soil temperature is below 10°C. Copper has not been effective against crown gall. In some areas, a biological control is available that has been used primarily with woody nursery plants such as roses. It features *Agrobacterium radiobacter*, which competes strongly with *A. tumefaciens* through the production of antibiotics (Moore and Putnam, 2021). Plants dipped into *A. radiobacter* Kerr 84 or the genetically modified, improved strain, K1026, are protected to a degree against infection by certain strains of *A. tumefaciens* (Ryder and Jones, 1991; Stockwell *et al.*, 1993). Other strains are not susceptible and will not be controlled by these products. *A. radiobacter* is not compatible with captan, mancozeb, PCNB, streptomycin, or fungicides containing etridiazole. *Pseudomonas fluorescens* is also listed as an antagonist of *A. tumefaciens*.

10.3.4 Phytoplasma diseases

Phytoplasmas are bacteria lacking a cell wall and living in phloem that cause a range of virus-like symptoms, including witches' brooms and flower abnormalities (Fig. 10.11). They are vectored by leaf hoppers and other insects in the order *Hemiptera*. Chrysanthemums are susceptible to the best-known of these, the aster yellows phytoplasma, *Candidatus* Phytoplasma asteris, which is in the 16Sr1 group (Raj *et al.*, 2007). Chrysanthemum slow decline, a disorder in which plants show leaf distortion and abnormal flowers, affected some cut flower cultivars over 30 years ago; researchers considered but rejected phytoplasmas as the cause (Dienelt and Lawson, 1988). A disease called chrysanthemum phloem necrosis was seen in certain cut flower cultivars in the Marble group and garden mums in the 1970s and 1980s; leaf necrosis symptoms were attributed to a phytoplasma (McGovern *et al.*, 1989) or manganese toxicity (Dienelt and Lawson, 1991). Recently, a survey in India found phytoplasmas in chrysanthemums from four different ribosomal groups, including 16SrI (Taloh *et al.*, 2020). To minimize phytoplasma disease losses, nearby weed hosts should be eliminated; highly susceptible crops may be protected from vectors with fine mesh nets (Babadoost, 1988).

Fig. 10.11. Symptoms of aster yellows phytoplasma in chrysanthemum, showing virescent flowers at the center. (Photo courtesy of G. P. Rao.)

10.4 Virus and Viroid Diseases

10.4.1 Chrysanthemum virus B (CVB)

Chrysanthemum virus B (CVB), a member of the genus *Carlavirus*, is globally distributed and can cause symptoms from mild mottle to crop-destroying flower effects (Wilson and Moran, 1983; Lin *et al.*, 2005). The particles are flexuous rods, 685 x 12 nm.

Symptoms and diagnosis: Plants may be symptomless, or they may be stunted with flowers underdeveloped; leaves may show yellow or necrotic mottling and vein clearing, as well as mosaic and surface distortion. CVB may be found in plant tissue by nested PCR (Guan *et al.*, 2017), ELISA, or immune electron microscopy (IEM). Detection in Taiwan was shown to be higher in winter, with temperatures 15–20°C, than under warmer conditions (Lin *et al.*, 2005).

Source, spread, and life history: The main source of CVB is infected cuttings. Spread is via vegetative propagation and by the green peach aphid (*Myzus persicae*) (Singh *et al.*, 2012). The virus has a very narrow host range, infecting only several *Nicotiana* spp., *Petunia hybrida*, *Vicia faba* and Marguerite daisy (*Argyranthemum frutescens*) in addition to chrysanthemum (Nakamura *et al.*, 1994).

Management: Only certified virus-free chrysanthemum cuttings should be grown. Treatment at 37°C combined with meristem culturing was tested as a means for getting virus-free chrysanthemums from cultivars infected with CVB. Meristem tips could be selected that were free from virus even without the heat treatment (Hakkaart and Quak, 1964).

10.4.2 Tomato aspermy virus

Tomato aspermy virus (TAV) belongs to the genus *Cucumovirus*; it is widely distributed across the globe. The name "aspermy" refers to the blockage of seed formation that is seen in some hosts. This virus was once known as Chrysanthemum aspermy virus (Hollings and Stone, 1971).

Symptoms and diagnosis: Symptoms of TAV are severe in chrysanthemum flowers: dwarfing, flower deformation including tubular florets, as well as bleaching and color break (Hollings, 1955; Hollings and Stone, 1971). Mixed infections with CVB may result in more intense symptoms. Some cultivars are symptomless carriers of TAV. Late-season infection may cause symptoms in only a few blooms, but systemic infection is apparent the following year (Hollings, 1955; Hollings and Stone, 1971). Diagnosis of TAV may be done by RT-PCR, direct antigen coating-enzyme-linked immunosorbent assay (DAC-ELISA), and double antibody sandwich-enzyme-linked immunosorbent assay (DAS-ELISA). TAV has spherical particles, with a diameter of about 25–30 nm (Hollings and Stone, 1971).

Source, spread, and life history: Mum cuttings are the most likely source of an outbreak, but the source may at times be other host plants. Spread of TAV is through vegetative propagation and by some species of aphids, including the green peach aphid (*Myzus persicae*), black chrysanthemum aphid (*Macrosiphoniella sanborni*), and foxglove aphid (*Aulacorthum solani*), which spread TAV nonpersistently. In research studies, transmission by a cutting knife was shown (at a low percentage), but seed transmission was not observed (Hollings, 1955). In an Australian survey, TAV was the second most frequently detected virus after CVB (Hill *et al.*, 1996). TAV has been more often a problem in field culture (Horst, 1985; Hill *et al.*, 1996). There are over 100 hosts of TAV, but natural infections have not been widely reported in ornamentals. TAV has been reported from *Chrysanthemum zawadskii* var. *latilobum*, a medicinal plant, in Korea (Kwak *et al.*, 2021).

Management: Only certified virus-free chrysanthemum cuttings should be grown. It is important to control aphids, especially in the early months of propagation when the chrysanthemum is most susceptible (Hollings, 1955). Virus-indexing programs and indoor cultivation of chrysanthemum greatly minimize the threat of TAV (Horst, 1985). Symptoms are most apparent during flowering and plants with any suspicious blooms should be rogued out. Meristem tip culture can be used to free valuable cultivars of the virus: plants were freed from a mixed infection of *Cucumber mosaic virus* and TAV by culturing 0.3-mm-long shoot meristems (Kumar *et al.*, 2009). Most strains of TAV can be eliminated by growing plants at 37°C for 4 weeks (Hollings, 1955).

10.4.3 Tomato spotted wilt orthotospovirus (TSWV)

Tomato spotted wilt orthotospovirus (TSWV) is an economically destructive disease around the globe, affecting many important crops including chrysanthemum. It is very damaging to some cultivars. High population of thrips vector leads to an unmanageable virus problem. The closely related *Impatiens necrotic spot orthotospovirus* may also affect chrysanthemum.

Symptoms and diagnosis: Symptoms of TSWV in chrysanthemum vary between cultivars. Stunting is common. Dark brown to black stem cankers, leaf necrosis, and internal stem necrosis mimic infection by fungal pathogens (Fig. 10.12). Chlorotic spotting, bronzing, yellow or brown ring spots or line patterns in leaves are more readily recognized as viral symptoms. Fresh weight, petal number, and number of flowers per stem for spray chrysanthemum were reduced by TSWV (Matteoni and Allen, 1989). Individual branches sometimes wilt due to TSWV infection. Infected chrysanthemums may also remain symptomless (Matteoni and Allen, 1989).

Detection of TSWV and *Impatiens necrotic spot virus* (INSV) in plants is challenging because of the notoriously uneven distribution of virus particles within the plant. Sample selection is thus all-important: symptomatic tissue should be tested where possible. DAS-ELISA has been used for TSWV detection (Matteoni and Allen, 1989); grower-friendly test kits are also available with lateral flow devices that can be used to test for the presence of TSWV or INSV by serology. A LAMP assay has also been developed for detection in the field (Fukuta *et al.*, 2004). A leaf-disc bioassay was found to be more reliable than whole-plant tests when evaluating TSWV susceptibility of cultivars after thrips-mediated inoculation (Van de Wetering *et al.*, 1999).

Source, spread, and life history: TSWV may originate from cuttings or, if a thrips vector is present, from weeds or crops in the vicinity. Vegetative propagation allows spread of TSWV; seed transmission is reported only rarely and is not known for chrysanthemum. Most commonly, problems with TSWV are directly linked to high populations of certain species of thrips, particularly the western flower thrips, *Frankliniella occidentalis*, the most efficient of its vectors. At least eight other Frankliniella thrips species (such as *Frankliniella bispinosa*, *F. cephalica*, *F. fusca*, *F. gemina*, *F. intonsa*, *F. schultzei*, *T. setosus*, and *T. tabaci*) may also transmit the virus (Chen *et al.*, 2019).

The virus has pleomorphic particles that are 80–120 nm in diameter (Adkins, 2000). It is acquired by thrips vectors when they feed as larvae on infected plants during the first and early second instar. TSWV is persistently transmitted by these thrips once they are winged adults. Thrips populations build up swiftly in greenhouses. If eggs are laid in virus-infected plants, there will be widespread dissemination of TSWV when the thrips developing from those eggs mature and fly about the greenhouse. Symptoms may take weeks to

Fig. 10.12. Leaf and stem lesions on chrysanthemum caused by *Tomato spotted wilt virus*. (Photo courtesy of Margery Daughtrey.)

appear after thrips feed upon a plant (Van de Wetering *et al.*, 1999). TSWV has over 800 plant hosts, including weeds, vegetables, and ornamentals (Adkins, 2000), many in the *Asteraceae* and *Solanaceae*. INSV has a similarly wide host range and the same life cycle as TSWV.

Management: Management of TSWV requires vigilance and an integrated approach. Cultivars known to be highly susceptible to the local thrips vector(s) and/or highly susceptible to TSWV should not be grown. Weeds that may be reservoirs for TSWV or INSV should be eliminated. Use yellow sticky cards to monitor for western flower thrips in greenhouses and note when thrips populations begin to increase. If biological control programs are being used for thrips management, monitor their effectiveness continually. Pesticide resistance has been a significant problem with western flower thrips (Van de Wetering *et al.*, 1999), complicating deployment of chemical insecticides and hastening the adoption of biological control programs. Insecticide applications must be made with care to rotate among effective active ingredients and slow the development of resistance. Thrips control measures, such as clean crop production areas, insect screening, thrips-resistant cultivars, and insecticide or biological control treatments, are essential for TSWV management. Most chrysanthemum cultivars are susceptible when tested by serology, but the percentage of cultivars showing symptoms is much lower. Chrysanthemums with both a high degree of vector resistance and a high degree of TSWV resistance are needed (Van de Wetering *et al.*, 1999).

10.4.4 Chrysanthemum stunt viroid (CSVd)

Viroids are composed of RNA, lacking the protein coat that is characteristic of viruses. Chrysanthemum stunt was described in the USA in 1947 as a new disease (Dimock, 1947); it is now distributed globally. At first the pathogen was thought to be a virus, but in 1973 it was shown to be a viroid (Diener and Lawson, 1973; Hollings and Stone, 1973). *Chrysanthemum stunt viroid* (CSVd), which causes this disease, is a widely distributed pospiviroid of high economic significance; excluding it from chrysanthemum production stock is critical. This very abbreviated organism is made up of only about 354–356 nucleotides in a circular molecule of single-stranded RNA (Yoon and Palukaitis, 2013), making it the smallest known pathogen of chrysanthemum (Tran *et al.*, 2016).

Symptoms and diagnosis: In some chrysanthemums and in other hosts, plants infected with CSVd are often symptomless (Palukaitis, 2017). Symptoms will vary with environmental conditions and cultivar (Handley and Horst, 1988). CSVd affects the development of flowers and leaves, as well as overall growth: roots and tops of plants are stunted. Flower numbers and size are reduced, and flowers may show petal bleaching, color break, distortion, and early opening (Palukaitis, 2017) (Fig. 10.13). Leaves may be stunted and chlorotic. Chlorotic spots, large necrotic blotches, vein clearing, vein banding, and distortion occur. Bioassays were once used for CSVd detection, but now RT-PCR and RT-LAMP assays are typically used (Palukaitis, 2017), with the LAMP assay potentially allowing detection in the field.

Source, spread, and life history: CSVd is often spread with cuttings. The viroid may be transmitted by handling, by foliage or root contact between plants, or by sap transfer on tools (Brierly and Smith, 1949). Transmission via seed has been documented (Chung and Pak, 2008), and there is a low level of transmission by pollen. Symptomless hosts complicate a grower's ability to guard against CSVd. Some genetic variability has been seen in isolates from a

Fig. 10.13. *Chrysanthemum stunt viroid* symptoms: stunting, distortion, and petal streaking on the flower at the right. (Photo courtesy of Jean Dunez.)

single host or a single country (Palukaitis, 2017). Symptom expression was highest between 26 and 29°C, and high light intensity increased symptoms in one study (Handley and Horst, 1988). In addition to cultivated chrysanthemum, CSVd is found naturally in various wild *Chrysanthemum* spp., as well as in ageratum, Marguerite daisy, dahlia, petunia, *Solanum jasminoides*, and *Vinca major* (Yoon and Palukaitis, 2013).

Management: Starting crops from germplasm certified as viroid-free is essential. If diseased material is detected, it must be rogued out. Standard meristem tip culture is insufficient to remove viroids, but using meristems free from leaf primordia can be successful (Palukaitis, 2017). Cross-protection has been noted when other viroids are present (Palukaitis, 2017). Combining psychrotherapy (cold treatment for 3 months at 4°C) with meristem tip culture and utilizing 50 or 100 ppm ribavirin as chemotherapy improved the recovery of plants free from viroids (Savitri *et al.*, 2013). Genetic resistance may occur naturally, and resistant transgenic plants have been developed (Tran *et al.*, 2016).

10.5 Nematodes

10.5.1 Root knot nematodes

High populations of root knot nematodes, *Meloidogyne* species, may damage chrysanthemums as well as other floral or vegetable crops. Economic loss has been documented in cut flower production in Colombia (Arbelaez, 1999) and in greenhouses in the Netherlands (DeLong *et al.*, 2021). Different *Meloidogyne* species are involved in different parts of the world, but the effects are similar.

Symptoms and diagnosis: Plants with root knot may be severely stunted and show nutrient-deficient foliage, which should lead growers to examine the roots. Root galls may be rounded or elongated, occasionally confluent so that the roots appear almost tuberous, but always integral to the root so that they cannot be broken off (Horst and Nelson, 1997). Larger galls are more typical of warm-climate species such as *M. javanica* and *M. incognita,* which are common problems in glasshouses in the Netherlands (DeLong *et al.*, 2021). In contrast, *M. hapla*, a cool-weather nematode, forms smaller, more discrete galls. Aboveground, chrysanthemums under attack by root knot grow poorly and show wilt or scorch, suggesting water or nutrient deficiency. An infested chrysanthemum bed may have a patchy, uneven appearance; plants may appear unresponsive to fertilization or extra-sensitive to herbicides. Enhanced root rot from other pathogens has been noted following root knot (Horst and Nelson, 1997), and *M. javanica* was reported to facilitate Fusarium wilt disease (Johnson and Littrell, 1969). Among plants grown as bedding plants in Florida, chrysanthemum was rated moderately susceptible to root knot nematode (Crow, 2003). Different cultivars vary in susceptibility and responses to different species of *Meloidogyne* will also vary.

Adult female root knot nematodes may be exposed in their feeding sites by dissection under a compound microscope, carefully peeling away the galled plant tissue to reveal the white, pear-shaped females within. For species-level identification, characteristics of the female perineal pattern are often used, which requires a compound microscope.

Root knot nematodes, of which there are approximately 100 species worldwide, are termed sedentary endoparasites because they set up a feeding site inside the root as juveniles and remain there, feeding from the xylem; only the egg masses are external to the root. The approximately 30 day life cycle begins with a larva (stage 2 juvenile) that emerges from an egg and travels through the soil to a root that it penetrates, moving to the cortex and then the stele (Horst and Nelson, 1997; DeLong *et al.*, 2021). There its hormonal secretions lead to the formation of three to five giant cells that are subsequently used for food absorption. Both males and females establish their own individual feeding sites; later, only males exit the root. After fertilization, the female extrudes a jelly-like mass holding 300–3000 eggs, which will hatch as larvae and begin the next cycle. The life cycles of *Meloidogyne* spp. are temperature dependent; there was no development at 13°C, and it took 87 days to complete the life cycle at 19°C and only 25 days for a life cycle at 23°C (Horst and Nelson, 1997).

Management: In outdoor chrysanthemum culture, crop rotation is often key for the management of root knot because steaming the soil is too expensive (Horst and Nelson, 1997), and the use of fumigants or carbamate or organophosphate nematicides that could contaminate groundwater is generally considered too harmful to the environment. Growing nonhost crops for one season is often effective for bringing the soil population of *Meloidogyne* to a level that will not appreciably affect yield; this benefit may last for 2–3 years. Growers should take care to avoid recontamination of beds, by introducing only cuttings free from nematodes.

The chemical tools used for root knot management are changing. Alternatives to the older organophosphate and carbamate materials are appearing, including two fungicides with activity against nematodes: fluopyram (FRAC Group 7) and fluopimomide (FRAC Group 43) (Chen *et al.*, 2020). Bacteria, including *Pasteuria penetrans* and *Bacillus* spp., and the fungi, *Arthrobotrys*, *Monacrosporium*, *Pochonia chlamydosporia*, and *Paecilomyces lilacinus*, are all antagonists of root knot nematodes. The challenge is to make these living controls affordable to produce and use (Mitkowski and Abawi, 2011). Botanical extracts are also being investigated: studies with garlic (*Allium sativum*) extract have shown some promise against root knot nematodes (Chalanska *et al.*, 2017).

In glasshouse culture of cut chrysanthemum in the Netherlands, periodic soil disinfestation by steaming, every five to six crop cycles, is the usual method for keeping nematodes and other pathogens in check (DeLong *et al.*, 2021). A recent study has explored the usefulness of alternative control measures with the goal of reducing the frequency of steaming events, which are energy intensive and harsh on the soil microbiome. Root knot nematode damage was reduced by treatments with garlic extract, chitosan HCl, and biostimulants based on sea minerals and plant oils (DeLong *et al.*, 2021).

10.5.2 Foliar nematodes

Species of *Aphelenchoides*, known as foliar nematodes, were a significant problem in chrysanthemum cultivation in the past. A recent review described the morphological traits that distinguish the genus (Handoo *et al.*, 2020). Now that soilless media is used for potted plants, the problem is rare in that production system but reports continue on cut flowers (Deimi *et al.*, 2008; Crozzoli *et al.*, 2008). *Aphelenchoides ritzemabosi* (chrysanthemum foliar nematode), *Aphelenchoides fragariae* (strawberry crimp nematode), and *Aphelenchoides besseyi* (rice white-tip nematode) have been found in *C. morifolium*. *A. besseyi* is more likely to be found in tropical climates, whereas *A. ritzemabosi* and *A. fragariae* are more commonly associated with ornamentals in temperate climates (Kohl, 2011). Nematodes identified earlier as *A. besseyi* may in some cases belong instead to the newly described cryptic species *A. pseudobesseyi*, which can be differentiated by molecular analysis (Subbotin *et al.*, 2021).

Symptoms and diagnosis: Brown V-shaped wedges surrounded by a chlorotic zone are seen in nematode-infested leaves (Fig. 10.14) and can be mistaken for Botrytis blight lesions. Lower foliage is typically infested first. The nematodes often feed within the leaf tissue, acting as endoparasites. In addition, foliar nematodes may act as ectoparasites, feeding externally on young leaves within the leaf buds to cause leaf scarring and deformity (Fig. 10.14).

Diagnosis: Diagnosis of a foliar nematode infestation may be made by tearing a leaf with a wedge-shaped lesion into smaller pieces and immersing the pieces in water in a transparent dish. Nematodes will begin to emerge from the margins of the lesion within minutes; they can be observed using a dissecting microscope with light transmitted from beneath and then identified using a compound microscope. There may be as many as 15,000 nematodes per leaf (Wallace, 1959), or 45–4380 nematodes per gram of leaf tissue (Crozzoli *et al.*, 2008). PCR has been used to identify foliar nematodes within plant tissue (McCuiston *et al.*, 2007).

Source, spread, and life history: Soil or cuttings are the primary sources of an infestation. *Aphelenchoides* spp. are also seedborne (CABI, 2019a, 2019b, 2020). Nematodes in the soil may move upwards along the stem in a film of moisture, crawl out onto the leaf, and enter it via the stomata. Nematodes mate and lay eggs within the leaf tissue. The larvae have four stages of development, the first one within the egg itself. Larvae are miniature versions of the adult but lack sexual organs. Feeding by the second-, third-, and fourth-stage larvae and adults leads to chlorosis and then necrosis of the tissue. Movement of the nematodes is restricted by larger veins, hence the wedge-shaped or vein-bounded pattern of lesions. Adult foliar nematodes can overwinter in a dry, anhydrobiotic state—this allows them to survive for several months up to 3 years without water (Kohl, 2011).

A wide host range is a characteristic of *A. fragariae* and *A. ritzemabosi*. The hundreds of hosts for *A. ritzemabosi* include many flower crops and ferns. Hosts

Fig. 10.14. Foliar nematode injury on chrysanthemum (*Aphelenchoides pseudobesseyi* inoculated from echinacea). Symptoms of leaf stunting and distortion after direct bud inoculation (ectoparasitic feeding) (A). Wedge-shaped foliar lesions (B). Flower symptoms (C). (Photo courtesy of William T. Crow.)

in the *Asteraceae* are most common; many common weeds are hosts (CABI, 2019b). *Aphelenchoides fragariae* is often found together with *A. ritzemabosi* on chrysanthemum, but it has a different host range: it is more likely than *A. ritzemabosi* to be found on ferns, Liliaceae, Primulaceae, and Ranunculaceae (CABI, 2019a). For *A. besseyi*, rice, strawberry, and various vegetable crops are hosts, in addition to ornamentals (Handoo *et al.*, 2020; CABI, 2020). *A. pseudobesseyi* has been found in *Dryopteris erythrosora*, *Echinacea* sp., and *Farfugium japonicum* and goes to chrysanthemum by inoculation (Subbotin *et al.*, 2021).

Management: Scouting for the symptoms of nematode infestation should be conducted regularly and diseased plants removed. Chrysanthemum cuttings can be freed from nematodes by immersion in hot water. A dip in water at 46°C for 5 min was used to control *A. ritzemabosi* on mum stock plants (Staniland, 2008). Reduce the time that plant surfaces are wet, avoid overhead irrigation, especially on stock plants, and space plants adequately. Additional research is needed on using nutritional factors to slow down nematode reproduction: effects of calcium and temperature on *A. ritzemabosi* in tissue culture have been reported (Dolliver *et al.*, 1962).

Resistant cultivars are an unwieldy solution in view of the versatility of foliar nematodes and the huge number of ever-changing cultivars of chrysanthemum. Previous chemical controls have been outlawed in many countries, as nematicides of the past have been highly toxic to humans and their solubility has made them contaminants of groundwater. Effectiveness of currently available chemical nematicides has been variable and is often limited by the thickness of the cuticle of the crop plants (Kohl, 2011). Results with abamectin, for example, may vary with host plant and with adjuvants that affect leaf penetration (LaMondia, 1999). A reduction in foliar nematode populations in anemone was seen with an extract of garlic (*Allium sativum*) applied in combination with abamectin (Chalanska *et al.*, 2017).

10.6 Physiological and Abiotic Disorders or Diseases

Chrysanthemums are subject to injury from environmental stresses such as excessive drought or waterlogged soil; either condition may result in stunted plants, slower growth with delayed flowering, or smaller flowers. These abiotic stresses may also predispose the plants to attack by parasites. Roots, for example, may be invaded by *Pythium* or *Fusarium* species following overfertilization. Leaves scorched from drought may be attacked by *Botrytis cinerea* or other opportunistic foliar fungi. Nontarget effects of herbicides or other pesticides used on or near chrysanthemums may cause the symptoms of phytotoxicity.

Chrysanthemums are also subject to nutrient deficiencies and toxicities. Macronutrients must be regularly delivered via fertilization, as chrysanthemums require moderate to heavy crop fertilization (Owen *et al.*, 2018).

Chrysanthemums with insufficient nitrogen have pale green leaves, while excessive nitrogen causes brittle, dark green foliage. Ammonium toxicity appears as leaf curl and marginal chlorosis followed by necrosis, beginning on lower leaves; this toxicity is seen most often at low growing temperatures, below 13°C (Horst and Nelson, 1997). Phosphorous deficiency caused by using P-free fertilizers or low growing temperature can result in older leaves turning purple during vegetative growth (Fig. 10.15). In contrast, upper leaf yellowing and necrosis will be noted if P supply is limited during flowering (Whipker, 2014). Potassium deficiency is initially seen as marginal leaf chlorosis mid-stem and then progresses to downcurled, necrotic lower leaves. Calcium deficiency may cause new growth to be chlorotic and strapped, and leaves will cup down and show necrotic margins (Horst and Nelson, 1997).

Micronutrient deficiencies are often associated with incorrect pH. In soilless media, ideal pH for chrysanthemums is considered to be pH 5.8–6.2 for both propagation and potting, while cut flower crops in soil are generally grown at a pH of 6.0–7.0 (Nau *et al.*, 2021). High pH (above pH 6.5 in soilless media) can be damaging to chrysanthemums: iron deficiency will be expressed in the youngest foliage as interveinal chlorosis (Horst and Nelson, 1997; Owen *et al.*, 2018). Insufficient manganese, also likely at high pH, will cause plants to produce new leaves that are lime green. Many nutrient deficiencies lead to reduced flowering; boron deficiency, in particular, negatively affects flower development, causing ray florets to die back from the tip (Horst and Nelson, 1997). A comprehensive chart of symptoms resulting from nutrient deficiencies and excesses for chrysanthemum is available in the compendium written by Horst and Nelson (1997), and a number of deficiencies are illustrated in the study of Owen *et al.* (2018). Foliar analysis data is extremely helpful for diagnosing nutrient stress problems.

10.7 Conclusions and Future Prospects

The management of diseases in chrysanthemum production remains a dynamic and challenging frontier in ornamental research and practice. This chapter has outlined the complex interaction of fungal, oomycete, bacterial, and viral pathogens that threaten the health and economic viability of chrysanthemum production worldwide. Despite significant advances in disease diagnostics and integrated disease management strategies, the emergence of new pathogens and the resurgence of older ones in novel geographic locales due to global trade emphasize the need for continuous vigilance and innovation. Future research should focus on the development of more resistant chrysanthemum cultivars through genetic enhancement, as well as the exploration of novel biocontrol agents that offer environmentally friendly alternatives to chemical fungicides. Additionally, advances in molecular techniques for rapid disease detection and monitoring will be critical in preempting outbreaks and mitigating their impacts. Collaboration across international borders will be essential in sharing best practices and coordinating responses to disease threats, ensuring the sustainability of chrysanthemum as a cherished ornamental species.

Fig. 10.15. Abnormal coloration indicating phosphorus deficiency in chrysanthemum. (Photo courtesy of Nora Catlin.)

References

Adkins, S. (2000) Tomato spotted wilt virus positive steps toward negative success. *Molecular Plant Pathology* 1(3), 151–157.

Alfieri, S.A. (1968) Septoria leaf spot of chrysanthemum. Florida Department of Agriculture, Plant Pathology Circular, no.73. Available at: https://ccmedia.fdacs.gov/content/download/11080/file/pp73.pdf

Arbelaez, G. (1999) Overview of the cut flower pathology in Colombia. *Acta Horticulturae* 482, 91–96.

Babadoost, M. (1988) Aster yellows. University of Illinois, department of crop sciences. Report on Plant Disease, no. 903. Available at: https://ipm.illinois.edu/diseases/rpds/903.pdf (accessed December 2022).

Bonkowski, J. (2021) Rhizoctonia web blight. (Vol. 22). Purdue University Landscaper Report. Available at: https://purduelandscapereport.org/article/rhizoctonia-web-blight/ (accessed December 2022).

Brierly, P. and Smith, F.F. (1949) Control measures effective against virus in florists' crop. *Florists' Review* 107, 27–30.

CABI (2019a) *Aphelenchoides fragariae* (strawberry crimp nematode). *Invasive Species Compendium*. Available at: https://www.cabi.org/isc/datasheet/6381 (accessed December 2022).

CABI (2019b) *Aphelenchoides ritzema-bosi* (chrysanthemum foliar eelworm). Available at: https://www.cabi.org/isc/datasheet/6384 (accessed December 2022).

CABI (2019c) *Rhizobium radiobacter* (crown gall). *Invasive Species Compendium.* Available at: https://www.cabi.org/isc/datasheet/3745#tonaturalEnemies (accessed December 2022).

CABI (2020) *Aphelenchoides besseyi* (rice leaf nematode). *Invasive Species Compendium.* Available at: https://www.cabi.org/isc/datasheet/6378 (accessed December 2022).

CABI (2021) *Puccinia horiana* (white rust of chrysanthemum). CABI Compendium. Available at: https://www.cabi.org/isc/datasheet/45806#REF-DDB-202162 (accessed December 2022).

Chalanska, A., Bogumil, A. and Labanowski, G. (2017) Management of foliar nematode *Aphelenchoides ritzemabosi* on *Anemone hupehensis* using plant extracts and pesticides. *Journal of Plant Diseases and Protection* 124, 437–443.

Chen, J., Li, Q.X. and Song, B. (2020) Chemical nematicides: Recent research progress and outlook. *Journal of Agricultural and Food Chemistry* 68, 12175–12188.

Chen, Y., Dessau, M., Rotenberg, D., Rasmussen, D.A. and Whitfield, A. (2019) Entry of bunyaviruses into plants and vectors. *Advances in Virus Research* 104, 65–96. DOI: 10.1016/bs.aivir.2019.07.001.

Chung, B.N. and Pak, H.S. (2008) Seed transmission of chrysanthemum stunt viroid in chrysanthemum. *Plant Pathology Journal* 24, 31–35.

Copes, W.E. and Scherm, H. (2005) Plant spacing effects on microclimate and rhizoctonia web blight development in container-grown azalea. *HortScience* 40(5), 1408–1412.

Copes, W.E. (2018) Sanitation for management of florists' crops diseases. In: McGovern, R.J. and Elmer, W.H. (eds) *Handbook of Florists' Crop Diseases*, Vol. 1. Handbook of Plant Disease Management, Springer AG, Cham, Switzerland, pp. 201–236.

Crow, B. (2003) Nematode management for bedding plants. Entomology and Nematology Department, University of Florida, IFAS Extension, Gainesville, ENY-052. Available at: https://edis.ifas.ufl.edu/publication/IN470 (accessed December 2022).

Crozzoli, R., Hurtado, T., Perichi, G. and Arcia, A. (2008) Characterization of a venezuelan population of *Aphelenchoides ritzemabosi* on chrysanthemum. *Nematologia Mediterranea* 36(1), 79–83.

Daughtrey, M., Gilrein, D., Helms, M., Mattson, N., Sanderson, J. *et al.* (2023) *2024–2025 Cornell Guide for the Integrated Management of Greenhouse Crops and Herbaceous Ornamentals.* Cornell Cooperative Extension, Ithaca, New York.

Deimi, A.M., Chitambar, J.J. and Maafi, Z.T. (2008) Nematodes associated with flowering ornamental plants in Mahallat, Iran. *Nematologia Mediterranea* 36(2), 115–123.

DeLong, J.R., Streminska, M.A., Persijn, A., Huisman, H.M. and van der Salm, C. (2021) Biological control of *Meloidogyne* spp. in glasshouse-grown chrysanthemum. *Journal of Nematology* 52, 1–125. DOI: 10.21307/jofnem-2020-125.

Demers, J.E., Crouch, J.A. and Castlebury, L.A. (2015) A multiplex real-time PCR assay for the detection of *Puccinia horiana* and *P. chrysanthemi* on chrysanthemum. *Plant Disease* 99, 195–200.

Dickens, J.S.W. (1990) Studies on the chemical control of chrysanthemum white rust caused by *Puccinia horiana*. *Plant Pathology* 39(3), 434–442.

Dienelt, M.M. and Lawson, R.H. (1988) Investigation of the etiology and cytology of chrysanthemum slow decline. *Acta Horticulturae* 234, 307–314. DOI: 10.17660/ActaHortic.1988.234.37.

Dienelt, M.M. and Lawson, R.H. (1991) Chrysanthemum foliar necrosis: Transmission electron microscopy of leaf lesions. *Phytopathology* 81, 1079–1087.

Diener, T.O. and Lawson, R.H. (1973) Chrysanthemum stunt: A viroid disease. *Virology* 51, 94–101.

Dimock, A.W. (1947) Chrysanthemum stunt. NY state flower growers. *Bulletin* 26, 2.

Dolliver, J.S., Hildebrandt, A.C. and Riker, A.J. (1962) Studies of reproduction of *Aphelenchoides ritzemabosi* (Schwanz) on plant tissues in culture. *Nematologica* 7(4), 294–300.

Engelhard, A.W. and Woltz, S.S. (1973) Fusarium wilt of chrysanthemum: Complete control of symptoms with an integrated fungicide-lime-nitrate regime. *Phytopathology* 63, 1256–1259.

Farr, D.F. and Rossman, A.Y. (2019) Fungal databases, U.S. National Fungus Collections, ARS, USDA. Available at: https://nt.ars-grin.gov/fungaldatabases/ (accessed December 2022).

Fukuta, S., Ohishi, K., Yoshida, K., Mizukami, Y., Ishida, A. *et al.* (2004) Development of immunocapture reverse transcription loop-mediated isothermal amplification for the detection of tomato spotted wilt virus from chrysanthemum. *Journal of Virological Methods* 121(1), 49–55.

Gardiner, D.C., Horst, R.K. and Nelson, P.E. (1987) Symptom enhancement of *Fusarium Wilt* of chrysanthemum by high temperatures. *Plant Disease* 71(12), 1106–1109.

Götz, M., Ulrich, R. and Werres, S. (2017) First detection of *Phytophthora chrysanthemi* on *Chrysanthemum indicum* in Germany. *New Disease Reports* 35, Article. DOI: 10.5197/j.2044-0588.2017.035.006.

Guan, Z., Wu, D., Song, A., Chen, F., Chen, S. *et al.* (2017) A highly sensitive method for the detection of Chrysanthemum virus B. *Electronic Journal of Biotechnology* 26, 64–68.

Hakkaart, F.A. and Quak, F. (1964) Effect of heat treatment of young plants on freeing chrysanthemums from virus B by means of meristem culture. *Netherlands Journal of Plant Pathology* 70, 154–157.

Handley, M.K. and Horst, R.K. (1988) The effect of temperature and light on chrysanthemum stunt viroid. *Acta Horticulturae* 234, 89–97.

Handoo, Z., Kantor, M. and Carta, L. (2020) Taxonomy and identification of principal foliar nematode species (*Aphelenchoides* and *Litylenchus*). *Plants* 9, 1490. DOI: 10.3390/plants9111490.

Hikichi, Y., Wali, U.M., Ohnishi, K. and Kiba, A. (2013) Mechanism of disease development caused by a multihost plant bacterium, *Pseudomonas cichorii*, and its virulence diversity. *Journal of General Plant Pathology* 79, 379–389.

Hill, M.F., Giles, R.J., Moran, J.R. and Hepworth, G. (1996) The incidence of chrysanthemum stunt viroid, chrysanthemum B carlavirus, tomato aspermy cucumovirus and tomato spotted wilt tospovirus in Australian chrysanthemum crops. *Australasian Plant Pathology* 25, 174–178.

Hollings, M. (1955) Investigation of chrysanthemum viruses 1. Aspermy flower distortion. *Annals of Applied Biology* 43(1), 86–102.

Hollings, M. and Stone, O.M. (1971) Tomato aspermy virus. *DPVweb.Net*. Available at: https://www.dpvweb.net/dpv/showdpv/?dpvno=79 (accessed November 2022).

Hollings, M. and Stone, O.M. (1973) Some properties of chrysanthemum stunt, a virus with characteristics of an uncoated ribonucleic acid. *Annals of Applied Biology* 74, 333–348.

Horst, R.K. (1985) Chrysanthemum diseases. In: Strider, D.L. (ed.) *Diseases of Floral Crops*. Praeger Publishers, New York, pp. 565–628.

Horst, R.K. and Nelson, P.E. (eds) (1997) *Compendium of Chrysanthemum Diseases*. American Phytopathological Society (APS Press), St. Paul, Minnesota.

Jewett, T.J. and Jarvis, W.R. (2001) Management of the greenhouse microclimate in relation to disease control: A review. *Agronomie* 21, 351–366.

Johnson, A.W. and Littrell, R.H. (1969) Effect of *Meloidogyne incognita, M. hapla* and *M. javanica* on the severity of *Fusarium wilt* of chrysanthemum. *Journal of Nematology* 1, 122–125.

Jones, J.B., Chase, A.R., Raju, B.C. and Harbaugh, B.K. (1984a) The roles of fertilizer rate, light intensity, and humidity as pre-inoculation factors in susceptibility of chrysanthemums to *Pseudomonas cichorii* leaf spot (Abstr.). *Phytopathology* 74, 628.

Jones, J.B., Raju, B.C. and Engelhard, A.W. (1984b) Effects of temperature and leaf wetness on development of bacterial spot of geraniums and chrysanthemums incited by *Pseudomonas cichorii*. *Plant Disease* 68, 248–251.

Kohl, L.M. (2011) Foliar nematodes: A summary of biology and control with a compilation of host range. *Plant Health Progress* 12, 23. DOI: 10.1094/PHP-2011-1129-01-RV.

Kumar, S., Khan, M.S., Raj, S.K. and Sharma, A.K. (2009) Elimination of mixed infection of Cucumber mosaic and Tomato aspermy virus from *Chrysanthemum morifolium* Ramat. cv. Pooja by shoot meristem culture. *Scientia Horticulturae* 119(2), 108–112.

Kwak, H.-R., Kim, J.-G., Kim, J.-E., Byun, H.-S., Choi, H.-S. *et al.* (2021) First report of tomato aspermy virus in *Chrysanthemum zawadskii* var. latilobum in Korea. *Journal of Plant Pathology* 103, 1005–1006.

LaMondia, J.A. (1999) Efficacy of insecticides for control of *Aphelenchoides fragariae* and *Ditylenchus dipsaci* in flowering perennial ornamentals. *Supplement to the Journal of Nematology* 31(4S), 644–649.

LaMondia, J.A. and Douglas, S.M. (1997) Sensitivity of *Botrytis cinerea* from Connecticut greenhouses to benzimidazole and dicarboximide fungicides. *Plant Disease* 81(7), 729–732.

Lin, M.-J., Chang, C.-A., Chen, C.-C. and Cheng, Y.-H. (2005) Occurrence of chrysanthemum virus B in Taiwan and preparation of its antibody against coat protein expressed in bacteria. *Plant Pathology Bulletin* 14, 191–202.

Lin, S., Martin, D.E., Taylor, N.J., Gabriel, C.K. and Peduto Hand, F. (2017) Occurrence of *Phytophthora chrysanthemi* causing root and stem rot on garden mums in the United States. *Plant Disease* 101(6), 1060. DOI: 10.1094/PDIS-12-16-1714-PDN.

Matic, S., Gilardi, G., Gullino, M.L. and Garibaldi, A. (2018) Evidence for an expanded host range of *Fusarium oxysporum* f. sp. *Chrysanthemi*. *Journal of Plant Pathology* 100, sp.

Matteoni, J.A. and Broadbent, A.B. (1988) Wounds caused by *Liriomyza trifolii* (Diptera: Agromyzidae) as sites for infection of chrysanthemum by *Pseudomonas cichorii*. *Canadian Journal of Plant Pathology* 10, 47–52.

Matteoni, J.A. and Allen, W.R. (1989) Symptomatology of tomato spotted wilt virus infection in florist's chrysanthemum. *Canadian Journal of Botany* 67, 373–380.

McCuiston, J.L., Hudson, L.C., Subbotin, S.A., Davis, E.L. and Warfield, C.Y. (2007) Conventional and PCR detection of *Aphelenchoides fragariae* in diverse ornamental host plant species. *Journal of Nematology* 39, 343–355.

McGovern, R.J., Horst, R.K. and Israel, H.W. (1989) Chrysanthemum phloem necrosis: Symptomatology and histopathology. *Canadian Journal of Botany* 67, 1014–1023.

McGovern, R.J., Horita, H. and Stiles, C.M. (2006) Host range of *Itersonilia perplexans* and management of Itersonilia petal blight of China aster. *Plant Health Progress* 7(1). DOI: 10.1094/PHP-2006-1018-02-RS.

Mitkowski, N.A. and Abawi, G.S. (2011) Root-knot nematodes. *The Plant Health Instructor*. Available at: https://doi.org/10.1094/PHI-I-2003-0917-01 (accessed November 2022).

Molnar, C., Nikolaeva, E., Kim, S., Olson, T., Bily, D. *et al.* (2020) Phytophthora diversity in pennsylvania nurseries and greenhouses inferred from clinical samples collected over four decades. *Microorganisms* 8, 1056.

Moore, L. and Putnam, M.L. (2021) Crown gall disease of nursery crops. In: Pscheidt, J.S. and Ocamb, C.M. (eds) *Pacific Northwest Plant Disease Management Handbook, Vol. Available at.* A Pacific Northwest Extension Publication, Corvallis, Oregon. Available at: https://pnwhandbooks.org/node/409

Naher, M., Motohash, K., Watanabe, H., Chikuo, Y., Senda, M. *et al.* (2011) *Phytophthora chrysanthemi* sp. nov., a new species causing root rot of chrysanthemum in Japan. *Mycological Progress* 10(1), 21–31.

Nakamura, Y., Fujisawa, I., Lee, K.H. and Uematsu, S. (1994) Cucumber mosaic virus and chrysanthemum virus B isolated from marguerite, *Chrysanthemum frutescens* L. In: *Proceedings of the Kanto-Tosan Plant Protection Society*, pp. 177–179 (Vol. 41).

Nau, J., Calkins, B. and Westbrook, A. (2021) *Ball RedBook*, 19th edn., Vol. 2. Crop Culture and Production. Ball Publishing, West Chicago, Illinois.

Newcombe, G. (2003) *Puccinia tanaceti*: Specialist or generalist. *Mycological Research* 107, 797–802.

Nishi, N., Muta, T., Ito, Y., Nakamura, M. and Tsukiboshi, T. (2009) Ray speck of chrysanthemum caused by *Stemphylium lycopersici* in Japan. *Journal of General Plant Pathology* 75, 80–82.

Osdaghi, E. (2020) *Pseudomonas cichorii* (bacterial blight of endive). In: *Invasive Species Compendium*. CABI, Wallingford, UK. Available at: https://doi.org/10.1079/ISC.44942.20210200740

Owen, W.G., Henry, J. and Whipker, B.E. (2018) Chrysanthemum, garden. *E-Gro Nutritional Monitoring*. Available at: http://www.fertdirtandsquirt.com/pdf/Chrysanthemum-Garden.pdf (accessed December 2022).

Palmer, C.L., Bonde, M.R., Nester, S.E., Revell, J.M. and Luster, D.G. (2015) Fungicide impact on *in vitro* germination of basidiospores of *Puccinia horiana*, the causal agent of chrysanthemum white rust. *Plant Health Progress* 16(2), 73–76. DOI: 10.1094/PHP-RS-14-0032.

Palukaitis, P. (2017) Chapter 17: Chrysanthemum stunt viroid. In: Hadidi, A., Flores, R., Randles, J.W. and Palukaitis, P. (eds) *Viroids and Satellites*. Elsevier, Cambridge, Massachusetts, pp. 181–190. DOI: 10.1016/B978-0-12-801498-1.00017-6.

Pscheidt, J.W. and Ocamb, C.M. (2022) *Pacific Northwest Plant Disease Management Handbook*. Oregon State University, Corvallis, Oregon. Available at: https://pnwhandbooks.org/plantdisease/host-disease/chrysanthemum-powdery-mildew

Pulawska, J., Willems, A. and Sobiczewski, P. (2012) *Rhizobium skierniewicense* sp. nov. isolated from tumours on chrysanthemum and cherry plum. *International Journal of Systematic and Evolutionary Microbiology* 62, 895–899. DOI: 10.1099/ijs.0.032532-0.

Punithalingam, E. and Wheeler, B.E.J. (1965) Septoria spp. occurring on species of chrysanthemum. *Transactions of the British Mycological Society* 48(3), 423–439.

Putnam, M. (2015) Crown gall: Still confounding scientists and growers alike. *Pest Management. Grower Talks*. Available at: https://www.growertalks.com/Article/?articleid=21821 (accessed December 2022).

Raj, S.K., Khan, M.S. and Kumar, S. (2007) Molecular identification of '*Candidatus* Phytoplasma asteris' associated with little leaf disease of *Chrysanthemum morifolium*. *Australasian Plant Disease Notes* 2, 21–22.

Randall-Schadel, B. (2016) *Phytophthora chrysanthemi* Naher, Hi Watan., Chikuo & Kageyama: crown and root rot of chrysanthemum. New Pest Advisory Group Report, USDA APHIS PPQ. Available at: http://www.canr.org/newsletter/PhytophthorachrysanthemiNPAGReport20160401R.pdf (accessed December 2022).

Ryder, M.H. and Jones, D.A. (1991) Biological control of crown gall using agrobacterium strains K84 and K1026. *Australian Journal of Plant Physiology* 18, 571–579.

Samson, R., Legendre, J.B., Christen, R., Fischer-Le Saux, M., Achouak, W. *et al.* (2005) Transfer of *Pectobacterium chrysanthemi* (Burkholder *et al.* 1953) Brenner *et al.* 1973 and *Brenneria paradisiaca* to the genus Dickeya gen. nov. as *Dickeya chrysanthemi* comb. nov. and *Dickeya paradisiaca* comb. nov. and delineation of four novel species, *Dickeya dadantii* sp. nov., *Dickeya dianthicola* sp. nov., *Dickeya dieffenbachiae* sp. nov. and *Dickeya zeae* sp. nov. *International Journal of Systematic and Evolutionary Microbiology* 55, 1415–1427.

Savitri, W.D., Park, K.I., Jeon, S.M., Chung, M.Y., Han, J.S. *et al.* (2013) Elimination of chrysanthemum stunt viroid (CSVd) from meristem tip culture combined with prolonged cold treatment. *Horticulture, Environment, and Biotechnology* 54, 177–182.

Singh, G. and Milne, K.S. (1974) Field evaluation of fungicides for the control of chrysanthemum flower blight. *New Zealand Journal of Experimental Agriculture* 2(2), 185–188. DOI: 10.1080/03015521.1974.10425759.

Singh, P.K. and Kumar, V. (2011) Biological control of *Fusarium wilt* of Chrysanthemum with Trichoderma and botanicals. *Journal of Agricultural Technology* 7(6), 1603–1613.

Singh, L., Hallan, V., Martin, D.P., Ram, R. and Zaidi, A.A. (2012) Genomic sequence analysis of four new Chrysanthemum virus B isolates: Evidence of RNA recombination. *Archives of Virology* 157, 531–537.

Staniland, L.N. (2008) Experiments on the control of chrysanthemum eelworm (*Aphelenchoides ritzema*-bosi Schwartz) by hot water treatment. *Annals of Applied Biology* 37(1), 11–18.

Stockwell, V.O., Moore, L.W. and Loper, J.E. (1993) Fate of agrobacterium radiobacter K84 in the environment. *Applied and Environmental Microbiology* 59(7), 2112–2120. DOI: 10.1128/aem.59.7.2112-2120.1993.

Subbotin, S.A., Oliveira, C.J., Alvarez-Ortega, S., Desaeger, J.A., Crow, W. *et al.* (2021) The taxonomic status of *Aphelenchoides besseyi* Christie, 1942 (Nematoda: Aphelenchoididae) populations from the southeastern USA, and description of *Aphelenchoides pseudobesseyi* sp. n. *Nematology* 381–413.

Taloh, A., Raju, D.V.S., Banyal, N., Kumar, G., Panda, P. *et al.* (2020) Genetic diversity of phytoplasma strains infecting chrysanthemum varieties in India and their possible natural reservoirs. *3 Biotech* 10(9), 411. DOI: 10.1007/s13205-020-02407.

Tran, D.H., Cho, S., Hoang, P.M., Kim, J., Kil, E.-J. *et al.* (2016) A codon-optimized nucleic acid hydrolyzing single-chain antibody confers resistance to chrysanthemums against chrysanthemum stunt viroid. *Plant Molecular Biology Reporter* 34, 221–232. DOI: 10.1007/s11105-015-0915-5.

Trolinger, J.C., McGovern, R.J., Elmer, W.H., Rechcigl, N.A. and Shoemaker, C.M. (2018) Diseases of chrysanthemum. In: McGovern, R.J. and Elmer, W.H. (eds) *Handbook of Florists' Crops Diseases*, Vol. 1. Springer, Cham, Switzerland, pp. 462–464.

USDA APHIS (2021) Chrysanthemum white rust. Online fact sheet. Available at: https://www.aphis.usda.gov/aphis/ourfocus/planthealth/plant-pest-and-disease-programs/pests-and-diseases/plant-disease/sa_cwr/ct_chrysanthemum_white_rust (accessed 16 August 2021).

Vaghefi, N., Hay, F.S., Pethybridge, S.J., Ford, R. and Taylor, P.W.J. (2016) Development of a multiplex PCR diagnostic assay for the detection of stagonosporopsis species associated with ray blight of asteraceae. *European Journal of Plant Pathology* 146, 581–595.

Van de Wetering, F., Posthuma, K., Goldbach, R. and Peters, D. (1999) Assessing the susceptibility of chrysanthemum cultivars to tomato spotted wilt virus. *Plant Pathology* 48, 693–699.

Vegh, A., Nemethy, Z., Salamon, P., Mandoki, Z. and Palkovics, L. (2014) First report of bacterial wilt on chrysanthemum caused by *Dickeya chrysanthemi* (syn. *Erwinia chrysanthemi*) in Hungary. *Plant Disease* 98(7), 988. DOI: 10.1094/PDIS-09-13-0948-PDN.

Waddell, H.T. and Weber, G.F. (1963) Physiology and pathology of Septoria species on chrysanthemum. *Mycologia* 55, 442–452.

Wallace, H.R. (1959) Movement of eelworms. V. observations on *Aphelenchoides ritzema-bosi* (Schwartz, 1912), Steiner, 1932 on florists' chrysanthemums. *Annals of Applied Biology* 47(2), 350–360.

Weisberg, A.J., Davis, E.J.I., Tabima, J., Belcher, M.S. and Miller, M. (2020) Unexpected conservation and global transmission of agrobacterial virulence plasmids. *Science* 368, 6495. DOI: 10.1126/science.aba5256.

Werres, S., Gotz, M., Tomic, Z., Ivic, D., Peduto Hand, F. *et al.* (2017) *Phytophthora chrysanthemi* (Naher, Hi. Watan., Chikuo & Kageyama). *JKI Data Sheets - Plant Diseases and Diagnosis* 1–14. DOI: 10.5073/jkidspdd.2017.002.

Whipker, B.E. (2014) *E-Gro Alert*, vol. 3, no. 62. Available at: https://www.e-gro.org/pdf/362.pdf

Wilson, J.M. and Moran, J.R. (1983) The incidence of chrysanthemum virus B, tomato aspermy virus and tomato spotted wilt virus in commercial chrysanthemum flower crops in Victoria. *Australasian Plant Pathology* 12(2), 17.

Yakabe, L.E., Parker, S.R. and Kluepfel, D.A. (2010) Effect of pre-plant soil fumigants on *Agrobacterium tumefaciens*, pythiaceous species, and subsequent soil recolonization by *A. tumefaciens*. *Crop Protection* 29, 583–590.

Yoon, J.-Y. and Palukaitis, P. (2013) Sequencing comparison of global chrysanthemum stunt viroid variants: Multiple polymorphic positions scattered through the viroid genome. *Virus Genes* 46, 97–104.

Young, J.M., Kuykendall, L.D., Martínez-Romero, E., Kerr, A. and Sawada, H. (2001) A revision of Rhizobium Frank 1889, with an emended description of the genus, and the inclusion of all species of Agrobacterium Conn 1942 and *Allorhizobium undicola* de Lajudie *et al.* 1998 as new combinations: *Rhizobium radiobacter*, *R.*

rhizogenes, *R. rubi*, *R. undicola* and *R. vitis*. *International Journal of Systematic and Evolutionary Microbiology* 51, 89–103.

Yourman, L.F. and Jeffers, S.N. (1999) Resistance to benzimidazole and dicarboximide fungicides in greenhouse isolates of *Botrytis cinerea*. *Plant Disease* 83(6), 569–575.

Zandvoort, R. (1968) Wind dispersal of *Puccinia horiana*. *Netherlands Journal of Plant Pathology* 74, 124–127.

Zhao, S., Chen, X., Deng, S., Dong, X., Song, A. *et al.* (2016) The effects of fungicide, soil fumigant, bio-organic fertilizer and their combined application on chrysanthemum *Fusarium wilt* controlling, soil enzyme activities and microbial properties. *Molecules* 21, 526. DOI: 10.3390/molecules2104052.

11

Postharvest Management of Chrysanthemum

John M. Dole[1]* and James E. Faust[2]

[1]*Department of Horticultural Science, North Carolina State University, Raleigh, NC 27695, USA*
[2]*Department of Plant and Environmental Sciences, Clemson University, E-143 Poole Agricultural Center, Clemson, SC 29634, USA*

Abstract

Chrysanthemums (*Chrysanthemum morifolium*) are one of the world's most important cut flowers, potted flowering plants, and garden plants due to their stunning array of flower colors, forms, and sizes as well as a long postharvest life. Many factors affect postharvest life, starting with selecting long-lasting cultivars, especially in regard to resistance to leaf yellow or necrosis. Optimum relative humidity, light and nutrition during production, and proper harvest stage are needed for a long shelf life. After harvest, several factors should be considered including minimizing ethylene, ensuring stems and plants are well hydrated, and preventing mechanical damage and diseases. Maintaining low temperatures, 0–1°C, from harvest to sale and using proper bucket and vase solutions is especially critical for cut flowers. The future is bright for chrysanthemums with continually improved breeding, sustainable production practices, improved transport logistics, such as sea transport, and new handling protocols, such as subzero storage of cut flowers.

Keywords

Chrysanthemums, ethylene sensitivity, flower senescence, flower supply chain, postharvest life, vase life

*Corresponding author: jmdole@ncsu.edu

DOI: 10.1079/9781789248289.0011

11.1 Introduction

The production of high-quality chrysanthemum cut flowers, potted flowering plants, and garden plants is an important first step in ensuring a high-quality product for consumers. Thereafter, plants must be properly harvested and handled all the way to the final customer. Optimum postharvest life is achieved with proper cultivar selection and the correct environmental conditions during production, as well as through the myriad of harvest, handling, and shipping steps. Fortunately, properly handled chrysanthemums will have a long postharvest life compared to many other floriculture crops. This characteristic, along with a stunning array of flower colors, forms, and sizes, has resulted in chrysanthemum becoming one of the most important flowers in the world.

11.2 Factors Affecting Postharvest Performance

11.2.1 Breeding and selection

The first step in the improvement of postharvest performance occurs during the breeding and selection process for new cultivar introductions (Baskaran *et al.*, 2010). For example, one current breeding effort is addressing storage-induced degreening of disk florets (van Geest *et al.*, 2017). The postharvest longevity of chrysanthemums is frequently limited by foliage, not flower, performance. Selection for insensitivity to ethylene has reduced leaf yellowing in the postharvest environment. However, leaf performance remains an issue for novel cut flower cultivars that have highly desirable and unusual flower colors or color patterns, for example, green, or bicolor flower petals, where leaf chlorosis or necrosis frequently terminates vase life prior to flower senescence. As a result of the collaborative efforts of plant physiologists and breeders, chrysanthemum postharvest performance continues to improve and is among the best of floriculture species.

11.2.2 Production (preharvest) environment

11.2.2.1 Relative humidity

The variation in cut flower postharvest life among cultivars is attributed to leaf stomatal characteristics (Fanourakis *et al.*, 2021). These stomatal characteristics and hence water loss from the foliage can be affected by the environment under which the stems were grown. Plants grown under high (90 ± 5%) relative humidity (low vapor pressure deficit) have larger stomata and higher stomatal densities than those grown under 55 ± 5% relative humidity (moderate vapor pressure deficit) (Aliniaeifard and van Meeteren, 2016). When harvested, stems grown under high humidity lose water more rapidly. Similar responses occurred even when the stems were subjected to high humidity for only a few days (five) prior to harvest, indicating the importance of increasing air movement during periods of foggy or rainy weather that can cause high humidity in the greenhouses.

Fig. 11.1. Shade curtain placed above a high-light-sensitive cut flower cultivar during flower development to prevent petal necrosis or burning.
(Photo courtesy of James E. Faust.)

11.2.2.2 Light

Light intensity during production affects postharvest performance. Nell *et al.* (1990) showed that higher light intensities (500 μmol m^{-2} s^{-1} compared to 100–300 μmol m^{-2} s^{-1}) during the final weeks of potted chrysanthemum production resulted in improved growth (larger flower size) and development (shorter time to flower) as well as improved postharvest longevity. Thus, acclimating plants to lower light intensities prior to harvest was not beneficial in the postharvest environment. However, cut flower chrysanthemum growers often place an additional shade screen above cultivars that are sensitive to high light intensities during flower petal development to avoid petal necrosis/burning (Fig. 11.1).

11.2.2.3 Nutrition

Chrysanthemums are vegetatively propagated, and cuttings may be shipped long distances and/or stored prior to propagation (Fig. 11.2). The amount of nitrogen delivered to the stock plant can affect rooting. For example, Drüge *et al.* (2000) reported that adventitious root number and length were positively correlated with initial total nitrogen in the cutting tissue. High nitrogen supplied

Fig. 11.2. Crate filled with plastic bags containing unrooted cuttings, stored in a cooler before propagation. (Photo courtesy of James E. Faust.)

to the stock plants also resulted in high sucrose concentrations and low starch levels, thus high nitrogen fertilization was associated with increased transport of nitrogen compounds and carbohydrates to the base of the cutting where root initiation occurs. During the last 5 weeks of potted chrysanthemum production, the ratio of ammonium-to-nitrate-nitrogen in the fertilizer solution affects the postharvest longevity (Roude *et al.*, 1991a). Longevity increased linearly as the ammonium-to-nitrate-nitrogen ratio increased from 1:0 to 0:1. Roude *et al.* (1991b) also reported that postharvest longevity increased with lower total nitrogen applied during potted chrysanthemum production. Some cut chrysanthemum growers using field soils in greenhouses will cease fertilization 1 or 2 weeks prior to harvest to reduce soil salinity, which can negatively affect the growth of the next crop planted in those same beds. However, this practice can increase the incidence of leaf chlorosis in the vase, so it may be a better practice to continue fertilization until the flowers are harvested and then leach the salts out of the soil with low-salinity water before planting the next crop. Foliar applications of humic acid, commencing every 15 days after transplanting, can improve chlorophyll content and net photosynthesis while also enhancing the activity of antioxidant enzymes in chrysanthemum cut flowers, which delays flower senescence and extends the vase life (Fan *et al.*, 2015).

11.3 Harvest and Postharvest Factors

11.3.1 Flower maturity

Flower maturity at harvest is most critical for cut flowers, where producers need to balance the degree of openness with ease of handling. Recently, open flowers typically have better color development than those harvested earlier, but petals may be damaged during harvesting and handling. On the other extreme, buds can be harvested, which can be subsequently opened after harvest using bucket solutions with carbohydrates, but flower color may not be well developed, and time will be required to open the buds.

Standard chrysanthemums are harvested when the outer petals are fully elongated (Dole *et al.*, 2017). To prevent petal damage on sensitive cultivars, especially those with long petals such as spiders, net sleeves may be placed on individual flowers near the end of the production phase (Fig. 11.3). Spray daisy chrysanthemums are harvested when the older flowers are open, but no pollen has been shed. Spray anemone types are cut when the older flowers are open but before disk flowers start to elongate, and spray decoratives are harvested when the centers of the older flowers are fully open. Thomas and Gollnow (2013) noted that at least five flowers should be open per spray stem and Kim *et al.* (2012) recommended three to four open flowers for the highest quality grade and five to six for second highest quality spray stems. Grades and standards for cut chrysanthemums have also been summarized by Nowak and Rudnicki (1990). For potted flowering plants, a broader stage of development can be used as flowers are more likely to continue to develop on the plant under proper postharvest conditions.

Potted plants for indoor use are marketed when 33–50% of the flowers are open and have developed pigmentation characteristic of the cultivar (Nell, 1993). The center of the flowers may turn brown on plants harvested

Fig. 11.3. Nets placed on the stems of standard cut flowers to be pulled over the open flowers at harvest, protecting them from physical damage during packing and shipping. (Photo courtesy of James E. Faust.)

when the flowers are too immature or when placed in excessively warm or dark conditions. The Produce Marketing Association and the Society of American Florists developed standards for potted flowering plants considering plant height and diameter, number of flower buds, stage of openness, and quality characteristics (PMA/SAF, n.d.). Potted chrysanthemums intended for outdoor use can be shipped when color is just showing as plants are likely to be sold from garden centers or other outdoor venues where the flowers can continue to develop. Customers are advised to purchase these plants before they are in full bloom to have a longer display period, but many customers are drawn to color and are more likely to purchase plants with fully open flowers.

11.3.2 Cut flower harvest

Cut stems are generally harvested as low as possible to maximize stem length; however, care should be taken not to cut too low on the stem where the tissue is woody, as that will restrict water uptake (Sacalis, 1993). Foliage is removed from the lower ½ to ⅓ of the stem.

11.3.3 Ethylene

Most references state that chrysanthemums are ethylene insensitive (Dole and Wilkins, 2005; Nell, 1993, 2003; Woltering and Van Doorn, 1988). However, ethylene can cause leaf yellowing (Doi *et al.*, 2003, 2004), which can be a major issue with cut chrysanthemums. Cultivar differences exist, with some being sensitive to ethylene at concentrations as low as 1 ppm for 24 hour, while other cultivars are insensitive. Application of 0.2 mM silver thiosulfate completely inhibited the leaf yellowing in sensitive cultivars and 1-MCP delayed it (Doi *et al.*, 2003). Fig. 11.4 demonstrates the use of a 1-MCP sachet in a box of cut chrysanthemums prior to packing.

11.3.4 Water

Water uptake is critical to maintaining cut flower vase life, with pH and electrical conductivity (EC) being the two most important factors. As with most species, low pH solutions increase chrysanthemum vase life due to improved water uptake and increased fresh weight (Carlson and Dole, 2013). Interestingly, chrysanthemums are quite tolerant of water EC with little difference in vase life noted at EC's ranging from 0.5 to 4.0 dS m^{-1}, although fresh weight was reduced and marginal necrosis slightly increased as the EC increased (Carlson and Dole, 2013). It should be noted that vase life was reduced at 0 dS m^{-1}, indicating the presence of a small amount of salt in the water is beneficial. In earlier work, Waters (1966) noted that chrysanthemum vase life decreased as the total soluble salt content increased and foliar deterioration increased, first as marginal, then general leaf chlorosis followed by necrosis. However, the foliage of modern cultivars is likely more resistant to salt stress than that of older cultivars. Dry-handled or dry-stored stems should be rehydrated in cold (0–5°C) water to improve water uptake (van Meeteren, 1992). Water temperature had little effect when stems were fully hydrated. Not surprisingly, the

Fig. 11.4. Sachets of 1-MCP placed in boxes before packing to protect stems from ethylene damage. Floral paper that releases chlorine dioxide is placed near the flowers to prevent fungal infections such as Botrytis. (Photo courtesy of James E. Faust.)

majority of water loss occurs through the foliage and stem and not through the flowers (Fanourakis *et al.*, 2021). As previously noted, water loss from the foliage during postharvest can be affected by the environment under which the chrysanthemums were grown.

11.3.5 Temperature

As with most floriculture crops, storage and transport temperatures are critical. Cut flowers should be stored at 0–1°C. Note that with proper equipment and monitoring, cut chrysanthemums can be stored at −0.6°C, allowing stems to be stored for longer (Sacalis, 1993; Kalinowski and Dole, 2024). Warmer cold storage temperatures, up to 4°C, can be tolerated, but vase life will be reduced. Potted flowering plants should be stored at 2–4°C for a maximum of 7 days (Nell, 1993). For longer storage, temperatures should be 1–2°C. Rooted chrysanthemum cuttings of some cultivars can be stored at 0–3°C for 4–6 weeks, but only for 1 week for weaker cultivars (Rajapakse *et al.*, 1996). Unrooted cuttings can be stored at temperatures near 0°C for up to 3 weeks (Dole and Gibson, 2006).

11.3.6 Nutrition

Potted flowering and garden plants should have a modest level of nutrition in the substrate at the time of harvest. Too high fertilizer concentration and EC in the substrate at harvest may reduce postharvest life (Nell *et al.*, 1989). Display life can be maximized by completely terminating fertilization at the disbud stage. Note that termination of fertilization assumes plants were properly fertilized prior to termination (Nell, 1993).

11.3.7 Light

For cut flowers, providing 10–20 $\mu mol\ m^{-2}\ s^{-1}$ of light will help to keep the foliage green and delay yellowing or browning (Sacalis, 1993). As cut stems are typically placed in boxes when stored for long durations, this practice is limited to short-duration storage in solutions and to those cultivars that are especially sensitive to leaf problems. Chrysanthemum flowers used for garlands last longer when exposed to 5 $\mu mol\ m^{-2}\ s^{-1}$ of white LED light (Zhou *et al.*, 2019). Potted plants are best held under a minimum of 10 $\mu mol\ m^{-2}\ s^{-1}$ during marketing and sales (Nell, 1993). Providing light at 23 $\mu mol\ m^{-2}\ s^{-1}$ to rooted cuttings during cold storage improves cutting quality (Rajapakse *et al.*, 1996).

11.3.8 Diseases

The most important postharvest disease is gray mold, caused by *Botrytis*. This common pathogen typically infects the plant material during production but may not be noticeable until after harvest. Botrytis is favored by warm temperatures, high humidity, and free water on the plant tissue, which are common conditions in the greenhouse, tunnel, or field and can occur during postharvest handling as well, especially after stems are packed in boxes or plants are sleeved. If boxes are allowed to get warm at any time during the storage and shipping process, the likelihood of significant damage from botrytis increases greatly. Control of botrytis is best accomplished during production to avoid infection altogether. In some cases, botrytis is so prevalent that producers will dip flowers in a fungicide solution after harvest. This can be problematic if the fungicide is not close to 100% effective, because the process puts free water on the petal tissue, which may encourage the disease. Floral paper that releases chlorine dioxide can be placed in packing boxes near the flowers to help protect the stems from fungal infections that may occur during transit.

Other diseases can occur during postharvest but are uncommon and, as with botrytis, they typically start during production and carry over into the postharvest period. Chrysanthemums are particularly susceptible to *Itersonilia* petal blight and *Stemphylium* leaf and flower spot. Chrysanthemums are sensitive to microbial growth in the vase water, which restricts water uptake and increases water stress (Put, 1990). Interestingly, freshly harvested chrysanthemum stems had higher levels of bacteria and fungi than roses, probably due to "the numerous tiny side leaves" on chrysanthemums, which harbored microbes.

Fig. 11.5. Standard (disbud) cut flower chrysanthemums protected with net sleeves. (Photo courtesy of James E. Faust.)

11.3.9 Mechanical damage

The petals of cut flowers can be easily damaged during harvest and handling, if care is not taken. Individual net sleeves can be used on large standard flowers to protect the petals during harvest and handling (Fig. 11.5). The cost of the nets and the labor to use them dictates that they are only used on the most valuable or sensitive flowers, especially spider or quill types with long delicate petals. While the focus is generally on the flowers, foliage is also susceptible to cracking due to mechanical damage during harvest, sleeving, and packing. Potted flowering plants and garden plants can be sensitive to stem breakage. Ethephon is commonly used to increase branching and prevent premature budding in garden chrysanthemums (Hayashi *et al.*, 2001). However, excessive application of ethephon during production can increase the likelihood of stem breakage. For large garden plants, producers may take the time and expense of staking and wrapping a string around the plants to prevent them from breaking apart. However, they will only do so when guaranteed a price that still allows a profit.

11.3.10 Disorders

The major limitations to the vase life of cut flowers or display life of potted plants are the failure of flower buds to open, browning of the central petals, and foliar chlorosis and necrosis. Flower buds may not open if cut stems or potted plants are subjected to low light levels during production and high temperatures during storage and shipping. High temperatures can also cause open flowers to develop discolored or brown centers. Cut flowers that are stored cold for more than 2–3 weeks may also exhibit brown centers. Foliar yellowing is a common issue of cut flowers, and the foliage often becomes unacceptable before the flowers senesce. Several causes have been identified including high temperatures, low light, ethylene, low nutrition, and the use of sugar in postharvest solutions. Temperature and light can interact in such a way that leaves may turn yellow when stored in dark at 5°C, but not at 1°C (Reid, 1997). Doi *et al.* (2004) showed that in ethylene-sensitive cultivars, the application of ethylene 1–3 days after harvest had no effect, while the plants were sensitive to applications of ethylene made 4–6 days after harvest. Thus, it is possible that ethylene sensitivity is related to the carbohydrate status, with plants becoming more sensitive as carbohydrate levels decrease. Of course, plants grown with insufficient nitrogen during production may exhibit leaf yellowing more quickly than those grown with sufficient fertilizer. Interestingly, sugars in the postharvest solutions may promote foliage yellowing; however, it should be noted that the benefits of the sugars in extending the vase life of the flowers outweigh the potential leaf yellowing (Sacalis, 1993). Recently, another foliage problem has been noted where the leaves turn black and desiccate after only a few days in the final vase solution (Faust and Dole, 2021) (Fig. 11.6). The cause or causes are not known, but there is variation among the cultivars with the green-flowered cultivars of spray forms being the most sensitive.

11.4 Postharvest Management Techniques

11.4.1 Harvesting, processing, grading, bunching, packing, and cooling of cut flowers

Flowers that will be processed immediately can be handled dry from the production site to the processing site. Fanourakis *et al.* (2021) noted that stems that lost 5% of their fresh weight were still able to fully rehydrate, while following a 10% loss, two cultivars fully rehydrated and four did not. Much of the processing of cut flower chrysanthemums can be done in the field where the lower foliage is removed and stems are graded and bunched, then the bunches are recut to even up the stems and sleeved. Field processing is an efficient way to process flowers if the crop is uniform. Bunches can be either transported dry to the packing area or placed in bucket solutions containing a biocide and an acidifier. Bucket solutions used to transport flowers from the greenhouse to the packing house are used four to five times before being replaced. This usually requires the solutions to be changed twice daily. Periodic testing of free chlorine concentration and pH is done to verify that the solutions are usable. Free chlorine should be 5 ppm, while pH should be less than 5.0. Bacteria counts are also done periodically to evaluate the operating procedures. If the stems are not going to be immediately packed, they are placed in a bucket

Fig. 11.6. Rapid leaf necrosis during the first week in the vase on sensitive cultivars, particularly novelty colors such as green in spray types. (Photo courtesy of James E. Faust.)

solution containing a biocide and an acidifier in the packing house until they are processed (Fig. 11.7). Stems are then graded, sorted, bunched, and sleeved.

Various grades and standards have been developed. Standard chrysanthemum stems are graded by stem length, minimum flower (head) size, and

Fig. 11.7. Cut flowers in a packaging house waiting to be processed, packed, and then placed in cold storage. (Photo courtesy of James E. Faust.)

quality (Reid, 1997). Spray chrysanthemums are graded by stem length, minimum number of florets, and quality. Kim *et al.* (2012) noted that the typical stem length for standard chrysanthemums is 80 and 70 cm for spray types. The standards for cut chrysanthemums are described in Table 11.1. For Australia, stems of 50–80 cm are used (Thomas and Gollnow, 2013). Stems to be used in mixed bouquets tend to be shorter than stems sold to wholesalers in boxes containing only chrysanthemums.

Cut stems are usually sorted and bunched in groups of ten and then sleeved. However, one, three, or other numbers of stems may be bunched and sleeved together if they are to be retailed in that package to the final consumer. It is best for flowers to be kept cold, less than 5°C, during the processing, grading, and bunching stages; although some parts of the process can be done at warmer temperatures, if they are accomplished quickly. Once flower stems are processed, they should be stored at 0–4°C, with close to 0°C being optimum and necessary if stems are to be stored for more than a few days. Large flowered standard mums, 13 cm in diameter, can be stored in wax paper-lined boxes dry for a maximum of 3 weeks; smaller, 7- to 13-cm-diameter flowers are stored for a maximum of only 2 weeks.

Precooling with a forced air cooler is an effective method for quickly removing residual heat from boxed flowers after they are packed (Fig. 11.8).

Table 11.1. Standards for cut chrysanthemums (Kim *et al.*, 2012).

Condition	Special grade (#1)	Good grade (#2)	Middle grade (#3)
Uniformity of size	No differences	Differences ≤ 5%	Differences ≤ 10%
Flowers	True to cultivar, vibrant leaf color		
Stem	Strong, unbowed stems and constant thickness		
Flowers			
Standard	Flower about ½ open	Flower about ⅔ open	Flower about ⅔ open
Spray	Three to four flowers open	Five to six flowers open	Less than good grade (#2)
Postharvest management	Dried and contaminated leaves completely removed		
Heavy defects	Nothing with heavy defects (weakness, bad shape)		
Light defects	Not exceeding 3%	Not exceeding 5%	Not exceeding 5%

Reid (2009) noted the following cooling times for a packed standard box of chrysanthemums:

- 1.25 static pressure: 38 l s^{-1} box^{-1} (liters/second/box), 62 min to 7/8th cooling
- 2.5 static pressure: 61 l s^{-1} box^{-1}, 58 min to 7/8th cooling
- 5.0 static pressure: 99 l s^{-1} box^{-1}, 54 min to 7/8th cooling

The standard box was defined as 57 × 21 × 12 cm with two 5-cm-diameter vent holes in each end and 7 kg gross weight. Newer box designs have rectangular flaps that can be opened and closed. Boxes packed with more material will be more difficult to move air through, that is, "9 kg box will allow 50% less air through the box." Due to the amount of floral and leaf mass, boxes of chrysanthemums tend to take longer to cool than those filled with other species such as *Dianthus*, *Gypsophila*, *Rosa*, or *Limonium*. Stems that are stored in the bud stage should be recut and placed in a bud-opening solution containing 2–3% sucrose at 15–20°C for 16 hour (Reid, 1997). In general, stems should be recut every time they are placed into a new solution in the postharvest environment. This will encourage continued uptake of solutions as stems are prone to blockage (Thomas and Gollnow, 2013).

11.4.2 Harvesting, packing, and transport of potted plants

Potted plants should be thoroughly irrigated prior to harvest but long enough before harvest that the foliage is dry. Plants should be groomed on the bench, removing unsightly leaves or the occasional broke or misshapen flower, and sleeved. Plants are then packed on trays and loaded onto carts. Each cart

Fig. 11.8. Boxes of cut flowers placed on a forced air cooler to remove the field heat prior to storage/shipping. Cold air from the surrounding cooler is sucked into the boxes through the round vents at the ends of each box. (Photo courtesy of James E. Faust.)

should be taken to the cooler as soon as it is loaded. If carts are to be loaded on trucks soon and not precooled, be sure the carts are out of the sunshine to prevent plants from becoming too warm and prevent moisture accumulation in the sleeves.

11.4.3 Bucket and vase solutions

Chrysanthemums respond well to postharvest hydration, prepackaging, holding, and vase solutions. Commercially prepared floral preservatives work well and have been heavily researched to be effective on chrysanthemums. However, firms producing a lot of cut flowers typically use large amounts of solutions and often find it more cost-effective to prepare their own solutions. Hydration solutions consist of an acidifier and biocide (sometimes one compound accomplishes both functions) and no sugars that can slow water uptake. The stems are then placed in prepackaging, holding, or shipping solutions as needed, and these typically include carbohydrates. The final treatment is the vase solution (also known as a "flower food"), which has the highest content of carbohydrates and is typically provided by the florist or the customer in the form of a small packet to be mixed with the vase water. Various products can be used as acidifiers, biocides, and carbohydrate sources. Citric acid can be used to adjust the pH of the water to 3.5. Several materials can be used as biocides including aluminum sulfate, isothiazolinone (methylchloroisothiazolinone + methylisothiazolinone), 8-hydroxyquinoline citrate or sulfate, and silver nitrate; the former two products are especially common in commercial solutions (Dole and Faust, 2020).

Many substances have been tested on mums in vase solutions and can extend chrysanthemum vase life including ammonium sulfate (100 mg l^{-3}), calcium nitrate (100–500 mg l^{-3}), and potassium nitrate (500 mg l^{-3}) (Souri *et al.*, 2018); amoxicillin (200 mg l^{-1}), rifampin (200 mg l^{-1}), and *Artemisia* oil (30%) (Hashemabadi *et al.*, 2012); copper nanoparticles (Rashidiani *et al.*, 2020b); ethanol (7%) and thyme oil (100 mg l^{-1}) (Bazaz *et al.*, 2015); *Eucalyptus* oil (3.3 ml l^{-1}), geranium oil (3.3–17 ml l^{-1}), and *Myrtus* oil (3.3 ml l^{-1}) (Bidarigh, 2015); clove oil (250 mg l^{-1}) and thyme oil (500 mg l^{-1}) (El-Sayed and El-Ziat, 2021); hydrogen sulfide (30 or 50 μM) (Wei *et al.*, 2021); salicylic acid and sodium nitroprusside (Mansouri, 2012; Mashhadian *et al.*, 2012); and silver nanoparticles (0.04g l^{-1}) (Rashidiani *et al.*, 2020a). Many research programs are looking for substances to extend vase life that are considered sustainable or can be certified organic. The main limitation in creating a completely organic floral solution is finding an effective biocide (Ahmad *et al.*, 2014).

11.4.4 Hormone solutions

Cytokinins can be applied as a pulse solution to cut flowers prior to storage and shipping to reduce leaf yellowing of yellowing-sensitive cultivars, which is associated with ethylene sensitivity (Reyes-Arribas *et al.*, 2000, 2001). The most studied cytokinins are benzyladenine (BA) and thidiazuron (TDZ). TDZ often provides a stronger response but is less commonly used commercially (Kaur and Singh, 2015). Chrysanthemum cultivars that are not sensitive to leaf yellowing are not as likely to benefit from the cytokinin treatments, although Ferrante *et al.* (2005) reported that even the nonsensitive cultivars showed improved water uptake when treated with TDZ. Pulse treatments are typically provided in a bucket solution in the packing house at ~20°C for up to 24 hour. BA treatments are often in the 10–50 mg l^{-1} range, while TDZ is provided at 1.1–2.2 mg l^{-1}. Gibberellic acid (GA_3) applied at a rate of 10–20 mg l^{-1} in combination with other beneficial compounds, for example, sucrose, STS, citric acid, 8-hydroxyquinoline citrate, and/or BA, has been reported to improve cut flower vase life (El-Ghait *et al.*, 2012; Abd-Allah *et al.*, 2019). Preharvest foliar sprays of GA_3 (100 mg l^{-1}) at the button stage for flower buds have been reported to extend vase life while also increasing flower size (Sajid *et al.*, 2018).

11.4.5 Logistics for air transport

Most cut flowers are shipped via airfreight since the shorter shipping time helps to maximize postharvest longevity and provide customers with the maximum vase life. Flight time is quite short, mostly less than 24 hour; however, the logistics from the farm to the airport and again from the importing airport to the local market results in most flowers being a minimum of 7 days from harvest to reaching the retail customer. Also, chrysanthemums may be stored for up to 2 weeks prior to exporting, so the maximum age for cut chrysanthemums is approximately 21–25 days from harvest. Regardless of shipping time, the vase life of chrysanthemums should be a minimum of 7 days, while 11–14 days are typical. Most farms regularly conduct postharvest evaluations of their products to assure that vase life is sufficient (Fig. 11.9), and humid chambers are used to evaluate the incidence and severity of botrytis infections. Temperatures vary throughout the 7 day delivery schedule due to various handling situations during transit. Variation occurs during loading and unloading of the cargo plane, during customs inspection, and during truck delivery. Leonard *et al.* (2011) recorded temperatures during 12 shipments of cut roses from Bogotá, Colombia to Gainesville, Florida, USA and reported a typical variation from 3 to 19°C during the 7 day shipment (Fig. 11.10).

11.4.6 Logistics for sea transport

The number of cut flower stems transported via sea container is increasing annually as this method reduces the transport costs and the carbon footprint of cut flowers. Sea transport is most common from the largest cut flower exporting countries, such as Colombia, Ecuador, Kenya, and Ethiopia, to the markets in North America, Europe, and Asia. A typical 40-foot sea container contains ~1,00,000 chrysanthemum stems, so only the largest exporters have the production capacity required for utilizing sea transport on a weekly basis. The duration of shipment ranges from 20 to 35 days, so only the most tolerant chrysanthemum cultivars can be shipped by sea. Sea containers provide excellent temperature control and are usually set at 1–3°C, to ship flowers packed in boxes. Thus, the longer shipping duration compared to air

Fig. 11.9. Evaluation rooms are used at most farms to assess plant quality and vase life. (Photo courtesy of James E. Faust.)

transport is compensated for somewhat by the fact that the cold chain is more easily maintained from start to finish. Controlled atmosphere containers can also manage humidity, carbon dioxide, and oxygen, but at an additional cost, so they are not yet commonly used for shipping flowers. Ultimately, the expectation is that vase life for cut flowers transported via container shipments is comparable to that of air shipments.

From a physiological perspective, the biggest limitations to sea transport are dehydration/wilting, while botrytis is the most common pathogen issue. Other important considerations that influence the feasibility of sea transport include proximity of the farm to a shipping port and the availability of cargo planes, for example, a country must have sufficient international airfreight imports to have planes available for flower export. Otherwise, bringing in empty planes adds significant costs to air transport.

11.5 Conclusion and Future Prospects

The importance of extending shelf life continues to drive innovations in the postharvest management of chrysanthemums. The extensive study of chrysanthemum has shown that many factors substantially influence

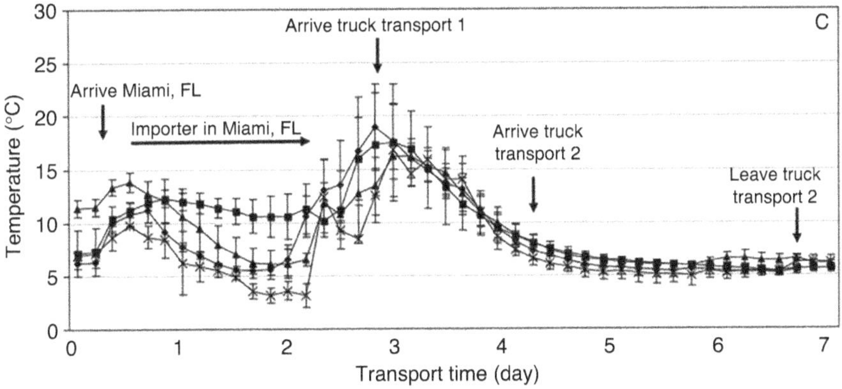

Fig. 11.10. Temperature recordings during transport of cut roses from Columbia, South America to the USA during a 7-day delivery, where flowers were transported via cargo plane from Bogotá to Miami, held at a Miami importer for 1.5 days, driven to a wholesaler in East Palatka, Florida (truck transport 1), and then driven to Gainesville, Florida (truck transport 2). Results are grouped by season: winter [◆] = December, January, February; spring [■] = March, April, May; summer [⅄] = June, July, August; and fall [✻] = September, October, November. Means ± SE. (Figure courtesy of Leonard *et al.*, 2011.)

the postharvest life and overall quality. Maximizing longevity starts with postharvest testing and proper selection of cultivars and preharvest care, including providing optimum relative humidity, light, and nutrition. After harvest, exposure to ethylene should be minimized, stems and plants should be well hydrated, and mechanical damage and diseases should be prevented. Maintaining low temperatures, 0–1°C, from harvest to sale and using proper bucket and vase solutions, is especially critical for cut flowers. Furthermore, the adoption of technologies such as controlled atmosphere containers and continued improvements in floral preservatives will be instrumental in further maintaining quality during transport and storage.

Looking forward, there is substantial potential for advancing chrysanthemum postharvest longevity through both traditional breeding and modern genetic engineering. Focusing on molecular biology could unveil genetic controls over senescence and ethylene sensitivity, leading to cultivars designed for enhanced postharvest performance. Moreover, integrating sustainable practices and developing eco-friendly treatments are essential to meet increasing consumer demands for sustainability without sacrificing flower quality. As global logistics systems evolve, optimizing the supply chain to minimize waste through improved packaging and cooling methods will be crucial. Such advancements will necessitate a collaborative approach among researchers, growers, and technologists, propelling innovation and sustainability within the floriculture industry.

References

Abd-Allah, A.E., Mohamed, T.A., Darwish, M.A., Khenizy, S.A.M. and Taha, M.H. (2019) Changes in post-harvest life of cut chrysanthemum as influenced by different holding solutions and two cultivars. *Middle East Journal of Agriculture* 8, 82–95.

Ahmad, I., Dole, J.M., Viloria, Z. and Blazich, F.A. (2014) Postharvest performance of cut carnation, chrysanthemum and rose as influenced by conventional and organic floral preservatives. *Biological Agriculture and Horticulture* 30, 109–118. DOI: 10.1080/01448765.2013.878880.

Aliniaeifard, A. and van Meeteren, U. (2016) Stomatal characteristics and desiccation response of leaves of cut chrysanthemum (*Chrysanthemum morifolium*) flowers grown at high air humidity. *Scientia Horticulturae* 205, 84–89. DOI: 10.1016/j.scienta.2016.04.025.

Baskaran, V., Jayanthi, R., Janakiram, T. and Abirami, K. (2010) Evaluation of post harvest quality of some cultivars of chrysanthemum. *Journal of Horticultural Science* 5, 81–83. DOI: 10.24154/jhs.v5i1.510.

Bazaz, A.M., Tehranifar, A. and Karizaki, A.R. (2015) Use of ethanol, methanol and essential oils to improve vase-life of chrysanthemum cut flowers. *International Journal of Sciences: Basic and Applied Research* 9, 1431–1436.

Bidarigh, S. (2015) Improvement vase life of chrysanthemum (*Dendranthema grandiflorum* L.) cut flowers using essential oils of Geranium, Eucalyptus and Myrtus. *Journal of Ornamental Plants* 5, 213–221. Available at: https://www.europub.co.uk/articles/-A-164457

Carlson, A.S. and Dole, J.M. (2013) Postharvest water quality affects vase life of cut *Dendranthema, Dianthus, Helianthus, and Zinnia*. *Scientia Horticulturae* 164, 277–286. DOI: 10.1016/j.scienta.2013.09.024.

Doi, M., Nakagawa, Y., Watabe, S., Aoe, K., Inamoto, K. *et al.* (2003) Ethylene-induced leaf yellowing in cut chrysanthemums (*Dendranthema grandiflora* Kitamura). *Journal of the Japanese Society for Horticultural Science* 72, 533–535. DOI: 10.2503/JJSHS.72.533.

Doi, M., Aoe, K., Watabe, S., Inamoto, K. and Imanishi, H. (2004) Leaf-yellowing of cut standard chrysanthemum (*Dendranthema grandiflora* Kitamura) 'Shuho-no-chikara' induced by ethylene and postharvest increase in ethylene sensitivity. *Journal of the Japanese Society for Horticultural Science* 73, 229–234. DOI: 10.2503/jjshs.73.229.

Dole, J.M. and Wilkins, H.F. (2005) *Floriculture - Principles and Species*, 2nd edn. Prentice Hall, Upper Saddle River, New Jersey.

Dole, J.M. and Gibson, J.L. (2006) *Cutting Propagation*. Ball Publishing, Batavia, Illinois.

Dole, J., Stamps, R., Carlson, A., Ahmad, I., Greer, L. *et al.* (2017) *Postharvest Handling of Cut Flowers and Greens*. Association of Specialty Cut Flower Growers, Oberlin, Ohio.

Dole, J.M. and Faust, J.E. (2020) Postharvest management. In: Faust, J.E. and Dole, J.M. (eds) *Cut Flowers and Foliages*. CAB International, pp. 342–368. DOI: 10.1079/9781789247602.0008.

Drüge, U., Zerche, S., Kadner, R. and Ernst, M. (2000) Relation between nitrogen status, carbohydrate distribution and subsequent rooting of chrysanthemum cuttings as affected by pre-harvest nitrogen supply and cold-storage. *Annals of Botany* 85, 687–701. DOI: 10.1006/anbo.2000.1132.

El-Sayed, I.M. and El-Ziat, R.A. (2021) Utilization of environmentally friendly essential oils on enhancing the postharvest characteristics of *Chrysanthemum morifolium* Ramat cut flowers. *Heliyon* 7, e05909. DOI: 10.1016/j.heliyon.2021.e05909.

El-Ghait, A.E.M., Gomaa, A.O, Youssef, A.S.M., and Mohamed, Y.F. (2012) Effect of some postharvest treatments on vase life and quality of chrysanthemum (*Dendranthema Grandiflorum Kitam*) cut flowers. *Research Journal of Agriculture and Biological Sciences* 8(2), 261–271.

Fan, H., Li, T., Sun, X., Sun, X. and Zheng, C. (2015) Effects of humic acid derived from sediments on the postharvest vase life extension in cut chrysanthemum flowers. *Postharvest Biology and Technology* 101, 82–87. DOI: 10.1016/j.postharvbio.2014.09.019.

Fanourakis, D., Papadopoulou, E., Valla, A., Tzanakakis, V.A. and Nektarios, P.A. (2021) Partitioning of transpiration to cut flower organs and its mediating role on vase life response to dry handling: A case study in chrysanthemum. *Postharvest Biology and Technology* 181, 111636. DOI: 10.1016/j.postharvbio.2021.111636.

Faust, J.E. and Dole, J.M. (2021) Major cut flowers. In: Faust, J.E. and Dole, J.M. (eds) *Cut Flowers and Foliages*. CAB International, Wallingford, UK, pp. 48–149.

Ferrante, A., Mensuali-Sod, A., Tognoni, F. and Serra, G. (2005) Postharvest studies on leaf yellowing of chrysanthemum cut flowers. *Advances in Horticultural Science* 19, 81-85.

Hashemabadi, D., Vand, S.H., Zarchini, M., Kaldeh, N.E., Ghaderi, A., Hajian, G. and Zarchini, S. (2012) Effect of Artemisia oil on vase life, flower opening index and fresh weight loss of cut chrysanthemum (*Dendranthema grandiflorum* L. cv. 'White'). *Annals of Biological Research* 3, 5399–5402.

Hayashi, T., Heins, R.D., Cameron, A.C. and Carlson, W.H. (2001) Ethephon influences flowering, height, and branching of several herbaceous perennials. *Scientia Horticulturae* 91, 305–324. DOI: 10.1016/S0304-4238(01)00225-4.

Kalinowski, J. and Dole, J.M. (2024) Extended storage of cut flowers using sub-zero temperature. *HortTechnology* 34, 101–115. DOI: 10.21273/HORTTECH05010-21.

Kim, S.J., Lee, S.K. and Kim, K.S. (2012) Current research trend of postharvest technology for chrysanthemum. *Korean Journal of Plant Research* 25, 156–168. DOI: 10.7732/KJPR.2012.25.1.156.

Kaur, P. and Kushal, S. (2015) Influence of growth regulators on physiology and senescence of cut stems of chrysanthemum (*Chrysanthemum morifolium* Ramat) var. Thai Ching Queen. *International Journal of Allied Practice, Research and Review* 2(2), 31–41.

Leonard, R.T., Alexander, A.M. and Nell, T.A. (2011) Postharvest performance of selected Colombian cut flowers after three transport systems to the United States. *HortTechnology* 21, 435–442. DOI: 10.21273/HORTTECH.21.4.435.

Mansouri, H. (2012) Salicylic acid and sodium nitroprusside improve postharvest life of chrysanthemums. *Scientia Horticulturae* 145, 29–33. DOI: 10.1016/j.scienta.2012.07.016.

Mashhadian, N.V., Tehranifar, A., Bayat, H. and Selahvarzi, Y. (2012) Salicylic acid and citric acid treatments improve the vase life of cut chrysanthemum flowers. *Journal of Agricultural Science Technology* 14, 879–887.

Nell, T.A. (1993) *Flowering Potted Plants, Prolonging Shelf Performance*. Ball Publishing, Batavia, Illinois.

Nell, T.A., Barrett, J.E. and Leonard, R.T. (1989) Fertilization termination influences postharvest performance of pot chrysanthemums. *HortScience* 24, 996–998. DOI: 10.21273/HORTSCI.24.6.996.

Nell, T.A., Leonard, R.T. and Barrett, J.E. (1990) Production and postproduction irradiance affects acclimatization and longevity of potted chrysanthemum and poinsettia. *Journal of the American Society for Horticultural Science* 115, 262–265. DOI: 10.21273/JASHS.115.2.262.

Nowak, J. and Rudnicki, R.M. (1990) *Postharvest Handling and Storage of Cut Flowers, Florist Greens and Potted Plants*. Timber Press, Portland, Oregon. DOI: 10.1007/978-94-009-0425-5.

Produce Marketing Association (PMA)/ Society of American Florists (SAF) (n.d.) Recommended grades and standards for potted flowering plants.

Put, H.M.C. (1990) Micro-organisms from freshly harvested cut flower stems and developing during the vase life of chrysanthemum, gerbera and rose cultivars. *Scientia Horticulturae* 43, 129–144. DOI: 10.1016/0304-4238(90)90044-F.

Rajapakse, N.C., Miller, W.B. and Kelly, J.W. (1996) Low-temperature storage of rooted chrysanthemum cuttings: Relationship to carbohydrate status of cultivars. *Journal of the American Society for Horticultural Science* 121, 740–745. DOI: 10.21273/JASHS.121.4.740.

Rashidiani, N., Nazari, F., Javadi, T. and Samadi, S. (2020a) Comparative postharvest responses of carnation and chrysanthemum to synthesized silver nanoparticles (AgNPs). *Advances in Horticultural Science* 34, 133–145. DOI: 10.13128/ahsc-7491.

Rashidiani, N., Nazari, F., Javadi, T. and Samadi, S. (2020b) Copper nanoparticles (CuNPs) increase the vase life of cut carnation and chrysanthemum flowers: Antimicrobial ability and morphophysiological improvements. *Ornamental Horticulture* 26, 225–235. DOI: 10.1590/2447-536x.v26i2.2156.

Reid, M.S. (1997) Ornamental produce facts: Chrysanthemum, florist mum. Available at: ostharvest.ucdavis.edu/Commodity_Resources/Fact_Sheets/Datastores/Ornamentals_English/?uid=11&ds=801 (accessed 18 December 2021).

Reid, M.S. (2009) Handling of cut flowers for export. *Proflora Bulletin*, 2009. Available at: ucanr.edu/sites/Postharvest_Technology_Center_/files/231308.pdf (accessed 18 December 2022).

Reyes-Arribas, T., Barrett, J.E., Nell, T.A. and Clark, D.G. (2000) Effect of ethylene, sucrose and benzyladenine on leaf senescence of two chrysanthemum cultivars 'Tara' and 'Boaldi'. *Acta Horticulturae* 518, 125–129. DOI: 10.17660/ActaHortic.2000.518.16.

Reyes-Arribas, T., Barrett, J.E., Huber, D.J., Nell, T.A. and Clark, D.G. (2001) Leaf senescence in a non-yellowing cultivar of chrysanthemum (*Dendranthema grandiflora*). *Physiologia Plantarum* 111, 540–544. DOI: 10.1034/j.1399-3054.2001.1110415.x.

Roude, N., Nell, T.A. and Barrett, J.E. (1991a) Nitrogen source and concentration, growing medium, and cultivar affect longevity of potted chrysanthemums. *HortScience* 26, 49–52. DOI: 10.21237/HORSCI.26.1.49.

Roude, N., Nell, T.A. and Barrett, J.E. (1991b) Longevity of potted chrysanthemums at various nitrogen and potassium concentrations and NH_4:NO_3 ratios. *HortScience* 26, 163–165. DOI: 10.21273/HORTSCI.26.2.163.

Sacalis, J.N. (1993) *Dendranthema grandiflora*. In: Seals, J.L. (ed.) *Cut Flowers, Prolonging Freshness*, 2nd edn. Ball Publishing, Batavia, Illinois, pp. 47–49.

Sajid, M., Rab, A., Khan, I.A., Jn, I., Amin, N. *et al.* (2018) The pre-harvest foliar application influenced the flower quality and vase life of chrysanthemum cultivars. *Horticulture International Journal* 2, 145–152. DOI: 10.15406/hij.2018.02.00043.
Souri, M.K., Goodarzizadeh, S., Ahmadi, M. and Hatamian, M. (2018) Characteristics of postharvest quality of chrysanthemum cut flowers under pretreatment with nitrogenous compounds. *Acta Scientiarum Polonorum Hortorum Cultus* 17, 83–90. DOI: 10.24326/ASPHC.2018.3.8.
Thomas, D. and Gollnow, B. (2013) What cut flower Is that? The essential care and handling guide for cut flower professionals. Rural Industries Research and Development Corporation, Publication No. 13/079, Canberra, Australia. Available at: agrifutures.com.au/product/what-cut-flower-is-that-the-essential-care-and-handling-guide-for-cut-flower-professionals/ (accessed 6 January 2022).
van Geest, G., Post, A., Arens, P., Visser, R.G.F. and van Meeteren, U. (2017) Breeding for postharvest performance in chrysanthemum by selection against storage-induced degreening of disk florets. *Postharvest Biology and Technology* 124, 45–53. DOI: 10.1016/j.postharvbio.2016.09.003.
van Meeteren, U. (1992) Role of air embolism and low water temperature in water balance of cut chrysanthemum flowers. *Scientia Horticulturae* 51, 275–284. DOI: 10.1016/0304-4238(92)90125-v.
Waters, W.E. (1966) Toxicity of certain florida waters to cut flowers. *Proceedings of the Florida State Horticultural Society* 79, 456–459.
Wei, L., Wang, C. and Weibiao, L. (2021) Hydrogen sulfide improves the vase life and quality of cut roses and chrysanthemums. *Journal of Plant Growth Regulation* 40, 2532–2547. DOI: 10.1007/s00344-021-10312-7.
Woltering, E.J. and Van Doorn, W.G. (1988) Role of ethylene in senescence of petals morphological and taxonomic relationships. *Journal of Experimental Botany* 39, 1605–1616. DOI: 10.1093/jxb/39.11.1605.
Zhou, F., Gu, S., Zuo, J., Gao, L., Wang, Q. *et al.* (2019) LED irradiation delays the postharvest senescence of garland chrysanthemum (*Chrysanthemum carinatum* Schousb.). *Journal of Food Measurement and Characterization* 13, 3005–3014. DOI: 10.1007/s11694-019-00221-5.

Value Addition of Chrysanthemum

12

Malleshaiah SharathKumar[1,2]*, Vikas Kumar Galipothu[3] and Visalakshi M.[4]

[1]Horticulture and Product Physiology, Department of Plant Sciences, Wageningen University and Research, Droevendaalsesteeg 1, 6708 PB Wageningen, The Netherlands
[2]Department of Breeding and Research, Deliflor Chrysanten B.V., Korte Kruisweg 163, 2676 BS Maasdijk, The Netherlands
[3]Department of Horticulture and Natural Resources, Kansas State University, Olathe, KS 66061, USA
[4]Department of Medicinal and Aromatic Crops, Tamil Nadu Agricultural University, Coimbatore, India

Abstract

Value addition in chrysanthemums enhances the intrinsic worth of these flowers through various modifications, making them more appealing and useful across different markets. This process is crucial in the floricultural and commercial sectors, aiming at maximizing profitability by transforming chrysanthemums into higher-value products. This chapter presents the most common value addition techniques for chrysanthemums. First, the dry flower technique preserves the natural beauty of flowers, extending their usability for a longer period. Methods such as air-drying, pressing, or chemical treatments of chrysanthemums make them suitable for decorative purposes year-round. Second, tinting enhances the economic and aesthetic value of chrysanthemums, particularly targeting simple white and light-colored flowers. By using food coloring dyes, these flowers are transformed into vibrant creations in a range of novel color shades, increasing their appeal in the floral market. Lastly, the culinary applications and the nutritional and medicinal value of edible chrysanthemums, such as garland chrysanthemum (*Chrysanthemum coronarium*) and chrysanthemum tea crafted from the dried flowers of *Chrysanthemum morifolium* and *Chrysanthemum*

*Corresponding author: sharath.malleshaiah@wur.nl; sharath.malleshaiah@deliflor.nl; sharath.malleshaiah@gmail.com

 Chrysanthemum (eds Malleshaiah SharathKumar and Ep Heuvelink)
DOI: 10.1079/9781789248289.0012

indicum, are explored. These applications not only utilize the flowers' aesthetic qualities but also capitalize on their health benefits.

Keywords
Dry flower technique, edible chrysanthemum, edible flowers, sustainability, tinting

12.1 Introduction

Chrysanthemums are among the most widely cultivated ornamental flowers globally, celebrated for their vibrant colors, diverse forms, and long-lasting blooms. Their versatility makes them popular as fresh-cut flowers, loose flowers, or potted ornamentals. Beyond their traditional role in ornamental horticulture, chrysanthemums hold immense potential for value addition, transforming their utility into various high-value products. The floriculture industry increasingly recognizes the economic and cultural importance of chrysanthemums, driving efforts to maximize their market potential through innovation and diversification. Value addition in chrysanthemums involves leveraging their inherent properties for diverse applications. This includes decorative uses, such as dry flower technology, which preserves the flowers' natural beauty for year-round purposes through techniques such as air-drying, pressing, or chemical treatments. Tinting is another innovative approach that enhances the aesthetic and economic value of chrysanthemums, especially for simple white and light-colored varieties. By introducing vibrant color shades using food dyes, these flowers become more appealing in the floral market, catering to evolving consumer preferences.

Additionally, chrysanthemums offer culinary and medicinal applications, particularly edible varieties such as *Chrysanthemum coronarium* and *C. morifolium*. Known for their bioactive compounds, including flavonoids and sesquiterpenes, these flowers provide antioxidant, anti-inflammatory, and antimicrobial benefits. Products such as chrysanthemum tea and essential oils cater to the growing demand for natural and wellness-oriented solutions, further expanding their market. Their cultural significance, particularly in Asia, adds another dimension to their value, as they are integral to traditional medicines, teas, and ceremonial uses. The concept of value addition not only enhances economic returns for growers but also aligns with sustainable agricultural practices by reducing postharvest losses and optimizing resource use. By exploring innovative processing techniques and targeting diverse markets, the chrysanthemum industry can transition from a traditional floriculture sector into a dynamic contributor to global agribusiness. This chapter delves into the strategies, applications, and future potential of value addition in chrysanthemums, discussing their role in fostering a diversified and sustainable floral economy.

12.2 Dry Flower Technology in Chrysanthemum

Dry flower technology offers extensive potential as an eco-friendly, durable, and versatile alternative to fresh flowers for various decorative and commercial purposes. Dried flowers extend usability beyond seasonal constraints, significantly contributing to global floriculture industry (Singh and Laishram, 2010; Kumari, 2022). The practice of drying flowers to preserve their beauty and extend their usability beyond the natural bloom period is a tradition with deep historical roots. The technique of drying flowers, known as "Oshana," in Japan, originates in Asia before spreading to Europe, revered for its ability to maintain the aesthetic qualities of flowers indefinitely (Bhutani, 1990; Joyce, 1998). In the floriculture sector, dry flower production has become a significant industry, particularly in India, which has emerged as one of the leading exporters of dry floral and plant products globally (Rajaram *et al.*, 2009). This industry not only caters to traditional uses such as herbariums and botanical identification but has also carved out a substantial niche in the decorative arts and commercial sectors, from home décor to fashion industry. In today's shift from eco-consciousness to sustainability, the use of natural plant products such as dried flowers, petals, leaves, stems, seeds, and roots has become a primary choice among urban populations for interior decoration, as well as for aesthetic and commercial purposes (Dhatt *et al.*, 2007). India's dry flower market has achieved remarkable growth, with a business volume reaching around $12 million, dominated by products such as potpourri that alone accounts for $6.6 million (Sindhuja *et al.*, 2017). The annual growth rate of 8–10% in demand for dry flowers signals expanding opportunities for entrepreneurs within the global floricultural trade (Singh, 2009).

Chrysanthemums, with their vibrant colors and intricate petal structures, are particularly well suited for dry flower preservation, enabling their use in everything from greeting cards to floral designs and festive decorations. Traditionally and spiritually, chrysanthemum flowers are revered in various religious contexts, often used in temples, churches, and gurdwaras, where their role in worship is significant. Due to their perishable nature, these flowers unfortunately do not retain their beauty for long and typically become waste after their ceremonial use. Recently, however, dry flower technology has been employed to extend their usefulness. By preserving worshipped flowers, this technology facilitates the creation of value-added products, thereby sustaining their spiritual and aesthetic significance beyond the initial ritual context (Vikas Kumar, 2022a). This section delves into how dried chrysanthemums are utilized in creating long-lasting floral arrangements, wreaths, and other decorative items, displaying their enduring appeal in home décor and fashion industries. As we explore the art and science of drying chrysanthemums, we will highlight their unique characteristics that make them ideal for preservation and their versatile use in various artistic and commercial applications. We will also discuss the various drying methods, including air, sun, oven, and microwave drying, as well as more sophisticated techniques such as freeze-drying and embedded drying (Bhattacharjee and De, 2003).

12.2.1 Selection criteria in chrysanthemum for drying purpose

Selecting chrysanthemums for drying involves considering several key criteria to ensure the flowers preserve well and maintain their aesthetic qualities. Here are the main factors to consider when selecting chrysanthemums for drying purposes:

12.2.1.1 Flower maturity/optimal harvesting stage

Choose flowers that are just at or slightly before peak bloom. Flowers that are too mature may lose petals during the drying process, while those that are not mature enough may not display their full color and shape. For air-drying, select flowers at the bud or partially opened stage, as these will continue to open during the drying process, which is crucial for achieving the best aesthetic in naturally shrunken and rustic-looking blooms.

12.2.1.2 Varietal characteristics

The variety of chrysanthemum affects how well the flowers will hold up and maintain their shape during drying.

- **Robustness and suitability:** Some cultivars are more robust and suitable for drying than others. Choose those known to respond well to your chosen drying method and avoid varieties prone to color fading or petal drop.
- **Color retention:** Some varieties of chrysanthemums retain their color better than others when dried. Typically, deeper and brighter colors yield more vivid results post-drying.
- **Petal density and structure:** Flowers with denser petals tend to dry better, retaining more of their original form and structure.
- **Shape and size:** Consider the shape and size of the flowers in relation to their intended use post-drying. Larger flowers may require longer drying times or specific methods such as freeze-drying to preserve their shape accurately.
- **Stem length:** Consider the final use of the dried flowers. If requiring long stems for versatility in arrangements, air-drying is preferable. If the design permits shorter stems, the silica gel method might be more appropriate due to its superior color and form retention.

12.2.1.3 Health of the flowers

Only healthy flowers free from pests, diseases, and blemishes should be chosen, as damaged or diseased flowers will not dry well and can deteriorate further during the process.

12.2.1.4 Moisture content

Flowers with lower initial moisture content will dry faster and are less likely to mold (surface growth of fungus, especially on damp or decaying matter). Avoid flowers that have been rained on recently unless they can be thoroughly dried before the process begins.

12.2.1.5 Harvesting time

Harvest chrysanthemums in the late morning after the dew has evaporated to reduce excess moisture and facilitate a more consistent drying process.

12.2.1.6 Overall aesthetics and appearance post-drying

Consider how the flower's appearance will change after drying and select varieties that are known to maintain their aesthetic appeal.

12.2.1.7 Drying method compatibility

Different drying methods may require specific flower traits. For example, flowers intended for microwave drying should be uniform in size and moisture content for uniform drying.

12.2.1.8 Intended use

The purpose for which the flowers are dried can also influence selection. Long stems provide more flexibility for floral arrangements and designs, while short stems are suitable for compact craft arrangements where the focus is on the bloom itself.

12.2.2 Different drying techniques for chrysanthemums to enhance preservation and quality

Dehydration is a crucial process for extending the storage life of flowers, enabling the preservation of their physical and biochemical properties. This process is essential for inhibiting enzymatic degradation and preventing microbial growth, thus maintaining the integrity of the flowers for longer periods (Ahrne *et al.*, 2007; Pereira *et al.*, 2007). Additionally, dehydration facilitates the extraction of valuable pigments. The period from harvesting to consumption can be lengthy, during which biochemical changes may affect the nutraceutical value of chrysanthemum flowers. Dehydration techniques significantly impact the retention of total anthocyanins and their antioxidant activities. Therefore, it is crucial to optimize various dehydration methods to maximize the retention of these beneficial compounds for edible or nutraceutical purposes. The quality and appearance of dried flowers are significantly influenced by the drying method employed (Pant and Singh, 2023). Each drying method affects the texture, color, and structural integrity of chrysanthemums differently, making the choice of technique crucial depending on the desired outcome of the drying process. Each drying method has its specific applications and benefits, and selecting the right one depends on factors such as the intended use of the dried flowers, desired quality of the product, and available resources. Recent studies and experiments have continuously explored and optimized these methods to enhance the ornamental and commercial value of dried chrysanthemums:

12.2.2.1 Air-drying

This method involves hanging plant materials on ropes or wires in either dark or sunny conditions for expedited drying. It requires a warm, clean, dark, and well-ventilated area with low humidity (Raghupathy *et al.*, 2000). Alternatively, flowers can also be spread over blotting sheets or newspapers under similar conditions (Datta, 1997). For chrysanthemums, it is advisable to start the drying process when the flowers are at the bud or partially opened stage, as they continue to open while drying (Sell, 1993).

12.2.2.2 Shade drying

Commonly used for producing powders for the food and feed industries, shade drying has drawbacks such as contamination risks, especially during rainy seasons, and prolonged drying times (Maskan, 2000).

12.2.2.3 Oven-drying

Hot air and microwave ovens provide faster drying, with controlled temperatures tailored to the specific plant species. Higher temperatures accelerate transpiration and moisture vaporization, potentially causing significant moisture loss in a short time and adversely affecting carotenoid content (Mayak and Halevy, 1980). Oven-drying at moderate temperatures (40°C) for 35–50 hour can preserve the color and structure of chrysanthemums without altering the physical dimensions of the floral parts. (Pandya *et al.*, 2001). Drying in a hot air oven or embedding in borax or sawdust are particularly effective methods for rapid drying and maintaining the aesthetic quality of dried chrysanthemum flowers (Mandal *et al.*, 2023).

12.2.2.4 Microwave drying

Microwave drying is the quickest method, though it may result in lower-quality flowers (Gurjar *et al.*, 2023). However, it is particularly effective for preserving key bioactive compounds in chrysanthemum flowers, maintaining high levels of anthocyanins and antioxidants crucial for enhancing the flowers' nutritional and pharmaceutical value. This method outperforms traditional methods such as shade drying and hot air oven-drying (Ullas *et al.*, 2018).

12.2.2.5 Press drying

Primarily used by botanists for herbarium preparations, this method involves placing flowers and foliage between newspaper sheets or blotting papers, spaced out to allow moisture to escape. Corrugated boards are incorporated between layers to facilitate vapor release. The process can be accelerated by placing the sheets in an oven at an appropriate temperature, typically around 50–60°C (Bhutani, 1990; Datta, 1997). Recently, Vikas Kumar (2022a) explored an innovative method of preserving chrysanthemum flowers that were collected after temple use. The flowers were dried using a wooden press for 4–5 days. During this process, the petals were evenly spread out on blotting paper, with additional layers of blotting paper placed between two wooden plates. This was done in a sandwich-like model, stacking approximately eight to ten wooden plates on top of each other to help preserve the original color of the flowers. The resulting flattened petals are ideal for artistic applications; they are tailored to fit printed images on various backdrop materials such as ivory sheets, canvas sheets, jute mats, photo paper, corrugated papers, and handmade papers. The dried florets are cut to match the shapes of the images and affixed with adhesive. To ensure longevity and protect against moisture, these floral images are encased in glass photo frames (Fig. 12.1). Properly sealed from moisture, these framed floral images can last more than 20 years.

12.2.2.6 Embedded drying

Various media, including sand, borax, silica gel, and sawdust, are used for embedding flowers. Sand and borax are cost-effective but require more time to effectively dry the flowers (Bhutani, 1990, 1995). A mix of borax and cornmeal has shown promising results for chrysanthemums, effectively retaining color and shape (Sell, 1993). Microwave oven-drying after embedding in silica gel has yielded the best color and shape retention (Bhalla *et al.*, 2006). Embedding chrysanthemum flowers in silica gel and drying them at 50°C for 48 hour in a hot air oven produces high-quality dried flowers (Dahiya *et al.*, 2003). Chrysanthemum flowers can also be preserved using a resin embedding technique (resin is a viscous substance that hardens to create a durable, glass-like finish, Vikas Kumar *et al.*, 2022b). Initially, the flowers are dried in silica gel for 3–4 days. After this drying phase, they are placed into silicone molds filled with a resin solution. This solution is made by mixing two parts resin with one part hardener, stirred for 3–4 min to ensure proper mixing, and then poured into the molds to achieve a self-leveling effect. The silicone molds are used to create various decorative items such as name boards, paperweights, pen stands, and alphabet decorations. In embedded drying, the resin solution in the molds should reach a semisolid state before the dried flowers are carefully positioned within it. Once the resin partially solidifies around the flowers, an additional layer of resin is added to fill the molds and left to set overnight. This method effectively encapsulates chrysanthemum flowers in resin, enhancing their durability and aesthetic appeal. The resin preservation maintains the flowers' beauty for years, making them suitable for long-lasting decorative products (Vikas Kumar *et al.*, 2022b; Fig. 12.2). Additionally, pretreating chrysanthemum flowers with a glycerol–water solution at a 1:3 ratio for 24 hour before embedded drying has proven beneficial.

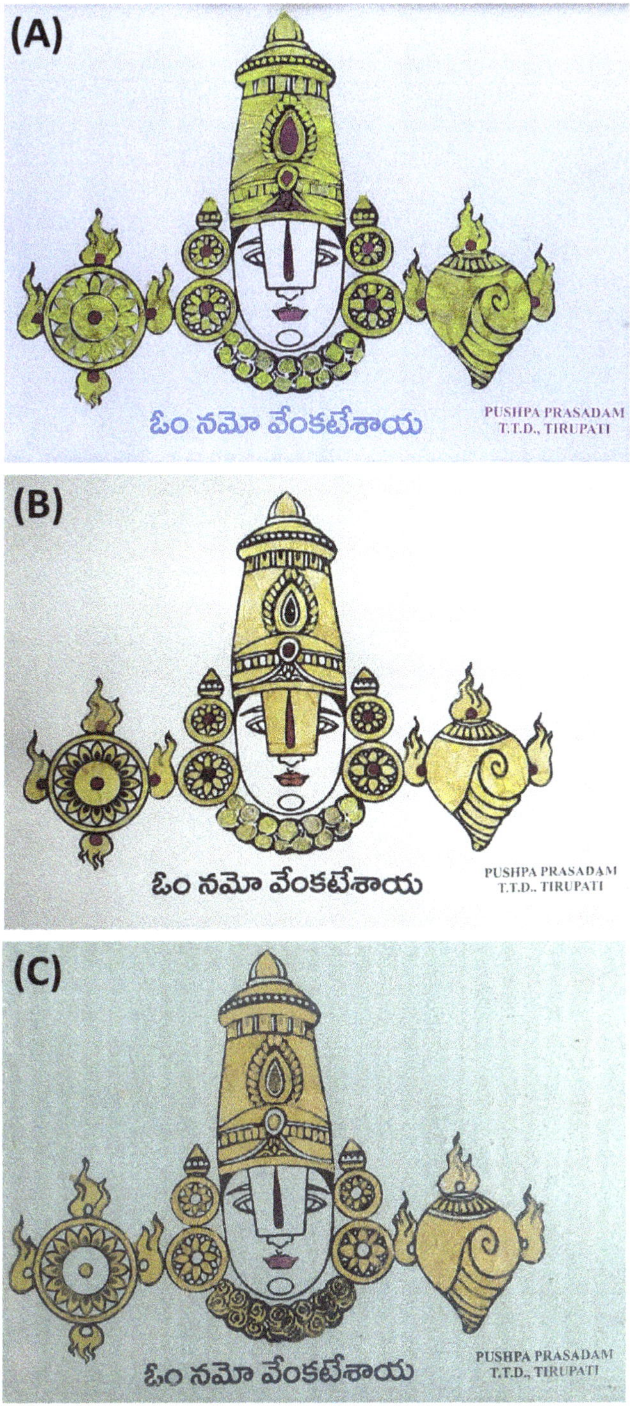

Fig. 12.1. Chrysanthemum petals pasted on various backdrop materials: ivory paper (A), canvas paper (B), and corrugated paper (C). (Photo courtesy of Vikas Kumar Galipothu.)

12.2.3 Handling, packing, and storing of dry chrysanthemums

Handling, packing, and storing dried chrysanthemums require careful attention to maintain their delicate structure and visual appeal. During handling, it is crucial to be gentle and minimize direct contact, using tools such as soft brushes to adjust their position in a clean, dust-free environment. After the drying process, each flower or bunch of flowers should be carefully wrapped in soft tissue paper to provide cushioning and protection from physical damage. For packing, it is important to use supportive and moisture-proof materials, such as rigid boxes with soft internal padding, and to seal the flowers in airtight containers with silica gel packets to avoid humidity. These wrapped flowers should then be stored in cardboard boxes or cupboard drawers, which help maintain a cool, dry environment essential for their preservation. This setup not only absorbs any residual moisture but also ensures the flowers are kept in a state that prevents degradation. Moreover, it is crucial to store dried flowers away from direct sunlight in a dark, well-ventilated area to maintain their color and structure.

12.2.4 Designs and arrangements made from dried chrysanthemums

Dried flower arrangement is an art form that significantly enhances the beauty of interior spaces, emphasizing the quality of the products used. Dried flowers can be arranged in vases similar to fresh-cut flowers, either alone or combined with fresh blooms to create diverse floral arrangements. Here are some popular uses and designs made from dried chrysanthemums (Fig. 12.3):

12.2.4.1 Potpourri

Potpourri is a mixture of dried, naturally fragrant plant materials, often arranged in decorative bowls or sachets to make room smell fresh.

12.2.4.2 Bouquets

Dried flowers can be arranged into bouquets that maintain their beauty over time, without the need for water.

12.2.4.3 Frames and wall hangings

Pressed flowers and leaves are often used to create beautiful, framed art pieces or wall hangings that serve as permanent decorations.

12.2.4.4 Wreaths

Crafting wreaths from dried flowers and foliage is a popular way to create seasonal or year-round decorations for doors and walls.

Fig. 12.2. Chrysanthemum flowers embedded in resin as paperweights. (Photo courtesy of Vikas Kumar Galipothu.)

12.2.4.5 Craft applications

Chrysanthemums offer unique potential for creative crafts. Pressed flowers can embellish greeting cards and bookmarks, adding a personal touch. Dried flowers embedded in resin create decorative paperweights, while laminated petals enhance table mats, blending functionality with aesthetic appeal (Fig. 12.2).

12.3 Tinting of Chrysanthemums

Tinting is a value-adding technique that involves the artificial addition of colors using food-grade dyes or color agents, particularly useful for species where natural color pigmentation is weak or absent. For chrysanthemums, globally recognized for their vivid and attractive colors, tinting offers a way to introduce a broader variety of hues, which can be specifically tailored to fit various decorative needs. Tinted flowers often command a premium on the market because they meet specific aesthetic requirements that natural colors cannot achieve (Fig. 12.4). Using dual colors in tinting allows the creation of two hues in a single bloom, adding a unique touch to bouquets, stage decorations, or general floral arrangements. Tinting chrysanthemums allows for creativity and innovation in floral design, making it possible to tailor the flowers to meet diverse and evolving consumer preferences. When opting for tinting in chrysanthemums, several factors must be considered. Color fastness is crucial, as some dyes may fade or leach when exposed to sunlight or water, affecting the flower's aesthetic appeal. Additionally, the health of the flower is important; tinting can reduce the longevity of the flowers as heavily dyed petals may clog the xylem, impairing the flower's ability to uptake water and sustain its freshness. Furthermore, while tinted chrysanthemums can command higher prices in the market, especially for

Fig. 12.3. The versatility of dried chrysanthemum flowers for creative expression in home décor: bunch of dried chrysanthemums (A), small dried whitish-pink chrysanthemum flowers for floral ornaments (B), and hairpins adorned with chrysanthemum flowers (C). (From https://www.etsy.com)

special events or in markets that value unique and vibrant colors, it is essential to balance the potential gains with the costs associated with the tinting process.

12.3.1 Dyes, methods of tinting, and considerations

Chrysanthemum flower stems treated with blue dye for 7.5 hour exhibited the highest vase life. Those treated with yellow dye for the same duration showed the maximum moisture content. Additionally, flowers tinted with blue, orange, or green dyes and placed in the dye for 10 hour received the highest consumer preference compared to quick dyeing due to their vibrant color and even distribution (Pavan *et al.*, 2022). Food-grade dyes are commonly used because they are safe and come in a variety of colors. The dyes are dissolved in water, sometimes with the addition of other chemicals to improve uptake and color fastness. Additionally, some florists use natural dyes derived from other plants or minerals, which can offer unique hues and are marketed as an eco-friendly option. The two primary methods of tinting involve, stem absorption and flower dipping. Both approaches allow for customizable hues, ranging from single colors to intricate combinations, creating unique and visually striking floral products.

12.3.1.1 Tinting via stem absorption

This is the most common method used for tinting chrysanthemums, particularly effective with primarily white or light-colored flowers. This technique involves

Fig. 12.4. Tinting of white spray chrysanthemums (top left) to achieve a diversified variety of hues, ranging from single to dual and multiple rainbow-colored chrysanthemums. (Photo courtesy of Malleshaiah SharathKumar.)

cutting the flower stems and placing them into a container with a warm solution of dye mixed with specific chemicals. To facilitate a higher uptake rate of the dye, the flowers are preconditioned by stressing them overnight at around 18°C to enhance the uptake of the dye solution. As the plants draw up the colored solution through the stems, the dye disperses uniformly throughout the petals, allowing for vibrant, even colors or intriguing gradient effects. It is essential to halt the dyeing process before the color reaches its peak intensity to prevent oversaturation, as residual dye in the stems continues to tint the flower when placed in vase solutions. This careful control ensures a precise and desirable tinting effect. Additionally, extended duration of tinting may result in discoloration of leaves, deteriorating the leaf quality of chrysanthemums.

12.3.1.2 Dipping flowers

This method is often employed for daisies and involves submerging the flower heads directly into a tinting solution made from aniline dyes dissolved in isopropanol. After dipping, the flowers are shaken to remove excess solution and then left to dry on a rack before they can be packed or stored. This method is quicker than tinting via stem absorption; however, it results in less uniformity but a strong coloration on the outer petals.

12.4 Edible Chrysanthemums

Edible chrysanthemums, known as "garland chrysanthemum" (*C. coronaria* or *Glebionis coronaria*) or "shungiku" in Japanese, involve various parts of the plant such as leaves, young shoots, flower buds, or petals. These are a versatile ingredient in Asian cuisines and have been cultivated for tea and functional foods because of their health-promoting properties (Fig. 12.5). Modern studies highlight chrysanthemum's pharmacological effects, including antibacterial, anti-inflammatory, antioxidant, antiviral, and antitumor properties, attributed to active ingredients such as flavonoids, phenolic acids, terpenoids, volatile oils, and polysaccharides (Yuan *et al.*, 2020; Liu *et al.*, 2024).

Edible chrysanthemums are among the most popular edible flowers globally (Zhang *et al.*, 2019) and exhibit a diversity in phytochemical profiles, which significantly contribute to their nutritional and medicinal functions (Yuan *et al.*, 2020; Lu *et al.*, 2022). The typical method of consuming chrysanthemums involves steeping the flowers in hot water (chrysanthemum tea) to directly extract phytochemicals. However, the phytochemical profiles vary significantly among different chrysanthemum varieties, which may influence their specific health benefits (Yang *et al.*, 2017; Chen *et al.*, 2021). Research highlights the inhibitory effects of chrysanthemum flowers on the formation of advanced glycation end-products (AGEs), which are linked to various chronic conditions such as diabetes, atherosclerosis, aging, and chronic inflammation (Tsuji-Naito *et al.*, 2009; Khan *et al.*, 2020; Sun *et al.*, 2020).

Particularly, the edible variety *Chrysanthemum* × *morifolium* Ramat. *forma esculentum* Makino, "Kotobuki," has been studied for its health-promoting flavonoids. These compounds, notably luteolin 7-O-(6″-O-malonyl)-glucoside, have shown significant potential in reducing key markers of liver injury, such as plasma aspartate aminotransferase and alanine aminotransferase, and exhibit strong radical scavenging activities, underscoring their effectiveness in mitigating liver damage (Suguwara and Igarashi, 2009).

12.4.1 Shungiku, Japanese-style

C. coronarium, known in Japan as shungiku, is renowned in Japanese cuisine for its distinctive flavor, which is slightly bitter, grassy, and herbal, with a tangy undertone and a succulent texture. The flavor of young shungiku leaves is mild,

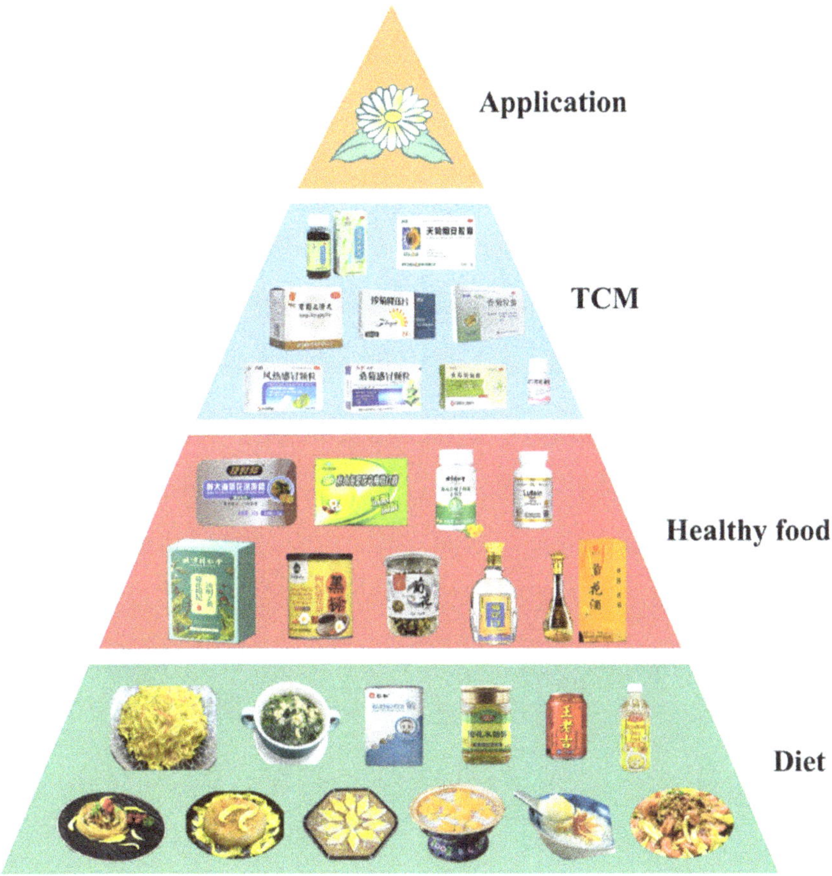

Fig. 12.5. Modern applications of chrysanthemum in TCM, healthy food, and diet. Note: TCM = traditional Chinese medicine. (Photo courtesy of Liu *et al.*, 2024.)

Fig. 12.6. *Chrysanthemum coronarium* used as a leafy vegetable (A), featuring Ohitashi, a dish made with edible chrysanthemums in Japanese cuisine (B). (Photo courtesy of (A) © State of New South Wales, CC BY 4.0—www.nsw.gov.au; (B) Ministry of Agriculture, Forestry and Fisheries website, CC BY 4.0—https://www.maff.go.jp)

but it becomes stronger and distinctly bitter as the plant matures and flowers, a taste reminiscent of coriander, making it a polarizing choice among diners. When young and tender, both the leaves and stalks of shungiku can be enjoyed raw in salads, with the ideal stalk diameter being no wider than 0.3 cm to avoid an overpowering bitterness. More mature greens, typically found in markets, are better suited for cooking, which mellows their flavor into a mildly grassy and herbaceous profile, with stalks that turn sweet and slightly crunchy when cooked. In Japanese dishes such as sukiyaki and shabu-shabu, shungiku is a praised addition and also enhances soups. The greens should be boiled briefly, no longer than 30 s to remain firm and vibrant. A popular way to serve them is blanched, cooled to room temperature, and dressed with a traditional sesame dressing. The versatility of shungiku extends to its preparation methods; it can be steamed, blanched, lightly boiled, stir-fried, or used in tempura. Visually, shungiku leaves resemble those from a florist, varying from slightly to deeply indented.

Shungiku thrives as a winter vegetable due to its hardiness and nutritional content. It is rich in vitamins A and B and a significant source of antioxidants, containing more potassium than an equivalent serving of bananas that are known for high potassium content. Its ability to tolerate light frosts allows it to be grown in open fields or unheated greenhouses from late autumn to early spring in mild climates. This adaptability, combined with its health benefits, makes shungiku a valuable addition to diets during the cooler months, offering a unique and healthful option for those seeking to enrich their culinary experiences.

12.4.2 Mottenohoka and Ohitashi: celebrating edible chrysanthemums in Yamagata cuisine

In Yamagata Prefecture of Japan, the culinary tradition of using edible chrysanthemums showcases a deep-rooted cultural appreciation and innovative applications in regional cuisine. This custom, dating back to the Edo period (1603–1868) and famously enjoyed by poet Matsuo Basho, remains robust today. One of the most cherished dishes is "Ohitashi of edible chrysanthemums," where chrysanthemum leaves and petals are boiled, chilled, and served with a soy sauce dressing, highlighting their delicate texture and natural flavors (Fig.12.6). The "Mottenohoka" variety, also known as "Enmeiraku," is particularly esteemed. This late-blooming, purple-colored chrysanthemum is noted for its fragrant aroma and exceptional flavor, described as almost outrageously delicious, hence its name "Mottenohoka," meaning "out of the question." Unique for its tubular, crunchy petals, this variety contrasts with the flatter petals of other chrysanthemums. Edible chrysanthemums from Yamagata are also used in "sunomono" (pickled vegetable dishes) and tempura, illustrating their seamless integration into local cuisine. These diverse applications highlight the distinctive qualities of different chrysanthemum varieties, affirming their status as celebrated ingredients in Yamagata's gastronomy (MAFF—https://www.maff.go.jp/e/policies/market/k_ryouri/search_menu/526/index.html).

12.4.3 Chrysanthemum tea

Chrysanthemum tea, a fragrant herbal infusion primarily made from the flowers of *C. morifolium* and *C. indicum*, is valued for both its culinary and medicinal properties. Renowned for its antioxidant, anticancer, anti-diabetic, and anti-inflammatory activities (Chen *et al*., 2019; Li *et al*., 2019), chrysanthemum tea has been consumed for its health benefits for millennia, with its use dating back over 3000 years to the Xia Dynasty in China (Hao *et al*., 2022). Chrysanthemum is classified into medicinal, tea, edible, and ornamental types, with its medicinal forms recognized in the Chinese Pharmacopoeia as Gongju, Hangju, Huaiju, Qiju, and Chuju. Among the 28 flower medicines listed in the Chinese Pharmacopoeia, eight are used for health-care purposes, including Juhua, the flower head of *C. morifolium*. Renowned for its traditional efficacies of "scattering cold," "clearing heat and toxins," and "brightening eyes," Juhua is used to treat conditions such as the common cold, eye pain, and dim-sightedness. It is effective against fever, headache, sore throat, and hypertension (Yuan *et al*., 2020). Beyond its dietary and medicinal uses, Juhua symbolizes Chinese culture and spirit, featuring prominently in poems and cultural works. Chrysanthemum tea is particularly popular in East Asian countries such as China, Korea, and Japan, where it is also used in various functional foods such as chrysanthemum wine, cake, and hot pot (Yang *et al*., 2019). Today, the medicinal and edible values of chrysanthemum are increasingly integrated, playing a significant role in daily life and health practices.

12.5 Conclusions and Future Prospects

The exploration of value addition in chrysanthemums highlights their multifaceted applications and significant potential in both traditional and innovative markets. Beyond ornamental value, chrysanthemums serve as a versatile crop used in traditional Chinese medicine, healthy foods, and diets. Technologies such as dry flower processing, tinting, and culinary uses have enhanced their economic value and practical applications. Innovations in drying techniques can further expand their use in pharmaceuticals, nutraceuticals, and functional foods. With growing interest in natural and sustainable products, chrysanthemums align well with eco-friendly consumer trends.

To advance the potential of value addition of chrysanthemums, future efforts should focus on developing advanced drying technologies to preserve their bioactive compounds, color, and structural integrity, thereby enhancing their applications in pharmaceuticals and nutraceuticals. Expanding market development strategies to promote chrysanthemum-based products in regions with a growing interest in natural health products is crucial, emphasizing their multifunctional benefits. Sustainability should be a priority, promoting eco-friendly farming practices and production methods. Comprehensive research on the health benefits of chrysanthemums can support the creation of new health supplements and functional foods. Culinary innovations should be explored to expand their presence in the food industry, creating novel products incorporating chrysanthemum petals, extracts, or flavors. Integrating chrysanthemums into modern health practices, especially within traditional Chinese medicine, can enhance their credibility and acceptance in contemporary health care. Finally, consumer education campaigns can increase awareness and drive demand, ensuring continued growth and innovation in the chrysanthemum industry.

References

Ahrne, L.M., Pereira, N.R., Staack, N. and Floberg, P. (2007) Microwave convective drying of plant foods at constant and variable microwave power. *Drying Technology* 25, 1149–1153. DOI: 10.1080/07373930701438436.

Bhalla, R., Dhiman, S.R., Moonaand Thakur, K.S. (2006) Standardization of drying techniques of chrysanthemum (*Dendranthema grandiflorum* tzvelev). *Journal of Ornamental Horticulture* 9(3), 159–163.

Bhattacharjee, S.K. and De, L.C. (2003) *Dried Flowers and Plant Parts*. Avishkar Publishers, Jaipur, pp. 162–173.

Bhutani, J.C. (1990) Capturing nature, a way with flower "Everlastings". *Indian Horticulture* 34(4), 15–19.

Bhutani, J.C. (1995) Drying of flowers and floral craft. *Advances in Horticulture Ornamental Plants* 12, 1053–1058.

Chen, M., Wang, K., Zhang, Y., Zhang, M., Ma, Y. *et al.* (2019) New insights into the biological activities of *Chrysanthemum morifolium*: Natural flavonoids alleviate diabetes by targeting alpha-glucosidase and the PTP-1B signaling pathway. *European Journal of Medicinal Chemistry* 178, 108–115. DOI: 10.1016/j.ejmech.2019.05.083.

Chen, S., Liu, J., Dong, G.Q., Zhang, X.T., Liu, Y. *et al.* (2021) Flavonoids and caffeoylquinic acids in *Chrysanthemum morifolium* Ramat flowers: A potentially rich source of bioactive compounds. *Food Chemistry* 344, 128733. DOI: 10.1016/j.foodchem.2020.128733.

Dahiya, D.S., Unnikrishnan, Gupta, A.K., Sehrawat, S.K and Siddiqui, S. (2003) Dehydration of annual chrysanthemum (*Chrysanthemum coronarium*). *Acta Horticulturae* 624, 358–388.

Datta, S.K. (1997) Dehydration of flowers and foliage and floral craft. *NBRI Bulletin* (3), 20.

Dhatt, K.K., Singh, K. and Kumar, R. (2007) Studies on methods of dehydration of rose buds. *Journal of Ornamental Horticulture* 10(4), 264–267.

Gurjar, A., Laishram, N., Singh, A., Pandey, R.K. and Sinha, B.K. (2023) Morphological and quality parameters of chrysanthemum flowers as influenced by different pre-drying treatments and drying methods. *Agriculture Association of Textile Chemical and Critical Reviews Journal* 11, 168–175. DOI: 10.58321/AATCCReview.2023.11.04.168.

Hao, D.C., Song, Y.J., Xiao, P.G., Zhong, Y., Wu, P.L. *et al.* (2022) The genus *Chrysanthemum*: Phylogeny, biodiversity, phytometabolites, and chemodiversity. *Frontiers in Plant Science* 13, 25. DOI: 10.3389/fpls.2022.973197.

Joyce, D.C. (1998) Dried and preserved ornamental plant material not new, but often overlooked and underrated. *Acta Horticulturae* 454, 133–145.

Khan, M., Liu, H., Wang, J. and Sun, B. (2020) Inhibitory effect of phenolic compounds and plant extracts on the formation of advance glycation end products: A comprehensive review. *Food Research International* 130, 108933. DOI: 10.1016/j.foodres.2019.

Kumari, P. (2022) *Dry flower technology*, training bulletin, CSIR-institute of himalayan bioresource technology, palampur.

Li, Y., Yang, P., Luo, Y., Gao, B., Sun, J. *et al.* (2019) Chemical compositions of chrysanthemum teas and their anti-inflammatory and antioxidant properties. *Food Chemistry* 286, 8–16. DOI: 10.1016/j.foodchem.2019.02.013.

Liu, Y., Lu, C., Zhou, J., Zhou, F., Gui, A. *et al.* (2024) *Chrysanthemum morifolium* as a traditional herb: A review of historical development, classification, phytochemistry, pharmacology and application. *Journal of Ethnopharmacology*. DOI: 10.1016/j.jep.2024.118198.

Lu, Y. F., Li, D. X., Zhang, R., Zhao, L. L., Qiu, Z. *et al.* (2022). Chemical antioxidant quality markers of *Chrysanthemum morifolium* using a spectrum-effect approach. *Frontiers in Pharmacology*, 13, 809482.

Mandal, M., Ashok, Y., Khan, A.M., Sarkar, I., Jana, J.C. *et al.* (2023) Application of different dehydration technology on chrysanthemum for quality dry flower production under Terai region of West Bengal, India. *International Journal of Plant & Soil Science* 35(18), 327–334. DOI: 10.9734/ijpss/2023/v35i183295.

Maskan, M. (2000) Microwave air and microwave finish drying of banana. *Journal of Food Engineering* 44(2), 71–78. DOI: 10.1016/S0260-8774(99)00167-3.

Mayak, S. and Halevy, A.H. (1980) Flower senescence. In: Thimann, K.V. (ed.) *Senescence in Plants*. CRC Press, Boca Raton, Florida, p. 132.

Pandya, H.P., Saxena, O.P., Nell, T.A. and Clark, D.G. (2001) Preservation of Chrysanthemum sp. by drying. *Acta Horticulturae* 543, 367–369.

Pant, P. and Singh, B. (2023) Refinement of technology on dry flowers: A review. *Agricultural Reviews* 44(4), 516–522. DOI: 10.18805/ag.R-2267.

Pavan, N.R., Kukanoor, L., Gorabal, K., Patil, S.R., Shiragur, M. *et al.* (2022) Standardization of tinting techniques in Chrysanthemum, Tuberose and Lilium. *The Pharma Innovation* 11(4), 713–717.

Pereira, N.R., Marsaioli, J.A. and Ahrne, L.M. (2007) Effect of microwave power, air velocity and temperature on the final drying of osmotically dehydrated bananas. *Journal of Food Engineering* 81(1), 79–87. DOI: 10.1016/j.jfoodeng.2006.09.025.

Raghupathy, R., Amuthan, G. and Kailappan, R. (2000) Dried flowers: Significance. *Kisan World* 28, 39.

Rajaram, P., Visalakshi, M., Jawaharlal, M. and Lakshmanan, V. (2009) Opportunities in dry flower industry for small and medium scale enterprises. In: *National Conference on Floriculture for Livelihood and Profitability*, IARI, New Delhi, pp. 36–38.

Sell, R. (1993) *Dried and fresh cut flowers*. NDSU Extension Service, North Dakota State University of Agriculture and Applied Science, Fargo, p. 62.

Sindhuja, S., Padmalatha, T., Padmavathamma, S.A. and Prasad, N.L.A. (2017) Studies on different methods of drying in carnation. *International Journal of Current Microbiology and Applied Sciences* 6(8), 870–882.

Singh, H.P. (2009) Floriculture industry in India: The bright future ahead. *Indian Horticulture* 54(1), 3–8.

Singh, A. and Laishram, N. (2010) Drying of flowers and other ornamental plant parts in India. *Floriculture and Ornamental Biotechnology* 4, 72–78.

Sun, C.D., Zhao, C., Guven, E.C., Paoli, P., Simal-Gandara, J. *et al.* (2020) Dietary polyphenols as antidiabetic agents: Advances and opportunities. *Food Frontiers* 1(1), 18–44. DOI: 10.1002/fft2.15.

Sugawara, T. and Igarashi, K. (2009). Identification of major flavonoids in petals of edible chrysanthemum flowers and their suppressive effect on carbon tetrachloride-induced liver injury in mice. *Food Science and Technology Research*, 15(5), 499–506.

Tsuji-Naito, K., Saeki, H. and Hamano, M. (2009) Inhibitory effects of Chrysanthemum species extract on formation of advanced glycation end products. *Food Chemistry* 116(4), 854–859. DOI: 10.1016/j.foodchem.200.

Ullas, P.S., Namita, N., Singh, K.P., Panwar, S., Kundu, A. *et al.* (2018) Influence of drying techniques on retention of anthocyanin and their antioxidant activities in chrysanthemum (*Chrysanthemum* x *Morifolium*) flowers. *The Indian Journal of Agricultural Sciences* 88(2), 228–233. DOI: 10.56093/ijas.v88i2.79189.

Vikas Kumar, G. (2022a) Studies on the retention of petal colour of worshipped flowers for making value added products through dry flower technology. Master's Thesis, Dr. Y.S.R. Horticultural University, Venkataramannagudem. Submitted November 2022.

Vikas Kumar, G., Nagaraju, R., Srinivas, P.T., Naik, M.R. and Ruth, C. (2022b) Value addition to *Chrysanthemum morifolium* through dry flower technology by standardizing different backdrop materials. *International Journal of Ecology and Environmental Sciences* 4(3), 133–135.

Yang, L., Nuerbiye, A., Cheng, P., Wang, J.H. and Li, H. (2017) Analysis of floral volatile components and antioxidant activity of different varieties of *Chrysanthemum morifolium*. *Molecules* 22(10), 1790. DOI: 10.3390/molecules22101790.

Yang, P.F., Yang, Y.N., He, C.Y., Chen, Z.F., Yuan, Q.S. *et al.* (2019) New caffeoylquinic acid derivatives and flavanone glycoside from the flowers of *Chrysanthemum morifolium* and their bioactivities. *Molecules* 24, 10. DOI: 10.3390/molecules24050850.

Yuan, H., Jiang, S., Liu, Y., Daniyal, M., Jian, Y. *et al.* (2020) The flower head of *Chrysanthemum morifolium* Ramat. (Juhua): A paradigm of flowers serving as Chinese dietary herbal medicine. *Journal of Ethnopharmacology* 261, 113043. DOI: 10.1016/j.jep.2020.113043.

Zhang, N., He, Z.J., He, S.Y. and Jing, P. (2019) Insights into the importance of dietary chrysanthemum flower (*Chrysanthemum morifolium* cv. Hangju) wolfberry (*Lycium barbarum* fruit) Jar combination in antioxidant and anti-inflammatory properties. *Food Research International* 116, 810–818. DOI: 10.1016/j.foodres.2018.09.

13

Festivals to FlowerTrials and Landscape Gardening of Chrysanthemum

Malleshaiah SharathKumar[1,2*]

[1]Horticulture and Product Physiology, Department of Plant Sciences, Wageningen University and Research, Droevendaalsesteeg 1, 6708 PB Wageningen, The Netherlands
[2]Department of Breeding and Research, Deliflor Chrysanten B.V., Korte Kruisweg 163, 2676 BS Maasdijk, The Netherlands

Abstract

Chrysanthemums, often affectionately referred to as "mums," have a rich history of cultivation dating back thousands of years. These ornamental beauties have captured the hearts of gardeners and played a significant role in various festivities and landscaping designs worldwide. From their importance in cultural celebrations to their role in creating stunning landscapes, chrysanthemums are versatile and elegant ornamental plants. In cultural festivals, chrysanthemums are cherished for their vibrant colors and symbolic meanings. They feature prominently in events such as the Chinese Chongyang Festival, Japanese Kiku Matsuri, and Korean Chrysanthemum Exhibition, symbolizing longevity, rejuvenation, and happiness. International flower festivals, including those at Longwood Gardens and the New York Botanical Garden, showcase chrysanthemums in elaborate displays, highlighting their global appeal. In landscaping, chrysanthemums are prized for their extensive color range, adaptability, and ability to extend the blooming season into late summer and autumn. Gardeners incorporate them into diverse designs, including mass plantings for dramatic visual impact, mixed borders for dynamic landscapes, and container gardening for flexible and decorative arrangements. This chapter explores the multifaceted allure of chrysanthemums, flowers that have captivated hearts and minds throughout the ages. Through their roles in festivals and landscaping, chrysanthemums have firmly established themselves as beloved flowers, enriching cultural traditions and enhancing the beauty of gardens worldwide.

*Corresponding author: sharath.malleshaiah@wur.nl; sharath.malleshaiah@deliflor.nl; sharath.malleshaiah@gmail.com

DOI: 10.1079/9781789248289.0013

Keywords

Celebrations, chrysanthemum, culture, festivals, floral displays, gardening, landscaping

13.1 Introduction

The chrysanthemum, often lovingly referred to as the "mum" by horticultural enthusiasts, holds a unique place in the world of ornamental gardening. Renowned for its diverse forms and vibrant colors, this flower has captivated people across the globe. Its versatility shines through in various aspects of our lives, from grand festivals to sophisticated landscaping. In the language of flowers, chrysanthemums often convey messages of longevity, love, and friendship. Different colors symbolize specific sentiments: red chrysanthemums signify love, white symbolizes purity and truth, purple denotes royalty and luxury, and yellow represents friendship, longevity, and prosperity (Arena Flowers, 2021; Flowwow, 2024). The chrysanthemum has a history that stretches back over two millennia. Originating in China, it was cultivated for both its aesthetic beauty and medicinal properties. By the 8th century, chrysanthemums had made their way to Japan, where they quickly became a symbol of Imperial power and prestige. The Chrysanthemum Throne remains the Japanese Imperial throne to this day, highlighting the flower's enduring cultural significance. In addition to Asia, the chrysanthemum found its way to Europe in the 17th century and the Americas in the 18th century. Its adaptability and stunning appearance led to its worldwide popularity. Embraced by different cultures, it has come to symbolize various meanings, ranging from longevity and vitality to purity and truth. Through its rich history and multifaceted uses, the chrysanthemum continues to enchant and inspire, solidifying its status as a beloved flower in gardens and festivals worldwide.

13.2 Cultural Significance of Chrysanthemums

Chrysanthemums carry deep cultural significance and hold a special place in many cultures around the world. In China, chrysanthemums symbolize nobility and integrity. They have long been associated with longevity and are considered one of the "Four Gentlemen" in traditional Chinese art, symbolizing the virtues of a noble character. Chrysanthemums have also been featured in Chinese poetry and painting for centuries, highlighting their esteemed status in Chinese culture. In Japan, chrysanthemums represent longevity and rejuvenation. The Imperial Seal of Japan features a stylized chrysanthemum, emphasizing its importance. As the national flower, the chrysanthemum holds a position of honor and is associated with the Imperial family, prominently featured on the country's Imperial crest. The term "Chrysanthemum Throne" or "Chrysanthemum Government" refers to the Japanese Imperial family, highlighting the flower's profound cultural significance (Amaranté London, 2022).

In India, chrysanthemums, referred to as "Shevanti," hold significant religious symbolism. The orange and yellow flowers are often used as garlands and offerings in pujas and temples. Chrysanthemums are commonly used to make garlands (malas) that are worn during religious ceremonies, weddings, and other significant events. These garlands are also used to adorn idols, pictures of deities, and altars. No religious function or ritual is complete without Shevanti flowers, and they form a prominent part of all celebrations. Frames and idols of gods are always decorated with these yellow and orange flowers, and garlands for frames of deities prominently feature Shevanti flowers. In Hindu religious practices, chrysanthemums are frequently used as offerings to deities. They are often part of floral decorations in temples and during various rituals and festivals. Their bright colors and freshness are considered auspicious. During festivals such as Diwali, Navratri, and Durga Puja, chrysanthemums are used extensively for decoration and worship. The presence of these flowers enhances the celebratory atmosphere and is believed to bring good fortune. Through their symbolism and representation in art, poetry, and national emblems, chrysanthemums have cemented their place as symbols of noble character, longevity, and rejuvenation in Chinese, Indian, and Japanese cultures.

13.3 Chrysanthemums in Landscape Gardening

The versatility of chrysanthemums extends to landscaping. In home and public gardens, and parks worldwide, chrysanthemums create stunning displays of color and form. Their many shapes, including daisy-like singles, pompoms, and spider-like petals, add texture and visual interest to garden designs. Chrysanthemums are used during festivals and landscaping to create visually striking and seasonally dynamic gardens. Through borders, topiary, bonsai, bushes, hedges, and other diverse applications, chrysanthemums demonstrate their exceptional adaptability and beauty in landscaping, contributing to visually appealing and dynamic outdoor spaces. Here are some landscaping ideas involving ornamental chrysanthemums:

- **Colorful chrysanthemum borders:** Planting chrysanthemums along walkways or garden borders can add color to your landscape. Mix chrysanthemum varieties with varying bloom colors and sizes for a vibrant display.
- **Chrysanthemum bonsai:** Chrysanthemum bonsai, with their miniature form and intricate pruning, is a testament to the artistry that can be achieved with this plant (Fig. 13.1). They add a touch of elegance and charisma to any garden or outdoor space. Bonsai, which translates to "plant growing in a tray," is an ancient Japanese art that nurtures dwarfed plants in containers. These bonsai are meticulously trained to reflect various natural landscapes, ranging from single-specimen trees to multiple-specimens displays that suggest expansive forest settings, showcasing the artistry and skill of bonsai cultivation.
- **Chrysanthemum bushes and hedges:** Chrysanthemum bushes and hedges create decorative borders and fences. The robust nature of

Fig. 13.1. The artistry of bonsai created using chrysanthemums, featuring the Cascade Bonsai style (A) and Informal Upright style (B and C), showcasing their miniature form and intricate training and pruning techniques. (Photo courtesy of Longwood Gardens, Pennsylvania, USA, with permission.)

chrysanthemums means they can withstand various weather conditions, making them an excellent choice for low-maintenance yet high-impact plants. Trimmed chrysanthemum hedges can serve as natural dividers in your garden, with dense foliage and colorful flowers providing an eye-catching and functional landscape feature.

- **Chrysanthemum topiary:** With careful pruning, chrysanthemums can be shaped into topiaries resembling animals, geometric shapes, or famous landmarks (Fig. 13.2). These living sculptures can be a unique addition to the garden.

Fig. 13.2. Display of yellow chrysanthemum topiaries shaped into spirals and flower domes, creating a picturesque effect of yellow blooms in conservatory. (Photo courtesy of Longwood Gardens, Pennsylvania, USA, with permission.)

- **Chrysanthemum water gardens:** Integrate chrysanthemums into your pond or water garden. Floating chrysanthemum blooms or potted plants at the water's edge can create a tranquil and stunning atmosphere.
- **Seasonal chrysanthemum displays:** Swap out your garden's floral displays with chrysanthemums each season to keep your landscape ever-changing and fresh. Spring, summer, and autumn chrysanthemums can be strategically planted to create a dynamic garden.

13.4 Festivals Celebrating Chrysanthemums

One of the most enchanting aspects of chrysanthemums is their role in festivals. In Asia, especially in China and Japan, chrysanthemums take center stage during fall festivals. In China, the Double Ninth Festival (Chongyang Festival) involves giving chrysanthemums to express good wishes and celebrate longevity. Chrysanthemums symbolize nobility and integrity, mirroring the festival's theme of respecting the elderly and seeking long life. In Japan, the Festival of Happiness, or "Kiku no Sekku," is celebrated with elaborate chrysanthemum displays. Various chrysanthemum cultivars are meticulously groomed and highlighted, demonstrating the flower's cultural significance and beauty. This festival highlights the Japanese appreciation for chrysanthemums as the symbols of rejuvenation and longevity. In the West, chrysanthemums are prominently featured in fall festivals, particularly in North America, where

they are often associated with Thanksgiving and Halloween decorations. Their warm and vibrant colors enhance the festive spirit of these celebrations, making them an emblem of the autumn season. Chrysanthemums' ability to thrive in cooler weather makes them a popular choice for autumnal displays. Chrysanthemum festivals often feature meticulously cultivated displays in various forms, such as cascading waterfalls of blooms, bonsai-style chrysanthemum trees, and intricate patterns created from chrysanthemum petals. These artistic presentations highlight the versatility and beauty of chrysanthemums, drawing visitors to admire and celebrate this remarkable flower. Through their prominent roles in festivals across different cultures, chrysanthemums symbolize celebration, good wishes, and the beauty of nature, reinforcing their status as a beloved and significant flower worldwide. Chrysanthemums are celebrated in several prominent festivals around the world, each highlighting the flower's cultural significance, beauty, and horticultural achievements. Some notable festivals dedicated to chrysanthemums are discussed in the following sections.

13.4.1 The art of Kiku: Japanese chrysanthemum masterpieces

The chrysanthemum, affectionately known as *kiku* in Japan, stands as a pinnacle of fall-flowering plants, celebrated for its captivating beauty and rich cultural significance. The name "chrysanthemum" is derived from the Greek words "chrysos" (gold) and "anthos" (flower), aptly describing the wild variety with its long, wiry stems and yellow daisy-like appearance. Kiku Matsuri (chrysanthemum festival) in Japan is a grand celebration of chrysanthemums held every November (Japan Society, 2008). The Tokyo Metropolitan Government Building often hosts stunning chrysanthemum exhibitions that attract thousands of visitors. In 1878, Emperor Meiji of Japan hosted an exclusive garden party, a remarkable gathering where chrysanthemums took center stage (New York Botanical Garden, 2018). This event ignited a tradition that would shape the world of chrysanthemum cultivation. Subsequently, a dedicated nursery was established in what is now the Shinjuku Gyoen National Garden in Tokyo, with the sole purpose of growing chrysanthemums for this annual event. In 1929, the garden party's venue shifted to Shinjuku Gyoen, where it continues to this day.

Over the past century, gardeners at Shinjuku Gyoen have perfected the art of growing and displaying these exquisite chrysanthemums. They employ intricate techniques that result in awe-inspiring creations, such as producing hundreds of beautiful flowers from a single stem. What was once a private event for the emperor's enjoyment is now open to the public and is known as "Kiku Kadan-ten." This 15-day exhibition, held each November, coincides with the chrysanthemum season. While the chrysanthemum exhibition at Shinjuku Gyoen lasts only for a limited time, the botanical Garden's exhibition runs for twice as long. To ensure an uninterrupted display, two groups of chrysanthemum plants are meticulously prepared. One group is showcased during the first half of the exhibition, while the other is held in reserve for the final weeks, guaranteeing the longevity of the spectacle.

13.4.1.1 Imperial styles of Kiku

The Enid A. Haupt Conservatory Courtyards at the Botanical Garden play host to four Imperial styles of kiku:

1. **Ozukuri ("Thousand Bloom")** This style involves meticulously training a single chrysanthemum plant to produce an impressive array of blooms on a single stem.
2. **Ogiku ("Single Stem")** In this style, the focus is on a single, magnificent chrysanthemum bloom, emphasizing its individual beauty.
3. **Kengai ("Cascade")** Chrysanthemum plants are artfully shaped to create cascading floral displays, creating an enchanting waterfall effect.
4. **Shino-tsukuri ("Driving Rain")** A new addition to the exhibition, this style mimics the beauty of chrysanthemums in the midst of a rain shower.

These chrysanthemum displays are housed in decorative Japanese garden pavilions known as uwaya. These intricately constructed structures, crafted from bamboo and cedar, are adorned with ceremonial drapery to protect, and frame the beauty of the kiku displays, enhancing their visual impact.

13.4.2 Chinese Chongyang Festival

The Chongyang Festival, also known as the Double Ninth Festival or the Senior Citizens' Festival, is celebrated on the ninth day of the ninth lunar month. This festival, which also has the name Chrysanthemum Festival, dates back over 2000 years to the Eastern Han Dynasty (25–220 AD) (Artmajeur Magazine, 2024). Chrysanthemum appreciation is a key element of the Chongyang Festival, deeply rooted in Chinese cultural and historical contexts. The tradition of appreciating chrysanthemums during this festival dates to the Jin Dynasty when the poet Tao Yuanming popularized it by enjoying the flowers and drinking chrysanthemum wine (Artmajeur Magazine, 2024). According to the Yin/Yang dichotomy in the Chinese worldview, Yin represents darkness, and Yang means light and brightness. In the book Yi Jing, the number six belongs to Yin, while the number nine is Yang. Therefore, the ninth day of the ninth lunar month, both being Yang, is named the Double Ninth Festival. As the number nine is associated with Yang, and the ninth day of the ninth month is a double yang day, the name "Chong Yang Festival" (Chong means "repeat" in Chinese). The ninth month also heralds the approach of winter, a time for providing warm clothing, including symbolic paper clothes burned as offerings for ancestors. On the Double Ninth Festival, people customarily climb mountains, appreciate chrysanthemum flowers, drink chrysanthemum wine, and eat double-ninth cakes. The festival is also known as the "Old Men Festival," and its activities aim to improve older people's health. It is a time for family gatherings, remembering ancestors, and acknowledging their sacrifices and hardships. Family outings often renew appreciation of nature and reaffirm love and concern for family members and close friends. Here are some specific aspects of chrysanthemum appreciation during the Chongyang Festival:

- **Viewing chrysanthemums:** Chrysanthemums have been revered in China for centuries, symbolizing longevity, nobility, and resilience. Parks and gardens across China host grand chrysanthemum exhibitions where people can admire various species in full bloom. These events often include competitions and displays of rare and exotic flower varieties (Jean, 2023).
- **Eating Chongyang cakes:** Chongyang cakes are made from ingredients such as chrysanthemum, jujubes, pecans, and other seasonal fruits. Celebrated around the harvest season, eating these cakes symbolizes the wish for a higher standard of living, as the word for cake sounds like "high" in Chinese.
- **Drinking chrysanthemum wine and tea:** Drinking chrysanthemum wine or tea is a traditional practice believed to promote health and longevity. The wine is often homemade during the festival to ward off evil spirits and diseases (Jean, 2023).
- **Decorating with chrysanthemums:** People decorate their homes with chrysanthemums during the festival. In ancient times, women wore chrysanthemum flowers in their hair, and the flowers were used to make wreaths and other decorative items.
- **Modern celebrations:** In contemporary times, chrysanthemum appreciation during the Chongyang Festival includes local festivals, floral art shows, and themed cultural events. These activities celebrate chrysanthemums' beauty and cultural heritage, bringing communities together to enjoy vibrant displays and partake in traditional customs (Li, 2023). By focusing on these aspects, the appreciation of chrysanthemums during the Chongyang Festival honors ancient traditions and fosters a sense of community and cultural continuity in modern China.

13.4.3 Chrysanthemum Festival at Longwood Gardens

The Chrysanthemum Festival at Longwood Gardens (Longwood Gardens, 2017), located in Kennett Square, Pennsylvania, is an annual event celebrated for its stunning displays and horticultural artistry (Sutton, 2017). Typically held from late October to mid-November, this festival transforms the garden's conservatories into a vibrant showcase of chrysanthemums. The chrysanthemums displayed at Longwood Gardens are primarily of Japanese origin and have been part of the garden's collection since the early 20th century. This festival celebrates the beauty of the flowers and their cultural heritage, incorporating traditional Asian gardening techniques.

13.4.3.1 Key features of the festival

- **Thousand bloom chrysanthemums:** One of the most remarkable displays is the "thousand bloom chrysanthemum," a single plant meticulously trained to produce over a thousand uniform blooms (Santos, 2023; Fig. 13.3). This horticultural feat involves over 1800 hour of work, including pruning, pinching, and training the chrysanthemum plant to grow into a large, symmetrical dome (Garden Design, 2024).
- **Cascade chrysanthemums:** Chrysanthemums are trained to grow along frames into shapes like spirals, columns, and curtains using stem-breaking techniques, allowing them to bend without damage (Sutton, 2017).
- **Diverse displays:** The festival includes a variety of mum styles, such as single stem, disbud, and spray mums, each contributing to the overall aesthetic. The displays showcase different sizes and forms, from small, intricate flowers to large, dramatic blooms.
- **Educational opportunities:** Visitors can learn about the 13 classifications of chrysanthemums, which are displayed with informational signage. This educational aspect helps visitors appreciate the diversity and complexity of the blooms.
- **Visiting the festival:** The chrysanthemum festival is one of the most horticulturally challenging exhibitions at Longwood Gardens, requiring extensive planning and cultivation. Visitors can expect thousands of chrysanthemums in full bloom, arranged in innovative and beautiful displays. The festival is a testament to the skill and dedication of Longwood Gardens' horticulturists, making it a must-visit event for flower enthusiasts and garden lovers alike (Longwood Gardens, 2017).

13.4.4 Kiku: The Art of Chrysanthemums at New York Botanical Garden

"Kiku: The Art of Chrysanthemums" is a celebrated annual exhibition at the New York Botanical Garden (NYBG), showcasing the exquisite art of Japanese chrysanthemum cultivation. This event in October transforms the Enid A. Haupt Conservatory into a vibrant display of meticulously trained chrysanthemums, known as "kiku" in Japanese. The exhibition highlights various traditional and contemporary styles, such as Ozukuri (thousand bloom), Kengai (cascade), and Ogiku (single stem). These chrysanthemums are cultivated using techniques learned through cultural exchanges with Japanese masters, making NYBG the first garden outside Japan to present kiku in the traditional Imperial style (Newman, 2016). In addition to the stunning floral displays, the event includes various cultural activities such as bonsai exhibitions, Japanese art displays, and hands-on activities for children. Visitors can also enjoy themed cocktails, guided tours, and special performances celebrating the chrysanthemum's cultural significance in Japanese style (Newman, 2016).

Fig. 13.3. Centerpiece of the thousand bloom chrysanthemum (*Chrysanthemum × morifolium* 'Susono-no-Hikari') displayed in the east conservatory of Longwood Gardens. Individual chrysanthemum flowers are supported with metal rings to achieve the uniformity and symmetry of the thousand blooms. (Photo courtesy of Longwood Gardens, Pennsylvania, USA, with permission.)

Fig. 13.4. Breeding companies exhibit chrysanthemum displays, showcasing their promising varieties through unique thematic presentations to highlight new breeds during chrysanthemum open days or flower trials for local and international customers. (Photo courtesy of Deliflor Chrysanten B.V.)

13.5 FlowerTrials

FlowerTrials is a prominent horticultural event held annually in the Netherlands and Germany in June, displaying the latest innovations and developments in the cut flowers, pot, and bedding plants industry (FlowerTrials®, 2024, https://www.flowertrials.com/en). The event is accessible to both industry professionals and hobby gardeners. Featuring 60 breeding companies across the Middle and South regions of the Netherlands and the Rhineland-Westphalia region in Germany (Hortinno, 2024). FlowerTrials presents the latest novelties and inspiring assortments of cut flowers, pot and bedding plants, including chrysanthemums. Leading chrysanthemum-breeding companies, such as Deliflor Chrysanten B.V., Dekker Chrysanten B.V., Floritech, Dümmen Orange, and Royal Van Zanten, participate by highlighting new assortments or varieties. Visitors can explore a variety of chrysanthemum displays, with each exhibitor presenting their promising new varieties through unique thematic displays (Fig. 13.4). FlowerTrials highlights novel and well-performing varieties in the market, synchronized to bloom during the event. The greenhouse exhibitions create a spectacular sea of chrysanthemums (Fig. 13.5), including spray, disbudded chrysanthemums, and Santini varieties, each distinctive in cultivation qualities, color, form, and resistance.

This event in the Netherlands is not confined to a single location but is spread across various venues, primarily in and around Aalsmeer and the Westland region, known as the "beating heart" of the event. This setup allows visitors to travel and explore different exhibitions, creating a multiday horticultural festival experience. FlowerTrials provides valuable opportunities

Fig. 13.5. Chrysanthemum beds featuring new and standard varieties displayed for local and international customers in a greenhouse setting, showcasing promising new varieties presented by a breeding company. (Photo courtesy of Deliflor Chrysanten B.V. and Malleshaiah SharathKumar.)

for learning and networking. Attendees can participate in talks and demonstrations covering cultivation techniques, retail concepts, and market trends, making it an essential event for industry professionals and plant enthusiasts. A similar concept has recently started in Colombia, known as Chrysanthemum Week. The chrysanthemum industry in Colombia is rapidly growing, with companies seeking robust varieties and novelties. Visitors from outside Colombia, including Bogota, Ecuador, and the USA, travel to Colombia for Chrysanthemum Week to discover new chrysanthemum varieties and innovations.

13.6 Conclusion and Future Prospects

Chrysanthemums continue to be closely associated with humanity, not only for their visual appeal but also for their deep-rooted significance in various cultures, adding color and elegance to settings ranging from festivals to garden landscapes. Their rich history and cultural significance make them a cherished flower in festivities and landscaping worldwide. From traditional celebrations to sophisticated garden designs, the chrysanthemum's charm and adaptability are unmatched. Its journey from ancient Asian countries to becoming a global symbol of beauty and resilience highlights its enduring appeal. Whether used in festive decorations or to enhance garden aesthetics, chrysanthemums continue to inspire and captivate enthusiasts worldwide.

Looking ahead, chrysanthemums hold significant potential for further innovation and integration into modern horticultural practices. Advanced breeding techniques and biotechnology offer opportunities to enhance the plant's resilience while expanding its color palette and bloom forms. As the demand for sustainable and eco-friendly gardening solutions grows, chrysanthemums' adaptability and low-maintenance requirements position them as ideal choices for urban and community gardens. Additionally, educational programs and public exhibitions can increase awareness of chrysanthemums' cultural heritage and horticultural benefits, fostering a deeper appreciation among future generations of gardeners and enthusiasts. By continuing to explore and expand the applications of chrysanthemums, we can ensure that this versatile and beautiful flower remains a beloved staple in gardens, festivities, and cultural celebrations worldwide.

References

Amaranté London (2022) The History of the Chrysanthemum, online. Available at: https://www.amarantelondon.com/blogs/blog/the-history-of-the-chrysanthemum (accessed 29 July 2024).

Arena Flowers (2021) The Meaning of Flowers: Chrysanthemum, online. Available at: https://www.arenaflowers.co.in/blogs/news/the-meaning-of-flowers-chrysanthemum (accessed 29 July 2024).

Artmajeur Magazine (2024) Chongyang Festival in Art: Honoring Bravery and Community Spirit, online. Available at: https://www.artmajeur.com/en/magazine/5-art-history/chongyang-festival-in-art-honoring-bravery-and-community-spirit/335587 (accessed 29 July 2024).

FlowerTrials® (2024) The Pot and Bedding Plant Event, online. Available at: https://www.flowertrials.com/en (accessed 29 July 2024).

Flowwow (2024) Chrysanthemum Flower Meaning, online. Available at: https://flowwow.co.uk/blog/en-chrysanthemum-flower-meaning/ (accessed 29 July 2024).

Garden Design (2024) Ancient Art form Lives on at Longwood Gardens. Available at: https://www.gardendesign.com/pennsylvania/longwood-gardens-chrysanthemum-festival.html (accessed 29 August 2024).

Hortinno (2024) FlowerTrials® 2024: What is It, When & Where?, online. Available at: https://www.hortinno.com/news/flowertrialsr-2024-what-it-when-where (accessed 29 July 2024).

Japan Society (2008) Kiku: The Art of the Japanese Chrysanthemum, online. Available at: https://japansociety.org/events/kiku-the-art-of-the-japanese-chrysanthemum/ (accessed 29 July 2024).

Jean (2023) Chongyang Festival. Mandarin Homeschool, online. Available at: https://mandarinhomeschool.com/chongyang-festival/ (accessed 29 July 2024).

Li, G. (2023) Chongyang Festival (Double Ninth Festival). Travel China Guide. Available at: https://www.travelchinaguide.com/essential/holidays/chongyang.htm (accessed 29 July 2024).

Longwood Gardens (2017) Get to Know Our Chrysanthemum Festival, online. Available at: https://longwoodgardens.org/blog/2017-11-01/get-know-our-chrysanthemum-festival (accessed 29 July 2024).

New York Botanical Garden (2018) Kiku: The Art of the Japanese Chrysanthemum, online. Available at: https://libguides.nybg.org/c.php?g=693023 (accessed 29 July 2024).

Newman, J. (2016) Kiku: The Art of the Chrysanthemums Opens at NYBG. Garden Collage, online. Available at: https://gardencollage.com/inspire/florals/kiku-art-chrysanthemums-opens-nybg/ (accessed 29 July 2024).

Santos, K. (2023) A Triumphant Return: Our Thousand Bloom Chrysanthemum is on display, with the rest of Chrysanthemum Festival, online. Available at: https://www.longwoodgardens.org/blog/2023/10/18/a-triumphant-return (accessed 29 July 2024).

Sutton, J. (2017) Get to Know Our Chrysanthemum Festival. Longwood Gardens, online. Available at: https://longwoodgardens.org/blog/2017-11-01/get-know-our-chrysanthemum-festival (accessed 29 July 2024).

Index

www.ingramcontent.com/pod-product-compliance
Lightning Source LLC
Chambersburg PA
CBHW041611120526
44952CB00026B/708